Analog Filter and Circuit Design Handbook

About the Author

Arthur B. Williams has over 45 years experience designing filters and analog circuitry. He is the author of 8 books on filters and analog design. He is also the holder of numerous Patents. Mr. Williams has been past chairman of the IEEE Circuits and Systems Society (Long Island) and the recipient of the IEEE Award and Honoree, 2007 "For Lifetime Advancement of the Technology of Electronic Filter Design". He is Chief Scientist at Telebyte Inc. where he is involved in the design of test equipment for the telecommunications industry.

Analog Filter and Circuit Design Handbook

Arthur B. Williams

New York Chicago San Francisco
Athens London Madrid
Mexico City Milan New Delhi
Singapore Sydney Toronto

Cataloging-in-Publication Data is on file with the Library of Congress.

McGraw-Hill Education books are available at special quantity discounts to use as premiums and sales promotions, or for use in corporate training programs. To contact a representative please visit the Contact Us page at www.mhprofessional.com.

Analog Filter and Circuit Design Handbook

1 2 3 4 5 6 7 8 9 0   DOC/DOC   1 2 0 9 8 7 6 5 4 3

ISBN 978-0-07-181671-7
MHID 0-07-181671-2

Parts of this book appeared previously in *Electronic Filter Design Handbook,* Fourth Edition, which was published by McGraw-Hill Education in 2006.

The pages within this book were printed on acid-free paper.

Sponsoring Editor
Michael Penn

Acquisitions Coordinator
Amy Stonebraker

Editorial Supervisor
David E. Fogarty

Project Manager
Charu Khanna, MPS Limited

Copy Editor
Lisa McCoy

Proofreader
MPS Limited

Production Supervisor
Richard C. Ruzycka

Composition
MPS Limited

Art Director, Cover
Jeff Weeks

This book is dedicated to my granddaughters
Leviah and Ilona Ehrlich

Contents

Preface

There have been four editions of the *Electronic Filter Design Handbook* where each edition had further enhancements over the previous edition. Prior to the book's introduction in 1981, the design of passive and active filters was considered a "black art" reserved for mathematicians and specialists in filter design, mainly because of the tedious calculations required and the skill and experience needed for the proper implementation of the filters. As a result of the book, filters of all types were being designed by individuals with a wide range of technical skills, including hobbyists, technicians, engineers, and scientists. This book became *the* reference for filters. It appeared on Wikipedia numerous times as the reference for filter-related searches, and was being used as a reference by many technical articles and papers as well as other books.

Although the previous *Electronic Filter Design Handbook* greatly simplified the design of passive and active filters, a void still existed, as filters are just a specialized segment of general analog design that has always been reserved for specialists. The need existed for a book that demystified analog design in general and expanded beyond filters. That is the function of this book, the first edition of the *Analog Filter and Circuit Design Handbook,* which not only simplifies filter design, but also breaks down the perception of analog design being a black art. Emphasis has been placed on using operational amplifiers as key building blocks to create working circuits that perform a variety of analog functions.

In addition to developing a strong foundation of understanding how op amps work and their limitations, circuit examples are given. Many of these will perform mathematical functions on analog signals in both a linear and nonlinear manner. Audio applications are shown, such as audio power amplifiers and crossover networks. Both voltage and current feedback amplifiers are covered. Analysis and the impact of nonideal amplifiers are presented. Waveform shaping and generation includes various types of oscillators, both sinusoidal and nonsinusoidal.

The filter-related material presented in the previous book has been revised in the *Analog Filter and Circuit Design Handbook,* which not only simplifies filter design, but includes new topics and enhances old ones. Chapter 1 has been expanded to further emphasize the pole-zero concept and its relationship to a transfer function and to modern network theory utilizing polynomials. The pole-zero concept can provide a useful tool for determining the feasibility of filters, as well as providing guidance as to their optimization. Various methods of synthesis are shown. The trade-offs between active and passive filter implementations are discussed. The frequency limitations of each type are considered.

The mathematical properties of standard filter response types are covered in Chap. 2, including Butterworth, Chebyshev, Bessel, linear phase with equiripple error, transitional, synchronously tuned, and constant delay with Chebyshev stopband. Extensive normalized curves for both frequency and time-domain parameters of these standard polynomial transfer functions are provided. Filter requirement normalization and filter scaling are shown. The highly efficient elliptic-function filter response is introduced in this chapter and emphasized throughout the handbook. Utilization of powerful programs, *Filter Solutions* (Book Version) from Nuhertz Technologies, and ELI 1.0, both used for the design of elliptic-function filters, is demonstrated. The "Optimum L" polynomials, also known as Legendre or Papoulis filters, are included within the various filter families with associated tables for poles, element values, and response curves.

In Chap. 3 the design of both passive and active low-pass filters is covered using normalized tables and the two programs provided with the book. Specialized passive low-pass filter design techniques are illustrated, such as designing for unequal terminations and compensating for the effects of component dissipation (low Q) using predistorted designs. Various active low-pass filter structures are covered for both all-pole and elliptic-function types. Techniques for designing low-pass filters with zero phase shift over a portion of their band are shown. This characteristic is required in many servo systems where a frequency shift should not result in a phase shift. Some new active low-pass filter active configurations are shown. A bidirectional active impedance converter is shown in conjunction with the GIC design approach to match loads of different impedances.

High-pass filters for both passive and active implementations are discussed in Chap. 4. The low-pass to high-pass mathematical transformation is demonstrated. Techniques are shown for making component values more practical using a variety of transformations. The generalized impedance converter (GIC) approach is applied to high-pass filter examples. A unique technique for the design of constant-delay high-pass filters using a delay line is presented.

Chapter 5 covers band-pass filters. Various passive filter transformations, approximations, and identities are shown to ensure practical element values, even for extreme conditions of center frequency, bandwidth, or impedance level. Some active band-pass implementations are given that exhibit low sensitivity at frequencies previously considered too high for active filters. The low-pass to band-pass transformation is presented. A Q multiplier approach is shown, where bandwidth can be adjustable. Some new band-pass active filter configurations are presented.

Techniques for band-reject filter design are presented in Chap. 6. Passive and active types are covered. The design of extremely steep band-reject filters is illustrated.

Chapter 7 covers the design of networks with properties best described in the time domain. All-pass delay and amplitude equalizers are discussed in detail. Delay equalization of filters is introduced. Methods are shown for the design of LC and active delay lines and wideband 90° phase shift networks. Topology for the design of very large-delay low-frequency LC delay lines using repetitious elements is shown, ranging into many milliseconds of delay, and formulas are provided for computing element values. This material has never been presented before in a book.

Refinements in LC filter design are covered in Chap. 8. Special techniques are presented to manipulate element values so that practical values can always be obtained. A variety of narrowband transformations are shown. New methods are illustrated for adding additional transmission zeros (nulls) to an existing design to enhance filter stopband performance with minimal effect on the passband. Equalizer circuits for amplitude equalization to compensate for inadequate Q are shown.

Careful component selection is critical for the successful operation of analog designs. Chapter 9 goes into detail on the properties of various types of inductors, capacitors, and resistors and the effects of their parasitic components.

Chapter 10 contains normalized tables for rapid design of both passive and active filters. In addition to the standard polynomial types, tables are provided for the unique constant-delay low-pass filters with Chebyshev stopband characteristics. This chapter is supplemented by two programs available for downloading with purchase of the book. A program called *Filter Solutions* from Nuhertz Technologies (www.nuhertz.com) enables the design of elliptic-function filters up to the 10th order. *Filter Solutions* can provide the transfer function, frequency response, group delay, time domain response, pole-zero plot, and schematic for up to a 10th-order low-pass elliptic-function passive filter. Also a Net List is available for Spice analysis. A second program, ELI 1.0, can be used for the design of odd-order elliptic-function filters up to a complexity of 15 nulls (transmission zeros), or the 31st order.

Chapter 11 covers switched-capacitor filters. The underlying theory behind this technology is presented. Some design examples are shown using standard building block ICs. Universal switched capacitor filter architectures are shown. A survey and updated convenient selection guide for IC switched capacitor filters is provided.

Both passive and active amplitude and delay equalizers are covered in Chap. 12. Configurations are provided that are both fixed and adjustable (tunable). Equalizers are in widespread use, mainly in audio graphic equalizers. Utilization applications are shown for the correction of amplitude and delay distortion. The pole-zero concept is related to equalizers.

Chapter 13 presents the theory of the voltage feedback operational amplifier and the analysis and effects of nonideal operational amplifiers. Negative and positive feedback, loop gain, and stability are covered. The impact of the open-loop response on the closed-loop response is analyzed. Practical methods of ensuring stability are shown. The feedback equation is explored in detail. Bode plots of amplitude and phase are analyzed for determining stability. Practical op-amp parameters, such as offset/bias voltage and current, input/output impedance, harmonic distortion, voltage and current noise, common-mode parameters, slewing, etc., are all explained and elaborated upon. Design methods for using a single supply to power amplifier circuits are covered. A voltage-feedback operational amplifier selection guide is provided.

Linear amplifier circuit applications are presented in Chap. 14. Differential input and output amplifiers, bridge amplifiers, weighted analog addition and subtraction, and other linear applications are shown. Sources of amplifier noise are discussed, along with their impact on the final circuit configuration. Techniques such as bootstrapping and use of a T-network to reduce resistor values are shown. Current-to-voltage and voltage-to-current converters, including the Howland current pump, are explained.

Chapter 15 covers nonlinear circuits, such as AGC amplifiers; one-, two-, and four-quadrant multipliers; analog dividers; squaring and square root functions; peak detectors; and sample and hold circuits. Half-wave and full-wave ideal rectifier circuits and absolute value circuits are presented. RMS-to-DC converters are explained. Logarithmic and antilog amplifier configurations are shown. Selection guides are provided.

Chapter 16 discusses various methods of shaping waveforms to convert them from sinusoidal to various levels that are amplitude-dependent on the sine wave. This includes differentiators, integrators, comparators, window comparators, limiters, and time-delay circuits. The concept of hysteresis is explained.

Chapter 17 presents waveform generators such as sine wave generators, square and triangular waveform generators, phase shift oscillators, sawtooth, Wien bridge sine wave oscillators, crystal oscillators, relaxation oscillators and other circuits, which can generate a variety of waveforms, along with their associated design equations. Consideration is given to methods of ensuring stability. The widely used 555 timer is explained, and applications are given.

The unique properties of current feedback amplifiers are introduced in Chap. 18 and illustrations are provided for the selection of current feedback versus voltage feedback amplifiers. Since current feedback amplifiers do not have the traditional structure of voltage feedback amplifiers, many engineers are uncomfortable using them. This chapter breaks down that barrier. The current feedback model is presented and compared to the voltage feedback model. The importance and selection of the feedback resistor is discussed, and how it controls gain, bandwidth, and stability is illustrated. A current feedback amplifier selection guide is provided.

Chapter 19 concentrates on large-signal output amplifiers, such as Class D audio power amplifiers, as well as other classes and circuits such as line drivers. Op-amp parameters that affect these classes of amplifiers are discussed and criteria presented to select the best device for a given application. The design of loudspeaker crossover networks is covered. Calculating heat sinking requirements for power amplifiers is demonstrated using the thermal resistance concept. Examples of choosing a heat sink are given. Active output impedance architectures using operational amplifiers are explained and the advantages and disadvantages are elaborated upon.

Arthur B. Williams

Analog Filter and
Circuit Design
Handbook

CHAPTER 1

Introduction to Modern Network Theory

In the early 1920s, a method of filter design was invented by Otto Zobel and George Campbell called *image parameter design*. This method provided a relatively simple way to design filters using a cascade of various building block circuits, each having an associated response. The final design required termination (source and load) by an image impedance. This is an impedance that is equal to the impedance seen looking into a network. However, the input and output impedance looking into a network containing inductors and capacitors is a complex function of frequency and is not simply resistive. In the real world, filters operate between fixed resistor values, so to terminate a network with a complex impedance is not practical. This book will concentrate on modern network theory, also known as the *insertion loss* method. This method starts with families of polynomials, which approximate a filter characteristic. A filter that operates between defined resistors can be synthesized from these polynomials. A few simple examples will be given in this chapter.

A generalized filter is shown in Fig. 1-1. The filter block may consist of inductors, capacitors, resistors, and possibly active elements, such as operational amplifiers and transistors. The terminations shown are a voltage source E_s, a source resistance R_s, and a load resistor R_L.

The circuit equations for the network of Fig. 1-1 can be written by using circuit-analysis techniques. Modern network theory solves these equations to determine the network values for optimum performance in some respect.

1.1 The Pole-Zero Concept

The frequency response of the generalized filter can be expressed as a ratio of two polynomials in s, where $s = j\omega$ ($j = \sqrt{-1}$, and ω, the frequency in radians per second, is $2\pi f$) and is referred to as a transfer function. This can be stated mathematically as:

$$T(s) = \frac{E_L}{E_s} = \frac{N(s)}{D(s)} \tag{1-1}$$

FIGURE **1-1**
A generalized
filter.

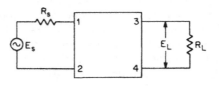

1

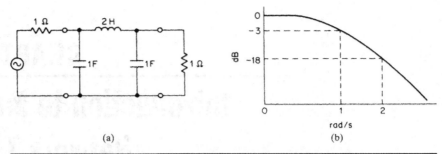

Figure 1-2   All-pole $n = 3$ low-pass filter: (a) filter circuit and (b) frequency response.

The roots of the denominator polynomial $D(s)$ are called poles, and the roots of the numerator polynomial $N(s)$ are referred to as zeros.

Deriving a network's transfer function could become quite tedious and is beyond the scope of this book. Also, knowledge of the pole-zero theory is not required for successful design of filters. However, the concept is important and allows designers to understand the differences among the various filter types and some of the sensitivities involved.

The following discussion explores the evaluation and representation of a relatively simple transfer function.

Analysis of the low-pass filter of Fig. 1-2a results in the following transfer function:

$$T(s) = \frac{1}{s^3 + 2s^2 + 2s + 1} \tag{1-2}$$

Let us now evaluate this expression at different frequencies after substituting $j\omega$ for s. The result will be expressed as the absolute magnitude of $T(j\omega)$ and the relative attenuation in decibels with respect to the response at DC.

$$T(j\omega) = \frac{1}{1 - 2\omega^2 + j(2\omega - \omega^3)} \tag{1-3}$$

ω	$|T(j\omega)|$	$20 \log |T(j\omega)|$
0	1	0 dB
1	0.707	−3 dB
2	0.124	−18 dB
3	0.0370	−29 dB
4	0.0156	−36 dB

The frequency-response curve is plotted in Fig. 1-2b.

Analysis of Eq. (1-2) indicates that the denominator of the transfer function has three roots or poles and the numerator has none. The filter is therefore called an all-pole type. Since the denominator is a third-order polynomial, the filter is also said to have an $n = 3$ complexity. The denominator poles are $s = -1$, $s = -0.500 + j0.866$, and $s = -0.500 - j0.866$.

These complex numbers can be represented as symbols on a complex-number plane. The abscissa is β, the real component of the root, and the ordinate is β, the imaginary part. Each pole is represented as the symbol X, and a zero is represented

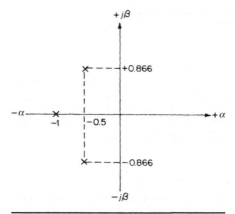

FIGURE 1-3 A complex-frequency plane representation of Eq. (1-2).

as 0. Figure 1-3 illustrates the complex-number plane representation for the roots of Eq. (1-2).

Figure 1-4 illustrates the general form of the complex-number plane, which we will call the S plane because it is based on roots of polynomials of s. Note that there are four quadrants. Because negative frequencies do not exist and are only a mathematical concept, we will only be concerned only with quadrants 1 and 2.

Let us now consider a somewhat more complex transfer function. This is given in the following equation:

$$T(s) = \frac{s+1}{(s^2 + 4s + 5)(s^2 + 2s + 5)} \tag{1-4}$$

The denominator of Eq. (1-4) can be factored into four terms as follows:

$$T(s) = \frac{s+1}{(s + 2 + j1)(s + 2 - j1)(s + 1 + j2)(s + 1 - j2)} \tag{1-5}$$

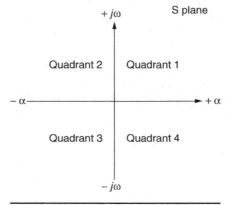

FIGURE 1-4 The S plane.

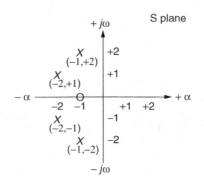

FIGURE 1-5   Poles and zeros of Eq. (1-5).

A zero in the S plane is determined at a point where the numerator is equal to zero. That would be at $s = -1$. A pole in the S plane would be determined at a point where the denominator would be zero. That would occur at the roots. The denominator has four roots, i.e., four values of s where the denominator would become zero. These are

$$s = -2 - j1$$
$$s = -2 + j1$$
$$s = -1 - j2$$
$$s = -1 + j2$$

If s were equal to any one of these four roots, the denominator would become zero and the value of $T(s)$ would be infinite. The poles and zeros of Eq. (1-5) are shown on the S plane in Fig. 1-5.

To determine the magnitude of the frequency response directly from the S plane, draw vectors from the frequency of interest to each pole and zero. This is shown in Fig. 1-6 for a frequency of interest of 0.75 rad/s. The magnitude of the frequency response is then the product of the length of all vectors from the frequency of interest to the zeros divided by the product of all vectors from the frequency of interest to the poles.

To compute the phase, all the angles of the vectors to the zeros are added, and then the sum of all the pole vectors is subtracted from that number.

Note that if the real part of a pole is small, the pole would have a dramatic effect on the frequency response in the vicinity of the imaginary part of the pole. The smaller the real part of the pole, the more influence it would have on the response and would cause a peaking effect as its imaginary part is approached on the $j\omega$ axis.

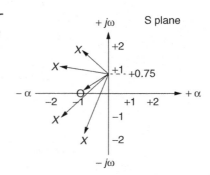

FIGURE 1-6   Evaluation of magnitude and phase directly from the S plane.

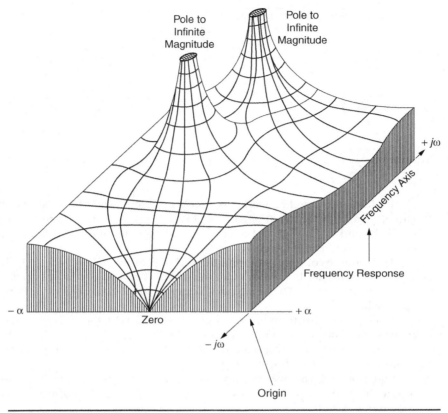

FIGURE 1-7   Circus tent analogy.

Figure 1-7 illustrates a "circus tent" analogy to poles and zeros in the S plane. Zeros are analogous to "tent stakes," where the tent would have zero height. Poles are like the "tent poles" but have infinite height. If you cut a segment away as shown in the figure, you will see a shape affected by the poles and zeros.

There are certain mathematical restrictions on the location of poles and zeros in order for the filter to be realizable. They must occur in pairs that are conjugates of each other, except for real-axis poles and zeroes, which may occur singly. Poles must also be restricted to the left plane (i.e., the real coordinate of the pole must be negative.). Zeros may occur in either plane. The numerator of the transfer function cannot be a higher order than the denominator. In other words there cannot be more zeros than poles. Also a pole cannot be on the $j\omega$ axis as that would imply an infinite value for $T(s)$, which would be an oscillatory condition; i.e., no input would produce an infinite output.

Let us consider one more S plane example. Band-pass filters are covered extensively in Chap. 5, but it would be useful to examine the pole-zero pattern and its effect on the behavior of a second-order band-pass filter here. A second-order band-pass transfer function can be expressed as:

$$T(s) = \frac{s}{s^2 + \dfrac{\omega_o}{Q}s + \omega_o{}^2} \tag{1-6}$$

FIGURE 1-8 Band-pass filter.   S plane

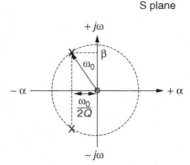

Q, also known as *selectivity factor*, is a term used to represent the "selectivity" of a pole and of a band-pass filter. (Sometimes Q is defined by the center frequency divided by the 3-dB bandwidth.) Figure 1-8 represents a two-pole band-pass filter with a zero at the origin as described by Eq. (1-6).

The center frequency is ω_o, and the real part of the pole α is $\omega_o/2Q$. Therefore Q is $\omega_o/2\alpha$. This implies that the smaller the real part of a pole (α), the higher will be the Q or selectivity factor. The imaginary component of the pole is β. For high Qs, ω_o approaches β.

It should be clear from Fig. 1-8 that as Q is increased, the real component of the pole α decreases. The pole will move closer to the $j\omega$ axis, and as a result will have a greater effect on the frequency response. In this band-pass case, a peaking will occur (increase in amplitude). This is shown in Fig. 1-9.

Equation (1-6) can also be shown in a normalized form, where the maximum magnitude of $T(s)$ is always 1. This would be Eq. (1-7), and the corresponding curves are shown in Fig. 1-10.

$$T(s) = \frac{\dfrac{\omega_o}{Q}s}{s^2 + \dfrac{\omega_o}{Q}s + \omega_o^2}$$

(1-7)

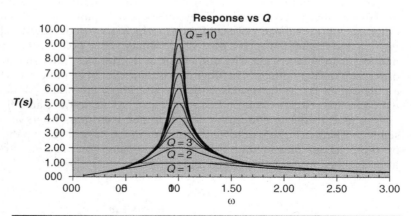

FIGURE 1-9 Effect of increasing Q in band-pass response.

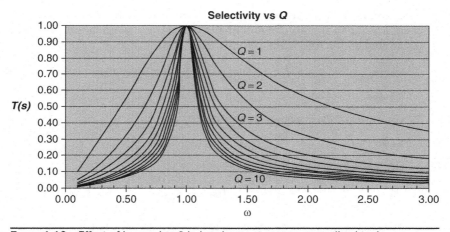

FIGURE 1-10   Effect of increasing Q in band-pass response normalized to 1.

1.2   Synthesis of Filters from Polynomials

Modern network theory has produced families of standard transfer functions that provide optimum filter performance in some desired respect. Synthesis is the process of deriving circuit component values from these transfer functions. Chapter **10** contains extensive tables of transfer functions and their associated component values so that design by synthesis is not required. Also, the downloaded computer programs (see App. A) simplify the design process. However, in order to gain some understanding as to how these values have been determined, we will now discuss a few methods of filter synthesis.

1.2.1   Synthesis by Expansion of Driving-Point Impedance

The input impedance to the generalized filter of Fig. 1-1 is the impedance seen looking into terminals 1 and 2 with terminals 3 and 4 terminated, and is referred to as the driving-point impedance or Z_{11} of the network. If an expression for Z_{11} could be determined from the given transfer function, this expression could then be expanded to define the filter.

A family of transfer functions describing the flattest possible shape and a monotonically increasing attenuation in the stop-band is known as the Butterworth low-pass response. These all-pole transfer functions have denominator polynomial roots, which all fall on a circle having a radius of unity from the origin of the $j\omega$ axis. The attenuation for this family is 3 dB at 1 rad/s.

The transfer function of Eq. (1-2) satisfies this criterion. It is evident from Fig. 1-3 that if a circle were drawn having a radius of 1, with the origin as the center, it would intersect the real root and both complex roots.

If R_s in the generalized filter of Fig. 1-1 is set to 1 Ω, a driving-point impedance expression can be derived in terms of the Butterworth transfer function as

$$Z_{11} = \frac{D(s) - s^n}{D(s) + s^n} \tag{1-8}$$

where $D(s)$ is the denominator polynomial of the transfer function, and n is the order of the polynomial.

After $D(s)$ is substituted into Eq. (1-8), Z_{11} is expanded using the continued fraction expansion. This expansion involves successive division and inversion of a ratio of two polynomials. The final form contains a sequence of terms, each alternately representing a capacitor and an inductor and finally the resistive termination. This procedure is demonstrated by the following example.

Example 1-1   Synthesis of $n = 3$ Butterworth Low-Pass Filter by Continued Fraction Expansion

Required:

A low-pass LC filter having a Butterworth $n = 3$ response.

Result:

(a) Use the Butterworth transfer function:

$$T(s) = \frac{1}{s^3 + 2s^2 + 2s + 1} \tag{1-2}$$

(b) Substitute $D(s) = s^3 + 2s^2 + 2s + 1$ and $s^n = s^3$ into Eq. (1-8), which results in

$$Z_{11} = \frac{2s^2 + 2s + 1}{2s^3 + 2s^2 + 2s + 1} \tag{1-9}$$

(c) Express Z_{11} so that the denominator is a ratio of the higher-order to the lower-order polynomial:

$$Z_{11} = \frac{1}{\dfrac{2s^3 + 2s^2 + 2s + 1}{2s^2 + 2s + 1}}$$

(d) Dividing the denominator and inverting the remainder results in

$$Z_{11} = \frac{1}{s + \dfrac{1}{\dfrac{2s^2 + 2s + 1}{s + 1}}}$$

(e) After further division and inversion, we get as our final expression:

$$Z_{11} = \frac{1}{s + \dfrac{1}{2s + \dfrac{1}{s + 1}}} \tag{1-10}$$

The circuit configuration of Fig. 1-11 is called a ladder network, since it consists of alternating series and shunt branches. The input impedance can be expressed as the following continued fraction:

$$Z_{11} = \frac{1}{Y_1 + \dfrac{1}{Z_2 + \dfrac{1}{Y_3 + \cdots \dfrac{1}{Z_{n-1} + \dfrac{1}{Y_n}}}}} \tag{1-11}$$

where $Y = sC$ and $Z = sL$ for the low-pass all-pole ladder except for a resistive termination where $Y_n = sC + 1/R_L$.

Figure 1-12 can then be derived from Eqs. (1-10) and (1-11) by inspection. This can be proved by reversing the process of expanding Z_{11}. By alternately adding admittances and impedances while working toward the input, Z_{11} is verified as being equal to Eq. (1-10).

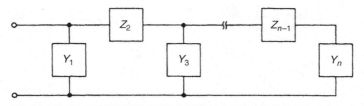

FIGURE 1-11 A general ladder network.

1.2.2 Synthesis for Unequal Terminations

If the source resistor is set equal to 1 Ω and the load resistor is desired to be infinite (unterminated), the impedance looking into terminals 1 and 2 of the generalized filter of Fig.1-1 can be expressed as

$$Z_{11} = \frac{D(s\ \text{even})}{D(s\ \text{odd})} \tag{1-12}$$

$D(s\ \text{even})$ contains all the even-power s terms of the denominator polynomial, and $D(s\ \text{odd})$ consists of all the odd-power s terms of any realizable all-pole low-pass transfer function. Z_{11} is expanded into a continued fraction, as was done in Example 1-1, to define the circuit.

Example 1-2 Synthesis of $n = 3$ Butterworth Low-Pass Filter for an Infinite Termination

Required:

Low-pass filter having a Butterworth $n = 3$ response with a source resistance of 1 Ω and an infinite termination.

Result:

(*a*) Use the Butterworth transfer function:

$$T(s) = \frac{1}{s^3 + 2s^2 + 2s + 1} \tag{1-2}$$

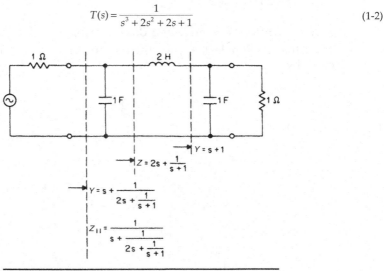

FIGURE 1-12 The low-pass filter for Eq. (1-10).

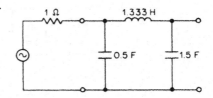

Figure 1-13 The low-pass filter of Example 1-2.

(*b*) Substitute $D(s \text{ even}) = 2s^2 + 1$ and $D(s \text{ odd}) = s^3 + 2s$ into Eq. (1-12):

$$Z_{11} = \frac{2s^2 + 1}{s^3 + 2s} \tag{1-13}$$

(*c*) Express Z_{11} so that the denominator is a ratio of the higher- to the lower-order polynomial:

$$Z_{11} = \frac{1}{\dfrac{s^3 + 2s}{2s^2 + 1}}$$

(*d*) Dividing the denominator and inverting the remainder results in

$$Z_{11} = \frac{1}{0.5s + \dfrac{1}{\dfrac{2s^2 + 1}{1.5s}}}$$

(*e*) Dividing and further inverting results in the final continued fraction:

$$Z_{11} = \frac{1}{0.5s + \dfrac{1}{1.333s + \dfrac{1}{1.5s}}} \tag{1-14}$$

The circuit is shown in Fig. 1-13.

1.2.3   Synthesis by Equating Coefficients

An active three-pole low-pass filter is shown in Fig. 1-14. Its transfer function is given by:

$$T(s) = \frac{1}{s^3 A + s^2 B + sC + 1} \tag{1-15}$$

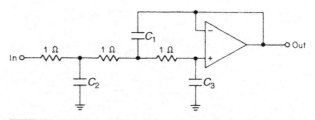

Figure 1-14 General $n = 3$ active low-pass filter.

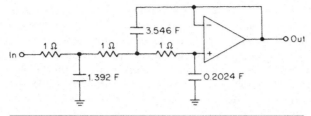

FIGURE 1-15   Butterworth $n = 3$ active low-pass filter.

where

$$A = C_1 C_2 C_3 \qquad (1\text{-}16)$$

$$B = 2C_3(C_1 + C_2) \qquad (1\text{-}17)$$

and

$$C = C_2 + 3C_3 \qquad (1\text{-}18)$$

If a Butterworth transfer function is desired, we can set Eq. (1-15) equal to Eq. (1-2):

$$T(s) = \frac{1}{s^3 A + s^2 B + sC + 1} = \frac{1}{s^3 + 2s^2 + 2s + 1} \qquad (1\text{-}19)$$

By equating coefficients, we obtain

$$A = 1$$
$$B = 2$$
$$C = 2$$

Substituting these coefficients in Eqs. (1-16) through (1-18) and solving for C_1, C_2, and C_3 results in the circuit of Fig. 1-15.

Synthesis of filters directly from polynomials offers an elegant solution to filter design. However, it also may involve laborious computations to determine circuit-element values. Design methods have been greatly simplified by the curves, tables, computer programs, and step-by-step procedures provided in this handbook, so design by synthesis can be left to the advanced specialist.

1.3   Active versus Passive Filters

The LC filters of Figs. 1-12 and 1-13 and the active filter of Fig. 1-15 all satisfy an $n = 3$ Butterworth low-pass transfer function. The filter designer is frequently faced with the sometimes difficult decision of choosing whether to use an active or LC design. A number of factors must be considered. Some of the limitations and considerations for each filter type will now be discussed.

1.3.1   Frequency Limitations

At subaudio frequencies, LC-filter designs require high values of inductance and capacitance along with their associated bulk. Active filters are more practical because they can be designed at higher impedance levels so that capacitor magnitudes are reduced.

Above 20 MHz or so, most commercial-grade operational amplifiers have insufficient open-loop gain for the average active filter requirement. However, amplifiers are available with extended bandwidth at an increased cost so that active filters at frequencies up to 100 MHz and even beyond are possible. LC filters, on the other hand, are practical at frequencies up to a few hundred megahertz. Beyond this range, filters become impractical to build in lumped form, and so distributed parameter techniques are used, such as stripline or microstrip, where a PC board functions as a distributed transmission line.

1.3.2   Size Considerations

Active filters are generally smaller than their LC counterparts since inductors are not required. Further reduction in size is possible with microelectronic technology. Surface-mount components for the most part have replaced hybrid technology, whereas in the past hybrids were the only way to reduce the size of active filters.

1.3.3   Economics and Ease of Manufacture

LC filters generally cost more than active filters because they use inductors. High-quality coils require efficient magnetic cores. Sometimes, special coil-winding methods are needed as well. These factors lead to the increased cost of LC filters.

Active filters have the distinct advantage that they can be easily assembled using standard off-the-shelf components. LC filters require coil-winding and coil-assembly skills. In addition, eliminating inductors prevents magnetic emissions, which can be troublesome.

1.3.4   Ease of Adjustment

In critical LC filters, tuned circuits require adjustment to specific resonances. Capacitors cannot be made variable unless they are below a few hundred picofarads. Inductors, however, can easily be adjusted, since most coil structures provide a means for tuning, such as an adjustment slug for a ferrite potcore.

Many active filter circuits are not easily adjustable, however. They may contain RC sections where two or more resistors in each section have to be varied in order to control resonance. These types of circuit configurations are avoided. The active filter design techniques presented in this handbook include convenient methods for adjusting resonances where required, such as for narrowband band-pass filters.

References

Guillemin, E. A. (1957). Introduction to Circuit Theory. *New York: John Wiley and Sons.*
Stewart, J. L. (1956). *Circuit Theory and Design.* New York: John Wiley and Sons.
White Electromagnetics (1963). *A Handbook on Electrical Filters.* White Electromagnetics, Inc.

CHAPTER 2

Selecting the Response Characteristic

2.1 Frequency-Response Normalization

Several parameters are used to characterize a filter's performance. The most commonly specified requirement is frequency response. When given a frequency-response specification, the engineer must select a filter design that meets these requirements. This is accomplished by transforming the required response to a normalized low-pass specification having a cutoff of 1 rad/s. This normalized response is compared with curves of normalized low-pass filters which also have a 1-rad/s cutoff. After a satisfactory low-pass filter is determined from the curves, the tabulated normalized element values of the chosen filter are transformed or denormalized to the final design.

Modern network theory has provided us with many different shapes of amplitude versus frequency which have been analytically derived by placing various restrictions on transfer functions. The major categories of these low-pass responses are:

- Butterworth **maximally flat amplitude**
- Chebyshev
- Bessel **maximally flat delay**
- Linear phase with equiripple error
- Transitional
- Synchronously tuned
- Elliptic-function
- Papoulis optimum "L"

With the exception of the elliptic-function family, these responses are all normalized to a 3-dB cutoff of 1 rad/s.

2.1.1 Frequency and Impedance Scaling

The basis for normalization of filters is the fact that a given filter's response can be scaled (shifted) to a different frequency range by dividing the reactive elements by a frequency-scaling factor (FSF). The FSF is the ratio of a reference frequency of the desired response to the corresponding reference frequency of the given filter. Usually

3-dB points are selected as reference frequencies of low-pass and high-pass filters, and the center frequency is chosen as the reference for band-pass filters. The FSF can be expressed as

$$\text{FSF} = \frac{\text{desired reference frequency}}{\text{existing reference frequency}} \tag{2-1}$$

The FSF must be a dimensionless number; so both the numerator and denominator of Eq. (2-1) must be expressed in the same units, usually radians per second. The following example demonstrates the computation of the FSF and frequency scaling of filters.

Example 2-1   Frequency Scaling of a Low-Pass Filter

Required:

A low-pass filter, either *LC* or active, with an $n = 3$ Butterworth transfer function having a 3-dB cutoff at 1,000 Hz.

Result:

Figure 2-1 illustrates the *LC* and active $n = 3$ Butterworth low-pass filters discussed in Chap. 1 and their response.

(*a*) Compute FSF, using Eq. (2-1).

$$\text{FSF} = \frac{2\pi 1,000 \text{ rad/s}}{1 \text{ rad/s}} = 6,280$$

(*b*) Dividing all the reactive elements by the FSF results in the filters of Fig. 2-2*a* and *b* and the response of Fig. 2-2*c*.

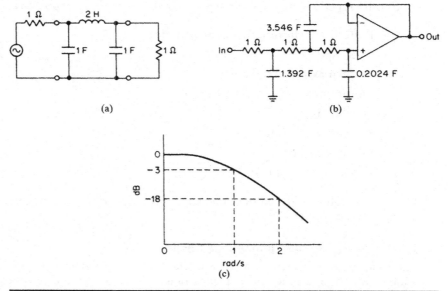

(a)   (b)

(c)

Figure 2-1 $n = 3$ Butterworth low-pass filter: (a) *LC* filter, (b) active filter, (c) frequency response.

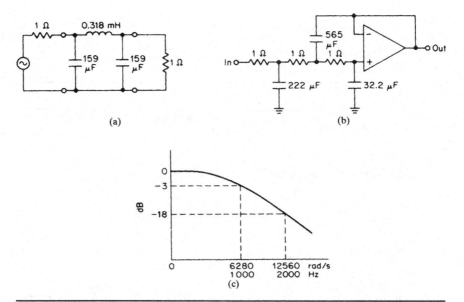

FIGURE 2-2 The denormalized low-pass filter of Example 2-1: (a) *LC* filter, (b) active filter, and (c) frequency response.

Note that all points on the frequency axis of the normalized response have been multiplied by the FSF. Also, since the normalized filter has its cutoff at 1 rad/s, the FSF can be directly expressed by $2\pi f_c$, where f_c is the desired low-pass cutoff frequency in hertz.

Frequency scaling a filter has the effect of multiplying all points on the frequency axis of the response curve by the FSF. Therefore, a normalized response curve can be directly used to predict the attenuation of the denormalized filter.

When the filters of Fig. 2-1 were denormalized to those of Fig. 2-2, the transfer function changed as well. The denormalized transfer function became

$$T(s) = \frac{1}{4.03 \times 10^{-12}\,s^3 + 5.08 \times 10^{-9}\,s^2 + 3.18 \times 10^{-4}\,s + 1} \tag{2-2}$$

The denominator has roots:

$$s = -6,280, \ s = -3,140 + j5,438, \text{ and } s = -3,140 - j5,438.$$

These roots can be obtained directly from the normalized roots by multiplying the normalized root coordinates by the FSF. Frequency scaling a filter also scales the poles and zeros (if any) by the same factor.

The component values of the filters in Fig. 2-2 are not very practical. The capacitor values are much too large and the 1-Ω resistor values are not very desirable. This situation can be resolved by impedance scaling. Any linear active or passive network maintains its transfer function if all resistor and inductor values are multiplied by an impedance-scaling factor Z, and all capacitors are divided by the same factor Z. This occurs because the Zs cancel in the transfer function. To prove this, let's investigate the transfer function of the simple two-pole low-pass filter of Fig. 2-3a, which is

$$T(s) = \frac{1}{s^2 LC + sCR + 1} \tag{2-3}$$

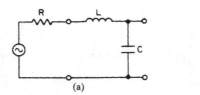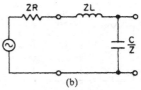

FIGURE **2-3**   A two-pole low-pass *LC* filter: (*a*) a basic filter and (*b*) an impedance-scaled filter.

Impedance scaling can be mathematically expressed as

$$R' = ZR \tag{2-4}$$

$$L' = ZL \tag{2-5}$$

$$C' = \frac{C}{Z} \tag{2-6}$$

where the primes denote the values after impedance scaling.

If we impedance-scale the filter, we obtain the circuit of Fig. 2-3*b*. The new transfer function then becomes

$$T(s) = \frac{1}{s^2 ZL \dfrac{C}{Z} + s \dfrac{C}{Z} ZR + 1} \tag{2-7}$$

Clearly, the *Z*s cancel, so both transfer functions are equivalent.

We can now use impedance scaling to make the values in the filters of Fig. 2-2 more practical. If we use impedance scaling with a *Z* of 1,000, we obtain the filters of Fig. 2-4. The values are certainly more suitable.

Frequency and impedance scaling are normally combined into one step rather than performed sequentially. The denormalized values are then given by

$$R' = R \times Z \tag{2-8}$$

$$L' = \frac{L \times Z}{\text{FSF}} \tag{2-9}$$

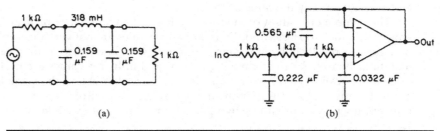

FIGURE **2-4**   The impedance-scaled filters of Example 2-1: (*a*) *LC* filter and (*b*) active filter.

$$C' = \frac{C}{\text{FSF} \times Z} \qquad (2\text{-}10)$$

where the primed values are both frequency- and impedance-scaled.

2.1.2 Low-Pass Normalization

In order to use normalized low-pass filter curves and tables, a given low-pass filter requirement must first be converted into a normalized requirement. The curves can now be entered to find a satisfactory normalized filter which is then scaled to the desired cutoff.

The first step in selecting a normalized design is to convert the requirement into a steepness factor A_s which can be defined as

$$A_s = \frac{f_s}{f_c} \qquad (2\text{-}11)$$

where f_s is the frequency having the minimum required stopband attenuation and f_c is the limiting frequency or cutoff of the passband, usually the 3-dB point. The normalized curves are compared with A_s, and a design is selected that meets or exceeds the requirement. The design is may have to be frequency scaled so that the selected passband limit of the normalized design occurs at f_c.

If the required passband limit f_c is defined as the 3-dB cutoff, the steepness factor A_s can be directly looked up in radians per second on the frequency axis of the normalized curves.

Suppose that we required a low-pass filter that has a 3-dB point at 100 Hz and more than 30-dB attenuation at 400 Hz. A normalized low-pass filter that has its 3-dB point at 1 rad/s and over 30-dB attenuation at 4 rad/s would meet the requirement if the filter were frequency-scaled so that the 3-dB point occurred at 100 Hz. Then there would be over 30-dB attenuation at 400 Hz, or four times the cutoff, because a response shape is retained when a filter is frequency scaled.

The following example demonstrates normalizing a simple low-pass requirement.

Example 2-2   Normalizing a Low-Pass Specification for a 3-dB Cutoff

Required:

Normalize the following specification:

   A low-pass filter
   3 dB at 200 Hz
   30-dB minimum at 800 Hz

Result:

(*a*) Compute A_s using Eq. (2-11):

$$A_s = \frac{f_s}{f_c} = \frac{800\ \text{Hz}}{200\ \text{Hz}} = 4$$

(*b*) Normalized requirement:

   3 dB at 1 rad/s
   30-dB minimum at 4 rad/s

In the event f_c does not correspond to the 3-dB cutoff, A_s can still be computed and a normalized design found that will meet the specifications. This is illustrated in the following example.

Example 2-3   Normalizing a Low-Pass Specification for a 1-dB Cutoff

Required:

Normalize the following specification:

A low-pass filter
1 dB at 200 Hz
30-dB minimum at 800 Hz

Result:

(a) Compute A_s, using Eq. (2-11):

$$A_s = \frac{f_s}{f_c} = \frac{800 \text{ Hz}}{200 \text{ Hz}} = 4$$

(b) Normalized requirement:

1 dB at K rad/s
30-dB minimum at 4 K rad/s
(where K is arbitrary)

A possible solution to Example 2-3 would be a normalized filter which has a 1-dB point at 0.8 rad/s and over 30-dB attenuation at 3.2 rad/s. The fundamental requirement is that the normalized filter makes the transition between the passband and stopband limits within a frequency ratio A_s.

2.1.3   High-Pass Normalization

A normalized $n = 3$ low-pass Butterworth transfer function was given in Sec. 1.1 as

$$T(s) = \frac{s^3}{s^3 + 2s^2 + 2s + 1} \tag{1-2}$$

and the results of evaluating this transfer function at various frequencies were as demonstrated in the following table.

ω	$|T(j\omega)|$	20 log $|T(j\omega)|$
0	1	0 dB
1	0.707	−3 dB
2	0.124	−18 dB
3	0.0370	−29 dB
4	0.0156	−36 dB

Let's now perform a high-pass transformation by substituting $1/s$ for s in Eq. (1-2). After some algebraic manipulations, the resulting transfer function becomes

$$T(s) = \frac{s^3}{s^3 + 2s^2 + 2s + 1} \tag{2-12}$$

If we evaluate this expression at specific frequencies, we can generate the following table:

ω	$|T(j\omega)|$	$20 \log |T(j\omega)|$
0.25	0.0156	−36 dB
0.333	0.0370	−29 dB
0.500	0.124	−18 dB
1	0.707	−3 dB
∞	1	0 dB

The response is clearly that of a high-pass filter. It is also apparent that the low-pass attenuation values now occur at high-pass frequencies that are exactly the reciprocals of the corresponding low-pass frequencies. A high-pass transformation of a normalized low-pass filter transposes the low-pass attenuation values to reciprocal frequencies and retains the 3-dB cutoff at 1 rad/s. This relationship is evident in Fig. 2-5, where both filter responses are compared.

The normalized low-pass curves could be interpreted as normalized high-pass curves by reading the attenuation as indicated and taking the reciprocals of the frequencies. However, it is much easier to convert a high-pass specification into a normalized low-pass requirement and use the curves directly.

To normalize a high-pass filter specification, calculate A_s, which in the case of high-pass filters is given by

$$A_s = \frac{f_c}{f_s} \tag{2-13}$$

Since the A_s, for high-pass filters is defined as the reciprocal of the A_s for low-pass filters, Eq. (2-13) can be directly interpreted as a low-pass requirement. A normalized low-pass filter can then be selected from the curves. A high-pass transformation is performed on the corresponding low-pass filter, and the resulting high-pass filter is scaled to the desired cutoff frequency.

FIGURE 2-5   A normalized low-pass high-pass relationship.

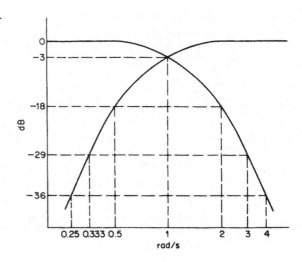

The following example shows the normalization of a high-pass filter requirement.

Example 2-4   Normalizing a High-Pass Specification

Required:

Normalize the following requirement:

A high-pass filter
3 dB at 200 Hz
30-dB minimum at 50 Hz

Result:

(a) Compute A_s, using Eq. (2-13)

$$A_s = \frac{f_c}{f_s} = \frac{200 \text{ Hz}}{50 \text{ Hz}} = 4$$

(b) Normalized equivalent low-pass requirement:

3 dB at 1 rad/s
30-dB minimum at 4 rad/s

2.1.4   Band-Pass Normalization

Band-pass filters fall into two categories: narrowband and wideband. If the ratio of the upper-cutoff frequency to the lower-cutoff frequency is over 2 (an octave), the filter is considered a wideband type.

Wideband Band-Pass Filters

Wideband filter specifications can be separated into individual low-pass and high-pass requirements, which are treated independently. The resulting low-pass and high-pass filters are then cascaded to meet the composite response.

Example 2-5   Normalizing a Wideband Band-Pass Filter

Required:

Normalize the following specification:

band-pass filter
3 dB at 500 and 1,000 Hz
40-dB minimum at 200 and 2,000 Hz

Result:

(a) Determine the ratio of upper cutoff to lower cutoff.

$$\frac{1,000 \text{ Hz}}{500 \text{ Hz}} = 2$$

wideband type

(b) Separate requirement into individual specifications.

High-pass filter:	Low-pass filter:
3 dB at 500 Hz	3 dB at 1,000 Hz
40-dB minimum at 200 Hz	40-dB minimum at 2000 Hz
$A_s = 2.5$ (2-13)	$A_s = 2.0$ (2-11)

FIGURE 2-6 The results of Example 2-5:
(a) cascade of low-pass and high-pass
filters and (b) frequency response.

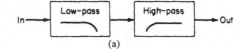

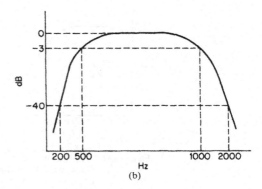

(b)

(c) Normalized high-pass and low-pass filters are now selected, scaled to the required cutoff frequencies, and cascaded to meet the composite requirements. Figure 2-6 shows the resulting circuit and response.

Narrowband Band-Pass Filters

Narrowband band-pass filters have a ratio of upper cutoff frequency to lower cutoff frequency of approximately 2 or less and cannot be designed as separate low-pass and high-pass filters. The major reason for this is evident from Fig. 2-7. As the ratio of upper cutoff to lower cutoff decreases, the loss at the center frequency will increase, and it may become prohibitive for ratios near unity.

If we substitute $s + 1/s$ for s in a low-pass transfer function, a band-pass filter results. The center frequency occurs at 1 rad/s, and the frequency response of the low-pass filter is directly transformed into the bandwidth of the band-pass filter at points of equivalent attenuation. In other words, the attenuation bandwidth ratios remain unchanged. This is shown in Fig. 2-8, which shows the relationship between a low-pass filter and its transformed band-pass equivalent. Each pole and zero of the low-pass filter is transformed into a *pair* of poles and zeros in the band-pass filter.

In order to design a band-pass filter, the following sequence of steps is involved.

1. Convert the given band-pass filter requirement into a normalized low-pass specification.

2. Select a satisfactory low-pass filter from the normalized frequency-response curves.

3. Transform the normalized low-pass parameters into the required band-pass filter.

The response shape of a band-pass filter is shown in Fig. 2-9, along with some basic terminology. The center frequency is defined as:

$$f_0 = \sqrt{f_L f_u} \tag{2-14}$$

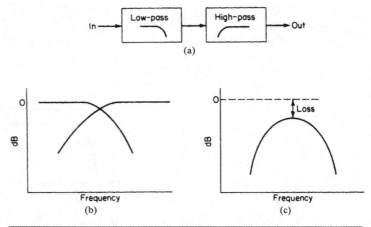

FIGURE 2-7 Limitations of the wideband approach for narrowband filters: (*a*) a cascade of low-pass and high-pass filters, (*b*) a composite response, and (*c*) algebraic sum of attenuation.

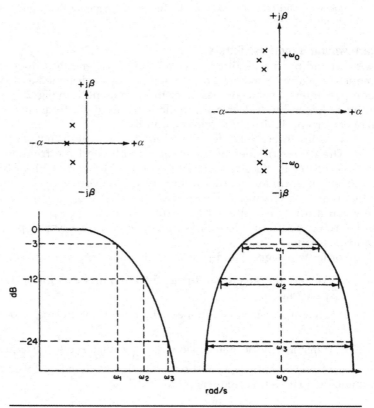

FIGURE 2-8 A low-pass to band-pass transformation.

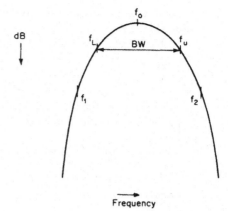

FIGURE 2-9 A general band-pass filter response shape.

where f_L is the lower passband limit and f_u is the upper passband limit, usually the 3-dB attenuation frequencies. For the more general case

$$f_0 = \sqrt{f_1 f_2} \tag{2-15}$$

where f_1 and f_2 are any two frequencies having equal attenuation. These relationships imply geometric symmetry; that is, the entire curve below f_0 is the mirror image of the curve above f_0 when plotted on a *logarithmic* frequency axis.

An important parameter of band-pass filters is the filter selectivity factor or Q, which is defined as

$$Q = \frac{f_0}{\text{BW}} \tag{2-16}$$

where BW is the passband bandwidth or $f_u - f_L$.

As the filter Q increases, the response shape near the passband approaches the arithmetically symmetrical condition which is mirror-image symmetry near the center frequency, when plotted using a *linear* frequency axis. For Qs of 10 or more, the center frequency can be redefined as the arithmetic mean of the passband limits, so we can replace Equation (2-14) with

$$f_0 = \frac{f_L + f_u}{2} \tag{2-17}$$

In order to utilize the normalized low-pass filter frequency-response curves, a given narrowband band-pass filter specification must be transformed into a normalized low-pass requirement. This is accomplished by first manipulating the specification to make it geometrically symmetrical. At equivalent attenuation points, corresponding frequencies above and below f_0 must satisfy

$$f_1 f_2 = f_0^2 \tag{2-18}$$

which is an alternate form of Eq. (2-15) for geometric symmetry. The given specification is modified by calculating the corresponding opposite geometric frequency for each stopband frequency specified. Each pair of stopband frequencies will result in two new frequency pairs. The pair having the *lesser* separation is retained, since it represents the more severe requirement.

A band-pass filter steepness factor can now be defined as

$$A_s = \frac{\text{stopband bandwidth}}{\text{passband bandwidth}} \qquad (2\text{-}19)$$

This steepness factor is used to select a normalized low-pass filter from the frequency-response curves that makes the passband to stopband transition within a frequency ratio of A_s.

The following example shows the normalization of a band-pass filter requirement.

Example 2-6   Normalizing a Band-pass Filter Requirement

Required:

Normalize the following band-Pass filter requirement:

A band-pass filter
A center frequency of 100 Hz
3 dB at ± 15Hz (85 Hz, 115 Hz)
40 dB at ± 30 Hz (70 Hz, 130 Hz)

Result:

(*a*) First, compute the center frequency f_0, using Eq. (2-14)

$$f_0 = \sqrt{f_L f_u} = \sqrt{85 \times 115} = 98.9 \text{ Hz}$$

(*b*) Compute two geometrically related stopband frequency pairs for each pair of stopband frequencies given, using Eq. (2-18).

Let $f_1 = 70$ Hz.

$$f_2 = \frac{f_0^2}{f_1} = \frac{(98.9)^2}{70} = 139.7 \text{ Hz}$$

Let $f_2 = 130$ Hz.

$$f_1 = \frac{f_0^2}{f_2} = \frac{(98.9)^2}{130} = 75.2 \text{ Hz}$$

The two pairs are

$$f_1 = 70 \text{ Hz}, f_2 = 139.7 \text{ Hz} \ (f_2 - f_1 = 69.7 \text{ Hz})$$

and

$$f_1 = 75.2 \text{ Hz}, f_2 = 130 \text{ Hz} \ (f_2 - f_1 = 54.8 \text{ Hz})$$

Retain the second frequency pair, since it has the lesser separation. Figure 2-10 compares the specified filter requirement and the geometrically symmetrical equivalent.

(*c*) Calculate A_s, using Eq. (2-19):

$$A_s = \frac{\text{stopband bandwidth}}{\text{passband bandwidth}} = \frac{54.8 \text{ Hz}}{30 \text{ Hz}} = 1.83 \qquad (2\text{-}19)$$

(*d*) A normalized low-pass filter can now be selected from the normalized curves. Since the passband limit is the 3-dB point, the normalized filter is required to have over 40 dB of rejection at 1.83 rad/s or 1.83 times the 1-rad/s cutoff.

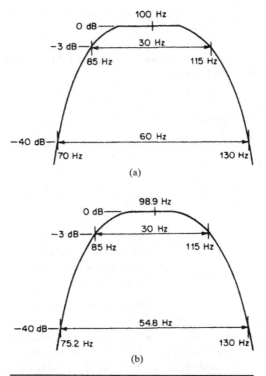

Figure 2-10 The frequency-response requirements of Example 2-6: (a) a given filter requirement and (b) a geometrically symmetrical requirement.

The results of Example 2-6 indicate that when frequencies are specified in an arithmetically symmetrical manner, the narrower stopband bandwidth can be directly computed by

$$BW_{stopband} = f_2 - \frac{f_0^2}{f_2} \qquad (2\text{-}20)$$

The narrower stopband bandwidth corresponds to the more stringent value of A_s, the steepness factor.

It is sometimes desirable to compute two geometrically related frequencies that correspond to a given bandwidth. Upon being given the center frequency f_0 and the bandwidth BW, the lower and upper frequencies are respectively computed by

$$f_1 = \sqrt{\left(\frac{BW}{2}\right)^2 + f_0^2} - \frac{BW}{2} \qquad (2\text{-}21)$$

$$f_2 = \sqrt{\left(\frac{BW}{2}\right)^2 + f_0^2} + \frac{BW}{2} \qquad (2\text{-}22)$$

Use of these formulas is illustrated in the following example.

Example 2-7   Determining Band-Pass Filter Bandwidths at Equal Attenuation Points

Required:

For a band-pass filter having a center frequency of 10 kHz, determine the frequencies corresponding to bandwidths of 100 Hz, 500 Hz, and 2,000 Hz.

Result:

Compute f_1 and f_2 for each bandwidth, using

$$f_1 = \sqrt{\left(\frac{BW}{2}\right)^2 + f_0^2} - \frac{BW}{2} \tag{2-21}$$

$$f_2 = \sqrt{\left(\frac{BW}{2}\right)^2 + f_0^2} + \frac{BW}{2} \tag{2-22}$$

BW, Hz	f_1, Hz	F_2, Hz
100	9,950	10,050
500	9,753	10,253
2,000	9,050	11,050

The results of Example 2-7 indicate that for narrow percentage bandwidths (1 percent) f_1 and f_2 are arithmetically spaced about f_0. For the wider cases, the arithmetic center of f_1 and f_2 would be slightly above the actual geometric center frequency f_0. Another and more meaningful way of stating the converse is that for a given pair of frequencies, the geometric mean is below the arithmetic mean.

Band-pass filter requirements are not always specified in an arithmetically symmetrical manner as in the previous examples. Multiple stopband attenuation requirements may also exist. The design engineer is still faced with the basic problem of converting the given parameters into geometrically symmetrical characteristics so that a steepness factor (or factors) can be determined. The following example demonstrates the conversion of a specification somewhat more complicated than the previous example.

Example 2-8   Normalizing a Nonsymmetrical Band-Pass Filter Requirement

Required:

Normalize the following band-pass filter specifications:

  Band-pass filter
  1-dB passband limits of 12 kHz and 14 kHz
  20-dB minimum at 6 kHz
  30-dB minimum at 4 kHz
  40-dB minimum at 56 kHz

Result:

(*a*) First, compute the center frequency, using Eq. (2-14):

$$f_L = 12 \text{ kHz} \qquad f_u = 14 \text{ kHz}$$

$$f_0 = 12.96 \text{ kHz}$$

FIGURE 2-11   The given and transformed responses of Example 2-7: (*a*) a given requirement and (*b*) geometrically symmetrical response.

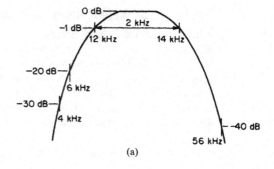

(a)

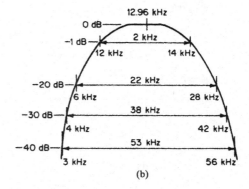

(b)

(*b*) Compute the corresponding geometric frequency for each stopband frequency given, using Eq. (2-18).

$$f_1 f_2 = f_0^2 \qquad (2\text{-}18)$$

Figure 2-11 illustrates the comparison between the given requirement and the corresponding geometrically symmetrical equivalent response.

f_1	f_2
6 kHz	28 kHz
4 kHz	42 kHz
3 kHz	56 kHz

(*c*) Calculate the steepness factor for each stopband bandwidth in Fig. 2-11*b*.

20 dB: $A_s = \dfrac{22 \text{ kHz}}{2 \text{ kHz}} = 11$ $\qquad (2\text{-}19)$

30 dB: $A_s = \dfrac{38 \text{ kHz}}{2 \text{ kHz}} = 19$

40 dB: $A_s = \dfrac{53 \text{ kHz}}{2 \text{ kHz}} = 26.5$

(*d*) Select a low-pass filter from the normalized tables. A filter is required that has over 20, 30, and 40 dB of rejection at, respectively, 11, 19, and 26.5 times its 1-dB cutoff.

2.1.5 Band-Reject Normalization

Wideband Band-Reject Filters

Normalizing a band-reject filter requirement proceeds along the same lines as for a band-pass filter. If the ratio of the upper cutoff frequency to the lower cutoff frequency is an octave or more, a band-reject filter requirement can be classified as wideband and separated into individual low-pass and high-pass specifications. The resulting filters are *paralleled* at the input and combined at the output. The following example demonstrates normalization of a wideband band-reject filter requirement.

Example 2-9 Normalizing a Wideband Band-Reject Filter

Required:

A band-reject filter
3 dB at 200 and 800 Hz
40-dB minimum at 300 and 500 Hz

Result:

(a) Determine the ratio of upper cutoff to lower cutoff, using

$$\frac{800 \text{ Hz}}{200 \text{ Hz}} = 4$$

wideband type

(b) Separate requirements into individual low-pass and high-pass specifications.

Low-pass filter:	High-pass filter:
3 dB at 200 Hz	3 dB at 800 Hz
40-dB minimum at 300 Hz	40-dB minimum at 500 Hz
$A_s = 1.5$ (2-11)	$A_s = 1.6$ (2-13)

(c) Select appropriate filters from the normalized curves and scale the normalized low-pass and high-pass filters to cutoffs of 200 Hz and 800 Hz, respectively. Figure 2-12 shows the resulting circuit and response.

The basic assumption of the previous example is that when the filter outputs are combined, the resulting response is the superimposed individual response of both filters. This is a valid assumption if each filter has sufficient rejection in the band of the other filter so that there is no interaction when the outputs are combined. Figure 2-13 shows the case where inadequate separation exists.

The requirement for a minimum separation between cutoffs of an octave or more is by no means rigid. Sharper filters can have their cutoffs placed closer together with minimal interaction.

Narrowband Band-Reject Filters

The normalized transformation described for band-pass filters where $s + 1/s$ is substituted into a low-pass transfer function can instead be applied to a high-pass transfer function to obtain a band-reject filter. Figure 2-14 shows the direct equivalence between a high-pass filter's frequency response and the transformed band-reject filter's bandwidth.

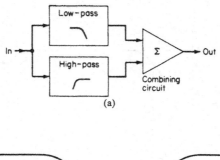

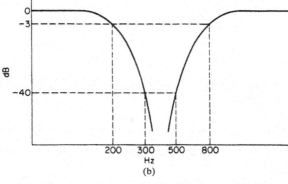

FIGURE 2-12 The results of Example 2-9: (*a*) combined low-pass and high-pass filters, and (*b*) a frequency response.

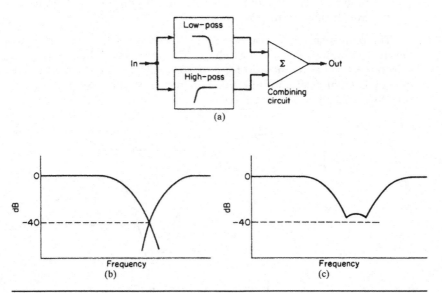

FIGURE 2-13 Limitations of the wideband band-reject design approach: (*a*) combined low-pass and high-pass filters (*b*) composite response; and (*c*) combined response by the summation of outputs.

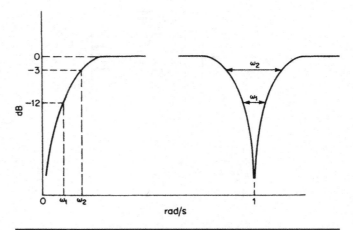

FIGURE 2-14 The relationship between band-reject and high-pass filters.

The design method for narrowband band-reject filters can be defined as follows:

1. Convert the band-reject requirement directly into a normalized low-pass specification.

2. Select a low-pass filter (from the normalized curves) that meets the normalized requirements.

3. Transform the normalized low-pass parameters into the required band-reject filter. This may involve designing the intermediate high-pass filter, or the transformation may be direct.

The band-reject response has geometric symmetry just as band-pass filters have. Figure 2-15 defines this response shape. The parameters shown have the same relationship to each other as they do for band-pass filters. The attenuation at the center frequency is theoretically infinite since the response of a high-pass filter at DC has been transformed to the center frequency.

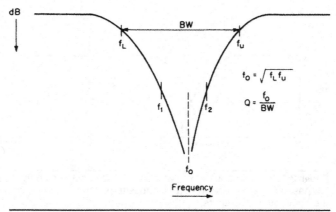

FIGURE 2-15 The band-reject response.

The geometric center frequency can be defined as

$$f_0 = \sqrt{f_L f_u} \tag{2-14}$$

where f_L and f_u are usually the 3-dB frequencies, or for the more general case:

$$f_0 = \sqrt{f_1 f_2} \tag{2-15}$$

The selectivity factor Q is defined as

$$Q = \frac{f_0}{\text{BW}} \tag{2-16}$$

where BW is $f_u - f_L$. For Qs of 10 or more, the response near the center frequency approaches the arithmetically symmetrical condition, so we can then state

$$f_0 = \frac{f_L + f_u}{2} \tag{2-17}$$

To use the normalized curves for the design of a band-reject filter, the response requirement must be converted to a normalized low-pass filter specification. In order to accomplish this, the band-reject specification should first be made geometrically symmetrical—that is, each pair of frequencies having equal attenuation should satisfy

$$f_1 f_2 = f_0^2 \tag{2-18}$$

which is an alternate form of Eq. (2-15). When two frequencies are specified at a particular attenuation level, two frequency pairs will result from calculating the corresponding opposite geometric frequency for each frequency specified. Retain the pair having the wider separation since it represents the more severe requirement. In the band-pass case, the pair having the lesser separation represented the more difficult requirement.

The band-reject filter steepness factor is defined by

$$A_s = \frac{\text{passband bandwidth}}{\text{stopband bandwidth}} \tag{2-23}$$

A normalized low-pass filter can now be selected that makes the transition from the passband attenuation limit to the minimum required stopband attenuation within a frequency ratio A_s.

The following example demonstrates the normalization procedure for a band-reject filter.

Example 2-10 Normalizing a Narrowband Band-Reject Filter

Required:

Band-reject filter
Center frequency of 1,000 Hz
3 dB at ± 300 Hz (700 Hz, 1,300 Hz)
40 dB at ± 200 Hz (800 Hz, 1,200 Hz)

Result:

(a) First, compute the center frequency f_0, using Eq. (2-14):

$$f_0 = \sqrt{f_L\, f_u} = \sqrt{700 \times 1,300} = 954 \text{ Hz}$$

(b) Using Eq. (2.18) compute two geometrically related stopband frequency pairs for each pair of stopband frequencies given:

Let $f_1 = 800$ Hz

$$f_2 = \frac{f_0^2}{f_1} = \frac{(954)^2}{800} = 1,138 \text{ Hz}$$

Let $f_2 = 1,200$ Hz

$$f_1 = \frac{f_0^2}{f_2} = \frac{(954)^2}{1,200} = 758 \text{ Hz}$$

The two pairs are

$$f_1 = 800 \text{ Hz}, f_2 = 1,138 \text{ Hz } (f_2 - f_1 = 338 \text{ Hz})$$

and

$$f_1 = 758 \text{ Hz}, f_2 = 1,200 \text{ Hz } (f_2 - f_1 = 442 \text{ Hz})$$

Retain the second pair since it has the *wider* separation and represents the more severe requirement. The given response requirement and the geometrically symmetrical equivalent are compared in Fig. 2-16.

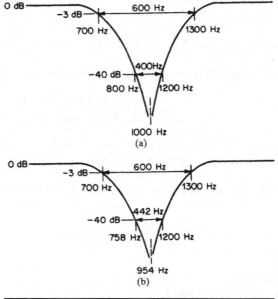

Figure 2-16 The response of Example 2-10: (a) given requirement; and (b) geometrically symmetrical response.

(*c*) Calculate A_s, using Eq. (2-23):

$$A_s = \frac{\text{passband bandwidth}}{\text{stopband bandwidth}} = \frac{600\,\text{Hz}}{442\,\text{Hz}} = 1.36$$

(*d*) Select a normalized low-pass filter from the normalized curves that makes the transition from the 3-dB point to the 40-dB point within a frequency ratio of 1.36. Since these curves are all normalized to 3 dB, a filter is required with over 40 dB of rejection at 1.36 rad/s.

2.2   Transient Response

In our previous discussions of filters, we have restricted our interest to frequency-domain parameters such as frequency response. The input forcing function was a sine wave. In real-world applications of filters, input signals consist of a variety of complex waveforms. The response of filters to these nonsinusoidal inputs is called *transient response.*

A filter's transient response is best evaluated in the time domain since we are usually dealing with input signals which are functions of time, such as pulses or amplitude steps. The frequency- and time-domain parameters of a filter are directly related through the Fourier or Laplace transforms.

2.2.1   The Effect of Nonuniform Time Delay

Evaluating a transfer function as a function of frequency results in both a magnitude and phase characteristic. Figure 2-17 shows the amplitude and phase response of a normalized $n = 3$ Butterworth low-pass filter. Butterworth low-pass filters have a phase shift of exactly n times $-45°$ at the 3-dB frequency. The phase shift continuously increases as the transition is made into the stopband and eventually approaches n times $-90°$ at frequencies far removed from the passband. Since the filter described by Fig. 2-17 has a complexity of $n = 3$, the phase shift is $-135°$ at the 3-dB cutoff and approaches $-270°$ in the stopband. Frequency scaling will transpose the phase characteristics to a new frequency range as determined by the FSF.

It is well known that a square wave can be represented by a Fourier series of odd harmonic components, as indicated in Fig. 2-18. Since the amplitude of each harmonic is reduced as the harmonic order increases, only the first few harmonics are of

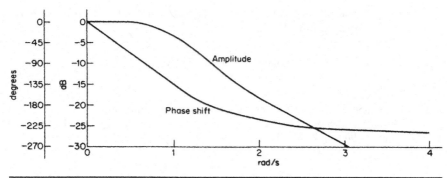

Figure 2-17   The amplitude and phase response of an $n = 3$ Butterworth low-pass filter.

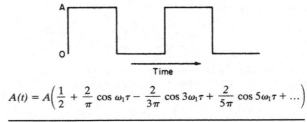

$$A(t) = A\left(\frac{1}{2} + \frac{2}{\pi} \cos \omega_1\tau - \frac{2}{3\pi} \cos 3\omega_1\tau + \frac{2}{5\pi} \cos 5\omega_1\tau + \ldots\right)$$

Figure 2-18 The frequency analysis of a square wave.

significance. If a square wave is applied to a filter, the fundamental and its significant harmonics must have a proper relative amplitude relationship at the filter's output in order to retain the square waveshape. In addition, these components must not be displaced in time with respect to each other. Let's now consider the effect of a low-pass filter's phase shift on a square wave.

If we assume that a low-pass filter has a linear phase shift between 0° at DC and n times −45° at the cutoff, we can express the phase shift in the passband as

$$\phi = -\frac{45nf_x}{f_c} \tag{2-24}$$

where f_x is any frequency in the passband, and f_c is the 3-dB cutoff frequency.

A phase-shifted sine wave appears displaced in time from the input waveform. This displacement is called *phase delay* and can be computed by determining the time interval represented by the phase shift, using the fact that a full period contains −360°. Phase delay can then be computed by

$$T_{pd} = \frac{\phi}{360} \frac{1}{f_x} \tag{2-25}$$

or, as an alternate form,

$$T_{pd} = -\frac{\beta}{\omega} \tag{2-26}$$

where β is the phase shift in radians (1 rad = 360/2π or 57.3°) and ω is the input frequency expressed in rad/s ($\omega = 2\pi f_x$).

Example 2-11   Effect of Nonlinear Phase on a Square Wave

Required:

Compute the phase delay of the fundamental and the third, fifth, seventh, and ninth harmonics of a 1 kHz square wave applied to an $n = 3$ Butterworth low-pass filter having a 3-dB cutoff of 10 kHz. Assume a linear phase shift with frequency in the passband.

Result:

Using Eqs (2-24) and (2-25), the table shown here can be computed.

Frequency	ϕ	T_{pd}
1 kHz	−13.5°	37.5 μs
3 kHz	−40.5°	37.5 μs
5 kHz	−67.5°	37.5 μs
7 kHz	−94.5°	37.5 μs
9 kHz	−121.5°	37.5 μs

The phase delays of the fundamental and each of the significant harmonics in Example 2-11 are identical. The output waveform would then appear nearly equivalent to the input except for a delay of 37.5 μs. If the phase shift is not linear with frequency, the ratio ϕ/f_x in Eq. (2-25) is not constant, so each significant component of the input square wave would undergo a different delay. This displacement in time of the spectral components, with respect to each other, introduces a distortion of the output waveform. Figure 2-19 shows some typical effects of a nonlinear phase shift upon a square wave. Most filters have nonlinear phase versus frequency characteristics, so some waveform distortion will usually occur for complex input signals.

Not all complex waveforms have harmonically related spectral components. An amplitude-modulated signal, for example, consists of a carrier and two sidebands, each sideband separated from the carrier by a modulating frequency. If a filter's phase characteristic is linear with frequency and intersects zero phase shift at zero frequency (DC), both the carrier and the two sidebands will have the same delay in passing through the filter—thus, the output will be a delayed replica of the input. If these conditions are not satisfied, the carrier and both sidebands will be delayed by different amounts. The carrier delay will be in accordance with the equation for phase delay:

$$T_{pd} = -\frac{\beta}{\omega} \tag{2-26}$$

(The terms *carrier delay* and *phase delay* are used interchangeably.)

A new definition is required for the delay of the sidebands. This delay is commonly called *group delay* and is defined as the derivative of phase versus frequency, which can be expressed as

$$T_{gd} = -\frac{d\beta}{d\omega} \tag{2-27}$$

Figure 2-19 The effect of a nonlinear phase: (*a*) an ideal square wave and (*b*) a distorted square wave.

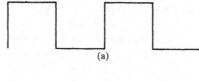
(a)

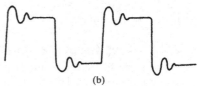
(b)

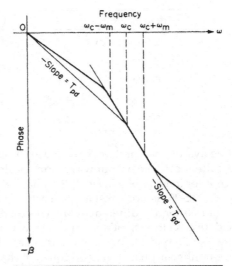

FIGURE 2-20 The nonlinear phase shift of a low-pass filter.

Linear phase shift results in constant group delay since the derivative of a linear function is a constant. Figure 2-20 illustrates a low-pass filter phase shift which is nonlinear in the vicinity of a carrier ω_c and the two sidebands: $\omega_c - \omega_m$ and $\omega_c + \omega_m$. The phase delay at ω_c is the negative slope of a line drawn from the origin to the phase shift corresponding to ω_c, which is in agreement with Eq. (2-26). The group delay at ω_c is shown as the negative slope of a line, which is tangent to the phase response at ω_c. This can be mathematically expressed as

$$T_{gd} = -\frac{d\beta}{d\omega}\bigg|_{\omega = \omega_c}$$

If the two sidebands are restricted to a region surrounding ω_c and having a constant group delay, the envelope of the modulated signal will be delayed by T_{gd}. Figure 2-21 compares the input and output waveforms of an amplitude-modulated signal applied to the filter depicted by Fig. 2-20. Note that the carrier is delayed by the phase delay, while the envelope is delayed by the group delay. For this reason, group delay is sometimes called *envelope delay*.

If the group delay is not constant over the bandwidth of the modulated signal, waveform distortion will occur. Narrow-bandwidth signals are more likely to encounter constant group delay than signals having a wider spectrum. It is common practice to use a group-delay variation as a criterion to evaluate phase nonlinearity and subsequent waveform distortion. The absolute magnitude of the nominal delay is usually of little consequence.

2.2.2 Step Response of Networks

If we were to define a hypothetical ideal low-pass filter, it would have the response shown in Fig. 2-22. The amplitude response is unity from DC to the cutoff frequency ω_c

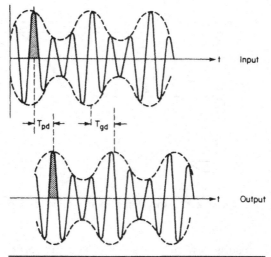

FIGURE 2-21   The effect of nonlinear phase on an AM signal.

and zero beyond the cutoff. The phase shift is a linearly increasing function in the pass-band, where n is the order of the ideal filter. The group delay is constant in the passband and zero in the stopband. If a unity amplitude step were applied to this ideal filter at $t = 0$ the output would be in accordance with Fig. 2-23. The delay of the half-amplitude point would be $n\pi/2\omega_c$, and the rise time, which is defined as the interval required to go from zero amplitude to unity amplitude with a slope equal to that at the half-amplitude point, would be equal to π/ω_c. Since rise time is inversely proportional to ω_c,

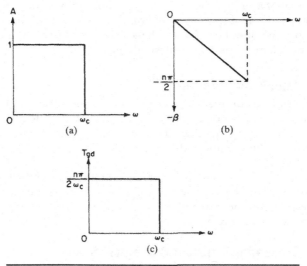

FIGURE 2-22   An ideal low-pass filter: (a) frequency response, (b) phase shift, and (c) group delay.

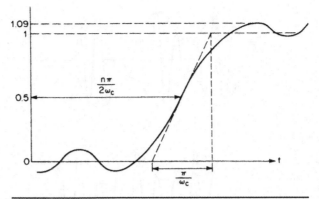

FIGURE 2-23   The step response of an ideal low-pass filter.

a wider filter results in reduced rise time. This proportionality is in agreement with a fundamental rule of thumb relating rise time to bandwidth, which is:

$$T_r \approx \frac{0.35}{f_c} \tag{2-28}$$

where T_r is the rise time in seconds, and f_c is the 3-dB cutoff in hertz.

A 9-percent overshoot exists on the leading edge. Also, a sustained oscillation occurs having a period of $2\pi/\omega_c$, which eventually decays, and then unity amplitude is established. This oscillation is called *ringing*. Overshoot and ringing occur in an ideal low-pass filter, even though we have linear phase. This is because of the abrupt amplitude roll-off at cutoff. Therefore, both linear phase and a prescribed roll-off are required for minimum transient distortion.

Overshoot and prolonged ringing are both very undesirable if the filter is required to pass pulses with minimum waveform distortion. The step-response curves provided for the different families of normalized low-pass filters can be very useful for evaluating the transient properties of these filters.

2.2.3   Impulse Response

A unit impulse is defined as a pulse which is infinitely high and infinitesimally narrow, and has an area of unity. The response of the ideal filter of Fig. 2-22 to a unit impulse is shown in Fig. 2-24. The peak output amplitude is ω_c/π, which is proportional to the filter's bandwidth. The pulse width $2\pi/\omega_c$, is inversely proportional to the bandwidth.

An input signal having the form of a unit impulse is physically impossible. However, a narrow pulse of finite amplitude will represent a reasonable approximation, so the impulse response of normalized low-pass filters can be useful in estimating the filter's response to a relatively narrow pulse.

2.2.4   Estimating Transient Characteristics

Group-delay, step-response, and impulse-response curves are given for the normalized low-pass filters discussed in the latter section of this chapter. These curves are

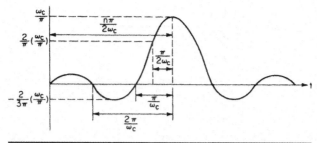

FIGURE 2-24   The impulse response of an ideal low-pass filter.

useful for estimating filter responses to nonsinusoidal signals. If the input waveforms are steps or pulses, the curves may be used directly. For more complex inputs, we can use the method of superposition, which permits the representation of a complex signal as the sum of individual components. If we find the filter's output for each individual input signal, we can combine these responses to obtain the composite output.

Group Delay of Low-Pass Filters

When a normalized low-pass filter is frequency-scaled, the delay characteristics are frequency-scaled as well. The following rules can be applied to derive the resulting delay curve from the normalized response:

1. Divide the delay axis by $2\pi f_c$ where f_c is the filter's 3-dB cutoff.
2. Multiply all points on the frequency axis by f_c.

The following example demonstrates the denormalization of a low-pass curve.

Example 2-12   Frequency Scaling the Delay of a Low-Pass Filter

Required:

Using the normalized delay curve of an $n = 3$ Butterworth low-pass filter given in Fig. 2-25a, compute the delay at DC and the delay variation in the passband if the filter is frequency-scaled to a 3-dB cutoff of 100 Hz.

Result:

To denormalize the curve, divide the delay axis by $2\pi f_c$ and multiply the frequency axis by f_c where f_c is 100 Hz. The resulting curve is shown in Fig. 2-25b. The delay at DC is 3.2 ms, and the delay variation in the passband is 1.3 ms.

The nominal delay of a low-pass filter at frequencies well below the cutoff can be estimated by the following formula:

$$T \approx \frac{125n}{f_c} \qquad\qquad (2\text{-}29)$$

where T is the delay in milliseconds, n is the order of the filter, and f_c is the 3-dB cutoff in hertz. Equation (2-29) is an approximation, which usually is accurate to within 25 percent.

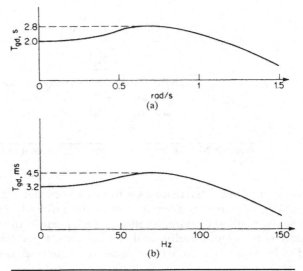

FIGURE 2-25 The delay of an $n = 3$ Butterworth low-pass filter: (a) normalized delay and (b) delay with $f_c = 100$ Hz.

Group Delay of Band-Pass Filters

When a low-pass filter is transformed to a narrowband band-pass filter, the delay is transformed to a nearly symmetrical curve mirrored about the center frequency. As the bandwidth increases from the narrow-bandwidth case, the symmetry of the delay curve is distorted approximately in proportion to the filter's bandwidth.

For the narrowband condition, the band-pass delay curve can be approximated by implementing the following rules:

1. Divide the delay axis of the normalized delay curve by πBW, where BW is the 3-dB bandwidth in hertz.

2. Multiply the frequency axis by BW/2.

3. A delay characteristic symmetrical around the center frequency can now be formed by generating the mirror image of the curve obtained by implementing steps 1 and 2. The total 3-dB bandwidth thus becomes BW.

The following example demonstrates the approximation of a narrowband band-pass filter's delay curve.

Example 2-13 Estimate the Delay of a Band-Pass Filter

Required:

Estimate the group delay at the center frequency and the delay variation over the passband of a band-pass filter having a center frequency of 1,000 Hz and a 3-dB bandwidth of 100 Hz. The band-pass filter is derived from a normalized $n = 3$ Butterworth low-pass filter.

Result:

The delay of the normalized filter is shown in Fig. 2-25a. If we divide the delay axis by πBW and multiply the frequency axis by BW/2, where BW = 100 Hz, we obtain the delay curve of Fig. 2-26a. We

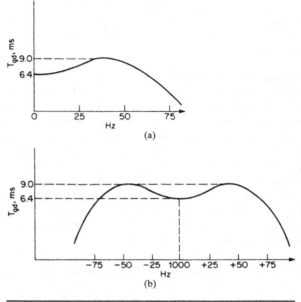

FIGURE 2-26 The delay of a narrow-band band-pass filter: (a) a low-pass delay and (b) a band-pass delay.

can now reflect this delay curve on both sides of the center frequency of 1,000 Hz to obtain Fig. 2-26b. The delay at the center frequency is 6.4 ms, whereas the delay variation over the passband is 2.6 ms.

The technique used in Example 2-13 to approximate a band-pass delay curve is valid for band-pass filter Qs of 10 or more (f_0/BW $\geq$ 10). As the fractional bandwidth increases, the delay becomes less symmetrical and peaks toward the low side of the center frequency, as shown in Fig. 2-27.

The delay at the center frequency of a band-pass filter can be estimated by

$$T \approx \frac{250n}{\text{BW}} \tag{2-30}$$

where T is the delay in milliseconds. This approximation is usually accurate to within 25 percent.

A comparison of Figs. 2-25b and 2-26b indicates that a band-pass filter has twice the delay of the equivalent low-pass filter of the same bandwidth. This results from the

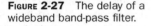

FIGURE 2-27 The delay of a wideband band-pass filter.

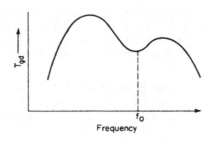

low-pass to band-pass transformation where a low-pass filter transfer function of order n always results in a band-pass filter transfer function with an order $2n$. However, a band-pass filter is conventionally referred to as having the same order n as the low-pass filter it was derived from.

Step Response of Low-Pass Filters

Delay distortion usually cannot be directly used to determine the extent of the distortion of a modulated signal. A more direct parameter would be the step response, especially where the modulation consists of an amplitude step or pulse.

The two essential parameters of a filter's step response are overshoot and ringing. Overshoot should be minimized for accurate pulse reproduction. Ringing should decay as rapidly as possible to prevent interference with subsequent pulses. Rise time and delay are usually less important considerations.

Step-response curves for standard normalized low-pass filters are provided in the latter part of this chapter. These responses can be denormalized by dividing the time axis by $2\pi f_c$ where f_c is the 3-dB cutoff of the filter. Denormalization of the step response is shown in the following example.

Example 2-14   Determining the Overshoot of a Low-Pass Filter

Required:

Determine the amount of overshoot of an $n = 3$ Butterworth low-pass filter having a 3-dB cutoff of 100 Hz. Also determine the approximate time required for the ringing to decay substantially—for instance, the settling time.

Result:

The step response of the normalized low-pass filter is shown in Fig. 2-28a. If the time axis is divided by $2\pi f_c$, where $f_c = 100$ Hz, the step response or Fig. 2-28b is obtained. The overshoot is slightly under 10 percent. After 25 ms, the amplitude will have almost completely settled.

If the input signal to a filter is a pulse rather than a step, the step-response curves can still be used to estimate the transient response, provided that the pulse width is greater than the settling time.

Example 2-15   Determining the Pulse Response of a Low-Pass Filter

Required:

Estimate the output waveform of the filter of Example 2-14 if the input is the pulse of Fig. 2-29a.

Result:

Since the pulse width is in excess of the settling time, the step response can be used to estimate the transient response. The leading edge is determined by the shape of the denormalized step response of Fig. 2-28b. The trailing edge can be derived by inverting the denormalized step response. The resulting waveform is shown in Fig. 2-29b.

The Step Response of Band-Pass Filters

The envelope of the response of a narrow band-pass filter to a step of the center frequency is almost identical to the step response of the equivalent low-pass filter having

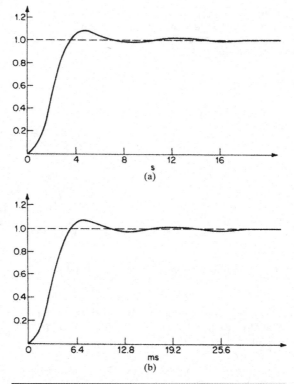

FIGURE 2-28 The step response of Example 2-14: (*a*) normalized step response and (*b*) denormalized step response.

FIGURE 2-29 The pulse response of Example 2-15: (*a*) input pulse and (*b*) output pulse.

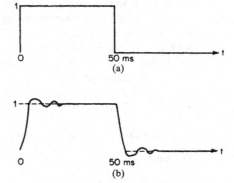

half the bandwidth. To determine this envelope shape, denormalize the low-pass step response by dividing the time axis by BW, where BW is the 3-dB bandwidth of the band-pass filter. The previous discussions of overshoot, ringing, and so on, can be applied to the carrier envelope.

FIGURE **2-30** The band-pass response to a center frequency step: (a) denormalized low-pass step response and (b) band-pass envelope response.

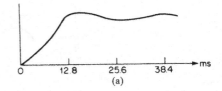

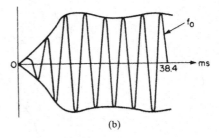

Example 2-16 Determining the Step Response of a Band-Pass Filter

Required:

Determine the envelope of the response to a 1,000 Hz step for an $n = 3$ Butterworth band-pass filter having a center frequency of 1,000 Hz and a 3-dB bandwidth of 100 Hz.

Result:

Using the normalized step response of Fig. 2-28*a*, divide the time axis by πBW, where BW = 100 Hz. The results are shown in Fig. 2-30.

The Impulse Response of Low-Pass Filters

If the duration of a pulse applied to a low-pass filter is much less than the rise time of the filter's step response, the filter's impulse response will provide a reasonable approximation to the shape of the output waveform.

Impulse-response curves are provided for the different families of low-pass filters. These curves are all normalized to correspond to a filter having a 3-dB cutoff of 1 rad/s, and have an area of unity. To denormalize the curve, multiply the amplitude by the FSF and divide the time axis by the same factor.

It is desirable to select a normalized low-pass filter having an impulse response whose peak is as high as possible. The ringing, which occurs after the trailing edge, should also decay rapidly to avoid interference with subsequent pulses.

Example 2-17 Determining the Impulse Response of a Low-Pass Filter

Required:

Determine the approximate output waveform if a 100-μs pulse is applied to an $n = 3$ Butterworth low-pass filter having a 3-dB cutoff of 100 Hz.

Result:

The denormalized step response of the filter is given in Fig. 2-28*b*. The rise time is well in excess of the given pulse width of 100 μs, so the impulse response curve should be used to approximate the output waveform.

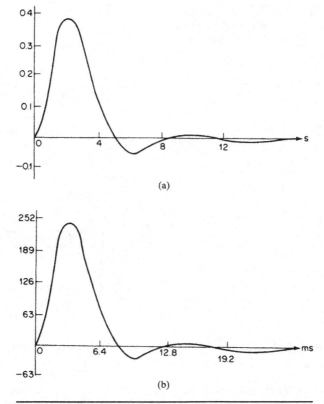

FIGURE 2-31   The impulse response for Example 2-17:
(a) normalized response and (b) denormalized response.

The impulse response of a normalized $n = 3$ Butterworth low-pass filter is shown in Fig. 2-31a. If the time axis is divided by the FSF and the amplitude is multiplied by this same factor, the curve of Fig. 2-31b results.

Since the input pulse amplitude of Example 2-17 is certainly not infinite, the amplitude axis is in error. However, the pulse shape is retained at a lower amplitude. As the input pulse width is reduced in relation to the filter rise time, the output amplitude decreases and eventually the output pulse vanishes.

The Impulse Response of Band-Pass Filters
The envelope of the response of a narrowband band-pass filter to a short tone burst of center frequency can be found by denormalizing the low-pass impulse response. This approximation is valid if the burst width is much less than the rise time of the denormalized step response of the band-pass filter. Also, the center frequency should be high enough so that many cycles occur during the burst interval.

To transform the impulse-response curve, multiply the amplitude axis by πBW and divide the time axis by the same factor, where BW is the 3-dB bandwidth of the band-pass filter. The resulting curve defines the shape of the envelope of the filter's response to the tone burst.

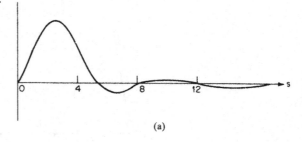

FIGURE 2-32 The results of Example 2-18: (a) normalized low-pass impulse response and (b) impulse response of band-pass filter.

(a)

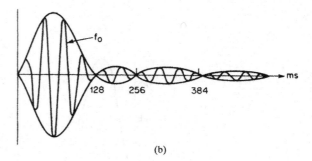

(b)

Example 2-18 Determining the Impulse Response of a Band-Pass Filter

Required:

Determine the approximate shape of the response of an $n = 3$ Butterworth band-pass filter having a center frequency of 1,000 Hz and a 3-dB bandwidth of 10 Hz to a tone burst of the center frequency having a duration of 10 ms.

Result:

The step response of a normalized $n = 3$ Butterworth low-pass filter is shown in Fig. 2-28a. To determine the rise time of the band-pass step response, divide the normalized low-pass rise time by πBW where BW is 10 Hz. The resulting rise time is approximately 120 ms, which well exceeds the burst duration. Also, 10 cycles of the center frequency occur during the burst interval, so the impulse response can be used to approximate the output envelope. To denormalize the impulse response, multiply the amplitude axis by πBW and divide the time axis by the same factor. The results are shown in Fig. 2-32.

Effective Use of the Group-Delay, Step-Response, and Impulse-Response Curves

Many signals consist of complex forms of modulation rather than pulses or steps, so the transient response curves cannot be directly used to estimate the amount of distortion introduced by the filters. However, the curves are useful as a figure of merit, since networks having desirable step- or impulse-response behavior introduce minimal distortion to most forms of modulation.

Examination of the step- and impulse-response curves in conjunction with group delay indicates that a necessary condition for good pulse transmission is a flat group delay. A gradual transition from the passband to the stopband is also required for low-transient distortion but is highly undesirable from a frequency-attenuation point of view.

In order to obtain a rapid pulse rise time, the higher-frequency spectral components should not be delayed with respect to the lower frequencies. The curves indicate that low-pass filters, that have a sharply increasing delay at higher frequencies, also have an impulse response which comes to a peak at a later time.

When a low-pass filter is transformed to a high-pass, a band-reject, or a wideband band-pass filter, the transient properties are not preserved. Lindquist and Zverev (see References) provide computational methods for the calculation of these responses.

2.3   Butterworth Maximally Flat Amplitude

The Butterworth approximation to an ideal low-pass filter is based on the assumption that a flat response at zero frequency is more important than the response at other frequencies. A normalized transfer function is an all-pole type having roots which all fall on a unit circle. The attenuation is 3 dB at 1 rad/s.

The attenuation of a Butterworth low-pass filter can be expressed by

$$A_{dB} = 10 \log \left[1 + \left(\frac{\omega_x}{\omega_c} \right)^{2n} \right] \tag{2-31}$$

where ω_x / ω_c is the ratio of the given frequency ω_x to the 3-dB cutoff frequency ω_c and n is the order of the filter.

For the more general case,

$$A_{dB} = 10 \log(1 + \Omega^{2n}) \tag{2-32}$$

where Ω is defined by the following table.

The value Ω is a dimensionless ratio of frequencies or a normalized frequency. $BW_{3\,dB}$ is the 3-dB bandwidth, and BW_x is the bandwidth of interest. At high values of Ω the attenuation increases at a rate of $6n$ dB per octave, where an octave is defined as a frequency ratio of 2 for the low-pass and high-pass cases, and a *bandwidth* ratio of 2 for band-pass and band-reject filters.

Filter Type	Ω
Low-pass	ω_x / ω_c
High-pass	ω_c / ω_x
Band-pass	$BW_x / BW_{3\,dB}$
Band-reject	$BW_{3\,dB} / BW_x$

The pole positions of the normalized filter all lie on a unit circle and can be computed by

$$-\sin \frac{(2K-1)\pi}{2n} + j \cos \frac{(2K-1)\pi}{2n}, \qquad K = 1, 2, \ldots, n \tag{2-33}$$

and the element values for an *LC* normalized low-pass filter operating between equal $1 - \Omega$ terminations can be calculated by

$$L_K \text{ or } C_K = 2 \sin \frac{(2K-1)\pi}{2n}, \qquad K = 1, 2, \ldots, n \tag{2-34}$$

where $(2K - 1)\pi/2n$ is in radians.

Equation (2-34) is exactly equal to twice the real part of the pole position of Eq. (2-33), except that the sign is positive.

Example 2-19   Calculating the Frequency Response, Pole Locations, and *LC* Element Values of a Butterworth Low-Pass Filter

Required:

Calculate the frequency response at 1, 2, and 4 rad/s, the pole positions, and the *LC* element values of a normalized $n = 5$ Butterworth low-pass filter.

Result:

(*a*) Using Eq. (2-32) with $n = 5$, the frequency-response table shown here can be derived.

Ω	Attenuation
1	3 dB
2	30 dB
4	60 dB

(*b*) The pole positions are computed using Eq. (2-33) as shown in this table.

K	$-\sin\dfrac{(2K-1)\pi}{2n}$	$j\cos\dfrac{(2K-1)\pi}{2n}$
1	−0.309	+j0.951
2	−0.809	+j0.588
3	−1	
4	−0.809	−j0.588
5	−0.309	−j0.951

(*c*) The element values can be computed by Eq. (2-34) and have the following values:

$$
\begin{array}{lll}
L_1 = 0.618 \text{ H} & & C_1 = 0.618 \text{ F} \\
C_2 = 1.618 \text{ F} & & L_2 = 1.618 \text{ H} \\
L_3 = 2 \text{ H} & \text{or} & C_3 = 2 \text{ F} \\
C_4 = 1.618 \text{ F} & & L_4 = 1.618 \text{ H} \\
L_5 = 0.618 \text{ H} & & C_5 = 0.618 \text{ F}
\end{array}
$$

The results of Example 2-19 are shown in Fig. 2-33.

Chapter 10 provides pole locations and element values for both *LC* and active Butterworth low-pass filters having complexities up to $n = 10$.

The Butterworth approximation results in a class of filters which have moderate attenuation steepness and acceptable transient characteristics. Their element values are more practical and less critical than those of most other filter types. The rounding of the frequency response in the vicinity of cutoff may make these filters undesirable where a sharp cutoff is required; nevertheless, they should be used wherever possible because of their favorable characteristics.

Figures 2-34 through 2-37 indicate the frequency response, group delay, impulse response, and step response for the Butterworth family of low-pass filters normalized to a 3-dB cutoff of 1 rad/s.

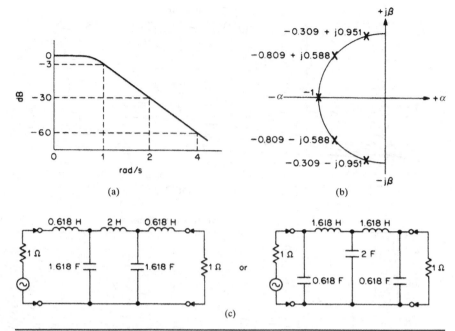

FIGURE 2-33 The Butterworth low-pass filter of Example 2-19: (a) frequency response, (b) pole locations, and (c) circuit configuration.

2.4   Chebyshev Response

If the poles of the normalized Butterworth low-pass transfer function were moved to right by multiplying the real parts of the pole position by a constant k_r, and the imaginary parts by a constant k_j, where both k's are <1, the poles would now lie on an ellipse instead of a unit circle. The frequency response would ripple evenly and have an attenuation at 1 rad/s equal to the ripple. The resulting response is called the Chebyshev or equiripple function.

The Chebyshev approximation to an ideal filter has a much more rectangular frequency response in the region near cutoff than the Butterworth family of filters. This is accomplished at the expense of allowing ripples in the passband.

The factors k_r and k_j are computed by

$$k_r = \sinh A \tag{2-35a}$$

$$k_j = \cosh A \tag{2-35b}$$

The parameter A is given by

$$A = \frac{1}{n} \sinh^{-1} \frac{1}{\varepsilon} \tag{2-36}$$

where

$$\varepsilon = \sqrt{10^{R_{dB}/10} - 1} \tag{2-37a}$$

and R_{dB} is the ripple in decibels.

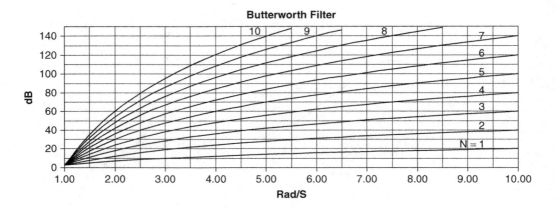

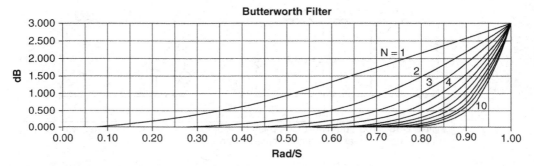

FIGURE 2-34   Attenuation characteristics for Butterworth filters.

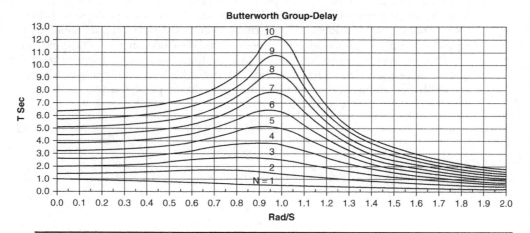

FIGURE 2-35   Group-delay characteristics for Butterworth filters.

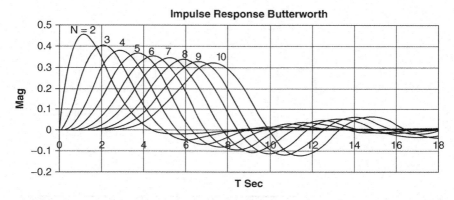

FIGURE 2-36   Impulse response for Butterworth filters.

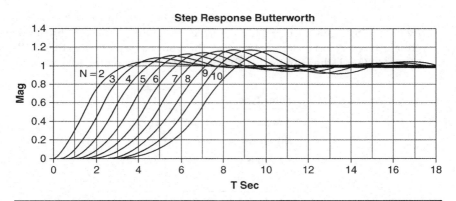

FIGURE 2-37   Step response for Butterworth filters.

Figure 2-38 compares the frequency response of an $n = 3$ Butterworth normalized low-pass filter and the Chebyshev filter generated by applying Eqs. (2-35a) and (2-35b). The Chebyshev filter response has also been normalized so that the attenuation is 3 dB at 1 rad/s. The actual 3-dB bandwidth of a Chebyshev filter computed using Eqs. (2-35a) and (2-35b) is $\cosh A_1$, where A_1 is given by

$$A_1 = \frac{1}{n}\cosh^{-1}\left(\frac{1}{\varepsilon}\right) \tag{2-37b}$$

The attenuation of Chebyshev filters can be expressed as

$$A_{\mathrm{dB}} = 10\log\left[1 + \varepsilon^2 C_n^2(\Omega)\right] \tag{2-38}$$

where $C_n(\Omega)$ is a Chebyshev polynomial whose magnitude oscillates between ±1 for $\Omega \le 1$. Table 2-1 lists the Chebyshev polynomials up to order $n = 10$.

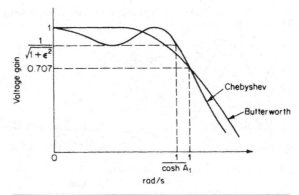

Figure 2-38   A comparison of Butterworth and Chebyshev low-pass filters.

At $\Omega = 1$, Chebyshev polynomials have a value of unity, so the attenuation defined by Eq. (2-38) would be equal to the ripple. The 3-dB cutoff is slightly above $\Omega = 1$ and is equal to cosh A_1. In order to normalize the response equation so that 3 dB of attenuation occurs at $\Omega = 1$, the Ω of Eq. (2-38) is computed by using the following table:

Filter Type	Ω
Low-pass	(cosh A_1) ω_x / ω_c
High-pass	(cosh A_1) ω_c / ω_x
Band-pass	(cosh A_1) $BW_x / BW_{3\,dB}$
Band-reject	(cosh A_1) $BW_{3\,dB} / BW_x$

Figure 2-39 compares the ratios of 3-dB bandwidth to ripple bandwidth (cosh A_1) for Chebyshev low-pass filters ranging from $n = 2$ to $n = 10$.

1. Ω
2. $2\Omega^2 - 1$
3. $4\Omega^3 - 3\Omega$
4. $8\Omega^4 - 8\Omega^2 + 1$
5. $16\Omega^5 - 20\Omega^3 + 5\Omega$
6. $32\Omega^6 - 48\Omega^4 + 18\Omega^2 - 1$
7. $64\Omega^7 - 112\Omega^5 + 56\Omega^3 - 7\Omega$
8. $128\Omega^8 - 256\Omega^6 + 160\Omega^4 - 32\Omega^2 + 1$
9. $256\Omega^9 - 576\Omega^7 + 432\Omega^5 - 120\Omega^3 + 9\Omega$
10. $512\Omega^{10} - 1,280\Omega^8 + 1,120\Omega^6 - 400\Omega^4 + 50\Omega^2 - 1$

Table 2-1   Chebyshev Polynomials

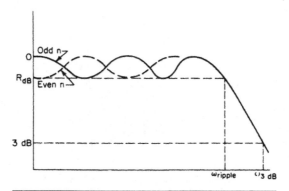

FIGURE 2-39 The ratio of 3-dB bandwidth to ripple bandwidth.

n	0.001 dB	0.005 dB	0.01 dB	0.05 dB
2	5.7834930	3.9027831	3.3036192	2.2685899
3	2.6427081	2.0740079	1.8771819	1.5120983
4	1.8416695	1.5656920	1.4669048	1.2783955
5	1.5155888	1.3510908	1.2912179	1.1753684
6	1.3495755	1.2397596	1.1994127	1.1207360
7	1.2531352	1.1743735	1.1452685	1.0882424
8	1.1919877	1.1326279	1.1106090	1.0673321
9	1.1507149	1.1043196	1.0870644	1.0530771
10	1.1215143	1.0842257	1.0703312	1.0429210
N	**0.10 dB**	**0.25 dB**	**0.50 dB**	**1.00 dB**
2	1.9432194	1.5981413	1.3897437	1.2176261
3	1.3889948	1.2528880	1.1674852	1.0948680
4	1.2130992	1.1397678	1.0931019	1.0530019
5	1.1347180	1.0887238	1.0592591	1.0338146
6	1.0929306	1.0613406	1.0410296	1.0234422
7	1.0680005	1.0449460	1.0300900	1.0172051
8	1.0519266	1.0343519	1.0230107	1.0131638
9	1.0409547	1.0271099	1.0181668	1.0103963
10	1.0331307	1.0219402	1.0147066	1.0084182

Odd-order Chebyshev *LC* filters have zero relative attenuation at DC. Even-order filters, however, have a loss at DC equal to the passband ripple. As a result, the even-order networks must operate between unequal source and load resistances to produce an impedance mismatch, whereas for odd *n*'s, the source and the load may be equal.

The element values for an *LC* normalized low-pass filter operating between equal 1-Ω terminations and having an odd *n* can be calculated from the following series of relations.

$$G_1 = \frac{2A_1 \cosh A}{Y} \qquad (2\text{-}39)$$

$$G_k = \frac{4A_{k-1}A_k \cosh^2 A}{B_{k-1}G_{k-1}} \quad k = 2, 3, 4, \ldots, n \tag{2-40}$$

where

$$Y = \sinh \frac{\beta}{2n} \tag{2-41}$$

$$\beta = \ln\left(\coth \frac{R_{dB}}{17.37} \right) \tag{2-42}$$

$$A_k = \sin \frac{(2k-1)\pi}{2n} \quad k = 1, 2, 3, \ldots, n \tag{2-43}$$

$$B_k = Y^2 + \sin^2\left(\frac{k\pi}{n} \right) \quad k = 1, 2, 3, \ldots, n \tag{2-44}$$

Coefficients G_1 through G_n are the element values.

An alternate form of determining LC element values can be done by synthesizing the driving-point impedance directly from the transfer function. Closed form formulas are given in Matthaei (see References). These methods include both odd- and even-order n's.

Example 2-20   Calculating the Pole Locations, Frequency Response, and LC Element Values of a Chebyshev Low-Pass Filter

Required:

Compute the pole positions, the frequency response at 1, 2, and 4 rad/s, and the element values of a normalized $n = 5$ Chebyshev low-pass filter having a ripple of 0.5 dB.

Result:

(a) To compute the pole positions, first solve for k_c as follows:

Using Eq. (2-37) $\varepsilon = \sqrt{10^{R_{dB}/10} - 1} = 0.349$

Using Eq. (2-38) $A = \frac{1}{n} \sinh^{-1} \frac{1}{\varepsilon} = 0.355$

Using Eq. (2-35a) $k_r = \sinh A = 0.3625$

Using Eq. (2-35b) $k_j = \cosh A = 1.0637$

Multiplication of the real parts of the normalized Butterworth poles of Example 2-19 by k_r and the imaginary parts by k_j results in

$$-0.1120 \pm j1.0116; -0.2933 \pm j0.6255; -0.3625$$

To denormalize these coordinates for 3 dB at 1 rad/s, divide all values by $\cosh A_1$, where A_1 [from Eq. (2-37b)] is given by

$$A_1 = 0.3428$$

so $\cosh A_1 = 1.0593$. The resulting pole positions are

$$-0.1057 \pm j0.9549; -0.2769 \pm j0.5905; -0.3422$$

(b) To calculate the frequency response, substitute a fifth-order Chebyshev polynomial and $\varepsilon = 0.349$ into Eq. (2-38). The following results are obtained:

Ω	A_{dB}
1.0	3 dB
2.0	45 dB
4.0	77 dB

(c) The element values are computed as follows:

Using Eq. (2-43)	$A_1 = 0.309$
Using Eq. (2-42)	$\beta = 3.55$
Using Eq. (2-41)	$Y = 0.363$
Using Eq. (2-39)	$G_1 = 1.81$
Using Eq. (2-40)	$G_2 = 1.30 \quad G_3 = 2.69 \quad G_4 = 1.30 \quad G_5 = 1.81$

Coefficients G_1 through G_5 represent the element values of a normalized Chebyshev low-pass filter having a 0.5-dB ripple and a 3-dB cutoff of 1 rad/s.

Figure 2-40 shows the results of this example.

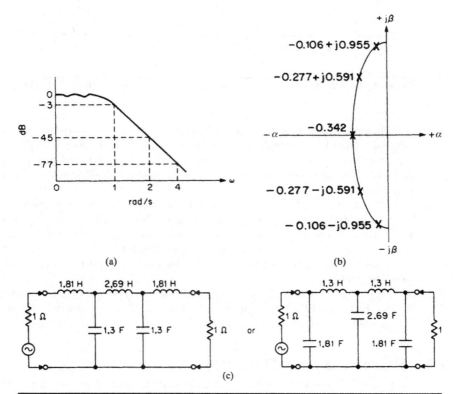

(a)

(b)

(c)

FIGURE 2-40 The Chebyshev low-pass filter of Example 2-20: (a) frequency response, (b) pole locations, and (c) circuit configuration.

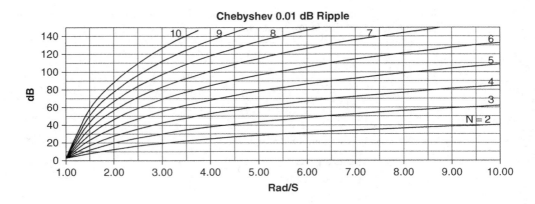

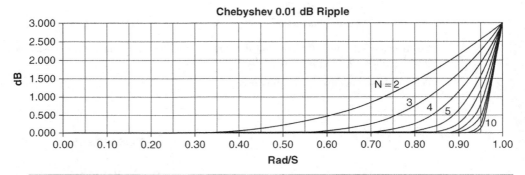

FIGURE 2-41   Attenuation characteristics for Chebyshev filters with 0.01-dB ripple.

Chebyshev filters have a narrower transition region between the passband and stopband than Butterworth filters but have more delay variation in their passband. As the passband ripple is made larger, the rate of roll-off increases, but the transient properties rapidly deteriorate. If no ripples are permitted, the Chebyshev filter degenerates to a Butterworth.

The Chebyshev function is useful where frequency response is a major consideration. It provides the maximum theoretical rate of roll-off of any all-pole transfer function for a given order. It does not have the mathematical simplicity of the Butterworth family, which should be evident from comparing Examples 2-20 and 2-19. Fortunately, the computation of poles and element values is not required since this information is provided in Chap. 10.

Figures 2-41 through 2-54 show the frequency and time-domain parameters of Chebyshev low-pass filters for ripples of 0.01, 0.1, 0.25, 0.5, and 1 dB, all normalized for a 3-dB cutoff of 1 rad/s.

2.5   Bessel Maximally Flat Delay

Butterworth filters have fairly good amplitude and transient characteristics. The Chebyshev family of filters offers increased selectivity but poor transient behavior. Neither approximation to an ideal filter is directed toward obtaining a constant delay in the passband.

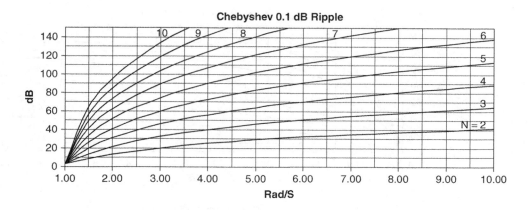

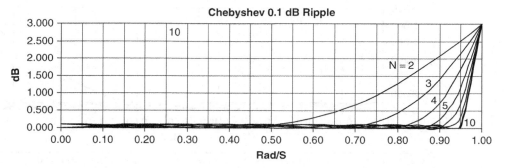

FIGURE 2-42 Attenuation characteristics for Chebyshev filters with 0.1-dB ripple.

The Bessel transfer function has been optimized to obtain a linear phase—in other words, a maximally flat delay. The step response has essentially no overshoot or ringing, and the impulse response lacks oscillatory behavior. However, the frequency response is much less selective than in the other filter types.

The low-pass approximation to a constant delay can be expressed as the following general transfer function:

$$T(s) = \frac{1}{\sinh s + \cosh s} \tag{2-45}$$

If a continued-fraction expansion is used to approximate the hyperbolic functions and the expansion is truncated at different lengths, the Bessel family of transfer functions will result.

A crude approximation to the pole locations can be found by locating all the poles on a circle and separating their imaginary parts by $2/n$, as shown in Fig. 2-55. The vertical spacing between poles is equal, whereas in the Butterworth case the angles were equal.

The relative attenuation of a Bessel low-pass filter can be approximated by

$$A_{\mathrm{dB}} = 3\left(\frac{\omega_x}{\omega_c}\right)^2 \tag{2-46}$$

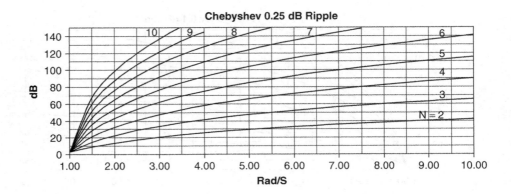

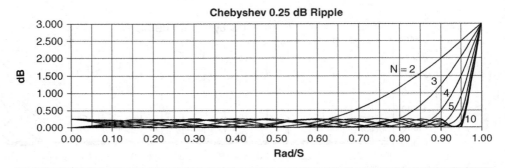

FIGURE 2-43 Attenuation characteristics for Chebyshev filters with 0.25-dB ripple.

This expression is reasonably accurate for ω_x/ω_c ranging between 0 and 2.

Figures 2-56 through 2-59 indicate that as the order n is increased, the region of flat delay is extended farther into the stopband. However, the steepness of roll-off in the transition region does not improve significantly. This restricts the use of Bessel filters to applications where the transient properties are the major consideration.

A similar family of filters is the Gaussian type. However, the Gaussian phase response is not as linear as the Bessel for the same number of poles, and the selectivity is not as sharp.

2.6 Linear Phase with Equiripple Error

The Chebyshev (equiripple amplitude) function is a better approximation of an ideal amplitude curve than the Butterworth. Therefore, it stands to reason that an equiripple approximation of a linear phase will be more efficient than the Bessel family of filters.

Figure 2-60 illustrates how a linear phase can be approximated to within a given ripple of ε degrees. For the same n, the equiripple-phase approximation results in a linear phase and, consequently, a constant delay over a larger interval than the Bessel approximation. Also the amplitude response is superior far from cutoff. In the transition region and below cutoff, both approximations have nearly identical responses.

As the phase ripple ε is increased, the region of constant delay is extended farther into the stopband. However, the delay develops ripples. The step response has slightly more overshoot than Bessel filters.

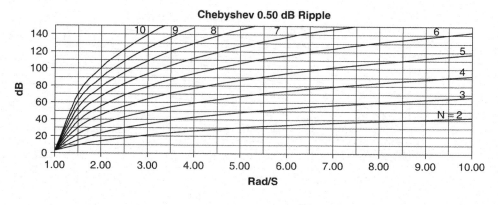

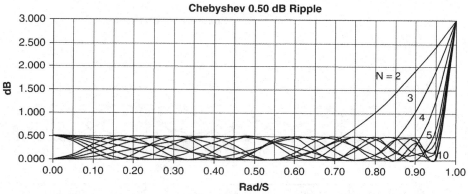

FIGURE 2-44   Attenuation characteristics for Chebyshev filters with 0.5-dB ripple.

A closed-form method for computation of the pole positions is not available. The pole locations tabulated in Chap. 10 were developed by iterative techniques. Values are provided for phase ripples of 0.05° and 0.5°, and the associated frequency and time-domain parameters are given in Figs. 2-61 through 2-68.

2.7   Transitional Filters

The Bessel filters discussed in Sec. 2.5 have excellent transient properties but poor selectivity. Chebyshev filters, on the other hand, have steep roll-off characteristics but poor time-domain behavior. A transitional filter offers a compromise between a Gaussian filter, which is similar to the Bessel family, and Chebyshev filters.

Transitional filters have a near linear phase shift and smooth amplitude roll-off in the passband. Outside the passband, a sharp break in the amplitude characteristics occurs. Beyond this breakpoint, the attenuation increases quite abruptly in comparison with Bessel filters, especially for the higher n's.

In the tables in Chap. 10, transitional filters are listed which have Gaussian characteristics to both 6 dB and 12 dB. The transient properties of the Gaussian to 6-dB filters are somewhat superior to those of the Butterworth family. Beyond the 6-dB point,

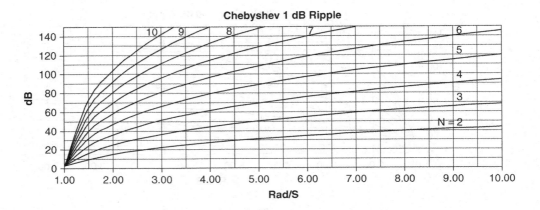

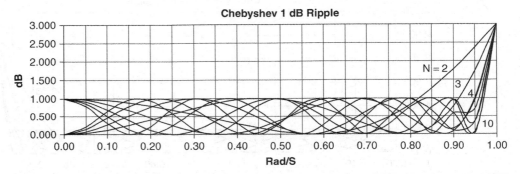

FIGURE 2-45   Attenuation characteristics for Chebyshev filters with 1-dB ripple.

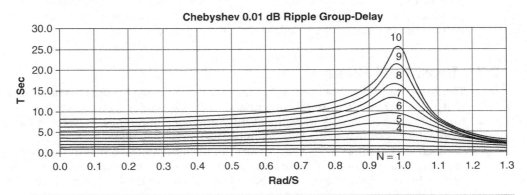

FIGURE 2-46   Group-delay characteristics for Chebyshev filters with 0.01-dB ripple.

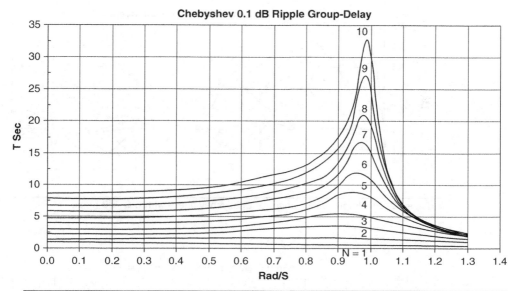

FIGURE 2-47 Group-delay characteristics for Chebyshev filters with 0.1-dB ripple.

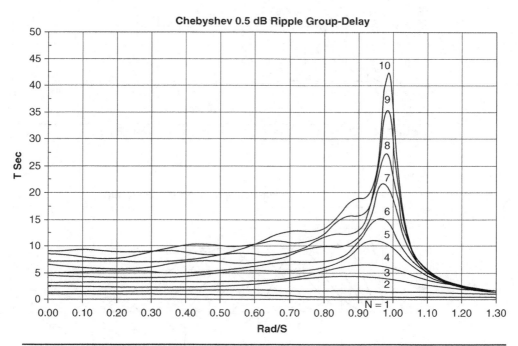

FIGURE 2-48 Group-delay characteristics for Chebyshev filters with 0.5-dB ripple.

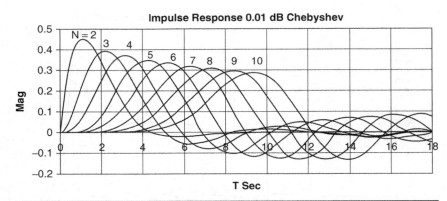

FIGURE 2-49   Impulse response for Chebyshev filters with 0.01-dB ripple.

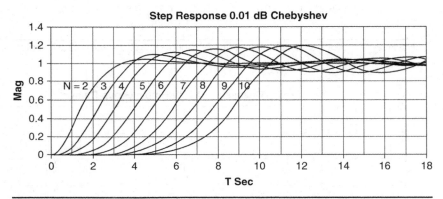

FIGURE 2-50   Step response for Chebyshev filters with 0.01-dB ripple.

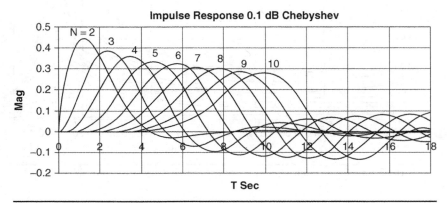

FIGURE 2-51   Impulse response for Chebyshev filters with 0.1-dB ripple.

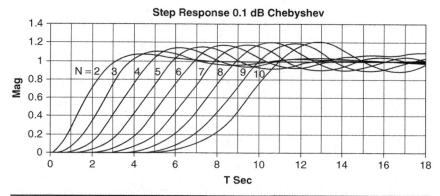

FIGURE 2-52 Step response for Chebyshev filters with 0.1-dB ripple.

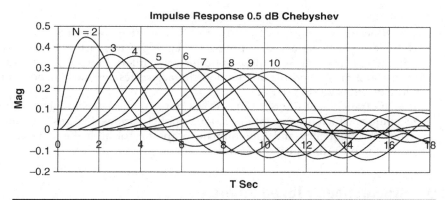

FIGURE 2-53 Impulse response for Chebyshev filters with 0.5-dB ripple.

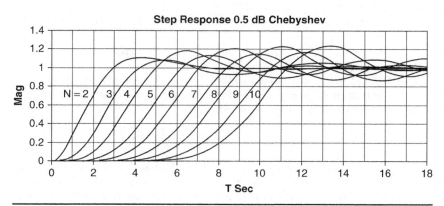

FIGURE 2-54 Step response for Chebyshev filters with 0.5-dB ripple.

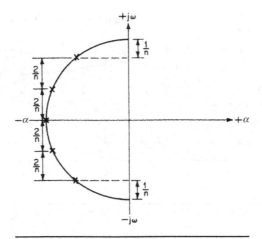

FIGURE 2-55 Approximate Bessel pole locations.

which occurs at approximately 1.5 rad/s, the attenuation characteristics are nearly comparable with Butterworth filters. The Gaussian to 12-dB filters have time-domain parameters far superior to those of Butterworth filters. However, the 12-dB breakpoint occurs at 2 rad/s, and the attenuation characteristics beyond this point are inferior to those of Butterworth filters.

The transitional filters tabulated in Chap. 10 were generated using mathematical techniques which involve interpolation of pole locations. Figures 2-69 through 2-76 indicate the frequency and time-domain properties of both the Gaussian to 6-dB and Gaussian to 12-dB transitional filters.

2.8 Synchronously Tuned Filters

Synchronously tuned filters are the most basic filter type and are the easiest to construct and align. They consist of identical multiple poles. A typical application is in the case of a band-pass amplifier, where a number of stages are cascaded, with each stage having the same center frequency and Q.

The attenuation of a synchronously tuned filter can be expressed as

$$A_{dB} = 10n \log[1 + (2^{1/n} - 1)\Omega^2]$$ (2-47)

Equation (2-47) is normalized so that 3 dB of attenuation occurs at $\Omega = 1$.

The individual section Q can be defined in terms of the composite circuit Q requirement using the following relationship:

$$Q_{section} = Q_{overall}\sqrt{2^{1/n} - 1}$$ (2-48)

Alternatively, we can state that the 3-dB bandwidth of the individual sections is reduced by the shrinkage factor $(2^{1/n} - 1)^{1/2}$. The individual section Q is less than the overall Q, whereas in the case of nonsynchronously tuned filters the section Qs may be required to be much higher than the composite Q.

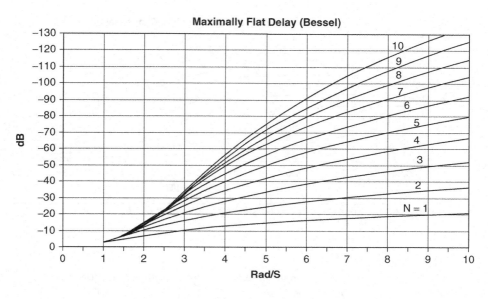

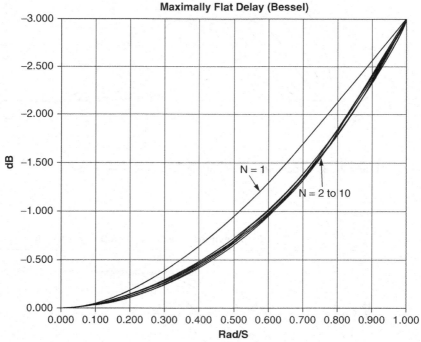

FIGURE 2-56 Attenuation characteristics for maximally flat delay (Bessel) filters.

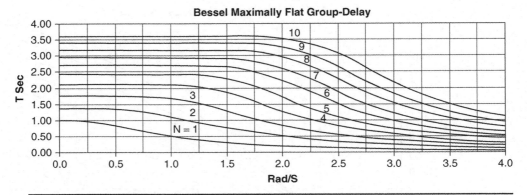

FIGURE 2-57   Group-delay characteristics for maximally flat delay (Bessel) filters.

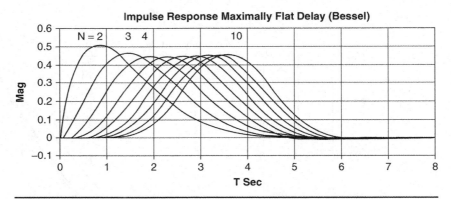

FIGURE 2-58   Impulse response for maximally flat delay (Bessel) filters.

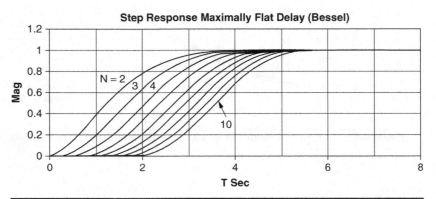

FIGURE 2-59   Step response for maximally flat delay (Bessel) filters.

FIGURE 2-60 An equiripple linear-phase approximation.

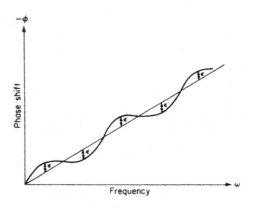

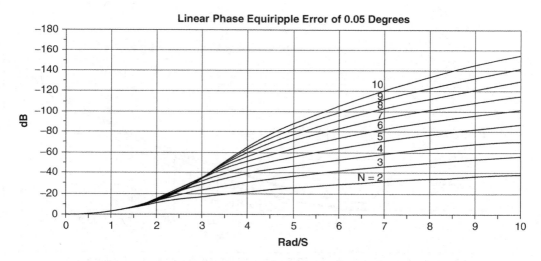

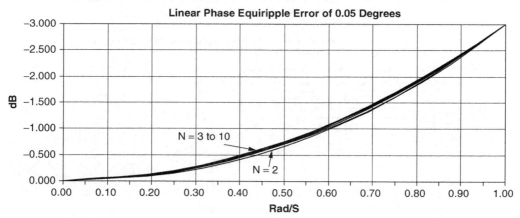

FIGURE 2-61 Attenuation characteristics for linear phase with equiripple error filters (phase error = 0.05°).

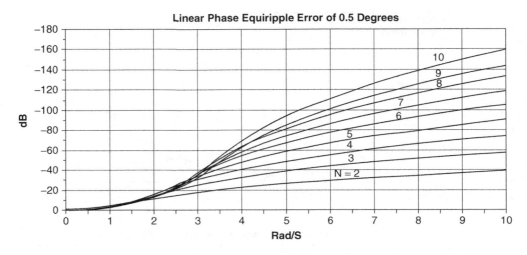

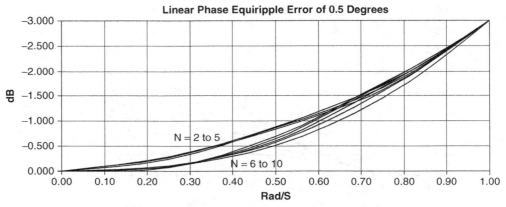

FIGURE 2-62 Attenuation characteristics for linear phase with equiripple error filters (Phase error = 0.5°).

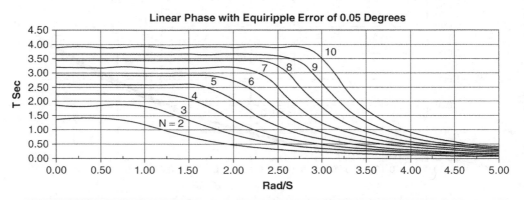

FIGURE 2-63 Group-delay characteristics for linear phase with equiripple error filters (phase error = 0.05°).

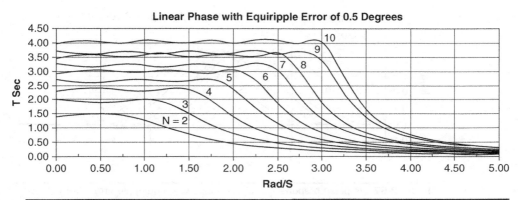

FIGURE 2-64   Group-delay characteristics for linear phase with equiripple error filters (phase error = 0.5°).

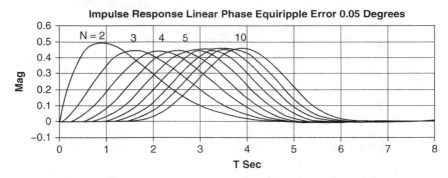

FIGURE 2-65   Impulse response for linear phase with equiripple error filters (phase error = 0.05°).

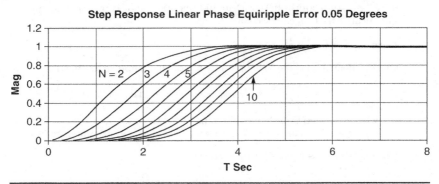

FIGURE 2-66   Step response for linear phase with equiripple error filters (phase error = 0.05°).

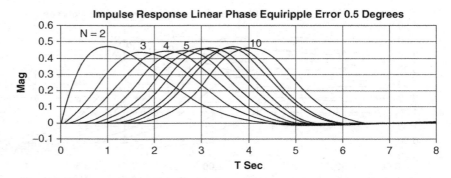

FIGURE 2-67   Impulse response for linear phase with equiripple error filters (phase error = 0.5°).

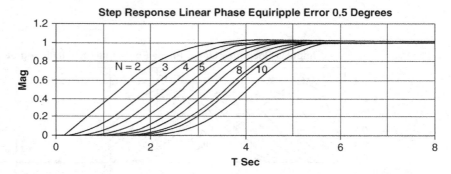

FIGURE 2-68   Step response for linear phase with equiripple error filters (phase error = 0.5°).

Example 2-21   Calculate the Attenuation and Section Q's of a Synchronously Tuned Band-Pass Filter

Required:

A three-section synchronously tuned band-pass filter is required to have a center frequency of 10 kHz and a 3-dB bandwidth of 100 Hz. Determine the attenuation corresponding to a bandwidth of 300 Hz, and calculate the Q of each section.

Result:

(*a*) The attenuation at the 300-Hz bandwidth can be computed, using Eq. (2-47) as

$$A_{dB} = 10n \log[1 + (2^{1/n} - 1)\Omega^2] = 15.7 \text{ dB}$$

where $n = 3$ and Ω the bandwidth ratio, is 300 Hz/100 Hz, or 3. (Since the filter is a narrowband type, conversion to a geometrically symmetrical response requirement was not necessary.)

(*b*) The Q of each section is [using Eq. (2-48)]

$$Q_{section} = Q_{overall} \sqrt{2^{1/n} - 1} = 51$$

where $Q_{overall}$ is 10 kHz/100 Hz, or 100.

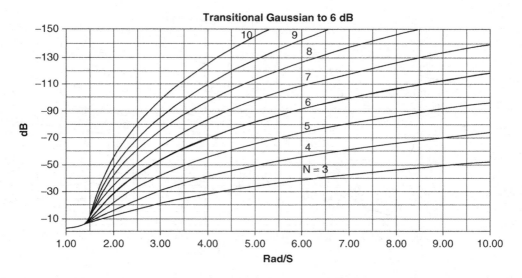

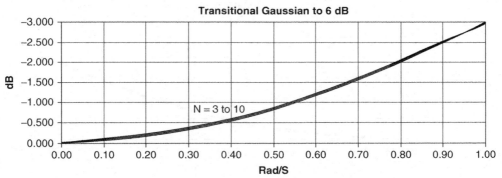

FIGURE 2-69 Attenuation characteristics for transitional filters (Gaussian to 6 dB).

The synchronously tuned filter of Example 2-21 has only 15.7 dB of attenuation at a normalized frequency ratio of 3, and for $n = 3$ Even the gradual roll-off characteristics of the Bessel family provide better selectivity than synchronously tuned filters for equivalent complexities.

The transient properties, however, are near optimum. The step response exhibits no overshoot at all and the impulse response lacks oscillatory behavior.

The poor selectivity of synchronously tuned filters limits their application to circuits requiring modest attenuation steepness and simplicity of alignment. The frequency domain characteristics are illustrated in Fig. 2-77.

2.9 Elliptic-Function Filters

All the filter types previously discussed are all-pole networks. They exhibit infinite rejection only at the extremes of the stopband. Elliptic-function filters have zeros as well as poles at finite frequencies. The location of the poles and zeros creates equiripple behavior in the passband similar to Chebyshev filters. Finite transmission zeros in the stopband reduce the transition region so that extremely sharp roll-off characteristics

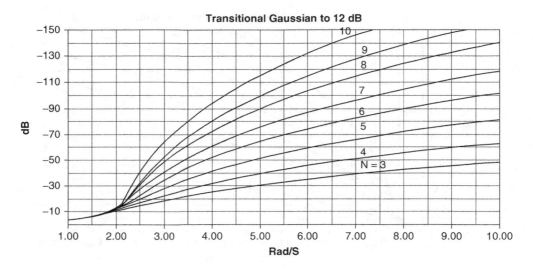

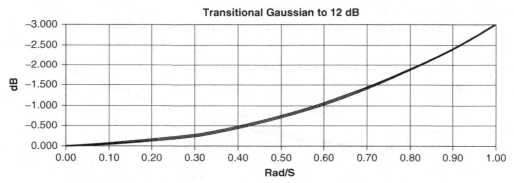

FIGURE 2-70 Attenuation characteristics for transitional filters (Gaussian to 12 dB).

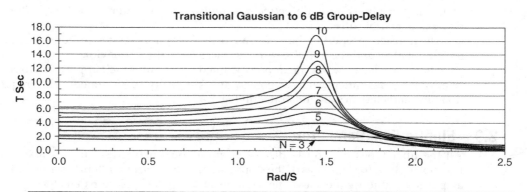

FIGURE 2-71 Group-delay characteristics for transitional filters (Gaussian to 6 dB).

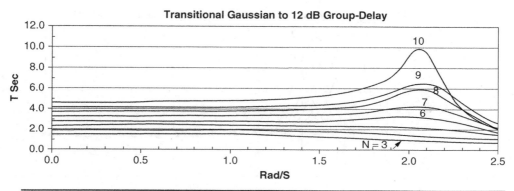

FIGURE 2-72   Group-delay characteristics for transitional filters (Gaussian to 12 dB).

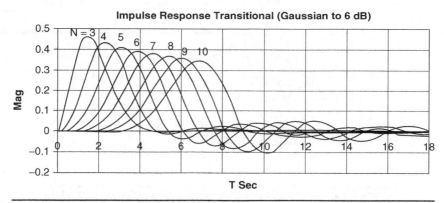

FIGURE 2-73   Impulse response for transitional filters (Gaussian to 6 dB).

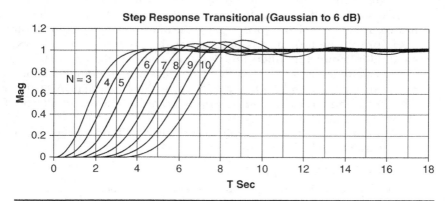

FIGURE 2-74   Step response for transitional filters (Gaussian to 6 dB).

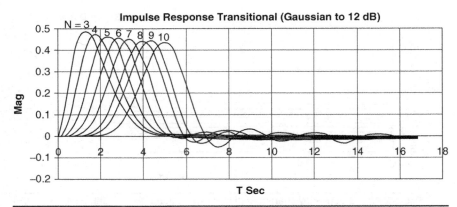

FIGURE 2-75 Impulse response for transitional filters (Gaussian to 12 dB).

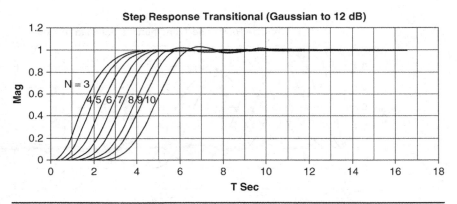

FIGURE 2-76 Step response for transitional filters (Gaussian to 12 dB).

can be obtained. The introduction of these transmission zeros allows the steepest rate of descent theoretically possible for a given number of poles.

Figure 2-78 compares a five-pole Butterworth, a 0.1-dB Chebyshev, and a 0.1-dB elliptic-function filter having two transmission zeros. Clearly, the elliptic-function filter has a much more rapid rate of descent in the transition region than the other filter types.

Improved performance is obtained at the expense of return lobes in the stopband. Elliptic-function filters are also more complex than all-pole networks. Return lobes usually are acceptable to the user, since a minimum stopband attenuation is required and the chosen filter will have return lobes that meet this requirement. Also, even though each filter section is more complex than all-pole filters, fewer sections are required.

The following definitions apply to normalized elliptic-function low-pass filters and are illustrated in Fig. 2-79:

R_{dB} = the passband ripple

A_{min} = the minimum stopband attenuation in decibels

Ω_s = the lowest stopband frequency at which A_{min} occurs

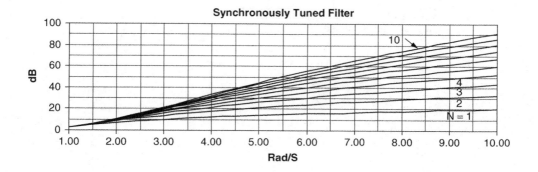

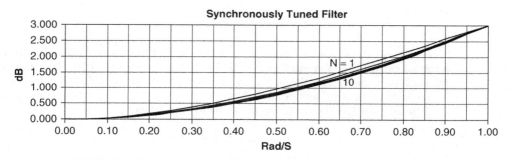

FIGURE 2-77   Attenuation characteristics for synchronously tuned filters.

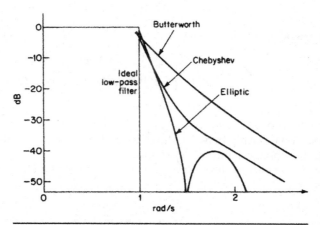

FIGURE 2-78   A comparison of $n = 5$ Butterworth, Chebyshev, and elliptic-function filters.

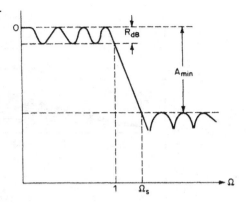

Figure 2-79 Normalized elliptic-function low-pass filter response.

The response in the passband is similar to that of Chebyshev filters except that the attenuation at 1 rad/s is equal to the passband ripple instead of 3 dB. The stopband has transmission zeros, with the first zero occurring slightly beyond Ω_s All returns (comebacks) in the stopband are equal to $A_{\min}$.

The attenuation of elliptic filters can be expressed as

$$A_{dB} = 10 \log\left[1 + \varepsilon^2 Z_n^2(\Omega)\right] \tag{2-49}$$

where ε is determined by the ripple [Eq. (2-37)] and $Z_n(\Omega)$ is an elliptic function of the nth order. Elliptic functions have both poles and zeros and can be expressed as

$$Z_n(\Omega) = \frac{\Omega\left(a_2^2 - \Omega^2\right)\left(a_4^2 - \Omega^2\right)\ldots\left(a_m^2 - \Omega^2\right)}{\left(1 - a_2^2\Omega^2\right)\left(1 - a_4^2\Omega^2\right)\ldots\left(1 - a_m^2\Omega^2\right)} \tag{2-50}$$

where n is odd and $m = (n-1)/2$, or

$$Z_n(\Omega) = \frac{\left(a_2^2 - \Omega^2\right)\left(a_4^2 - \Omega^2\right)\ldots\left(a_m^2 - \Omega^2\right)}{\left(1 - a_2^2\Omega^2\right)\left(1 - a_4^2\Omega^2\right)\ldots\left(1 - a_m^2\Omega^2\right)} \tag{2-51}$$

where n is even and $m = n/2$.

The zeros of Z_n are $a_2, a_4, \ldots, a_m$, whereas the poles are $1/a_2, 1/a_4, \ldots, 1/a_m$. The reciprocal relationship between the poles and zeros of Z_n results in equiripple behavior in both the stopband and the passband.

The values for a_2 through a_m are derived from the elliptic integral, which is defined as

$$K_e = \int_0^{\pi/2} \frac{d\theta}{\sqrt{1 - k^2 \sin^2 \theta}} \tag{2-52}$$

Numerical evaluation may be somewhat difficult. Glowatski (see References) contains tables specifically intended for determining the poles and zeros of $Z_n(\Omega)$.

Elliptic-function filters have been extensively tabulated by Saal and Zverev (see Bibliography). The basis for these tabulations was the order n and the parameters θ (degrees) and reflection coefficient ρ (percent).

θ, degrees	Ω_s
0	∞
10	5.759
20	2.924
30	2.000
40	1.556
50	1.305
60	1.155
70	1.064
80	1.015
90	1.000

TABLE 2-2 Ω_s VS. θ

Elliptic-function filters are sometimes called *Cauer filters* in honor of network theorist Professor Wilhelm Cauer. They were tabulated using the following convention

$$C\,n\,\rho\,\theta$$

where C represents Cauer, n is the filter order, ρ is the reflection coefficient, and θ is the modular angle. A fifth-order filter having a ρ of 15 percent and a θ of 29° would be described as CO5 15 $\theta = 29°$.

The angle θ determines the steepness of the filter and is defined as

$$\theta = \sin^{-1}\frac{1}{\Omega_s} \tag{2-53}$$

or, alternatively, we can state

$$\Omega_s = \frac{1}{\sin\theta} \tag{2-54}$$

Table 2-2 gives some representative value of θ and Ω_s.

The parameter ρ, the reflection coefficient, can be derived from

$$\rho = \frac{VSWR - 1}{VSWR + 1} = \sqrt{\frac{\varepsilon^2}{1 + \varepsilon^2}} \tag{2-55}$$

where VSWR is the standing-wave ratio and ε is the ripple factor (see Sec. 2.4 on the Chebyshev response). The passband ripple and reflection coefficient are related by

$$R_{dB} = -10\log(1 - \rho^2) \tag{2-56}$$

Table 2-3 interrelates these parameters for some typical values of the reflection coefficient, where ρ is expressed as a percentage.

As the parameter θ approaches 90° the edge of the stopband Ω_s approaches unity. For θs near 90° extremely sharp roll-offs are obtained. However, for a fixed n, the stopband attenuation A_{mni} is reduced as the steepness increases. Figure 2-80 shows the

ρ, %	R_{dB}	VSWR	ε (ripple factor)
1	0.0004343	1.0202	0.0100
2	0.001738	1.0408	0.0200
3	0.003910	1.0619	0.0300
4	0.006954	1.0833	0.0400
5	0.01087	1.1053	0.0501
8	0.02788	1.1739	0.0803
10	0.04365	1.2222	0.1005
15	0.09883	1.3529	0.1517
20	0.1773	1.5000	0.2041
25	0.2803	1.6667	0.2582
50	1.249	3.0000	0.5774

TABLE 2-3 ρ vs. R_{dB} VSWR, and ε

frequency response of an $n = 3$ elliptic filter for a fixed ripple of 1 dB ($\rho = 50$ percent) and different values of θ.

For a given θ and order n, the stopband attenuation parameter A_{min} increases as the ripple is made larger. Since the poles of elliptic-function filters are approximately located on an ellipse, the delay curves behave in a similar manner to those of the Chebyshev family. Figure 2-81 compares the delay characteristics of $n = 3$, 4, and 5 elliptic filters, all having an A_{min} of 60 dB. The delay variation tends to increase sharply with increasing ripple and filter order n.

The factor ρ determines the input impedance variation with frequency of LC elliptic filters, as well as the passband ripple. As ρ is reduced, a better match is achieved between the resistive terminations and the filter impedance. Figure 2-82 illustrates the

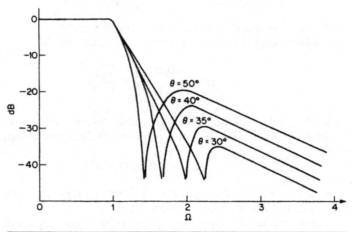

FIGURE 2-80 The elliptic-function low-pass filter response for $n = 3$ and $R_{dB} = 1$ dB.

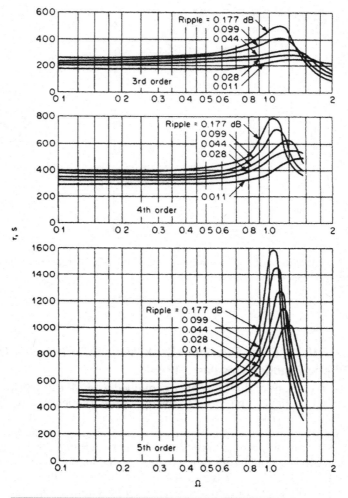

FIGURE 2-81 Delay characteristics of elliptic-function filters n = 3, 4, 5, and with an A_{min} of 60 dB.

input impedance variation with frequency of a normalized n = 5 elliptic-function low-pass filter. At DC, the input impedance is 1 Ω resistive. As the frequency increases, both positive and negative reactive components appear. All maximum values are within the diameter of a circle whose radius is proportional to the reflection coefficient ρ. As the complexity of the filter is increased, more gyrations occur within the circle.

The relationship between ρ and filter input impedance is defined by

$$|\rho|^2 = \left| \frac{R - Z_{11}}{R + Z_{11}} \right|^2 \tag{2-57}$$

where R is the resistive termination and Z_{11} is the filter input impedance.

FIGURE 2-82   Impedance variation in the passband of a normalized $n = 5$ elliptic-function low-pass filter.

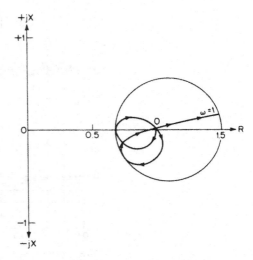

The closeness of matching between R and Z_{11} is frequently expressed in decibels as a returns loss, which is defined as

$$A_\rho = 20 \log \left| \frac{1}{\rho} \right| \tag{2-58}$$

2.9.1   Using Filter Solutions (Book Version) Software for Design of Elliptic Function Low-Pass Filters

Previous editions of this book have contained extensive numerical tables of normalized values which have to be scaled to the operating frequencies and impedance levels during the design process. This is no longer the case.

A program called "Filter Solutions Book Version" is included in the download (see App. A). This program is limited to Elliptic Function LC filters (up to $N = 10$) and is a subset of the complete program which is available from Nuhertz Technologies® (www.nuhertz.com). The reader is encouraged to obtain the full version, which in addition to passive implementations covers many filter polynomial types and includes distributed, active, switched capacitor, and digital along with many very powerful features, and integrates with the popular Microwave Office® design software from AWR® Corporation, CST Studio Suite®, and Sonnet® software."

The program is quite intuitive and self-explanatory; thus, readers are encouraged to explore its many features on their own. Nevertheless, all design examples using this program will elaborate on its usage and provide helpful hints. Extract and install this program by running *Nuhertz_FB2_Install.*

Filter Solutions can be installed by running *Filter_Solutions_FB2_Install* which is part of the download (see App. A).

Example 2-22   Determining the Order of an Elliptic Function Filter using *Filter Solutions*

Required:

Determine the order of an elliptic-function filter having a passband ripple less than 0.2 dB up to 1,000 Hz, and a minimum rejection of 60 dB at 1,300 Hz and above. Use *Filter Solutions*.

Result:

(*a*) Open *Filter Solutions.*
    Check the *Stop Band Freq* box.
    Enter 0.2 in the *Pass Band Ripple(dB)* box.
    Enter **1,000** in the *Pass Band Freq* box.
    Enter **1,300** in the *Stop Band Freq* box.
    Check the *Frequency Scale Hertz* box.

(*b*) Click the *Set Order* control button to open the second panel.
    Enter **60** for the *Stop band Attenuation (dB)*
    Click the *Set Minimum Order* button and then click *Close.*
    7 Order is displayed on the main control panel.

(*c*) The result is that a 7th order elliptic-function low-pass filter provides the required attenuation. By comparison, a 27th-order Butterworth low-pass filter would be needed to meet the requirements of Example 2-22, so the elliptic-function family is a *must* for steep filter requirements.

2.9.2   Using the ELI 1.0 Program for the Design of Odd-Order Elliptic-Function Low-Pass Filters up to the 31st Order

This program allows the design of odd-order elliptic function *LC* low-pass filters up to a complexity of 15 nulls (transmission zeros), or the 31st order. It is based on an algorithm developed by Amstutz.

The program inputs are passband edge (Hz), stopband edge (Hz), number of nulls (up to 15), stopband rejection in dB, and source and load terminations (which are always equal). The output parameters are critical Q (theoretical minimum Q), passband ripple (dB), nominal 3-dB cutoff and a list of component values along with resonant null frequencies.

To install the program, first copy **ELI1** from the companion website (see App. A) to the desktop and then create a folder called ELI1. Place the program in the folder.

To run the program, double-click "ELI1." and enter inputs as requested. Upon completing the execution, a dataout.text file will appear in the folder which can open using Notepad and containing the resulting circuit description.

If the number of nulls is excessive for the response requirements (indicated by zero passband ripple) the final capacitor may have a negative value as a result of the algorithm. Reduce the number of nulls, increase the required attention, define a steeper filter, or do a combination of these.

2.10   Maximally Flat Delay With Chebyshev Stopband

The Bessel, linear phase with equiripple error, and transitional filter families all exhibit either maximally flat or equiripple-delay characteristics over most of the passband and, except for the transitional type, even into the stopband. However, the amplitude versus frequency response is far from ideal. The passband region in the vicinity of the cutoff is very rounded, while the stopband attenuation in the first few octaves is poor.

Elliptic-function filters have an extremely steep rate of descent into the stopband because of transmission zeros. However, the delay variation in the passband is unacceptable when the transient behavior is significant.

The maximally flat delay with Chebyshev stopband filters is derived by introducing transmission zeros into a Bessel-type transfer function. The constant delay properties in the passband are retained. However, the stopband rejection is significantly improved because of the effectiveness of the transmission zeros.

The step response exhibits no overshoot or ringing, and the impulse response has essentially no oscillatory behavior. Constant delay properties extend well into the stopband for higher-order networks.

Normalized tables of element values for the maximally flat delay with the Chebyshev stopband family of filters are provided in Table 10-56. These tables are normalized so that the 3-dB response occurs at 1 rad/s. The tables also provide the delay at DC and the normalized frequencies corresponding to a 1-percent and 10-percent deviation from the delay at DC. The amplitude response below the 3-dB point is identical to the attenuation characteristics of the Bessel filters shown in Fig. 2-56.

2.11   Papoulis Optimum "L" Filter

Professor A. Papoulis published two papers in 1958 and 1959, which introduced a new set of filters based on Legendre polynomials. This class of filters is named after Papoulis and may also be known as "Optimum L" with the "L" representing Legendre. These are all-pole filters that have the steepest rate of descent into the stopband without a

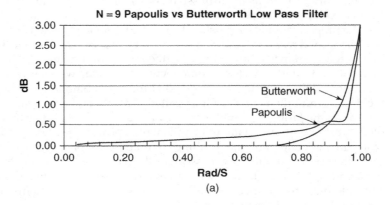

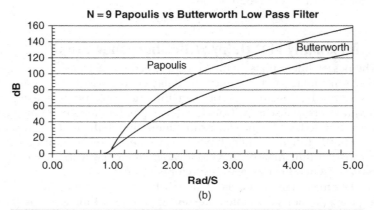

Figure 2-83 $N = 9$ Butterworth compared to Papoulis low-pass filter: (a) Response below 1 Rad/S, (b) stopband response.

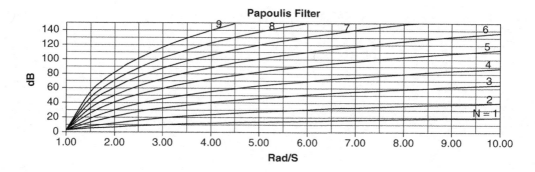

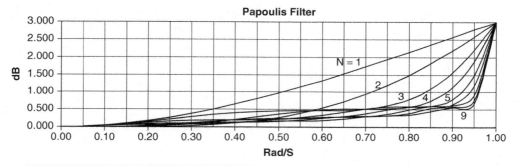

Figure 2-84 Attenuation characteristics Papoulis filter.

passband ripple and without transmission zeros. Their response is always monotonic. As the order increases a "step" appears to form in the response as the corner is approached but the response never reverses direction. Figure 2-83 compares an $N = 9$ Butterworth and Papoulis low-pass filter. The apparent advantage of the Papoulis response over the Butterworth is evident.

However, it must be said that the transient behavior of the Papoulis filer is poorer than the Butterworth.

Attenuation curves for orders' 1 through 9 are provided in Fig. 2-84.

Normalized pole locations and element values are given in Chap. 10.

References

Amstutz, P. "Elliptic Approximation and Elliptic Filter Design on Small Computers." *IEEE Transactions on Circuits and Systems* CAS-25, No.12 (December, 1978).

Feistel, V. K., and R. Unbehauen. "Tiefpasse mit Tschebyscheff—Charakter der Betriebsdampfung im Sperrbereich und Maximal geebneter Laufzeit." *Frequenz* 8 (1965).

Glowatski, E. "Sechsstellige Tafel der Cauer-Parameter." *Verlag der Bayr*, Akademie der Wissenchaften (1955).

Lindquist, C. S. *Active Network Design*. California: Steward and Sons, 1977.

Matthaei, G. L., Young, L., and E. M. T. Jones. *"Microwave Filters, Impedance-Matching Networks, and Coupling Structures."* Massachusetts: Artech House, 1980.

Saal, R. "Der Entwurf von Filtern mit Hilfe des Kataloges Normierter Tiefpasse." *Telefunken GMBH* (1963).

White Electromagnetics. *A Handbook on Electrical Filters*. White Electromagnetics Inc., 1963.

Zverev, A. I. *Handbook of Filter Synthesis*. New York: John Wiley and Sons, 1967.

CHAPTER 3

Low-Pass Filter Design

3.1 *LC* Low-Pass Filters

3.1.1 All-Pole Filters

LC low-pass filters can be designed from the tables provided in Chap. 10 or the software that can be downloaded as indicated in App. A. A suitable filter must first be selected using the guidelines established in Chap. 2, however. The chosen design is then frequency- and impedance-scaled to the desired cutoff and impedance level when using the tables, or directly designed when using the software.

Example 3-1 Design of an LC Low-Pass Filter from the Tables

Required:

An *LC* low-pass filter
3 dB at 1,000 Hz
20-dB minimum at 2,000 Hz
$R_s = R_L = 600\ \Omega$

Result:

(*a*) To normalize the low-pass requirement, compute A_s, using Eq. (2.11):

$$A_s = \frac{f_s}{f_c} = \frac{2,000\ \text{Hz}}{1,000\ \text{Hz}} = 2$$

(*b*) Choose a normalized low-pass filter from the curves of Chap. 2 having at least 20 dB of attenuation at 2 rad/s.

Examination of the curves indicates that an $n = 4$ Butterworth or third-order 0.1-dB Chebyshev satisfies this requirement. Let us select the latter, since fewer elements are required.

(*c*) Table 10-28 contains element values for normalized 0.1-dB Chebyshev *LC* filters ranging from $n = 2$ through $n = 10$. The circuit corresponding to $n = 3$ and equal source and load resistors ($R_s = 1\ \Omega$) is shown in Fig. 3-1*a*.

(*d*) Using Eqs. (2-8) to (2-10) denormalize the filter using a Z of 600 and a frequency-scaling factor (FSF) of $2\pi f_c$ or 6,280.

$$R'_s = R'_L = 600\ \Omega$$

$$L'_2 = \frac{L \times Z}{\text{FSF}} = \frac{1.5937 \times 600}{6,280} = 0.152\ \text{H}$$

85

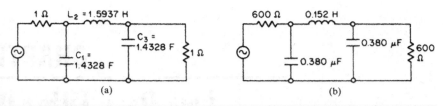

FIGURE 3-1 The results of Example 3-1: (a) normalized filter from Table 10-28, (b) frequency- and impedance-scaled filter.

$$C_1' = C_3' = \frac{C}{\text{FSF} \times Z} = \frac{1.4328}{6,280 \times 600} = 0.380 \ \mu\text{F}$$

The resulting filter is shown in Fig. 3-1b.

The normalized filter used in Example 3-1 is shown in Table 10-28 (in Chap. 10) as having a current source input with a parallel resistor of 1 Ω. The reader will recall that Thévenin's theorems permit the replacement of this circuit with a voltage source having an equivalent series source resistance.

3.1.2 Elliptic-Function Filters

Elliptic Function Low-Pass Filters Using the Filter Solutions Program

The following example illustrates the design of an elliptic-function low-pass filter using the Filter Solutions program introduced in Sec. 2.9.

Example 3-2 Design of an Elliptic Function Low-Pass Filter Using Filter Solutions Program

Required:

LC low-pass filter
0.25-dB maximum ripple DC to 100 Hz
60-dB minimum at 132 Hz
$R_s = R_L = 900 \ \Omega$

Result:

(a) Open *Filter Solutions*.
Check the *Stop Band Freq* box.
Enter **0.18** in the *Pass Band Ripple (dB)* box.
Enter **100** in the *Pass Band Freq* box.
Enter **132** in the *Stop Band Freq* box.
Check the *Frequency Scale Hertz* box.
Enter **900** for *Source Res* and *Load Res*.

(b) Click the *Set Order* control button to open the second panel.
Enter **60** for *Stop band Attenuation (dB)*.
Click the *Set Minimum Order* button and then click *Close*.
7 Order is displayed on the main control panel.

(c) Click the *Synthesize Filter* button.

Two schematics are presented and shown in Fig. 3-2. The circuit of Fig. 3-2a has a shunt capacitor as its first element, and the circuit of Fig. 3-2b has a series inductor as its first element. Normally, the user would select the configuration having less inductors, which is the first circuit. The frequency response is shown in fig 3-2c.

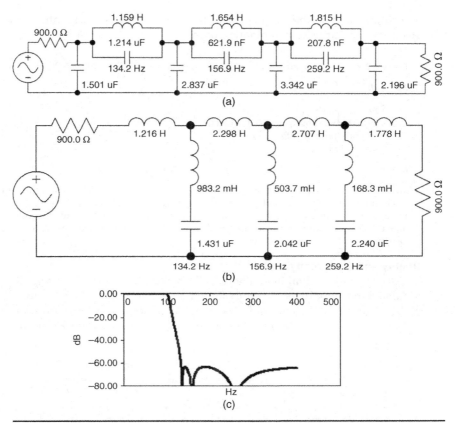

Figure 3-2 Filters of Example 3-2: (a) first element shunt capacitor, (b) first element series inductor, and (c) frequency response.

Note: All examples in the book using *Filter Solutions* are based on starting with program default settings. To restore these settings click the **Initialize** button, then **Default**, and then click **Save**.

Using the ELI 1.0 Program for Designing Odd-Order Elliptic Function Low-Pass Filters up to the 31st Order

The following example illustrates the design of an elliptic-function low-pass filter using the **ELI1.0** program first introduced in Sec. 2.9. This program allows the design of odd-order elliptic function LC low-pass filters up to a complexity of 15 nulls (transmission zeros) or the 31st order. It is based on an algorithm developed by Amstutz (see References).

The program inputs are passband edge (Hz), stopband edge (Hz), number of nulls (up to 15), stopband rejection in dB, and source and load terminations (which are always equal). The output parameters are critical Q (theoretical minimum Q), passband ripple (dB), nominal 3 dB cutoff, and a list of component values along with resonant null frequencies.

If the number of nulls is excessive for the response requirements (indicated by zero passband ripple), the final capacitor may have a negative value as a result of the algorithm. Reduce the number of nulls, increase the required attention, or define a steeper filter—or use a combination of these.

Example 3-3 Design of an Elliptic Function Low-Pass Filter Using ELI 1.0 Program

Required:

An *LC* low-pass filter
0.25-dB maximum ripple DC to 100 Hz
35-dB minimum at 105 Hz
$R_s = R_L = 10$ kΩ

Result:

To run, double-click the "eli1" shortcut and enter inputs as requested. Upon completing execution, a dataout.text file will be created in the ELI 1.0 folder (as shown in Fig. 3-3) and will contain the resulting circuit description. Note that the capacitors are all listed in one column, the inductors in another, and the corresponding resonant frequencies in a third column lined up with the parallel tuned circuits.

Duality and Reciprocity

A network and its dual have identical response characteristics. Each all-pole *LC* filter tabulated in Chap. 10 has an equivalent dual network. The circuit configuration shown at the bottom of each table, and the bottom set of nomenclature, corresponds to the dual of the upper filter. For elliptic filters using *Filter Solutions*, a checkmark in **1st Ele Shunt**

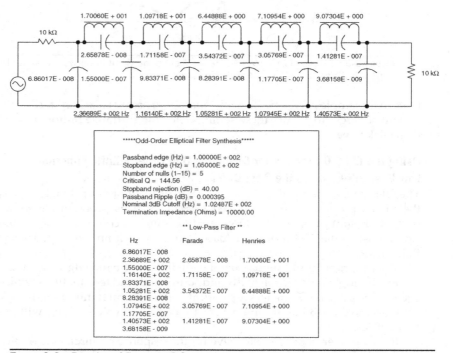

FIGURE 3-3 Results of Example 3-3.

and **1st Ele Series** will give you dual networks in the normalized case of equal 1-ohm source and load terminations.

Any ladder-type network can be transformed into its dual by implementing the following rules:

1. Convert every series branch into a shunt branch and every shunt branch into a series branch.
2. Convert circuit branch elements in series to elements in parallel, and vice versa.
3. Transform each inductor into a capacitor, and vice versa. The values remain unchanged—for instance, 4 H becomes 4 F.
4. Replace each resistance with a conductance—for example, 3 Ω becomes 3 mhos or 1/3 Ω.
5. Change a voltage source into a current source, and vice versa.

Figure 3-4 shows a network and its dual.

The theorem of reciprocity states that if a voltage located at one point of a linear network produces a current at any other point, the same voltage acting at the second point results in the same current at the first point. Alternatively, if a current source at one point of a linear network results in a voltage measured at a different point, the same current source at the second point produces the same voltage at the first point. As a result, the response of an *LC* filter is the same regardless of which direction the signal flows in, except for a constant multiplier. It is perfectly permissible to turn a filter schematic completely around with regard to its driving source, provided that the source- and load-resistive terminations are also interchanged.

The laws of duality and reciprocity are used to manipulate a filter to satisfy termination requirements or to force a desired configuration.

Designing for Unequal Terminations

Tables of all-pole filter *LC* element values are provided in Chap. 10 for both equally terminated and unequally terminated networks. A number of different ratios of source-to-load resistance are tabulated, including the impedance extremes of infinity and zero.

To design an unequally terminated filter, first determine the desired ratio of R_s/R_L. Select a normalized filter from the table that satisfies this ratio. The reciprocity

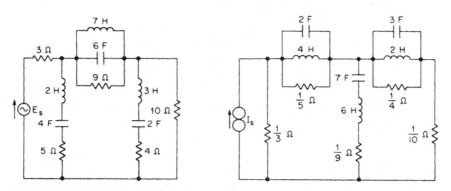

Figure 3-4 An example of dual networks.

theorem can be applied to turn a network around end for end and the source and load resistors can be interchanged. The tabulated impedance ratio is inverted if the dual network given by the lower schematic is used. The chosen filter is then frequency- and impedance-scaled.

For unequally terminated elliptic filters, you can enter the required source and load terminations in the **Source Res** and **Load Res** boxes of *Filter Solutions* before clicking the *Synthesize Filter* button.

Example 3-4   Design of an LC Low-Pass Filter for Unequal Terminations

Required:

>An *LC* low-pass filter
>1 dB at 900 Hz
>20-dB minimum at 2,700 Hz
>$R_s = 1 \text{ k}\Omega$
>$R_L = 5 \text{ k}\Omega$

Result:

(*a*) Compute A_s, using Eq. (2.11):

$$A_s = \frac{f_s}{f_c} = \frac{2,700 \text{ Hz}}{900 \text{ Hz}} = 3$$

(*b*) Normalized requirement:

>1 dB at X rad/s
>20-dB minimum at $3X$ rad/s
>(where X is arbitrary)

(*c*) Select a normalized low-pass filter that makes the transition from 1 dB to at least 20 dB over a frequency ratio of 3:1. A Butterworth $n = 3$ design satisfies these requirements since Fig. 2-34 indicates that the 1-dB point occurs at 0.8 rad/s and that more than 20 dB of attenuation is obtained at 2.4 rad/s. Table 10-2 provides element values for normalized Butterworth low-pass filters for a variety of impedance ratios. Since the ratio of R_s/R_L is 1:5, we will select a design for $n = 3$, corresponding to $R_s = 0.2 \ \Omega$, and use the upper schematic. (Alternatively, we could have selected the lower schematic corresponding to $R_s = 5 \ \Omega$ and turned the network end for end, but an additional inductor would have been required.)

(*d*) The normalized filter from Table 10-2 is shown in Fig. 3-5*a*. Since the 1-dB point is required to be 900 Hz, the FSF is calculated using Eq. (2-1):

$$\text{FSF} = \frac{\text{desired reference frequency}}{\text{existing reference frequency}}$$

$$= \frac{2\pi 900 \text{ rad/s}}{0.8 \text{ rad/s}} = 7,069$$

Making use of Eqs. (2-8) to (2-10) and using a Z of 5,000 and an FSF of 7,069, the denormalized component values are

$$R'_s = R \times Z = 1 \text{ k}\Omega$$

$$R'_L = 5 \text{ k}\Omega$$

$$C'_1 = \frac{C}{\text{FSF} \times Z} = \frac{2.6687}{7,069 \times 5,000} = 0.0755 \ \mu\text{F}$$

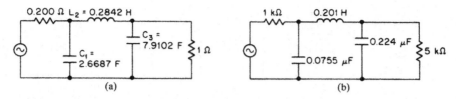

FIGURE 3-5   Low-pass filter with unequal terminations: (a) normalized low-pass filter, (b) frequency- and impedance-scaled filter.

$$C_3' = 0.22 \ \mu F$$

$$L_2 = \frac{L \times Z}{\text{FSF}} = \frac{0.2842 \times 5,000}{7,069} = 0.201 \ \text{H}$$

The scaled filter is shown in Fig. 3-5b.

If an infinite termination is required, a design having an R_s of infinity is selected. When the input is a current source, the configuration is used as given. For an infinite load impedance, the entire network is turned end for end.

If the design requires a source impedance of 0 Ω, the dual network is used corresponding to $1/R_s$ of infinity or $R_s = 0 \ \Omega$.

In practice, impedance extremes of near zero or infinity are not always possible. However, for an impedance ratio of 20 or more, the load can be considered infinite in comparison with the source, and the design for an infinite termination is used. Alternatively, the source may be considered zero with respect to the load and the dual filter corresponding to $R_s = 0 \ \Omega$ may be used. When n is odd, the configuration having the infinite termination has one less inductor than its dual.

An alternate method of designing filters to operate between unequal terminations involves partitioning the source or load resistor between the filter and the termination. For example, a filter designed for a 1-kΩ source impedance could operate from a 250-Ω source if a 750-Ω resistor were placed within the filter network in series with the source. However, this approach would result in a higher insertion loss.

Bartlett's Bisection Theorem   A filter network designed to operate between equal terminations can be modified for unequal source and load resistors if the circuit is symmetrical. Bartlett's bisection theorem states that if a symmetrical network is bisected and one half is impedance-scaled, including the termination, the response shape will not change. All tabulated odd-order Butterworth and Chebyshev filters having equal terminations satisfy the symmetry requirement.

Example 3-5   Design of an *LC* Low-Pass Filter for Unequal Terminations Using Bartlett's Bisection Theorem

Required:

An *LC* low-pass filter
3 dB at 200 Hz
15-dB minimum at 400 Hz
$R_s = 1 \ \text{k}\Omega$
$R_L = 1.5 \ \text{k}\Omega$

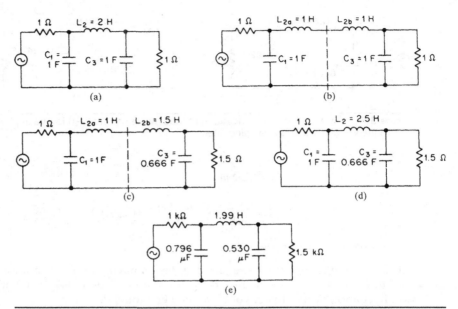

FIGURE 3-6 An example of Bartlett's bisection theorem: (a) normalized filter having equal terminations, (b) bisected filter, (c) impedance-scaled right-half section, (d) recombined filter, and (e) final scaled network.

Result:

(a) Compute A_s using Eq. (2-11)

$$A_s = \frac{f_s}{f_c} = \frac{400}{200} = 2$$

(b) Figure 2-34 indicates that an $n = 3$ Butterworth low-pass filter provides 18-dB rejection at 2 rad/s. Normalized LC values for Butterworth low-pass filters are given in Table 10-2. The circuit corresponding to $n = 3$ and equal terminations is shown in Fig. 3-6a.

(c) Since the circuit of Fig. 3-6a is symmetrical, it can be bisected into two equal halves, as shown in Fig. 3-6b. The requirement specifies a ratio of load-to-source resistance of 1.5 (1.5 kΩ/1 kΩ), so we must impedance-scale the right half of the circuit by a factor of 1.5. The circuit of Fig. 3-6c is thus obtained.

(d) The recombined filter of Fig. 3-6d can now be frequency- and impedance-scaled using an FSF of $2\pi200$ or 1,256 and a Z of 1,000.

$$R'_s = 1 \text{ k}\Omega$$

$$R'_L = 1.5 \text{ k}\Omega$$

Using Eqs. (2-9) and (2-10):

$$C'_1 = \frac{C}{\text{FSF} \times Z} = \frac{1}{1,256 \times 1,000} = 0.796 \ \mu\text{F}$$

$$C'_3 = 0.530 \ \mu\text{F}$$

$$L'_2 = \frac{L \times Z}{\text{FSF}} = \frac{2.5 \times 1,000}{1,256} = 1.99 \text{ H}$$

The final filter is shown in Fig. 3-6e.

3.1.3 Effects of Dissipation

Filters designed using the tables of *LC* element values in Chap. 10 require lossless coils and capacitors to obtain the theoretical response predicted in Chap. 2. In the practical world, capacitors are usually obtainable that have low losses, but inductors are generally lossy, especially at low frequencies. Losses can be defined in terms of Q, the figure of merit or quality factor of a reactive component.

If a lossy coil or capacitor is resonated in parallel with a lossless reactance, the ratio of resonant frequency to 3-dB bandwidth of the resonant circuit's impedance (in other words, the band over which the magnitude of the impedance remains within 0.707 of the resonant value) is given by

$$Q = \frac{f_0}{\mathrm{BW}_{3\,\mathrm{dB}}} \tag{3-1}$$

Figure 3-7 gives the low-frequency equivalent circuits for practical inductors and capacitors. Their Qs can be calculated by

Inductors:
$$Q = \frac{\omega L}{R_L} \tag{3-2}$$

Capacitor:
$$Q = \omega C R_c \tag{3-3}$$

where ω is the frequency of interest, in radians per second.

Using elements having a finite Q in a design intended for lossless reactances has the following mostly undesirable effects:

- At the passband edge, the response shape becomes more rounded. Within the passband, the ripples are diminished and may completely vanish.

- The insertion loss of the filter is increased. The loss in the stopband is maintained (except in the vicinity of transmission zeros), so the relative attenuation between the passband and the stopband is reduced.

Figure 3-8 shows some typical examples of these effects on all-pole and elliptic function low-pass filters.

The most critical problem caused by the finite element Q is the effect on the response shape near cutoff. Estimating the extent of this effect is somewhat difficult without extensive empirical data. The following variations in the filter design

Figure 3-7 Low-frequency equivalent circuits of practical inductors and capacitors.

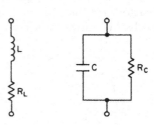

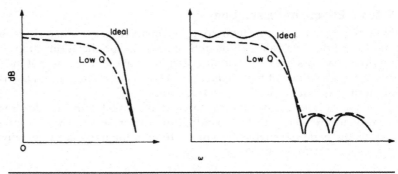

Figure 3-8   The effects of finite Q.

parameters will cause *increased* rounding of the frequency response near cutoff for a fixed Q:

- Going to a larger passband ripple
- Increasing the filter order n
- Decreasing the transition region of elliptic-function filters

Changing these parameters in the opposite direction, of course, reduces the effects of dissipation.

Filters can be designed to have the responses predicted by modern network theory using finite element Qs. Figure 3-9 shows the minimum Qs required at the cutoff for different low-pass responses. If elements used are having Qs slightly above the minimum values given in Fig. 3-9, the desired response can be obtained provided that certain predistorted element values are used. However, the insertion loss will be prohibitive. It is therefore highly desirable that element Qs be several times higher than the values indicated.

The effect of low Q on the response near cutoff can usually be compensated for by going to a higher-order network or a steeper filter and using a larger design bandwidth to allow for rounding. However, this design approach does not always result in satisfactory results, since the Q requirement may also increase. A method of compensating for low Q by using amplitude equalization is discussed later in Sec. 8.4.

The insertion loss of low-pass filters can be computed by replacing the reactive elements with resistances corresponding to their Qs since at DC the inductors become short circuits and capacitors become open, which leaves the resistive elements only.

Figure 3-10a shows a normalized third-order 0.1-dB Chebyshev low-pass filter where each reactive element has a Q of 10 at the 1-rad/s cutoff. The series and shunt resistors for the coil and capacitors are calculated using Eqs. (3-2) and (3-3), respectively. At 1 rad/s these equations can be simplified and reexpressed as

$$R_L = \frac{L}{Q} \tag{3-4}$$

$$R_c = \frac{Q}{C} \tag{3-5}$$

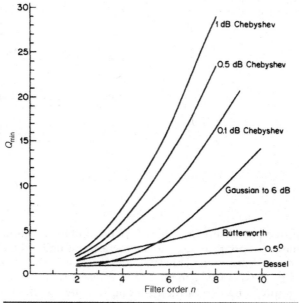

FIGURE 3-9 Minimum Q requirements for low-pass filters.

The equivalent circuit at DC is shown in Fig. 3-10b. The insertion loss is 1.9 dB. The actual loss calculated was 7.9 dB, but the 6-dB loss caused by the source and load terminations is normally not considered a part of the filter's insertion loss because it would also occur in the event that the filter was completely lossless.

3.1.4 Using Predistorted Designs

The effect of finite element Q on an LC filter transfer function is to increase the real components of the pole positions by an amount equal to the dissipation factor d, where

$$d = \frac{1}{Q} \tag{3-6}$$

Figure 3-11 shows this effect. All poles are displaced to the left by an equal amount.

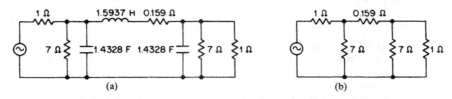

FIGURE 3-10 Calculation of insertion loss: (a) third-order 0.1-dB Chebyshev low-pass filter with Q = 10, (b) equivalent circuit at DC.

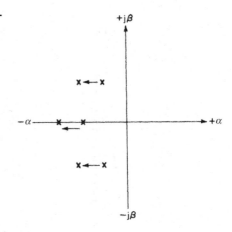

Figure 3-11 The effects of dissipation on the pole pattern.

If the desired poles were first shifted to the right by an amount equal to d, the introduction of the appropriate losses into the corresponding LC filter would move the poles back to the desired locations. This technique is called *predistortion*. Predistorted filters are obtained by predistorting the required transfer function for a desired Q and then synthesizing an LC filter from the resulting transfer function. When the reactive elements of the filter have the required losses added, the response shape will correspond to the original transfer function.

The maximum amount that a group of poles can be displaced to the right in the process of predistortion is equal to the smallest real part among the poles given that further movement corresponds to locating a pole in the right half plane, which is an unstable condition. The minimum Q therefore is determined by the highest Q pole (in other words, the pole having the smallest real component). The Qs shown in Fig. 3-9 correspond to $1/d$, where d is the real component of the highest Q pole.

Tables are provided in Chap. 10 for all-pole predistorted low-pass filters. These designs are all singly terminated with a source resistor of $1\,\Omega$ and an infinite termination. Their duals turned end for end can be used with a voltage source input and a $1\text{-}\Omega$ termination.

Two types of predistorted filters are tabulated for various ds. The uniform dissipation networks require uniform losses in both the coils and the capacitors. The second type are the Butterworth lossy-L filters, where only the inductors have losses, which closely agrees with practical components. It is important for both types that the element Qs are closely equal to $1/d$ at the cutoff frequency. In the case of the uniform dissipation networks, losses must usually be added to the capacitors.

Example 3-6   Design of a Predistorted Lossy-L LC Low-Pass Filter

Required:

An LC low-pass filter
3 dB at 500 Hz
24-dB minimum at 1,200 Hz
$Rs = 600\,\Omega$
$RL = 100\ \text{k}\Omega$ minimum
Inductor Qs of 5 at 500 Hz
Lossless capacitors

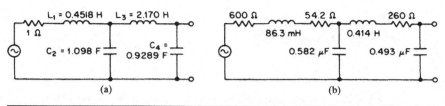

FIGURE 3-12 The lossy-*L* low-pass filter of Example 3-6: (*a*) normalized filter and (*b*) scaled filter.

Result:

(*a*) Compute A_s, using Eq. (2-11):

$$A_s = \frac{1,200}{500} = 2.4$$

(*b*) The curves of Fig. 2-34 indicate that an $n = 4$ Butterworth low-pass filter has over 24 dB of rejection at 2.4 rad/s. Table 10-14 contains the element values for Butterworth lossy-*L* network where $n = 4$. The circuit corresponding to $d = 0.2$ ($d = 1/Q$) is shown in Fig. 3-12*a*.

(*c*) Using Eq. (2-9), the normalized filter can now be frequency- and impedance-scaled using an FSF of $2\pi500 = 3,142$ and a *Z* of 600.

$$R_s' = 600 \ \Omega$$

$$L_1' = \frac{L \times Z}{\text{FSF}} = \frac{0.4518 \times 600}{3,142} = 86.3 \ \text{mH}$$

$$L_3' = 0.414 \ \text{H}$$

$$C_2' = \frac{C}{\text{FSF} \times Z} = \frac{1.098}{3,142 \times 600} = 0.582 \ \mu\text{F}$$

$$C_4' = 0.493 \,\mu\text{F}$$

(*d*) Using Eq. (3-2), the resistive coil losses are

$$R_1 = \frac{\omega L}{Q} = 54.2 \ \Omega$$

and

$$R_3 = 260 \ \Omega$$

where

$$\omega = 2\pi f_c = 3142$$

The final filter is given in Fig. 3-12*b*.

Example 3-7   Design of a Uniform Distortion *LC* Low-Pass Filter

Required:

An *LC* low-pass filter
3 dB at 100 Hz
58-dB minimum at 300 Hz
$R_s = 1 \ \text{k}\Omega$
$R_L = 100 \ \text{k}\Omega$ minimum
Inductor *Q*s of 11 at 100 Hz
Lossless capacitors

Result:

(a) Compute A_s, using Eq. (2-11)

$$A_s = \frac{300}{100} = 3$$

(b) Figure 2-42 indicates that a fifth-order 0.1-dB Chebyshev has about 60 dB of rejection at 3 rad/s. Table 10-32 provides *LC* element values for 0.1-dB Chebyshev uniform dissipation networks. The available inductor Q of 11 corresponds to a *d* of 0.091 ($d = 1/Q$). Values are tabulated for an $n = 5$ network having a *d* of 0.0881, which is sufficiently close to the requirement. The corresponding circuit is shown in Fig. 3-13a.

(c) Making use of Eq. (2-10), the normalized filter is frequency-and impedance-scaled using an FSF of $2\pi100 = 628$ and a Z of 1,000.

$$C_1' = \frac{C}{\text{FSF} \times Z} = \frac{1.1449}{628 \times 1,000} = 1.823\ \mu F$$

$$C_3' = 3.216\ \mu F$$

$$C_5' = 1.453\ \mu F$$

Using Eq. (2-9)

$$L_2' = \frac{L \times Z}{\text{FSF}} = \frac{1.8416 \times 1,000}{628} = 2.932\ H$$

$$L_4' = 2.681\ H$$

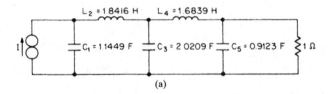

(a)

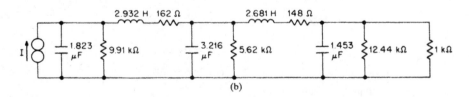

(b)

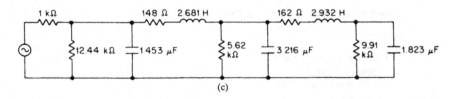

(c)

Figure 3-13 The design of the uniform dissipation network from Example 3-7: (a) normalized fifth-order 0.1-dB Chebyshev with $d = 0.0881$, (b) frequency- and impedance-scaled filter including losses, and (c) final network.

(*d*) Using Eq. (3-3), the shunt resistive losses for capacitors C_1', C_3', and C_5' are

$$R_1 = \frac{Q}{\omega C} = 9.91 \text{ k}\Omega$$

$$R_3' = 5.62 \text{ k}\Omega$$

$$R_5' = 12.44 \text{ k}\Omega$$

Using Eq. (3-2), the series resistive inductor losses are

$$R_1 = \frac{\omega L}{Q} = 162 \ \Omega$$

$$R_4' = 148 \ \Omega$$

where

$$Q = \frac{1}{d} = \frac{1}{0.0881} = 11.35$$

and

$$\omega = 2\pi f_c = 2\pi 100 = 628$$

The resulting circuit, including all losses, is shown in Fig. 3-13*b*. This circuit can be turned end for end so that the requirement for a 1-kΩ source resistance is met. The final filter is given in Fig. 3-13*c*.

It is important to remember that uniform dissipation networks require the presence of losses in both the coils and capacitors, thus resistors must usually be added. Component *Q*s within 20 percent of 1/*d* are usually sufficient for satisfactory results.

Resistors can sometimes be combined to eliminate components. In the circuit of Fig. 3-13*c*, the 1-kΩ source and the 12.44-kΩ resistor can be combined, which results in a 926-Ω equivalent source resistance. The network can then be impedance scaled to restore a 1-kΩ source.

3.2 Active Low-Pass Filters

Active low-pass filters are designed using a sequence of operations similar to the design of *LC* filters. The specified low-pass requirement is first normalized and a particular filter type of the required complexity is selected using the response characteristics given in Chap. 2. Normalized tables of active filter component values are provided in Chap. 10 for each associated transfer function. The corresponding filter is denormalized by frequency and impedance scaling.

Active filters can also be designed directly from the poles and zeros. This approach sometimes offers some additional degrees of freedom and will also be covered.

3.2.1 All-Pole Filters

The transfer function of a passive *RC* network has poles that lie only on the negative real axis of the complex frequency plane. In order to obtain the complex poles required by the all-pole transfer functions of Chap. 2, active elements must be introduced. Integrated circuit operational amplifiers are readily available that have nearly ideal properties, such as high gain. However, these properties are limited to frequencies below a few MHz, so active filters beyond this range are difficult.

Unity-Gain Single-Feedback Realization

Figure 3-14 shows two active low-pass filter configurations. The two-pole section provides a pair of complex conjugate poles, whereas the three-pole section produces a pair of complex conjugate poles and a single real-axis pole. The operational amplifier is configured in the voltage-follower configuration, which has a closed-loop gain of unity, very high-input impedance, and nearly zero output impedance.

The two-pole section has the transfer function

$$T(s) = \frac{1}{C_1 C_2 s^2 + 2C_2 s + 1} \tag{3-7}$$

A second-order low-pass transfer function can be expressed in terms of the pole locations as

$$T(s) = \frac{1}{\dfrac{1}{\alpha^2 + \beta^2} s^2 + \dfrac{2\alpha}{\alpha^2 + \beta^2} s + 1} \tag{3-8}$$

Equating coefficients and solving for the capacitors results in

$$C_1 = \frac{1}{\alpha} \tag{3-9}$$

$$C_2 = \frac{\alpha}{\alpha^2 + \beta^2} \tag{3-10}$$

where α and β are the real and imaginary coordinates of the pole pair.

The transfer function of the normalized three-pole section was discussed in Sec. 1.2 and given by

$$T(S) = \frac{1}{s^3 A + s^2 B + sC + 1} \tag{1-15}$$

where $$A = C_1 C_2 C_3 \tag{1-16}$$

$$B = 2C_3(C_1 + C_2) \tag{1-17}$$

and $$C = C_2 + 3C_3 \tag{1-18}$$

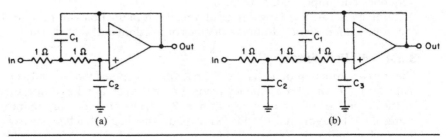

Figure 3-14   Unity-gain active low-pass configurations: (a) two-pole section and (b) three-pole section.

The solution of these equations to find the values of C_1, C_2, and C_3 in terms of the poles is somewhat laborious and is best accomplished with a digital computer.

If the filter order n is an even order, $n/2$ two-pole filter sections are required. Where n is odd, $(n-3)/2$, two-pole sections and a single three-pole section are necessary. This occurs because even-order filters have complex poles only, whereas an odd-order transfer function has a single real pole in addition to the complex poles.

At DC, the capacitors become open circuits; so the circuit gain becomes equal to that of the amplifier, which is unity. This can also be determined analytically from the transfer functions given by Eqs. (3-7) and (1-17). At DC, $s = 0$ and $T(s)$ reduces to 1. Within the passband of a low-pass filter, the response of individual sections may have sharp peaks and some corresponding gain.

All resistors are 1 Ω in the two normalized filter circuits of Fig. 3-14. Capacitors C_1, C_2, and C_3 are tabulated in Chap. 10. These values result in the normalized all-pole transfer functions of Chap. 2 where the 3-dB cutoff occurs at 1 rad/s.

To design a low-pass filter, a filter type is first selected from Chap. 2. The corresponding active low-pass filter values are then obtained from Chap. 10. The normalized filter is denormalized by dividing all the capacitor values by FSF $\times$ Z, which is identical to the denormalization formula for LC filters, as shown in the following

$$C' = \frac{C}{\text{FSF} \times Z} \tag{2-10}$$

where FSF is the frequency-scaling factor $2\pi f_c$ and Z is the impedance-scaling factor. The resistors are multiplied by Z, which results in equal resistors throughout, having a value of $Z\Omega$.

The factor Z does not have to be the same for each filter section, since the individual circuits are isolated by the operational amplifiers. The value of Z can be independently chosen for each section so that practical capacitor values occur, but the FSF must be the same for all sections. The sequence of the sections can be rearranged if desired.

The frequency response obtained from active filters is usually very close to theoretical predictions, provided that the component tolerances are small and that the amplifier has satisfactory properties. The effects of low Q, which occurs in LC filters, do not apply, so the filters have no insertion loss and the passband ripples are well-defined.

Example 3-8   Design of an Active All-Pole Low-Pass Filter

Required:

    An active low-pass filter
    3 dB at 100 Hz
    70-dB minimum at 350 Hz

Result:

(a) Compute the low-pass steepness factor A_s, using Eq. (2-11):

$$A_s = \frac{f_s}{f_c} = \frac{350}{100} = 3.5$$

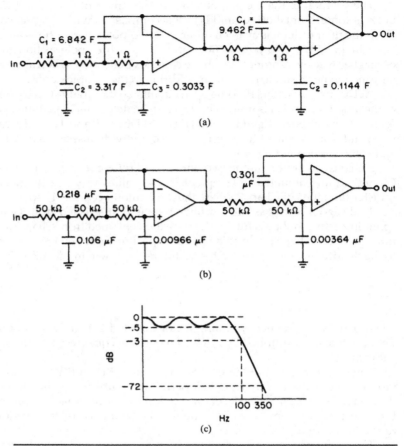

(a)

(b)

(c)

FIGURE 3-15   The low-pass filter of Example 3-8: (a) normalized fifth-order 0.5-dB Chebyshev low-pass filter, (b) denormalized filter, and (c) frequency response.

(b) The response curve of Fig. 2-44 indicates that a fifth-order 0.5-dB Chebyshev low-pass filter meets the 70-dB requirement at 3.5 rad/s.

(c) The normalized values can be found later in Table 10-39. The circuit consists of a three-pole section followed by a two-pole section and is shown in Fig. 3-15a.

(d) Let us arbitrarily select an impedance-scaling factor of 5×10^4. Using an FSF of $2\pi f_c$ or 628, the resulting new values are

Three-pole section:

$$C_1' = \frac{C}{\text{FSF} \times Z} = \frac{6.842}{628 \times 5 \times 10^4} = 0.218 \ \mu\text{F} \tag{2-10}$$

$$C_2' = 0.106 \ \mu\text{F}$$

$$C_3' = 0.00966 \ \mu\text{F}$$

Two-pole section:

$$C_1' = \frac{C}{FSF \times Z} = \frac{9.462}{628 \times 5 \times 10^4} = 0.301 \ \mu F$$

$$C_2' = 0.00364 \ \mu F$$

The resistors in both sections are multiplied by Z, resulting in equal resistors throughout of 50-kΩ. The denormalized circuit is given in Fig. 3-15b, having the frequency response of Fig. 3-15c.

The first section of the filter should be driven by a voltage source having a source impedance much less than the first resistor of the section. The input must have a DC return to ground if a blocking capacitor is present. Since the filter's output impedance is low, the frequency response is independent of the terminating load, provided that the operational amplifier has sufficient driving capability.

Real-Pole Configurations

All odd-order low-pass transfer functions have a single real-axis pole. This pole is realized as part of the $n = 3$ section of Fig. 3-14b when the tables of active all-pole low-pass values in Chap. 10 are used. If an odd-order filter is designed directly from the tabulated poles, the normalized real-axis pole can be generated using one of the configurations given in Fig. 3-16.

The most basic form of a real pole is the circuit of Fig. 3-16a. The capacitor C is defined by

$$C = \frac{1}{\alpha_0} \tag{3-11}$$

where α_0 is the normalized real-axis pole. The circuit gain is unity with a high-impedance termination.

If gain is desirable, the circuit of Fig. 3-16a can be followed by a noninverting amplifier, as shown in Fig. 3-16b, where A is the required gain. When the gain must be inverting, the circuit of Fig. 3-16c is used.

The chosen circuit is frequency- and impedance-scaled in a manner similar to the rest of the filter. The value R in Fig. 3-16b is arbitrary since only the ratio of the two feedback resistors determines the gain of the amplifier.

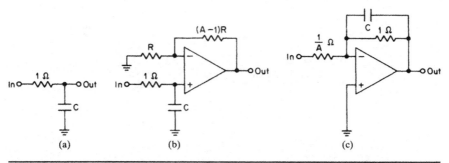

Figure 3-16 The first-order pole configurations: (a) a basic RC section, (b) a noninverting gain configuration, and (c) an inverting gain circuit.

Example 3-9   Design of an Active All-Pole Low-Pass filter with a Separate Real-Pole Section

Required:

An active low-pass filter
3 dB at 75 Hz
15-dB minimum at 150 Hz
A gain of 40 dB ($A = 100$)

Result:

(*a*) Compute the steepness factor using Eq. (2-11):

$$A_s = \frac{f_s}{f_c} = \frac{150}{75} = 2$$

(*b*) Figure 2-34 indicates that an $n = 2$ Butterworth low-pass response satisfies the attenuation requirement. Since a gain of 100 is required, we will use the $n = 2$ section of Fig. 3-14a, followed by the $n = 1$ section of Fig. 3-16b, which provides the gain. The circuit configuration is shown in Fig. 3-17a.

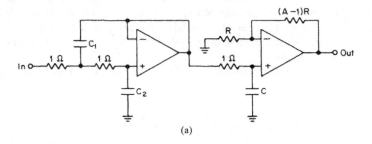

(a)

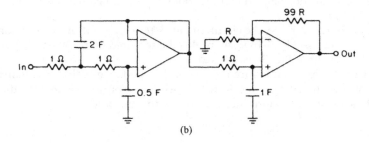

(b)

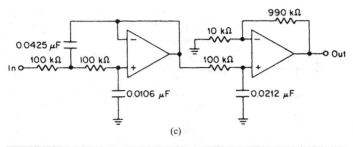

(c)

Figure 3-17   The low-pass filter of Example 3-9: (*a*) circuit configuration, (*b*) normalized circuit, and (*c*) scaled filter.

(c) The following pole locations of a normalized $n = 3$ Butterworth low-pass filter are obtained from Table 10-1:

Complex pole $\qquad \alpha = 0.5000 \qquad\qquad\qquad \beta = 0.8660$
Real pole $\qquad\quad\; \alpha = 0.1.0000$

The component values for the $n = 2$ section are

$$C_1 = \frac{1}{\alpha} = \frac{1}{0.5} = 2 \text{ F} \tag{3-9}$$

$$C_2 = \frac{\alpha}{\alpha^2 + \beta^2} = \frac{0.5}{0.5^2 + 0.866^2} = 0.5 \text{ F} \tag{3-10}$$

The capacitor in the $n = 1$ circuit is computed by

$$C = \frac{1}{\alpha_0} = \frac{1}{1.0} = 1 \text{ F} \tag{3-11}$$

Since $A = 100$, the feedback resistor is $99R$ in the normalized circuit shown in Fig. 3-17b.

(d) Using an FSF of $2\pi f_c$ or 471 and selecting an impedance-scaling factor of 10^5, the denormalized capacitor values are

n = 2 section:

$$C_1' = \frac{C}{\text{FSF} \times Z} = \frac{2}{471 \times 10^5} = 0.0425 \text{ } \mu\text{F} \tag{2-10}$$

$$C_2' = 0.0106 \text{ } \mu\text{F}$$

n = 1 section:

$$C' = 0.0212 \text{ } \mu\text{F}$$

The value R for the $n = 1$ section is arbitrarily selected at 10 kΩ. The final circuit is given in Fig. 3-17c.

Although these real-pole sections are intended to be part of odd-order low-pass filters, they can be independently used as an $n = 1$ low-pass filter, and have the transfer function

$$T(s) = K \frac{1}{sC + 1} \tag{3-12}$$

where $K = 1$ for Fig. 3-16a, $K = A$ for Fig. 3-16b, and $K = -A$ for Fig. 3-16c. If $C = 1$F, the 3-dB cutoff occurs at 1 rad/s.

The attenuation of a first-order filter can be expressed as

$$A_{\text{dB}} = 10 \log \left[1 + \left(\frac{\omega_x}{\omega_c} \right)^2 \right] \tag{3-13}$$

where ω_x / ω_c is the ratio of a given frequency to the cutoff frequency. The normalized frequency response corresponds to the $n = 1$ curve of the Butterworth low-pass filter response curves of Fig. 2-34. The step response has no overshoot, and the impulse response does not have any oscillatory behavior.

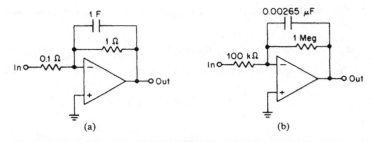

Figure 3-18   The $n = 1$ low-pass filter of Example 3-10: (a) normalized filter and (b) frequency- and impedance-scaled filter.

Example 3-10   Design of an Active All-Pole Low-Pass filter with a Gain of 10 dB

Required:

> An active low-pass filter
> 3 dB at 60 Hz
> 12-dB minimum attenuation at 250 Hz
> A gain of 20 dB with inversion

Result:

(a) Compute A_s, using Eq. (2-11)

$$A_s = \frac{f_s}{f_c} = \frac{250}{60} = 4.17$$

(b) Figure 2-34 indicates that an $n = 1$ filter provides over 12 dB attenuation at 4.17 rad/s. Since an inverting gain of 20 dB is required, the configuration of Fig. 3-16c will be used. The normalized circuit is shown in Fig. 3-18a, where $C = 1F$ and $A = 10$, corresponding to a gain of 20 dB.

(c) Using an FSF of $2\pi60$ or 377 and an impedance-scaling factor of 10^6, the denormalized capacitor is

$$C' = \frac{C}{FSF \times Z} = \frac{1}{377 \times 10^6} = 0.00265 \ \mu F \tag{2-10}$$

The input and output feedback resistors are 100 kΩ and 1 MΩ, respectively. The final circuit is shown in Fig. 3-18b.

Second-Order Section with Gain

If an active low-pass filter is required to have a gain higher than unity and the order is even, the $n = 1$ sections of Fig. 3-16 cannot be used since a real pole is not contained in the transfer function.

The circuit of Fig. 3-19 realizes a pair of complex poles and provides a gain of $-A$. The element values are computed using the following formulas:

$$C_1 = (A + 1)\left(1 + \frac{\beta^2}{\alpha^2}\right) \tag{3-14}$$

$$R_1 = \frac{\alpha}{A(\alpha^2 + \beta^2)} \tag{3-15}$$

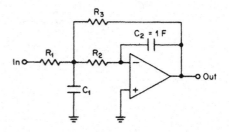

Figure 3-19 Second-order section with gain.

$$R_2 = \frac{AR_1}{A+1} \tag{3-16}$$

$$R_3 = AR_1 \tag{3-17}$$

This section is used in conjunction with the $n = 2$ section of Fig. 3-14a to realize even-order low-pass filters with gain. This is shown in Example 3-11.

Example 3-11   Design of an Active All-Pole Low-Pass filter with a Gain of 2

Required:

An active low-pass filter
3 dB at 200 Hz
30-dB minimum at 800 Hz
No step-response overshoot
A gain of 6 dB with inversion ($A = 2$)

Result:

(a) Compute A_s, using Eq. (2-11):

$$A_s = \frac{f_s}{f_c} = \frac{800}{200} = 4 \tag{2-11}$$

(b) Since no overshoot is permitted, a Bessel filter type will be used. Figure 2-56 indicates that a fourth-order network provides over 30 dB of rejection at 4 rad/s. Since an inverting gain of 2 is required and $n = 4$, the circuit of Fig. 3-19 will be used, followed by the two-pole section of Fig. 3-14a. The basic circuit configuration is given in Fig. 3-20a.

(c) The following pole locations of a normalized $n = 4$ Bessel low-pass filter are obtained from Table 10-41:

$$\alpha = 1.3596 \quad \beta = 0.4071$$

and

$$\alpha = 0.9877 \quad \beta = 1.2476$$

The normalized component values for the first section are determined by the following formulas, where $\alpha = 1.3596$, $\beta = 0.4071$, and $A = 2$:

$$C_1 = (A+1)\left(1 + \frac{\beta^2}{\alpha^2}\right) = 3\left(1 + \frac{0.4071^2}{1.3596^2}\right) = 3.27 \ F \tag{3-14}$$

$$R_1 = \frac{\alpha}{A(\alpha^2 + \beta^2)} = \frac{1.3596}{2(1.3596^2 + 0.4071^2)} = 0.3375 \ \Omega \tag{3-15}$$

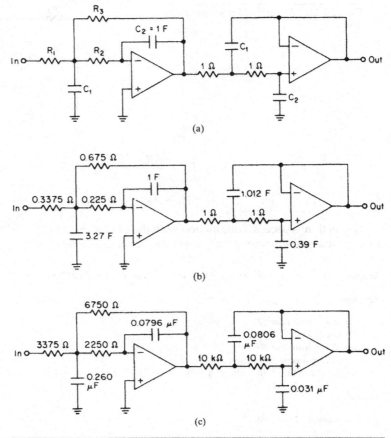

FIGURE 3-20   The $n = 4$ Bessel low-pass filter of Example 3-11: (a) circuit configuration, (b) normalized filter, and (c) frequency- and impedance-scaled filter.

$$R_2 = \frac{AR_1}{A+1} = \frac{2 \times 0.3375}{3} = 0.225 \ \Omega \tag{3-16}$$

$$R_3 = AR_1 = 2 \times 0.3375 = 0.675 \ \Omega \tag{3-17}$$

The remaining pole pair of $\alpha = 0.9877$ and $\beta = 1.2476$ is used to compute the component values of the second section.

$$C_1 = \frac{1}{\alpha} = \frac{1}{0.9877} = 1.012 \ \text{F} \tag{3-9}$$

$$C_2 = \frac{\alpha}{\alpha^2 + \beta^2} = \frac{0.9877}{0.9877^2 + 1.2476^2} = 0.39 \ \text{F} \tag{3-10}$$

The normalized low-pass filter is shown in Fig. 3-20b.

(d) Using Eqs. (2-8) and (2-10), an FSF $2\pi f_c$ of or 1256 and an impedance-scaling factor of 10 for both sections, the denormalized values are

n = 2 section with A = 2:

$$R_1' = R \times Z = 0.3375 \times 10^4 = 3,375 \ \Omega$$

$$R_2' = 2,250 \ \Omega$$

$$R_3' = 6,750 \ \Omega$$

$$C_1' = \frac{C}{FSF \times Z} = \frac{3.27}{1,256 \times 10^4} = 0.260 \ \mu F$$

$$C_2' = 0.0796 \ \mu F$$

n = 2 section having unity gain:

$$R' = 10 \ k\Omega$$

$$C_1' = \frac{C}{FSF \times Z} = \frac{1.012}{1,256 \times 10^4} = 0.0806 \ \mu F$$

$$C_2' = 0.0310 \ \mu F$$

The final circuit is shown in Fig. 3-20c.

3.2.2 *VCVS* Uniform Capacitor Structure

The unity-gain $n = 2$ all-pole configuration of Fig. 3-14a requires unequal capacitor values and noninteger capacitor ratios. The inconvenience usually results in either the use of nonstandard capacitor values or the paralleling of two or more standard values.

An alternate configuration is given in this section. This structure features equal capacitors. However, the circuit sensitivities are somewhat higher than the previously discussed configuration. Nevertheless, the more convenient capacitor values may justify its use in many instances where higher sensitivities are tolerable.

The $n = 2$ low-pass circuit of Fig. 3-21 features equal capacitors and a gain of 2. The element values are computed as follows:

Select C.

Then

$$R_1 = \frac{1}{2\alpha'C} \qquad\qquad (3\text{-}18)$$

and

$$R_2 = \frac{2\alpha'}{C(\alpha'^2 + \beta'^2)} \qquad\qquad (3\text{-}19)$$

FIGURE 3-21   An all-pole configuration.

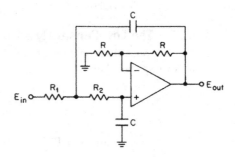

where α' and β' are the denormalized real and imaginary pole coordinates. R may be conveniently chosen.

Example 3-12   Design of an Active All-Pole Low-Pass filter Using Uniform Capacitor Values

Required:

Design a fourth-order 0.1-dB Chebyshev active low-pass filter for a 3-dB cutoff of 100 Hz using 0.01 μF capacitors throughout.

Result:

(*a*) The pole locations for a normalized 0.1-dB Chebyshev low-pass filter are obtained from Table 10-23 and are as follows:

$$\alpha = 0.2177 \quad \beta = 0.9254$$

and

$$\alpha = 0.5257 \quad \beta = 0.3833$$

(*b*) Two sections of the filter of Fig. 3-21 will be cascaded. The value of C is 0.01 μF, and R is chosen at 10 kΩ. Use Eqs. (3-18) and (3-19).

Section 1:

$$\alpha = 0.2177 \quad \alpha' = \alpha \times \mathrm{FSF} = 136.8$$

$$\beta = 0.9254 \quad \beta' = \beta \times \mathrm{FSF} = 581.4$$

where FSF $2\pi f_c = 628.3$

$$R_1 = \frac{1}{2\alpha' C} = 365.5 \text{ k}\Omega$$

$$R_2 = \frac{2\alpha'}{C(\alpha'^2 + \beta'^2)} = 76.7 \text{ k}\Omega$$

Section 2:

$$\alpha = 0.5257 \quad \alpha' = 330.0$$

$$\beta = 0.3833 \quad \beta' = 240.8$$

$$R_1 = \frac{1}{2\alpha' C} = 151.4 \text{ k}\Omega$$

$$R_2 = \frac{2\alpha'}{C(\alpha^2 + \beta^2)} = 395.4 \text{ k}\Omega$$

The final filter is shown in Fig. 3-22 using standard resistor values. The gain is 2^2 or 4.

3.2.3   The Low-Sensitivity Second-Order Section

The low-pass filter section of Fig. 3-23 realizes a second-order transfer function which can be expressed as

$$T(s) = \frac{1}{\tau_1 \tau_2 s^2 + \tau_2 s + 1} \tag{3-20}$$

where $\tau_1 = R_1 C_1$ and $\tau_1 = R_2 C_2$.

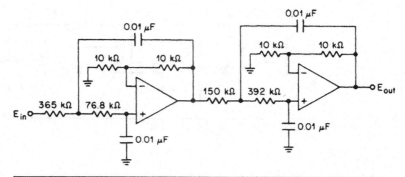

FIGURE 3-22 The equal capacitor circuit of Example 3-12.

If we first equate Eq. (3-20) with Eq. (3-7), the general form for a second-order transfer function, and then solve for R_1 and R_2, we obtain

$$R_1 = \frac{1}{2\alpha C_1} \tag{3-21}$$

and
$$R_2 = \frac{2\alpha}{(\alpha^2 + \beta^2)C_2} \tag{3-22}$$

Two important observations can be made from Fig. 3-23 and the associated design equations. Since both operational amplifiers are configured as voltage followers, the circuit sensitivity to amplifier open-loop gain is not as severe as for the previous circuit which requires a gain of 2. Secondly, both the transfer function and design equations clearly indicate that the circuit operation is dictated by two time constants: R_1C_1 and R_2C_2. Thus, C_1 and C_2 can be independently selected for convenient values or made equal, as desired.

Example 3-13 illustrates the application of this configuration.

Example 3-13   Design of an Active All-Pole Low-Pass filter Using Low-Sensitivity Second-Order Sections

Required:

Design a fourth-order 0.1-dB Chebyshev low-pass filter for a 3-dB cutoff frequency of 10 kHz using the low-sensitivity second-order section.

FIGURE 3-23   The low-sensitivity second-order section.

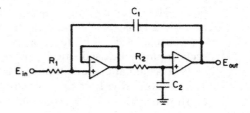

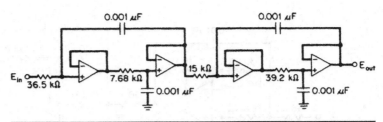

FIGURE 3-24   The circuit of Example 3-13.

Result:

(a) The pole locations for a normalized 0.1-dB Chebyshev low-pass filter (given later in Table 10-23) are as follows:

$$\alpha = 0.2177 \quad \beta = 0.9254$$

and

$$\alpha = 0.5257 \quad \beta = 0.3833$$

(b) Denormalizing the pole locations (multiply α and β by the FSF):

$$\alpha' = 13,678 \quad \beta' = 58,145$$

and

$$\alpha' = 33,031 \quad \beta' = 24,083$$

(c) Using Eqs. (3-21) and (3-22), compute the component values as follows:

Section 1:

$$\alpha' = 13,678 \quad \beta' = 58,145$$

Let

$$C_1 = C_2 = 0.001 \; \mu\text{F}$$

$$R_1 = \frac{1}{2\alpha'C} = 36.56 \; \text{k}\Omega$$

$$R_2 = \frac{2\alpha'}{(\alpha'^2 + \beta'^2)C_2} = 7.667 \; \text{k}\Omega$$

Section 2:

$$\alpha' = 33,031 \quad \beta' = 24,083$$

$$R_1 = \frac{1}{2\alpha'C} = 15.14 \; \text{k}\Omega$$

$$R_2 = \frac{2\alpha'}{(\alpha^2 + \beta^2)C_2} = 39.53 \; \text{k}\Omega$$

The resulting circuit is shown in Fig. 3-24 using standard resistor values. The overall gain is unity.

3.2.4   Elliptic-Function VCVS Filters

Elliptic-function filters were first discussed in Sec. 2.9. They contain zeros as well as poles. The zeros begin just outside the passband and force the response to decrease rapidly as Ω_s is approached. (Refer to Fig. 2-82 for frequency-response definitions.)

Because of these finite zeros, the active filter circuit configurations of the previous section cannot be used since they are restricted to the realization of poles only.

The schematic of an elliptic-function low-pass filter section is shown in Fig. 3-25*a*. This section provides a pair of complex conjugate poles and a pair of imaginary zeros, as shown in Fig. 3-25*b*. The complex pole pair has a real component of α and an

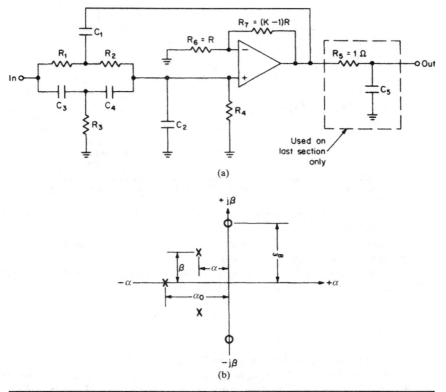

FIGURE 3-25 The elliptic-function low-pass filter section: (*a*) VCVS circuit configuration for *K* > 1 and (*b*) pole-zero pattern.

imaginary coordinate of β. The zeros are located at $\pm j\omega_\infty$. The RC section consisting of R_5 and C_5 introduces a real pole at α_0.

The configuration contains a VCVS as the active element and is frequently referred to as a VCVS realization. Although this structure requires additional elements when compared with other VCVS configurations, it has been found to yield more reliable results and has lower sensitivity factors.

The normalized element values are determined by the following relations:
First calculate

$$a = \frac{2\alpha'}{\sqrt{\alpha'^2 + \beta'^2}} = \frac{1}{Q} \tag{3-23}$$

$$b = \frac{\omega'^2_\infty}{\alpha'^2 + \beta'^2} = \frac{\omega^2_\infty}{\omega^2_0} \tag{3-24}$$

$$c = \sqrt{\alpha'^2 + \beta'^2} = \omega_0 \tag{3-25}$$

where α', β', and ω'_∞ are the denormalized pole-zero coordinates. The second forms of Eqs. (3-23) through (3-25) involving Q and ω_0 are used when these parameters are directly provided already denormalized by the *Filter Solutions* program.

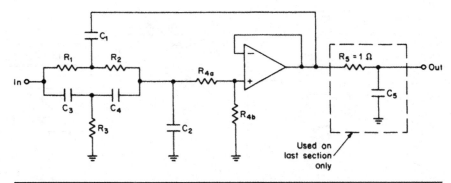

Figure 3-26 The elliptic-function VCVS low-pass filter section for $K < 1$.

A controlled amplification of K is required between the noninverting amplifier input and the section output. Since the gain of a noninverting operational amplifier is the ratio of the feedback resistors plus 1, R_6 and R_7 are R and $(K - 1)$ R, respectively, where R can be any convenient value.

In the event that K is less than 1, the amplifier is reconfigured as a voltage follower and R_4 is split into resistors R_{4a} and R_{4b} where

$$R_{4a} = (1 - K)R_4 \tag{3-26}$$

$$R_{4b} = KR_4 \tag{3-27}$$

The modified circuit is shown in Fig. 3-26.

The design of active elliptic-function filters utilizes the *Filter Solutions* program (downloaded as described in App. A) for obtaining pole-zero locations which are already denormalized. The design method proceeds as if a passive elliptic low-pass filter is being designed. However, once a design is completed, the **Transfer Function** button is depressed and then the **Casc** box is checked. The poles and zeros are displayed in cascaded form (rather than rectangular form), which must now be utilized to compute **a**, **b**, and **c** of Eqs. (3-23), (3-24), and (3-25), as follows:

$$a = \frac{2\alpha'}{\sqrt{\alpha'^2 + \beta'^2}} = \frac{1}{Q} \tag{3-23}$$

$$b = \frac{\omega'^2_\infty}{\alpha'^2 + \beta'^2} = \frac{\omega^2_\infty}{\omega^2_0} \tag{3-24}$$

$$c = \sqrt{\alpha'^2 + \beta'^2} = \omega_0 \tag{3-25}$$

For odd-order filters, the real pole α_0 is presented as $(S + \alpha_0)$ in the denominator. The element values are computed as follows:

Select C

Then
$$C_1 = C \tag{3-28}$$

$$C_3 = C_4 = \frac{C_1}{2} \tag{3-29}$$

let
$$C_2 \geq \frac{C_1(b-1)}{4} \tag{3-30}$$

$$R_3 = \frac{1}{cC_1\sqrt{b}} \tag{3-31}$$

$$R_1 = R_2 = 2R_3 \tag{3-32}$$

$$R_4 = \frac{4\sqrt{b}}{cC_1(1-b) + 4cC_2} \tag{3-33}$$

$$K = 2 + \frac{2C_2}{C_1} - \frac{a}{2\sqrt{b}} + \frac{2}{C_1\sqrt{b}}\left(\frac{1}{cR_4} - aC_2\right) \tag{3-34}$$

$$\text{Section gain} = \frac{bKC_1}{4C_2 + C_1} \tag{3-35}$$

Capacitor C_5 is determined from the denormalized real pole by

$$C_5 = \frac{1}{R_5\alpha_0'} \tag{3-36}$$

where both R and R_5 can be arbitrarily chosen and α_0' is $\alpha_0 \times$ FSF.

Odd-order elliptic-function filters are more efficient than even-order since maximum utilization is made of the number of component elements used. Since the circuit of Fig. 3-25 provides a single pole pair (along with a pair of zeros), the total number of sections required for an odd-order filter is determined by $(n-1)/2$, where n is the order of the filter. Because an odd-order transfer function has a single real pole, R_5 and C_5 appear on the output section only.

In the absence of a detailed analysis, it is a good rule of thumb to pair poles with their nearest zeros when allocating poles and zeros to each active section. This applies to high-pass, band-pass, and band-reject filters as well.

Example 3-14   Design of an Active Elliptic Function Low-Pass Filter using the VCVS Structure

Required:

Design an active elliptic-function low-pass filter corresponding to a 0.177-dB ripple, a cutoff of 100 Hz and a minimum attenuation of 37 dB at 292.4 Hz using the VCVS structure of Fig. 3-25.

Result:

(a) Open *Filter Solutions.*
    Check the *Stop Band Freq* box.
    Enter **0.177** in the *Pass Band Ripple (dB)* box.
    Enter **100** in the *Pass Band Freq* box.
    Enter **292.4** in the *Stop Band Freq* box.
    Check the *Frequency Scale Hertz* box.

(b) Click the Set Order control button to open the second panel.
Enter 37 for *Stop band Attenuation (dB)*.
Click the *Set Minimum Order* button and then click *Close*.
3 Order is displayed on the main control panel.

(c) Click the *Transfer Function* button.
Check the *Casc* box.

The following is displayed:

Continuous Transfer Function

$$Wn = 2105$$

$$\frac{71.83 \quad (S^2 + 4.432e+06)}{(S^2 + 485.5*S + 5.713e+05) \quad (S + 557.4)}$$

$$Wo = 755.8$$
$$Q = 1.557$$

3rd Order Low Pass Elliptic

Pass Band Frequency = 100.0 Hz   Stop Band Ratio = 2.924
Pass Band Ripple = 177.0 mdB   Stop Band Frequency = 292.4 Hz
                                  Stop Band Attenuation = 37.43 dB

(d) The design parameters are summarized as follows:

$$\text{Section } Q = 1.557$$
$$\text{Section } \omega_0 = 755.8$$
$$\text{Section } \omega_\infty = 2,105$$
$$\alpha_0 = 557.4 \text{ (from the denominator)}$$

(e) The element values are computed using Eqs. (3-23) to (3-25) and Eqs. (3-28) to (3-36) as follows:

$$a = 0.6423$$
$$b = 7.7569$$
$$c = 775.8$$

Select
$$C_1 = C = 0.1 \ \mu F$$
$$C_3 = C_4 = 0.054 \ \mu F$$
$$C_2 \geq 0.169 \ \mu F$$

Let
$$C_2 = 0.22 \ \mu F$$
$$R_3 = 4,751 \ \Omega$$
$$R_1 = R_2 = 9,502 \ \Omega$$
$$R_4 = 72.2 \ k\Omega$$
$$K = 5.402$$

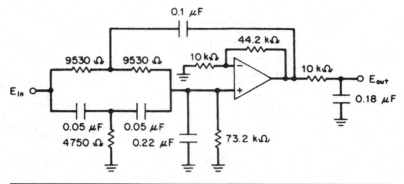

FIGURE 3-27 The circuit for the active elliptic-function low-pass filter.

Let $\qquad\qquad\qquad\qquad$ $R = R_5 = 10 \text{ k}\Omega$

then $\qquad\qquad\qquad\qquad$ $C_5 = 0.180 \text{ }\mu\text{F}$

The resulting circuit is shown in Fig. 3-27 using standard values.

3.2.5 State-Variable Low-Pass Filters

The poles and zeros of the previously discussed active filter configurations cannot be easily adjusted because of the interaction of circuit elements. For most industrial requirements, sufficient accuracy is obtained by specifying 1-percent resistors and 1- or 2-percent capacitors. In the event that greater precision is required, the state-variable approach features independent adjustment of the pole and zero coordinates. Also, the state-variable configuration has a lower sensitivity to many of the inadequacies of operational amplifiers such as finite bandwidth and gain.

All-Pole Configuration

The circuit of Fig. 3-28 realizes a single pair of complex poles. The low-pass transfer function is given by

$$T(s) = \frac{1}{R_2 R_4 C^2} \; \frac{1}{s^2 + \dfrac{1}{R_1 C} s + \dfrac{1}{R_2 R_3 C^2}} \tag{3-37}$$

If we equate Eq. (3-37) to the second-order low-pass transfer function expressed by Eq. (3-8) and solve for the element values, after some algebraic manipulation we obtain the following design equations:

$$R_1 = \frac{1}{2\alpha C} \tag{3-38}$$

$$R_2 = R_3 = R_4 = \frac{1}{C\sqrt{\alpha^2 + \beta^2}} \tag{3-39}$$

where α and β are the real and imaginary components, respectively, of the pole locations and C is arbitrary. The value of R in Fig. 3-28 is also optional.

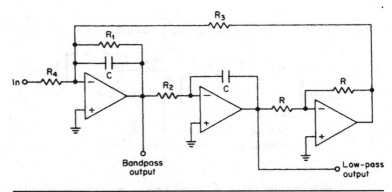

FIGURE 3-28   The state-variable all-pole low-pass configuration.

The element values computed by Eqs. (3-38) and (3-39) result in a DC gain of unity. If a gain of $-A$ is desired, R_4 can instead be defined by

$$R_4 = \frac{1}{AC\sqrt{\alpha^2 + \beta^2}} \tag{3-40}$$

Sometimes it is desirable to design a filter directly at its cutoff frequency instead of calculating the normalized values and then frequency- and impedance-scaling the normalized network. Equations (3-38) and (3-39) result in the denormalized values if α and β are first denormalized by the frequency-scaling factor FSF as follows:

$$\alpha' = \alpha \times \mathrm{FSF} \tag{3-41}$$

$$\alpha' = \beta \times \mathrm{FSF} \tag{3-42}$$

Direct design of the denormalized filter is especially advantageous when the design formulas permit the arbitrary selection of capacitors and all network capacitors are equal. A standard capacitance value can then be chosen.

Figure 3-28 indicates that a band-pass output is also provided. Although a discussion of band-pass filters will be deferred until Chap. 5, this output is useful for tuning of the low-pass filter. To adjust the low-pass real and imaginary pole coordinates, first compute the band-pass resonant frequency:

$$f_0 = \frac{\sqrt{(\alpha')^2 + (\beta')^2}}{2\pi} \tag{3-43}$$

Trim the value of R_3 until resonant conditions occur at the band-pass output with f_0 applied. Resonance can be determined by exactly 180° of phase shift between input and output or by peak output amplitude. The 180° phase shift method normally results in more accuracy and resolution. By connecting the vertical channel of an oscilloscope to the section input and the horizontal channel to the band-pass output, a *Lissajous pattern* is obtained. This pattern is an ellipse that will collapse to a straight line (at a 135° angle) when the phase shift is 180°.

For the final adjustment, trim R_1 for a band-pass Q (for example, f_0/3-dB bandwidth) equal to

$$Q = \frac{\pi f_0}{\alpha'} \tag{3-44}$$

Resistor R_1 can be adjusted for a measured band-pass output gain at f_0 equal to the computed ratio of R_1/R_4. Amplifier phase shift creates a Q-enhancement effect where the Q of the section is increased. This effect also increases the gain at the band-pass output, so adjustment of R_1 for the calculated gain will usually restore the desired Q. Alternatively, the 3-dB bandwidth can be measured and the Q computed. Although the Q measurement approach is the more accurate method, it certainly is slower than a simple gain adjustment.

Example 3-15   Design of an Active State-Variable All-Pole Low-Pass Filter

Required:

An active low-pass filter
3 dB ± 0.25 dB at 500 Hz
40-dB minimum at 1375 Hz

Result:

(*a*) Compute A_s using Eq. (2-11)

$$A_s = \frac{1,375}{500} = 2.75$$

(*b*) Figure 2-42 indicates that a fourth-order 0.1-dB Chebyshev low-pass filter has over 40 dB of rejection at 2.75 rad/s. Since a precise cutoff is required, we will use the state-variable approach so the filter parameters can be adjusted if necessary.

The pole locations for a normalized fourth-order 0.1-dB Chebyshev low-pass filter are obtained from Table 10-23 and are as follows:

$$\alpha = 0.2183 \quad \beta = 0.9262 \quad \text{and} \quad \alpha = 0.5271 \quad \beta = 0.3836$$

(*c*) Two sections of the circuit of Fig. 3-28 will be cascaded. The denormalized filter will be designed directly. The capacitor value C is chosen to be 0.01uF , and R is arbitrarily selected to be 10 kΩ.

Section 1:

$$\alpha = 0.2183 \quad \alpha' = \alpha \times \text{FSF} = 685.5 \tag{3-41}$$

$$\beta = 0.9262 \quad \beta' = \beta \times \text{FSF} = 2,908 \tag{3-42}$$

where $\text{FSF} = 2\pi f_c = 2\pi 500 = 3,140$

$$R_1 = \frac{1}{2\alpha'C} = 72.94 \text{ k}\Omega \tag{3-38}$$

$$R_2 = R_3 = R_4 = \frac{1}{C\sqrt{(\alpha')^2 + (\beta')^2}} = 33.47 \text{ k}\Omega \tag{3-39}$$

Section 2:

$$\alpha = 0.5271 \quad \alpha' = \alpha \times \text{FSF} = 1,655 \tag{3-41}$$

$$\beta = 0.3836 \quad \beta' = \beta \times \text{FSF} = 1,205 \tag{3-42}$$

where FSF = 3140

$$R_1 = \frac{1}{2\alpha'C} = 30.21 \text{ k}\Omega \tag{3-38}$$

$$R_2 = R_3 = R_4 = \frac{1}{C\sqrt{(\alpha')^2 + (\beta')^2}} = 48.85 \text{ k}\Omega \tag{3-39}$$

(d) The resulting filter is shown in Fig. 3-29. The resistor values have been modified as in previous examples so that standard 1-percent resistors are used and adjustment capability is provided. The band-pass resonant frequency and Q are

Section 1:

$$f_0 = \frac{\sqrt{(\alpha')^2 + (\beta')^2}}{2\pi} = 476 \text{ Hz} \tag{3-43}$$

$$Q = \frac{\pi f_0}{\alpha'} = 2.18 \tag{3-44}$$

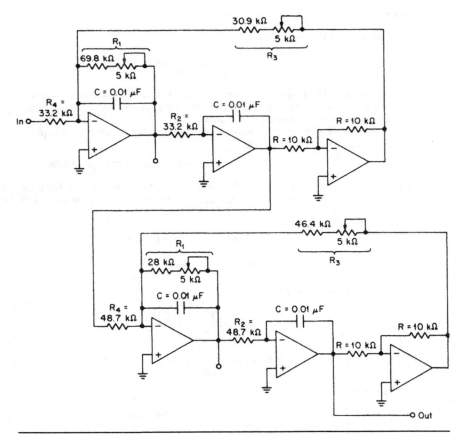

FIGURE 3-29   The state-variable low-pass filter of Example 3-15.

Section 2:

$$f_0 = 326 \text{ Hz} \tag{3-43}$$

$$Q = 0.619 \tag{3-44}$$

Elliptic-Function Configuration

When precise control of the parameters of elliptic-function filters is required, a state-variable elliptic-function approach is necessary. This is especially true in the case of very sharp filters where the location of the poles and zeros is highly critical.

The circuit in Fig. 3-30 has the transfer function

$$T(s) = -\frac{R_6}{R}\left[\frac{s^2 + \dfrac{1}{R_2R_3C^2}\left(1 + \dfrac{R_3R}{R_4R_5}\right)}{s^2 + \dfrac{1}{R_1C}s + \dfrac{1}{R_2R_3C^2}}\right] \tag{3-45}$$

where $R_1 = R_4$ and $R_2 = R_3$. The numerator roots result in a pair of imaginary zeros, and the denominator roots determine a pair of complex poles. Since both the numerator and denominator are second-order, this transfer function form is frequently referred to as *biquadratic*, while the circuit is called a *biquad*. The zeros are restricted to frequencies beyond the pole locations—for example, the stopband of elliptic-function low-pass filters. If R_5 in Fig. 3-30 were connected to node 2 instead of node 1, the zeros would occur below the poles as in high-pass elliptic-function filters.

The design of active elliptic-function filters using biquads utilizes the *Filter Solutions* program (downloaded as described in App. A) for obtaining pole-zero locations,

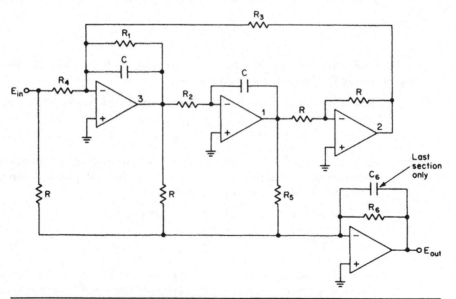

Figure 3-30 The state-variable (biquad) configuration for elliptic-function low-pass filters.

which are already denormalized. The design method proceeds as if a passive elliptic low-pass filter is being designed. However once a design is completed the *Transfer Function* button is depressed and then the *Casc* box is checked. The poles and zeros are displayed in cascaded form (rather than rectangular form), which can then be used in the design equations.

The parameters obtained from *Filter Solutions* are ω_∞, ω_0, Q, and α_0, which are *already denormalized.*

First, compute:

$$\alpha = \frac{\omega_0}{2Q} \tag{3-46}$$

The component values are

$$R_1 = R_4 = \frac{1}{2\alpha C} \tag{3-47}$$

$$R_2 = R_3 = \frac{1}{\omega_0 C} \tag{3-48}$$

$$R_5 = \frac{2\alpha\omega_0 R}{(\omega_\infty)^2 - (\omega_0)^2} \tag{3-49}$$

$$R_6 = \left(\frac{\omega_0}{\omega_\infty}\right)^2 AR \tag{3-50}$$

where C and R are arbitrary and A is the desired low-pass gain at DC.

Since odd-order elliptic-function filters contain a real pole, the last section of a cascade of biquads should contain capacitor C_6 in parallel with R_6. To compute C_6

$$C_6 = \frac{1}{\alpha_0 R_6} \tag{3-51}$$

The poles and zeros of the biquad configuration of Fig. 3-30 can be adjusted by implementing the following sequence of steps:

1. *Resonant frequency:* The band-pass resonant frequency is defined by

$$f_0 = \frac{\omega_0}{2\pi} \tag{3-52}$$

If R_3 is made adjustable, the section's resonant frequency can be tuned to f_0 by monitoring the band-pass output at node 3. The 180° phase shift method is preferred for the determination of resonance.

2. *Q adjustment:* The band-pass Q is given by

$$Q = \frac{\pi f_0}{\alpha} \tag{3-53}$$

Adjustment of R_1 for unity gain at f_0 measured between the section input and the band-pass output at node 3 will usually compensate for any Q enhancement resulting from amplifier phase shift.

3. *Notch frequency:* The notch frequency is given by

$$f_\infty = \omega_\infty/2\pi \tag{3-54}$$

Adjustment of f_∞ usually is not required if the circuit is first tuned to f_0, since f_∞ will then fall in. However, if independent tuning of the notch frequency is desired, R_5 should be made adjustable. The notch frequency is measured by determining the input frequency where E_{out} is nulled.

Example 3-16 Design of an Active Elliptic-Function Low-Pass Filter Using the State-Variable Configuration

Required:

Design an active elliptic-function low-pass filter corresponding to a 0.18-dB ripple, a cutoff of 1,000 Hz and a minimum attenuation of 18 dB at 1556 Hz using the state-variable (biquad) configuration for elliptic-function low-pass filters of Fig. 3-30.

Result:

(*a*) Open *Filter Solutions.*
   Check the *Stop Band Freq* box.
   Enter **.18** in the *Pass Band Ripple (dB)* box.
   Enter **1,000** in the *Pass Band Freq* box.
   Enter **1,556** in the *Stop Band Freq* box.
   Check the *Frequency Scale Hertz* box.

(*b*) Click the *Set Order* control button to open the second panel.
   Enter **18** for *Stop band Attenuation (dB)*.
   Click the Set *Minimum Order* button and then click *Close.*
   3 *Ord*er is displayed on the main control panel.

(*c*) Click the *Transfer Function* button.
   Check the *Casc* box.

The following is displayed:

Continuous Transfer Function

Wn = 1.095e+04

$$\frac{3040 \quad (S^2 + 1.199e+08)}{(S^2 + 3573*S + 5.511e+07) \quad (S + 6613)}$$

Wo = 7423
Q = 2.077

3rd Order Low Pass Elliptic

Pass Band Frequency = 1.000 KHz   Stop Band Ratio = 1.556
Pass Band Ripple = 180.0 mdB   Stop Band Frequency = 1.556 KHz
   Stop Band Attenuation = 18.63 dB

(d) The design parameters are summarized as follows:

$$\text{Section } Q = 2.077$$

$$\text{Section } \omega_0 = 7,432$$

$$\text{Section } \omega_\infty = 10,950$$

$$\alpha_0 = 6613 \text{ (from the denominator)}$$

(e) A single section is required. Let $R = 100\text{k}\Omega$ and $C = 0.1$, and let the gain equal unity ($A = 1$). The values are computed as follows:

$$\alpha = \frac{\omega_0}{2Q} = 1,787 \tag{3-46}$$

$$R_1 = R_4 = \frac{1}{2\alpha C} = 2,798 \ \Omega \tag{3-47}$$

$$R_2 = R_3 = \frac{1}{\omega_0 C} = 1347 \ \Omega \tag{3-48}$$

$$R_5 = \frac{2\alpha\omega_0 R}{(\omega_\infty)^2 - (\omega_0)^2} = 40.94 \ \text{k}\Omega \tag{3-49}$$

$$R_6 = \left(\frac{\omega_0}{\omega_\infty}\right)^2 AR = 45.96 \ \text{k}\Omega \tag{3-50}$$

Since a real pole is required, C_6 is introduced in parallel with R_6 and is calculated by

$$C_6 = \frac{1}{\alpha_0 R_6} = 3,260 \ \text{pF} \tag{3-51}$$

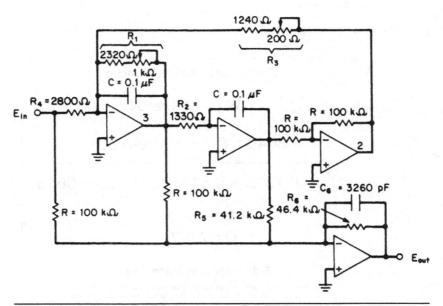

FIGURE 3-31 The elliptic-function low-pass filter of Example 3-16.

(f) The resulting filter is shown in Fig. 3-31. The resistor values are modified so that standard 1-percent values are used and the circuit is adjustable. The sections f_0 and Q are computed by

$$f_0 = \frac{\omega_0}{2\pi} = 1,181\,\text{Hz}$$

(3-52)

$$Q = \frac{\pi f_0}{\alpha} = 2.076$$

(3-53)

The frequency of infinite attenuation is given by

$$f_\infty = \omega_\infty/2\pi = 1,743\,\text{Hz}$$

(3-54)

3.2.6 Generalized Impedance Converters

The circuit of Fig. 3-32 is known as a generalized impedance converter (GIC). The driving-point impedance can be expressed as

$$Z_{11} = \frac{Z_1 Z_3 Z_5}{Z_2 Z_4}$$

(3-55)

By substituting RC combinations of up to two capacitors for Z_1 through Z_5, a variety of impedances can be simulated. If, for instance, Z_4 consists of a capacitor having an impedance $1/sC$, where $s = j\omega$ and all other elements are resistors, the driving-point impedance is given by

$$Z_{11} = \frac{sCR_1R_3R_5}{R_2}$$

(3-56)

The impedance is proportional to frequency and is therefore identical to an inductor, having a value of

$$L = \frac{CR_1R_3R_5}{R_2}$$

(3-57)

as shown in Fig. 3-33.

If two capacitors are introduced for Z_1 and Z_3, and Z_2, Z_4, and Z_5 are resistors, the resulting driving-point impedance expression can be expressed in the form of

$$Z_{11} = \frac{R_5}{s^2 C^2 R_2 R_4}$$

(3-58)

FIGURE 3-32 A generalized impedance converter (GIC).

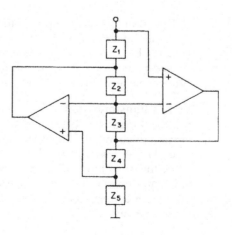

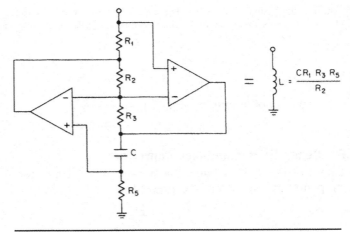

FIGURE 3-33 A GIC inductor simulation.

An impedance proportional to $1/s^2$ is called a D element, whose driving point impedance is given by

$$Z_{11} = \frac{1}{s^2 D} \tag{3-59}$$

Equation (3-58) therefore defines a D element having the value

$$D = \frac{C^2 R_2 R_4}{R_5} \tag{3-60}$$

If we let $C = 1$ F, $R_2 = R_5 = 1\Omega$, and $R_4 = R$, Eq. (3-64) simplifies to $D = R$.

In order to gain some insight into the nature of this element, let us substitute $s = j\omega$ into Eq. (3-58). The resulting expression is

$$Z_{11} = -\frac{R_5}{\omega^2 C^2 R_2 R_4} \tag{3-61}$$

Equation (3-61) corresponds to a frequency-dependent negative resistor (FDNR).

A GIC in the form of a normalized D element and its schematic designation are shown in Fig. 3-34. Bruton (see References) has shown how the FDNR or D element can be used to generate active filters directly from the LC normalized low-pass prototype values. This technique will now be described.

The 1/s Transformation

If all the impedances of an LC filter network are multiplied by $1/s$, the transfer function remains unchanged. This operation is equivalent to impedance-scaling a filter by the factor $1/s$ and should not be confused with the high-pass transformation which involves the substitution of $1/s$ for s. Section 2.1 under "Frequency and Impedance Scaling" demonstrated that the impedance scaling of a network by any factor Z does not change the frequency response, since the Zs cancel in the transfer function, so the validity of this transformation should be apparent.

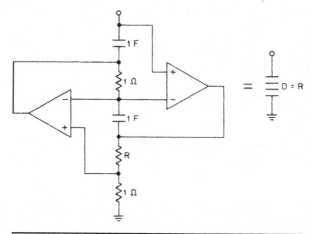

FIGURE 3-34   A GIC realization of a normalized D element.

When the elements of a network are impedance-scaled by $1/s$, they undergo a change in form. Inductors are transformed into resistors, resistors into capacitors, and capacitors into D elements, which are summarized in Table 3-1. Clearly, this design technique is extremely powerful. It enables us to design active filters directly from passive LC circuits. Knowledge of the pole and zero locations is unnecessary.

The design method proceeds by first selecting a normalized low-pass LC filter. All capacitors must be restricted to the shunt arms only, since they will be transformed into D elements which are connected to ground. The dual LC filter (defined by the lower schematic in the tables of Chap. 10 for the all-pole case) is usually chosen to be transformed to minimize the number of D elements. The circuit elements are modified by the $1/s$ transformation, and the D elements are realized using the normalized GIC circuit of Fig. 3-34. The transformed filter is then frequency- and impedance-scaled in the conventional manner. The following example demonstrates the design of an all-pole active low-pass filter using the $1/s$ impedance transformation and the GIC.

Element	Impedance	Transformed Element	Transformed Impedance
L	sL	L	L
C	$\dfrac{1}{sC}$	C	$\dfrac{1}{s^2C}$
R	R	$\dfrac{1}{R}$	$\dfrac{R}{s}$

TABLE 3-1   The $1/s$ Impedance Transformation

Example 3-17 Design of an Active All-Pole Low-Pass Filter Using a D Element

Required:

An active low-pass filter
3 dB to 400 Hz
20-dB minimum at 1,200 Hz
Minimal ringing and overshoot

Result:

(*a*) Compute the steepness factor using Eq. (2-11):

$$A_s = \frac{f_s}{f_c} = \frac{1,200 \text{ Hz}}{400 \text{ Hz}} = 3$$

(*b*) Since low transient distortion is desired, a linear phase filter with a phase error of 0.5° will be selected. The curves of Fig. 2-62 indicate that a filter complexity of $n = 3$ provides over 20 dB of attenuation at 3 rad/s.

The $1/s$ transformation and a GIC realization will be used.

(*c*) The normalized LC low-pass filter from Table 10-47 corresponding to $n = 3$ is shown in Fig. 3-35*a*. The dual circuit has been selected so that only a single D element will be required.

(*d*) The normalized filter is transformed in accordance with Table 3-1, resulting in the circuit of Fig. 3-35*b*. The D element is realized using the normalized GIC configuration of Fig. 3-34, as shown in Fig. 3-35*c*.

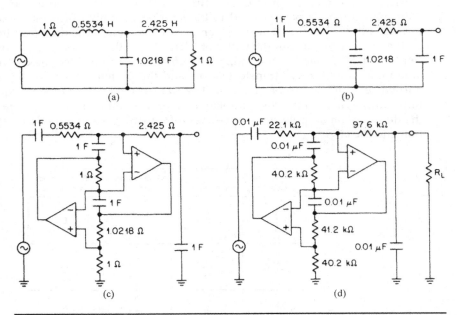

FIGURE 3-35 The network of Example 3-17: (a) normalized low-pass prototype; (b) normalized circuit after 1/s transformation; (c) realization of D element; and (d) final circuit.

(*e*) Since all normalized capacitors are 1 F, it would be desirable if they were all denormalized to a standard value such as 0.01 μF. Using an FSF of $2\pi f_c$ or 2513 and a C' of 0.01 μF, the required impedance-scaling factor can be found by solving Eq. (2-10) for Z as follows:

$$Z = \frac{C}{\text{FSF} \times C'} = \frac{1}{2{,}513 \times 0.01 \times 10^{-6}} = 39{,}800$$

Using Eq. (2-10), an FSF of $2\pi f_c$ or 2,513 and an impedance-scaling factor Z of 39.8×10^3, the normalized filter is scaled by dividing all capacitors by $Z \times \text{FSF}$ and multiplying all resistors by Z. The final circuit is given in Fig. 3-35*d*. The resistor values were modified for standard 1-percent values. The filter loss is 6 dB, corresponding to the loss due to the resistive source and load terminations of the *LC* filter.

The *D* elements are usually realized with dual operational amplifiers which are available as a matched pair in a single package. In order to provide a bias current for the noninverting input of the upper amplifier, a resistive termination to ground must be provided. This resistor will cause a low-frequency roll-off, so true DC-coupled operation will not be possible. However, if the termination is made much larger than the nominal resistor values of the circuit, the low-frequency roll-off can be made to occur well below the frequency range of interest. If a low-output impedance is required, the filter can be followed by a voltage follower or an amplifier if gain is also desired. The filter input should be driven by a source impedance much less than the input resistor of the filter. A voltage follower or amplifier could be used for input isolation.

Elliptic-Function Low-Pass Filters Using the GIC

The $1/s$ transformation and GIC realization are particularly suited for the realization of active high-order elliptic-function low-pass filters. These circuits exhibit low sensitivity to component tolerances and amplifier characteristics. They can be made tunable, and are less complex than the state-variable configurations. The following example illustrates the design of an elliptic-function low-pass filter using the GIC as a *D* element.

Example 3-18   Design of an Active Elliptic-Function Low-Pass Filter using *D* Elements

Required:

An active low-pass filter
0.18-dB ripple at 260 Hz
45-dB minimum at 270 Hz

Result:

Note: The passive elliptic low-pass filter will be designed for a 1 rad/sec cutoff and 1 Ω terminations to obtain the initial low-pass filter prototype.

(*a*) Compute the steepness factor using Eq. (2-11):

$$A_s = \frac{f_s}{f_c} = \frac{270 \text{ Hz}}{260 \text{ Hz}} = 1.0385$$

(*b*) Open **Filter Solutions**.
Check the **Stop Band Freq** box.

Enter **0.18** in the *Pass Band Ripple (dB)* box.
Enter **1** in the *Pass Band Freq* box.
Enter **1.0385** in the *Stop Band Freq* box.
The *Frequency Scale Rad/Sec* box should be checked.
Enter **1** for *Source Res* and *Load Res*.

(c) Click the *Set Order* control button to open the second panel.
Enter **45** for *Stop band Attenuation (dB)*.
Click the *Set Minimum Order* button and then click *Close*.
9 Order is displayed on the main control panel.

(d) Click the *Synthesize Filter* button.

Two schematics are presented by *Filter Solutions*. Select the one representing the dual (Lumped Filter 2), which is shown in Fig. 3-36a.

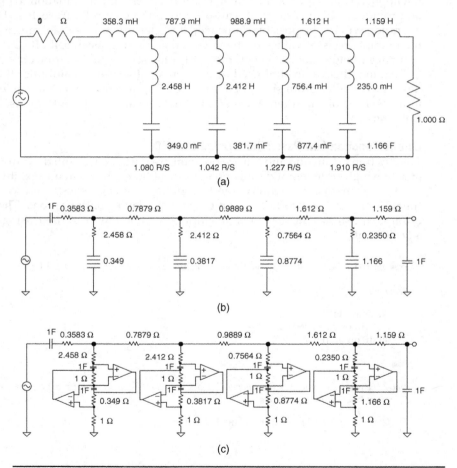

FIGURE 3-36 The filter of Example 3-18: (a) normalized low-pass filter; (b) circuit after 1/s transformation; (c) normalized configuration using GICs for D elements; (d) denormalized filer; and (e) frequency response.

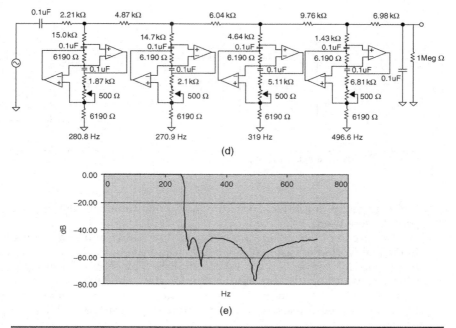

Figure 3-36 (Continued).

(e) The $1/s$ impedance transformation modifies the elements in accordance with Table 3-1, resulting in the circuit of Fig. 3-36b. The D elements are realized using the GIC of Fig. 3-34, as shown in Fig. 3-36c.

(f) The normalized circuit can now be frequency- and impedance-scaled. Since all normalized capacitors are equal, it would be desirable if they could all be scaled to a standard value such as 0.1 μF. The required impedance-scaling factor can be determined from Eq. (2-10) by using an FSF of $2\pi f_c$ or 1634, corresponding to a cutoff of 260 Hz. Therefore, using Eq. (2-10):

$$Z = \frac{C}{FSF \times C'} = \frac{1}{1{,}634 \times 0.1 \times 10^{-6}} = 6{,}120$$

Frequency and impedance scaling by dividing all capacitors by $Z \times$ FSF and multiplying all resistors by Z results in the final filter circuit of Fig. 3-36d having the frequency response of Fig. 3-36e. The resistor values have been modified so that standard 1-percent values are used and the transmission zeros can be adjusted. The frequency of each zero was computed by multiplying each zero in rad/sec of Fig. 3-36a by f_c. A resistive termination is provided so that bias current can be supplied to the amplifiers.

The transmission zeros generated by the circuit of Fig. 3-36d occur because at specific frequencies the value of each FDNR is equal to the positive resistor in series, therefore creating cancellations or null in the shunt branches. By adjusting each D element, these nulls can be tuned to the required frequencies. These adjustments are usually sufficient to obtain satisfactory results. State-variable filters permit the adjustment of the poles and zeros directly for greater accuracy. However, the realization is more complex—for instance, the filter of Example 3-18 would require

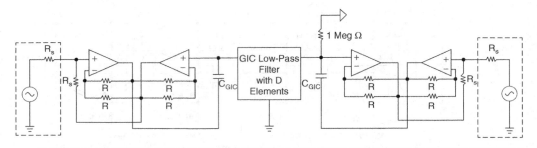

Figure 3-37 Bidirectional active impedance converter.

twice as many amplifiers, resistors, and potentiometers if the state-variable approach were used.

The circuit of Fig. 3-36 requires a capacitive source and a capacitive load (0.1 μF). In the real world the vast majority of applications require a resistive source and resistive load. This conversion can be accomplished by the circuit of Fig. 3-37. R_s is the source resistor and also the load, assuming they are equal. R is a convenient value typically 10 KΩ. Capacitors' C_{GIC} are the source and load capacitors of Fig. 3-36. As in Fig. 3-36 a 1 Meg Ω resistor is required from the output of the GIG low-pass filter to ground.

3.3 Minimal Phase-Shift Filters

In some applications, such as servo systems, filters are required having essentially no phase shift over a portion of their passband. Although minimal phase shift low-pass filters are discussed in this section, they can be transformed into band-pass filters as we will discuss in Chap. 5. However they should have a relatively narrow bandwidth (10 percent or less) to retain the minimal phase-shift properties.

If a Chebyshev low-pass filter is designed to have an extremely high ripple, a unique effect occurs. In the valley of each ripple the filters' phase shift slope is reduced to the point where it appears there's no phase shift over the band of most of the valleys. The circuit of Fig. 3-38 has a ripple of 17.37 dB and is normalized for a 17.37 dB cut-off of 1 rad/s. The design equations of Sec. 2.4 were used to develop these element values.

The attenuation and phase-shift curves are shown in Fig. 3-39.

Figure 3-38 Minimal phase-shift 17.37-dB ripple Chebyshev filter.

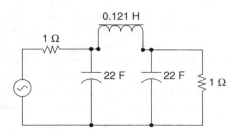

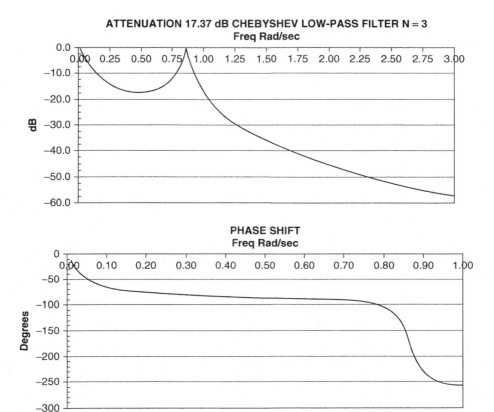

ATTENUATION 17.37 dB CHEBYSHEV LOW-PASS FILTER N = 3

FIGURE 3-39   (a) Attenuation of n = 3 17.37 dB Chebyshev low-pass filter; (b) corresponding phase shift.

References

Amstutz, P. "Elliptic Approximation and Elliptic Filter Design on Small Computers." *IEEE Transactions on Circuits and Systems* CAS-25, No.12 (December, 1978).

Bruton, L. T. "Active Filter Design Using Generalized Impedance Converters." *EDN* (February, 1973).

Bruton, L. T. "Network Transfer Functions Using the Concept of Frequency-Dependent Negative Resistance." *IEEE Transactions on Circuit Theory* CT-16 (August, 1969): 406–408.

Christian, E., and E, Eisenmann. *Filter Design Tables and Graphs*. New York: John Wiley Sons, 1966.

Geffe, P. *Simplified Modern Filter Design*. New York: John F. Rider, 1963.

Huelsman, L. P. *Theory and Design of Active RC Circuits*. McGraw-Hill, 1968.

Saal, R., and E. Ulbrich. "On the Design of Filters by Synthesis." *IRE Transactions on Circuit Theory* (December, 1958).

Shepard, B. R. "Active Filters Part 12." *Electronics* (August 18, 1969): 82–91.

Thomas, L. C. "The Biquad: Part I—Some Practical Design Considerations." *IEEE Transactions on Circuit Theory* CT-18 (May, 1971): 350–357.

Tow, J. "A Step-by-Step Active Filter Design." *IEEE Spectrum* vol. 6 (December, 1969) 64–68.

Williams, A. B. "Design Active Elliptic Filters Easily from Tables." *Electronic Design* 19, no. 21 (October 14, 1971): 76–79.

Williams, A. B. *Active Filter Design*. Dedham, Massachusetts: Artech House, 1975.

Zverev, A. I. *Handbook of Filter Synthesis*. New York: John Wiley and Sons, 1967.

CHAPTER 4

High-Pass Filter Design

4.1 LC High-Pass Filters

LC high-pass filters can be directly designed by mapping the values of a normalized LC low-pass filter into a high-pass filter. This allows use of existing tables of normalized low-pass values to create high-pass filters.

4.1.1 The Low-Pass to High-Pass Transformation

If $1/s$ is substituted for s in a normalized low-pass transfer function, a high-pass response is obtained. The low-pass attenuation values will now occur at high-pass frequencies, which are the reciprocal of the corresponding low-pass frequencies. This was demonstrated in Sec. 2.1.

A normalized LC low-pass filter can be transformed into the corresponding high-pass filter by simply replacing each coil with a capacitor and vice versa, using reciprocal element values. This can be expressed as

$$C_{\text{hp}} = \frac{1}{L_{\text{Lp}}} \tag{4-1}$$

and

$$L_{\text{hp}} = \frac{1}{C_{\text{Lp}}} \tag{4-2}$$

The source and load resistive terminations are unaffected.

The transmission zeros of a normalized elliptic-function low-pass filter are also reciprocated when the high-pass transformation occurs. Therefore,

$$\omega_\infty(\text{hp}) = \frac{1}{\omega_\infty(\text{Lp})} \tag{4-3}$$

To minimize the number of inductors in the high-pass filter, the dual low-pass circuit defined by the lower schematic in the tables of Chap. 10 is usually chosen to be transformed except for even-order all-pole filters, where either circuit may be used. For elliptic-function high-pass filters, the *Filter Solutions* program is used to obtain a low-pass filter prototype normalized to 1 radian/sec and 1-Ω. Thus, the circuit representing the filters' dual, labeled as "Passive Lumped Filter 2," is selected.

The objective is to start with a low-pass prototype containing more inductors than capacitors, since after the low-pass to high-pass transformation, the result will contain more capacitors than inductors.

After the low-pass to high-pass transformation, the normalized high-pass filter is frequency- and impedance-scaled to the required cutoff frequency. The following two examples demonstrate the design of high-pass filters.

Example 4-1 Design of an All-Pole LC High-Pass Filter from a Normalized Low-Pass Filter

Required:

An LC high-pass filter
3 dB at 1 MHz
28-dB minimum at 500 kHz
$R_s = R_L = 300\ \Omega$

Result:

(*a*) To normalize the requirement, use Eq. (2-13) to compute the high-pass steepness factor A_s.

$$A_s = \frac{f_c}{f_s} = \frac{1\ \text{MHz}}{500\ \text{kHz}} = 2$$

(*b*) Select a normalized low-pass filter that offers over 28 dB of attenuation at 2 rad/s.

Inspection of the curves of Chap. 2 indicates that a normalized $n = 5$ Butterworth low-pass filter provides the required attenuation. Table 10-2 contains element values for the corresponding network. The normalized low-pass filter for $n = 5$ and equal terminations is shown in Fig. 4-1*a*. The dual circuit as defined by the lower schematic of Table 10-2 was chosen.

(*c*) To transform the normalized low-pass circuit to a high-pass configuration, replace each coil with a capacitor and vice versa, using reciprocal element values, as shown in Fig. 4-1*b*.

(*d*) Using Eqs. (2-10) and (2-9), denormalize the high-pass filter using a Z of 300 and a frequency-scaling factor (FSF) of $2\pi f_c$ of 6.28×10^6.

$$C'_1 = \frac{C}{\text{FSF} \times Z} = \frac{\dfrac{1}{0.618}}{6.28 \times 10^6 \times 300} = 858\ \text{pF}$$

$$C'_3 = 265\ \text{pF}$$

$$C'_5 = 858\ \text{pF}$$

$$L'_2 = \frac{L \times Z}{\text{FSF}} = \frac{\dfrac{1}{1.618} \times 300}{6.28 \times 10^6} = 29.5\ \mu\text{H}$$

$$L'_4 = 29.5\ \mu\text{H}$$

The final filter is given in Fig. 4-1*c*, with the frequency response shown in Fig. 4-1*d*.

Example 4-2 Design of an Elliptic-Function LC High-Pass Filter Using *Filter Solutions*

Required:

An LC high-pass filter
2-dB maximum at 3,220 Hz
52-dB minimum at 3,020 Hz
$R_s = R_L = 300\ \Omega$

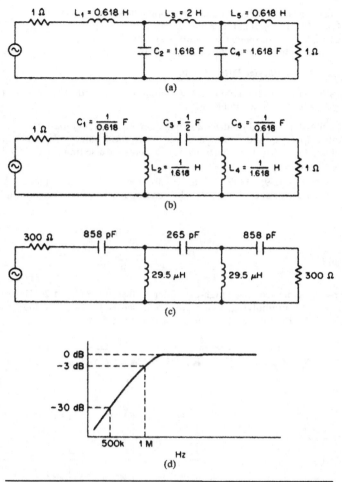

FIGURE 4-1 The high-pass filter of Example 4-1: (a) normalized low-pass filter; (b) high-pass transformation; (c) frequency- and impedance-scaled filter; and (d) frequency response.

Result:

(a) Using Eq. (2-13), compute the high-pass steepness factor A_s.

$$A_s = \frac{f_c}{f_s} = \frac{3,220 \text{ Hz}}{3,020 \text{ Hz}} = 1.0662$$

(b) Since the filter requirement is very steep, an elliptic function will be selected.
   Open *Filter Solutions*.
   Check the *Stop Band Freq* box.
   Enter .2 in the *Pass Band Ripple (dB)* box.
   Enter 1 in the *Pass Band Freq* box.
   Enter **1.0662** in the *Stop Band Freq* box.
   The *Frequency Scale Rad/Sec* box should be checked.
   Enter 1 for the *Source Res* and *Load Res*.

(c) Click the **Set Order** control button to open the second panel.
Enter **52** for **Stop Band Attenuation (dB)**.
Click the **Set Minimum Order** button and then click **Close**.
9 Order is displayed on the main control panel

(d) Click the **Synthesize Filter** button.

Two schematics are presented by **Filter Solutions**. Select the one representing the dual (Lumped Filter 2), which is shown in Fig. 4-2a.

(e) To transform the normalized low-pass circuit into a high-pass configuration, convert inductors into capacitors and vice versa, using reciprocal values. The transformed high-pass filter is illustrated in Fig. 4-2b. The transmission zeros are also transformed by converting them to reciprocal values.

(f) Using Eqs. (2-9) and (2-10), denormalize the high-pass filter using a Z of 300 and a frequency-scaling factor of $2\pi f_c$ or 20,232. The denormalized elements are computed by

$$L' = \frac{L \times Z}{\text{FSF}}$$

and

$$C' = \frac{C}{\text{FSF} \times Z}$$

The resulting denormalized high-pass filter is illustrated in Fig. 4-2c. The frequency of each zero was computed by multiplying each zero in rad/sec of Fig. 4-2b by the design cutoff frequency of $f_c = 3,220$ Hz. The frequency response is given in Fig. 4-2d.

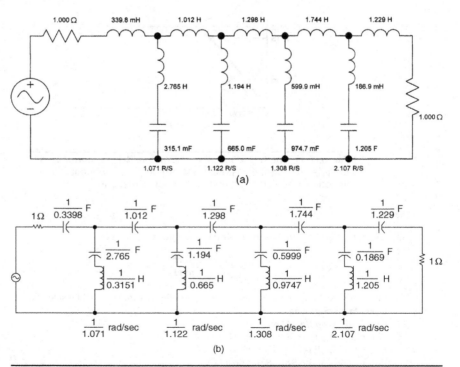

FIGURE 4-2 The high-pass filter of Example 4-2: (a) normalized low-pass filter from *filter solutions*; (b) transformed high-pass filter; (c) frequency- and impedance-scaled high-pass filter; and (d) frequency response.

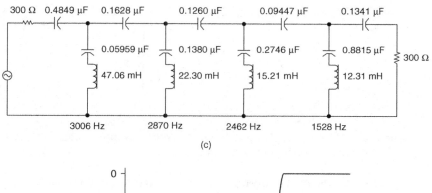

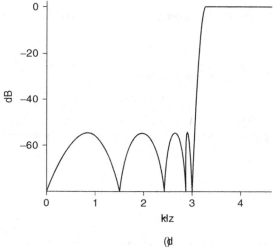

FIGURE 4-2 (*Continued*).

4.1.2   The T-to-Pi Capacitance Conversion

When the elliptic-function high-pass filters are designed for audio frequencies and at low impedance levels, the capacitor values tend to be large. The T-to-pi capacitance conversion will usually restore practical capacitor values.

The two circuits of Fig. 4-3 have identical terminal behavior and are, therefore, equivalent if

$$C_a = \frac{C_1 C_2}{\Sigma C} \tag{4-4}$$

$$C_b = \frac{C_1 C_3}{\Sigma C} \tag{4-5}$$

$$C_c = \frac{C_2 C_3}{\Sigma C} \tag{4-6}$$

Figure 4-3 The T-to-pi capacitance transformation.

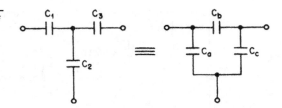

where $\Sigma C = C_1 + C_2 + C_3$. The following example demonstrates the effectiveness of this transformation in reducing large capacitances.

Example 4-3   The T-to-Pi Capacitance Transformation to Reduce Capacitor Values

Required:

The high-pass filter of Fig. 4-2c contains a 0.8815-μF capacitor in the last shunt branch. Use the T-to-pi transformation to provide some relief.

Result:

The circuit of Fig. 4-2c is repeated in Fig. 4-4a showing a "T" of capacitors, including the undesirable 0.8815 μF capacitor. Using Eqs. (4-4) to (4-6), the T-to-pi transformation results in

$$C_a = \frac{C_1 C_2}{\Sigma C} = 0.0750 \mu F$$

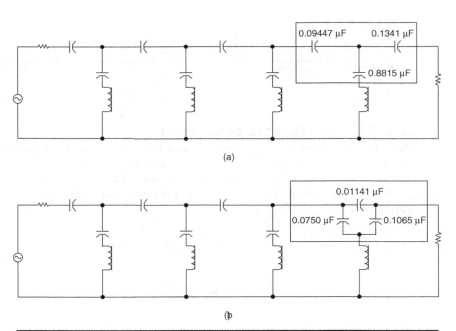

Figure 4-4 The T-to-pi transformation of Example 4-3: (a) the high-pass filter of Example 4-2; and (b) the modified configuration.

$$C_b = \frac{C_1 C_3}{\Sigma C} = 0.01141 \mu F$$

and

$$C_c = \frac{C_2 C_3}{\Sigma C} = 0.1065 \mu F$$

where $C_1 = 0.09447 \ \mu F$, $C_2 = 0.8815 \ \mu F$, and $C_3 = 0.1341 \ \mu F$. The transformed circuit is given in Fig. 4-4b, where the maximum capacitor value has undergone a more than 8:1 reduction.

4.2 Active High-Pass Filters

4.2.1 The Low-Pass to High-Pass Transformation

Active high-pass filters can be derived directly from the normalized low-pass configurations by employing a suitable transformation in a similar manner to LC high-pass filters. To make the conversion, replace each resistor with a capacitor having the reciprocal value and vice versa, as follows:

$$C_{hp} = \frac{1}{R_{Lp}} \tag{4-7}$$

$$R_{hp} = \frac{1}{C_{Lp}} \tag{4-8}$$

It is important to recognize that only the resistors that are a part of the low-pass RC networks are transformed into capacitors by Eq. (4-7). Feedback resistors that strictly determine operational amplifier gain, such as R_6 and R_7 in Fig. 3-20a, are omitted from the transformation.

After the normalized low-pass configuration is transformed into a high-pass filter, the circuit is frequency- and impedance-scaled in the same manner as in the design of low-pass filters. The capacitors are divided by $Z \times FSF$, and the resistors are multiplied by Z. A different Z can be used for each section, but the FSF must be uniform throughout the filter.

4.2.2 All-Pole High-Pass Filters

Active two-pole and three-pole low-pass filter sections were shown in Fig. 3-13 and correspond to the normalized active low-pass values obtained from relevant tables in Chap. 10. These circuits can be directly transformed into high-pass filters by replacing the resistors with capacitors and vice versa, using reciprocal element values, and then frequency- and impedance-scaling the filter network. The filter gain is unity at frequencies well into the passband corresponding to unity gain at DC for low-pass filters. The source impedance of the driving source should be much less than the reactance of the capacitors of the first filter section at the highest passband frequency of interest. The following example demonstrates the design of an all-pole high-pass filter.

Example 4-4   Design of an Active All-Pole High-Pass Filter

Required:

An active high-pass filter
3 dB at 100 Hz
75-dB minimum at 25 Hz

Result:

(*a*) Compute the high-pass steepness factor using Eq. (2-13).

$$A_s = \frac{f_c}{f_s} = \frac{100}{25} = 4$$

(*b*) A normalized low-pass filter that makes the transition from 3 to 75 dB within a frequency ratio of 4:1 must first be selected. The curves of Fig. 2-44 indicate that a fifth-order 0.5-dB Chebyshev filter is satisfactory. The corresponding active filter consists of a three-pole section and a two-pole section whose values are obtained from Table 10-39 and are shown in Fig. 4-5*a*.

(*c*) To transform the normalized low-pass filter into a high-pass filter, replace each resistor with a capacitor, and vice versa, using reciprocal element values. The normalized high-pass filter is given in Fig. 4-5*b*.

(*d*) Since all normalized capacitors are equal, the impedance-scaling factor Z will be computed so that all capacitors become 0.015 μF after denormalization. Since the cutoff frequency is 100 Hz, the FSF is $2\pi f_c$ or 628, so that using Eq. (2-10) results in

$$Z = \frac{C}{\text{FSF} \times C'} = \frac{1}{628 \times 0.015 \times 10^{-6}} = 106.1 \times 10^3$$

If we frequency- and impedance-scale the normalized high-pass filter by dividing all capacitors by $Z \times$ FSF and multiplying all resistors by Z, the circuit of Fig. 4-5*c* is obtained. The resistors were rounded off to standard 1 percent values.

4.2.3   Elliptic-Function High-Pass Filters

High-pass elliptic-function filters can be designed using the elliptic-function voltage-controlled voltage source (VCVS) configuration discussed in Sec. 3.2. This structure can introduce transmission zeros either above or below the pole locations and is, therefore, suitable for elliptic-function high-pass requirements.

The normalized low-pass poles and zeros must first be transformed into the high-pass form. Each complex low-pass pole pair consisting of a real part α and imaginary part β is transformed into a normalized high-pass pole pair as follows:

$$\alpha_{\text{hp}} = \frac{\alpha}{\alpha^2 + \beta^2} = \frac{1}{2Q\omega_0} \tag{4-9}$$

$$\beta_{\text{hp}} = \frac{\beta}{\alpha^2 + \beta^2} = \frac{1}{\omega_0} \sqrt{1 - \frac{1}{4Q^2}} \tag{4-10}$$

The second forms of Eqs. (4-9) and (4-10) involving Q and ω_0 are used when these parameters are provided by the *Filter Solutions* program.

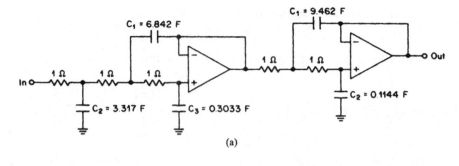

(a)

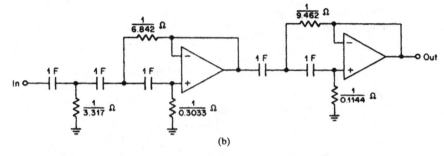

(b)

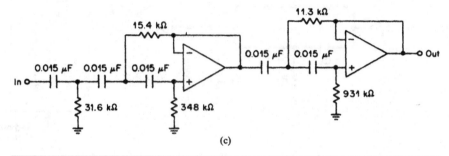

(c)

FIGURE 4-5   The all-pole high-pass filter of Example 4-4: (a) normalized low-pass filter; (b) high-pass transformation; and (c) frequency- and impedance-scaled high-pass filter.

The transformed high-pass pole pair can be denormalized by

$$\alpha'_{hp} = \alpha_{hp} \times FSF \tag{4-11}$$

$$\beta'_{hp} = \beta_{hp} \times FSF \tag{4-12}$$

where FSF is the frequency scaling factor $2\pi f_c$. If the pole is real, the normalized pole is transformed by

$$\alpha_{0,hp} = \frac{1}{\alpha_0} \tag{4-13}$$

The denormalized real pole is obtained from

$$\alpha'_{0,\text{hp}} = \alpha_{0,\text{hp}} \times \text{FSF}$$

(4-14)

To transform zeros, we first use Eq. (4-3) to compute

$$\omega_\infty(\text{hp}) = \frac{1}{\omega_\infty(\text{Lp})}$$

Denormalization occurs by

$$\omega'_\infty(\text{hp}) = \omega_\infty(\text{hp}) \times \text{FSF}$$

(4-15)

The elliptic-function VCVS circuit of Sec. 3.2 is repeated in Fig. 4-6. The elements are then computed as follows:

First, calculate

$$a = \frac{2\alpha'_{\text{hp}}}{\sqrt{(\alpha'_{\text{hp}})^2 + (\beta'_{\text{hp}})^2}}$$

(4-16)

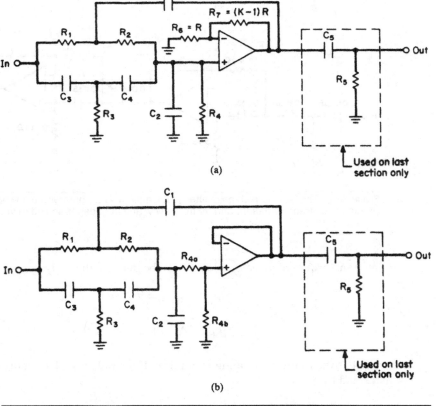

(a)

(b)

Figure 4-6   VCVS elliptic-function high-pass section: (a) circuit for $K > 1$; and (b) circuit for $K < 1$.

$$b = \frac{[\omega'_\infty(\text{hp})]^2}{(\alpha'_{\text{hp}})^2 + (\beta'_{\text{hp}})^2} \tag{4-17}$$

$$c = \sqrt{(\alpha'_{\text{hp}})^2 + (\beta'_{\text{hp}})^2} \tag{4-18}$$

where α'_{hp}, β'_{hp}, and $\omega'_\infty(\text{hp})$ are the denormalized high-pass pole-zero coordinates.
Select C

$$C_1 = C \tag{4-19}$$

Then

$$C_3 = C_4 = \frac{C_1}{2} \tag{4-20}$$

Let

$$C_2 \geq \frac{C_1(b-1)}{4} \tag{4-21}$$

$$R_3 = \frac{1}{cC_1\sqrt{b}} \tag{4-22}$$

$$R_1 = R_2 = 2R_3 \tag{4-23}$$

$$R_4 = \frac{4\sqrt{b}}{cC_1(1-b) + 4cC_2} \tag{4-24}$$

$$K = 2 + \frac{2C_2}{C_1} - \frac{a}{2\sqrt{b}} + \frac{2}{C_1\sqrt{b}}\left(\frac{1}{cR_4} - aC_2\right) \tag{4-25}$$

$$R_6 = R \tag{4-26}$$

$$R_7 = (K-1)R \tag{4-27}$$

where R can be arbitrarily chosen. If K is less than 1, the circuit of Fig. 4-6b is used. Then

$$R_{4a} = (1-K)R_4 \tag{4-28}$$

$$R_{4b} = KR_4 \tag{4-29}$$

$$\text{Section gain} = \frac{bKC_1}{4C_2 + C_1} \tag{4-30}$$

R_5 is determined from the denormalized real pole by

$$R_5 = \frac{1}{C_5\alpha'_{0,\text{hp}}} \tag{4-31}$$

where C_5 can be arbitrarily chosen and $\alpha'_{0,\text{hp}}$ is $\alpha'_{0,\text{hp}} \times \text{FSF}$. The design procedure is illustrated in Example 4-5.

Example 4-5 Design of an Active Elliptic-Function High-Pass Filter Using the VCVS Configuration

Required:

An active high-pass filter
0.2-dB maximum at 3,000 Hz
35-dB minimum rejection at 1,026 Hz

Result:

(*a*) Compute the high-pass steepness factor using Eq. (2-13).

$$A_s = \frac{f_c}{f_s} = \frac{3,000}{1,026} = 2.924$$

(*b*) Open *Filter Solutions*.
    Check the *Stop Band Freq* box.
    Enter .177 in the *Pass Band Ripple (dB)* box.
    Enter 1 in the *Pass Band Freq* box.
    Enter 2.924 in the *Stop Band Freq* box.
    Check the *Frequency Scale Radians* box.

(*c*) Click the *Set Order* control button to open the second panel.

    Enter 35 for *Stop Band Attenuation (dB)*.
    Click the *Set Minimum Order* button and then click *Close*.
    3 Order is displayed on the main control panel.

(*d*) Click the *Transfer Function* button.

    Check the *Casc* box.

The following is displayed:

Continuous Transfer Function

Wn = 3.351

$$\frac{.1143 \quad (S^2 + 11.23)}{(S^2 + .7727{*}S + 1.447) \quad (S + .8871)}$$

Wo = 1.203
Q = 1.557

3rd Order Low Pass Elliptic

Pass Band Frequency = 1.000 Rad/Sec   Stop Band Ratio = 2.924
Pass Band Ripple = 177.0 mdB   Stop Band Frequency = 2.924 Rad/Sec
    Stop Band Attenuation = 37.43 dB

(e) The design parameters are summarized as follows:

$$\text{Section } Q = 1.557$$

$$\text{Section } \omega_0 = 1.203$$

$$\text{Section } \omega_\infty = 3.351$$

$$\alpha_0 = 0.8871 \text{ (from the denominator)}$$

(f) To compute the element values, first transform the low-pass poles and zeros to the high-pass form using Eqs. (4-9) and (4-10).

Complex pole:

$$\alpha_{hp} = \frac{1}{2Q\omega_0} = 0.2670$$

$$\beta_{hp} = \frac{1}{\omega_0} \sqrt{1 - \frac{1}{4Q^2}} = 0.7872$$

The pole-pair is denormalized using Eqs. (4-11) and (4-12):

$$\alpha'_{hp} = 5,033$$

$$\beta'_{hp} = 14,838$$

where $\text{FSF} = 2\pi \times 3,000$. Using Eqs. (4-13), (4-14), (4-3), and (4-15), we have:

Real pole:

$$\alpha_0 = 0.8871$$

$$\alpha_{0,hp} = 1.127$$

$$\alpha'_{0,hp} = 21,248$$

Zero:

$$\omega_\infty = 3.351$$

$$\omega_\infty(hp) = 0.2984$$

$$\omega'_\infty(hp) = 5625$$

(g) The results of (c) are summarized as follows:

$$\alpha'_{hp} = 5,033$$

$$\beta'_{hp} = 14,838$$

$$\alpha'_{0,hp} = 21,248$$

$$\omega'_\infty(hp) = 5625$$

The element values can now be computed using Eqs. (4-16) to (4-18):

$$a = 0.6424$$

$$b = 0.1289$$

$$c = 15,668$$

Let $\qquad\qquad C = 0.02 \ \mu F$

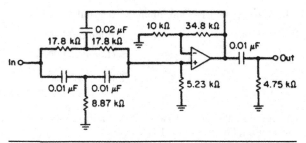

Figure 4-7   The circuit of Example 4-5.

then, using Eqs. (4-19) to (4-21) $C_1 = 0.02\ \mu F$

$$C_3 = C_4 = 0.01\ \mu F$$

$$C_2 \geq -0.00436 \mu F$$

Let $C_2 = 0$

Use Eqs. (4-22) to (4-25): $R_3 = 8.89\ k\Omega$

$$R_1 = R_2 = 17.8\ k\Omega$$

$$R_4 = 5.26\ k\Omega$$

$$K = 4.48$$

Using Eqs. (4-26) and (4-27), let $R_6 = R = 10\ k\Omega$

$$R_7 = 34.8\ k\Omega$$

Using Eq. (4-31), let $C_5 = 0.01\ \mu F$

$$R_5 = 4.70\ k\Omega$$

The resulting circuit is illustrated in Fig. 4-7.

4.2.4   State-Variable High-Pass Filters

The all-pole and elliptic-function active high-pass filters previously discussed cannot be easily adjusted. If the required degree of accuracy results in unreasonable component tolerances, the state-variable or biquad approach will permit independent adjustment of the filter's pole and zero coordinates. Another feature of this circuit is the reduced sensitivity of the response to many of the amplifier limitations such as finite bandwidth and gain.

All-Pole Configuration

In order to design a state-variable all-pole high-pass filter, the normalized low-pass poles must first undergo a low-pass to high-pass transformation. Each low-pass pole pair consisting of a real part α and imaginary part β is transformed into a normalized high-pass pole pair as follows using Eqs. (4-9) and (4-10):

$$\alpha_{hp} = \frac{\alpha}{\alpha^2 + \beta^2}$$

$$\beta_{hp} = \frac{\beta}{\alpha^2 + \beta^2}$$

The transformed high-pass pole pair can now be denormalized by using Eqs. (4-11) and (4-12):

$$\alpha'_{hp} = \alpha_{hp} \times FSF$$

$$\beta'_{hp} = \beta_{hp} \times FSF$$

where FSF is the frequency-scaling factor $2\pi f_x$.

The circuit in Fig. 4-8 realizes a high-pass second-order biquadratic transfer function. The element values for the all-pole case can be computed in terms of the high-pass pole coordinates as follows:

First, compute

$$\omega'_0 = \sqrt{(\alpha'_{hp})^2 + (\beta'_{hp})^2} \tag{4-32}$$

The component values are

$$R_1 = R_4 = \frac{1}{2\alpha'_{hp}C} \tag{4-33}$$

$$R_2 = R_3 = \frac{1}{\omega'_0 C} \tag{4-34}$$

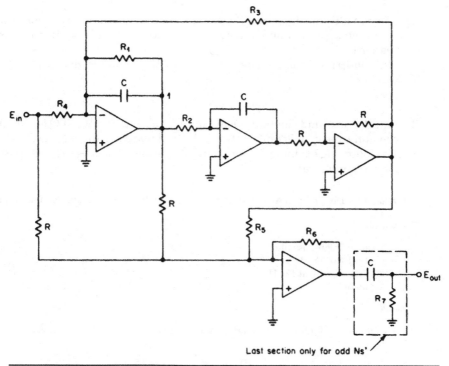

Last section only for odd Ns'

FIGURE 4-8   The biquadratic high-pass configuration.

$$R_5 = \frac{2\alpha'_{hp}}{\omega'_0} R \tag{4-35}$$

$$R_6 = AR \tag{4-36}$$

where C and R are arbitrary and A is the section gain.

   If the transfer function is of an odd order, a real pole must be realized. To transform the normalized low-pass real pole α_0, use Eq. (4-13) to compute

$$\alpha_{0,hp} = \frac{1}{\alpha_0}$$

Then denormalize the high-pass real pole by using Eq. (4-14):

$$\alpha'_{0,hp} = \alpha_{0,hp} \times FSF$$

The last section of the filter is followed by an RC network, as shown in Fig. 4-8. The value of R_7 is computed by

$$R_7 = \frac{1}{\alpha'_{0,hp}C} \tag{4-37}$$

where C is arbitrary.

   A bandpass output is provided at node 1 for tuning purposes. The bandpass resonant frequency is given by

$$f_0 = \frac{\omega'_0}{2\pi} \tag{4-38}$$

R_3 can be made adjustable and the circuit tuned to resonance by monitoring the phase shift between E_{in} and node 1 and adjusting R_3 for 180° of phase shift at f_0 using a Lissajous pattern.

   The bandpass Q can then be monitored at node 1 and is given by

$$Q = \frac{\pi f_0}{\alpha'_{hp}} \tag{4-39}$$

R_1 controls the Q and can be adjusted until either the computed Q is obtained or, more conveniently, the gain is unity between E_{in} and node 1 with f_0 applied.

   The following example demonstrates the design of an all-pole high-pass filter using the biquad configuration.

Example 4-6   Design of an Active All-Pole High-Pass Filter Using a State-Variable Approach

Required:

   An active high-pass filter
   3 ± 0.1 dB at 300 Hz
   30-dB minimum at 120 Hz
   A gain of 2

Result:

(*a*) Compute the high-pass steepness factor using Eq. (2-13).

$$A_s = \frac{f_c}{f_s} = \frac{300}{120} = 2.5$$

(b) The curves of Fig. 2-45 indicate that a normalized third-order 1-dB Chebyshev low-pass filter has over 30 dB of attenuation at 2.5 rad/s. Since an accuracy of 0.1 dB is required at the cutoff frequency, a state-variable approach will be used so that an adjustment capability is provided.

(c) The low-pass pole locations are found in Table 10-26 and are as follows:

Complex pole $\alpha = 0.2257$ $\beta = 0.8822$
Real pole $\alpha_0 = 0.4513$

Complex pole-pair realization:

The complex low-pass pole pair is transformed to a high-pass pole pair as follows using Eqs. (4-9) and (4-10):

$$\alpha_{hp} = \frac{\alpha}{\alpha^2 + \beta^2} = \frac{0.2257}{0.2257^2 + 0.8822^2} = 0.2722$$

and

$$\beta_{hp} = \frac{\beta}{\alpha^2 + \beta^2} = \frac{0.8822}{0.2257^2 + 0.8822^2} = 1.0639$$

The transformed pole pair is then denormalized by using Eqs. (4-11) and (4-12):

$$\alpha'_{hp} = \alpha_{hp} \times FSF = 513$$

$$\beta'_{hp} = \beta_{hp} \times FSF = 2,005$$

where FSF is $2\pi f_c$ or 1,885 since $f_c = 300$ Hz. If we choose $R = 10 \text{ k}\Omega$ and $C = 0.01 \ \mu\text{F}$, the component values are calculated by Eq. (4-32) as follows:

$$\omega'_0 = \sqrt{(\alpha'_{hp})^2 + (\beta'_{hp})^2} = 2,070$$

then using Eqs. (4-33) to (4-36), we have $R_1 = R_4 = \dfrac{1}{2\alpha'_{hp}C} = 97.47 \text{ k}\Omega$

$$R_2 = R_3 = \frac{1}{\omega'_0 C} = 48.31 \text{ k}\Omega$$

$$R_5 = \frac{2\alpha'_{hp}}{\omega'_0} R = 4,957\,\Omega$$

$$R_6 = AR = 20 \text{ k}\Omega$$

where $A = 2$.

The bandpass resonant frequency and Q are calculated as follows, using Eqs. (4-38) and (4-39):

$$f_0 = \frac{\omega'_0}{2\pi} = \frac{2,070}{2\pi} = 329 \text{ Hz}$$

$$Q = \frac{\pi f_0}{\alpha'_{hp}} = \frac{\pi 329}{513} = 2.015$$

Real-pole realization:

Transform the real pole using Eq. (4-13):

$$\alpha_{0,\text{hp}} = \frac{1}{\alpha_0} = \frac{1}{0.4513} = 2.216$$

To denormalize the transformed pole, compute the following using Eq. (4-14):

$$\alpha'_{0,\text{hp}} = \alpha_{0,\text{hp}} \times \text{FSF} = 4,177$$

Using $C = 0.01\ \mu\text{F}$, the real pole section resistor is given by Eq. (4-37):

$$R_7 = \frac{1}{\alpha'_{0,\text{hp}}} = 23.94\ \text{k}\Omega$$

The final filter configuration is shown in Fig. 4-9. The resistors were rounded off to standard 1 percent values, and R_1 and R_3 were made adjustable.

Elliptic-Function Configuration

The biquadratic configuration of Fig. 4-8 can also be applied to the design of elliptic-function high-pass filters. The design of active high-pass elliptic-function filters using biquads utilizes the *Filter Solutions* program, available from the companion website (see App. A), for obtaining normalized low-pass pole-zero locations, which are then converted into high-pass pole-zero locations, denormalized, and used to compute the component values. The parameters obtained from *Filter Solutions* are ω_∞, ω_0, Q, and α_0.

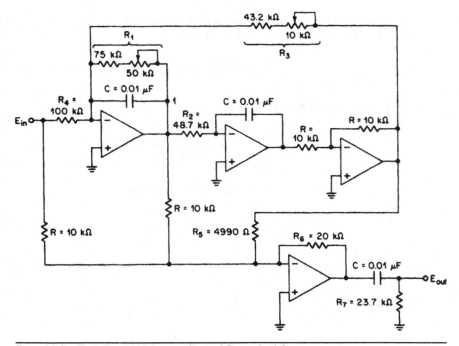

FIGURE 4-9 The all-pole high-pass filter of Example 4-6.

The normalized low-pass poles and zeros must first be transformed to the high-pass form. Each complex low-pass pole pair consisting of a real part α and imaginary part β is transformed into a normalized high-pass pole pair as follows using Eqs. (4-9) and (4-10):

$$\alpha_{hp} = \frac{\alpha}{\alpha^2 + \beta^2} = \frac{1}{2Q\omega_0}$$

$$\beta_{hp} = \frac{\beta}{\alpha^2 + \beta^2} = \frac{1}{\omega_0} \sqrt{1 - \frac{1}{4Q^2}}$$

The second forms of Eqs. (4-9) and (4-10) involving Q and ω_0 are used when these parameters are provided by the *Filter Solutions* program.

The transformed high-pass pole pair can be denormalized by using Eqs. (4-11) and (4-12) as shown:

$$\alpha'_{hp} = \alpha_{hp} \times \text{FSF}$$

$$\beta'_{hp} = \beta_{hp} \times \text{FSF}$$

where FSF is the frequency scaling factor $2\pi f_c$. If the pole is real, the normalized pole is transformed by Eq. (4-13) as follows:

$$\alpha_{0,hp} = \frac{1}{\alpha_0}$$

The denormalized real pole is obtained from Eq. (4-14) as shown:

$$\alpha'_{0,hp} = \alpha_{0,hp} \times \text{FSF}$$

To transform zeros, we compute the following using Eq. (4-3):

$$\omega_\infty(hp) = \frac{1}{\omega_\infty(Lp)}$$

The component values are computed by using the same formulas as for the all-pole case, except for R_5, which is given by

$$R_5 = \frac{2\alpha'_{hp}\omega'_0}{(\omega'_0)^2 - [\omega'_\infty(hp)]^2} R \tag{4-40}$$

where $\omega'_\infty(hp)$ is the denormalized high-pass transmission zero, which is obtained from

$$\omega'_\infty(hp) = \omega_\infty(hp) \times \text{FSF} \tag{4-41}$$

As in the all-pole circuit, the bandpass resonant frequency f_0 is controlled by R_3 and the bandpass Q is determined by R_1. In addition, the section notch can be adjusted if R_5 is made variable. However, this adjustment is usually not required if the circuit is first tuned to f_0, since the notch will then usually fall in.

The following example illustrates the design of an elliptic-function high-pass filter using the biquad configuration of Fig. 4-8.

Example 4-7   Design of an Active Elliptic-Function High-Pass Filter Using a State-Variable Approach

Required:

An active high-pass filter
0.3-dB maximum ripple above 1,000 Hz
18-dB minimum at 643 Hz

Result:

(a) Using Eq. (2-13), compute the high-pass steepness factor.

$$A_s = \frac{f_c}{f_s} = \frac{1,000}{643} = 1.556$$

(b) Open *Filter Solutions*.
Check the *Stop Band Freq* box.
Enter .18 in the *Pass Band Ripple (dB)* box.
Enter 1 in the *Pass Band Freq* box.
Enter 1.556 in the *Stop Band Freq* box.
Check the *Frequency Scale Radians* box.

(c) Click the *Set Order* control button to open the second panel.

Enter 18 for *Stop Band Attenuation (dB)*.
Click the *Set Minimum Order* button and then click *Close*.
3 *Order* is displayed on the main control panel.

(d) Click the *Transfer Function* button.
Check the *Casc* box.

The following is displayed:

Continuous Transfer Function

$$Wn = 1.743$$

$$\frac{.4838 \quad (S^2 + 3.037)}{(S + 1.053) \quad (S^2 + .5687*S + 1.396)}$$

$$Wo = 1.181$$
$$Q = 2.077$$

3rd Order Low Pass Elliptic

Pass Band Frequency = 1.000 Rad/Sec   Stop Band Ratio = 1.556
Pass Band Ripple = 180.0 mdB   Stop Band Frequency = 1.556 Rad/Sec
Stop Band Attenuation = 18.63 dB

The design parameters are summarized as follows:

$$\text{Section } Q = 2.077$$

$$\text{Section } \omega_0 = 1.181$$

$$\text{Section } \omega_\infty = 1.743$$

$$\alpha_0 = 1.053 \text{ (from the denominator)}$$

(e) To compute the element values, first transform the normalized low-pass poles and zeros to the high-pass form, using Eqs. (4-9), (4-10), and (4-3), respectively.

Complex pole:

$$\alpha_{hp} = \frac{1}{2Q\omega_0} = 0.2038$$

$$\beta_{hp} = \frac{1}{\omega_0} \sqrt{1 - \frac{1}{4Q^2}} = 0.8218$$

Zero:

$$\omega_\infty(hp) = \frac{1}{\omega_\infty(Lp)} = \frac{1}{1.743} = 0.5737$$

(f) The poles and zeros are denormalized as follows, using Eqs. (4-11), (4-12), and (4-41):

$$\alpha'_{hp} = \alpha_{hp} \times FSF = 1,280$$

$$\beta'_{hp} = \beta_{hp} \times FSF = 5,163$$

and

$$\omega'_\infty(hp) = \omega_\infty(hp) \times FSF = 3,605$$

where FSF = $2\pi f_c$ or 6,283.

(g) If we arbitrarily choose C = 0.01 μF and R = 100 kΩ, the component values can be obtained by using Eqs. (4-32), (4-33), (4-34), (4-40), and (4-36) as follows:

$$\omega'_0 = \sqrt{(\alpha'_{hp})^2 + (\beta'_{hp})^2} = 5,319$$

then

$$R_1 = R_4 = \frac{1}{2\alpha'_{hp}C} = 39.1 \text{ k}\Omega$$

$$R_2 = R_3 = \frac{1}{\omega'_0 C} = 18.8 \text{ k}\Omega$$

$$R_5 = \frac{2\alpha'_{hp}\omega'_0}{(\omega'_0)^2 - [\omega'_\infty(hp)]^2} R = 89.0\text{k}\Omega$$

and

$$R_6 = AR = 100 \text{ k}\Omega$$

where the gain A is unity.

The bandpass resonant frequency and Q are determined from Eqs. (4-38) and (4-39) as shown:

$$f_0 = \frac{\omega'_0}{2\pi} = 847 \text{ Hz}$$

and

$$Q = \frac{\pi f_0}{\alpha'_{hp}} = 2.077$$

The notch frequency occurs at $\omega_\infty(hp) \times f_c$ or 574 Hz.

(*h*) The normalized real low-pass pole is transformed to a high-pass pole using Eq. (4-13):

$$\alpha_{0,hp} = \frac{1}{\alpha_0} = 0.950$$

and is then denormalized by using Eq. (4-14):

$$\alpha'_{0,hp} = \alpha_{0,hp} \times FSF = 5967$$

Resistor R_7 is found by using Eq. (4-37):

$$R_7 = \frac{1}{\alpha'_{0,hp}C} = 16.8 \text{ k}\Omega$$

where $\qquad\qquad\qquad C = 0.01 \ \mu F.$

The final circuit is given in Fig. 4-10*a* using standard 1 percent values with R_1 and R_3 made adjustable. The frequency response is illustrated in Fig. 4-10*b*.

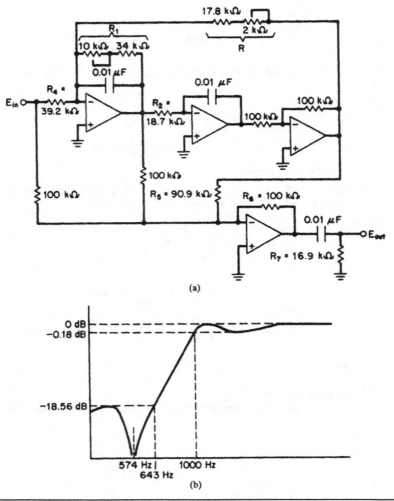

(a)

(b)

FIGURE 4-10   The elliptic-function high-pass filter of Example 4-7: (*a*) filter using the biquad configuration; and (*b*) frequency response.

4.2.5 High-Pass Filters Using the GIC

The generalized impedance converter (GIC) was first introduced in Sec. 3.2. This versatile device is capable of simulating a variety of different impedance functions. The circuit of Fig. 3-28 simulated an inductor whose magnitude was given by Eq. (3-57) as shown:

$$L = \frac{CR_1R_3R_5}{R_2}$$

If we set R_1 through R_3 equal to 1 Ω and $C = 1$ F, a normalized inductor is obtained where L 5 R_5. This circuit is shown in Fig. 4-11.

An active realization of a grounded inductor is particularly suited for the design of active high-pass filters. If a passive LC low-pass configuration is transformed into a high-pass filter, shunt inductors to ground are obtained, which can be implemented using the GIC. The resulting normalized filter can then be frequency- and impedance-scaled. If R_5 is made variable, the equivalent inductance can be adjusted. This feature is especially desirable in the case of steep elliptic-function high-pass filters, since the inductors directly control the location of the critical transmission zeros in the stopband.

The following example illustrates the design of an active all-pole high-pass filter directly from the LC element values using the GIC as a simulated inductance.

Example 4-8 Design of an Active All-Pole High-Pass Filter Using a GIC Approach

Required:

> An active high-pass filter
> 3 dB at 1,200 Hz
> 35-dB minimum at 375 Hz

Result:

(*a*) Compute the high-pass steepness factor using Eq. (2-13):

$$A_s = \frac{f_c}{f_s} = \frac{1,200}{375} = 3.2$$

> The curves of Fig. 2-45 indicate that a third-order 1-dB Chebyshev low-pass filter provides over 35 dB of attenuation at 3.2 rad/s. For this example, we will use a GIC to simulate the inductor of an $n = 3$ LC high-pass configuration.

FIGURE 4-11 A normalized inductor using the GIC.

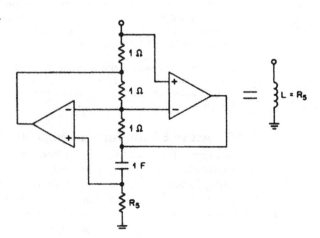

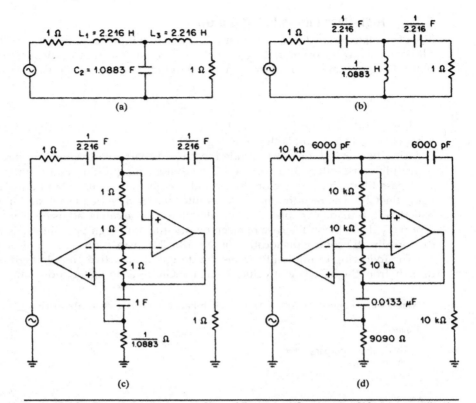

FIGURE 4-12   The all-pole high-pass filter of Example 4-8 using the GIC: (a) normalized low-pass filter; (b) transformed high-pass filter; (c) active inductor realization; and (d) the final network after scaling.

(b) The normalized low-pass filter is obtained from Table 10-31 and is shown in Fig. 4-12a. The dual filter configuration is used to minimize the number of inductors in the high-pass filter.

(c) To transform the normalized low-pass filter into a high-pass configuration, replace the inductors with capacitors, and vice versa, using reciprocal element values. The normalized high-pass filter is shown in Fig. 4-12b. The inductor can now be replaced by the GIC of Fig. 4-11, resulting in the high-pass filter of Fig. 4-12c.

(d) The filter is frequency- and impedance-scaled. Using an FSF of $2\pi f_c$ or 7,540 and a Z of 10^4, divide all capacitors by Z × FSF and multiply all resistors by Z. The final configuration is shown in Fig. 4-12d using standard 1 percent resistor values.

4.2.6   Active Elliptic-Function High-Pass Filters Using the GIC

Active elliptic-function high-pass filters can also be designed directly from normalized elliptic-function low-pass filters using the GIC. This approach is much less complex than a high-pass configuration involving biquads, and still permits adjustment of the transmission zeros. Design of an elliptic-function high-pass filter using the GIC and the *Filter Solutions* program is demonstrated in the following example.

Example 4-9   Design of an Active Elliptic-Function High-Pass Filter Using a GIC Approach

Required:

An active high-pass filter
0.5-dB maximum at 2,500 Hz
60-dB minimum at 1,523 Hz

Result:

(*a*) Compute the high-pass steepness factor using Eq. (2-13):

$$A_s = \frac{f_c}{f_s} = \frac{2,500}{1,523} = 1.641$$

Open *Filter Solutions*.
Check the *Stop Band Freq* box.
Enter .18 in the *Pass Band Ripple (dB)* box.
Enter 1 in the *Pass Band Freq* box.
Enter 1.641 in the *Stop Band Freq* box.
The *Frequency Scale Rad/Sec* box should be checked.
Enter 1 for the *Source Res* and *Load Res*.

(*b*) Click the *Set Order* control button to open the second panel.
Enter 60 for the *Stopband Attenuation (dB)*.
Click the *Set Minimum Order* button and then click *Close*.
6 *Order* is displayed on the main control panel.
Check the *Even Order Mod* box.

(*c*) Click the *Synthesize Filter* button.

Two schematics are presented by *Filter Solutions*. Select the one representing the dual circuit (Lumped Filter 2), which is shown in Fig. 4-13*a*.

(*d*) To transform the network into a normalized high-pass filter, replace each inductor with a capacitor that has a reciprocal value, and vice versa. The zeros are also reciprocated. The normalized high-pass filter is given in Fig. 4-13*b*.

(*e*) The inductors can be replaced using the GIC inductor simulation of Fig. 4-11, resulting in the circuit of Fig. 4-13*c*.

To scale the network, divide all capacitors by $Z \times$ FSF and multiply all resistors by Z, which is arbitrarily chosen at 10^4 and FSF is $2\pi f_c$ or 15,708. The final filter is shown in Fig. 4-13*d*. The stopband peaks were computed by multiplying each normalized high-pass transmission zero by $f_c = 2,500$ Hz, resulting in the frequencies indicated.

4.2.7   Constant-Delay High-Pass Filters

The constant-delay properties of low-pass filter families with linear phase properties in the passband do not get transformed into the passband of a high-pass filter. In general, the delay curve of a low-pass filter is severely distorted during the low-pass to high-pass transformation. The delay rapidly increases as the frequency approaches the stopband. Intuitively, a high-pass filter has no delay at infinite frequency, since all capacitors become shorted and all inductors become open. As the frequency is lowered, the delay will increase and then rapidly ascend to a high value as the band edge is approached. This can be illustrated by an extreme case of the high-pass filter in Example 4-2.

Figure 4-14*a* expands the passband frequency response of the low-pass filter in Example 4-2. Its group delay is shown in Fig. 4-14*b*.

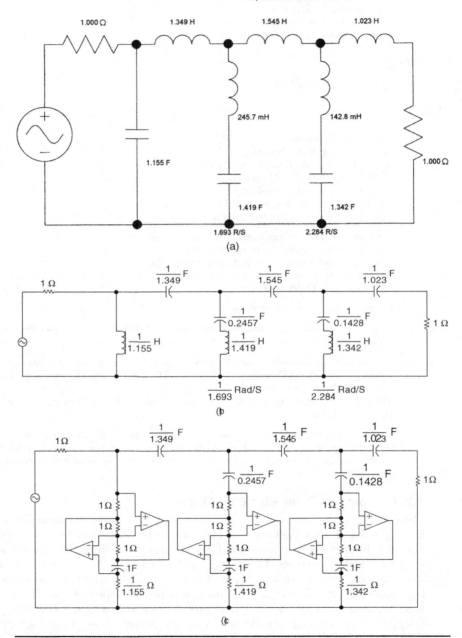

6th Order Low Pass Elliptic

Pass Band Frequency = 1.000 Rad/Sec   Stop Band Ratio = 1.641
Pass Band Ripple = 180.0 mdB   Stop Band Frequency = 1.641 Rad/Sec
                                        Stop Band Attenuation = 63.00 dB

Figure 4-13   The elliptic-function high-pass filter of Example 4-9: (a) normalized low-pass filter; (b) transformed high-pass filter; (c) high-pass filter using the GIC; and (d) frequency- and impedance-scaled network.

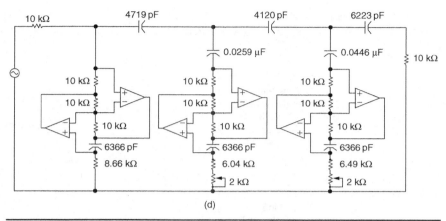

(d)

FIGURE **4-13** (*Continued*).

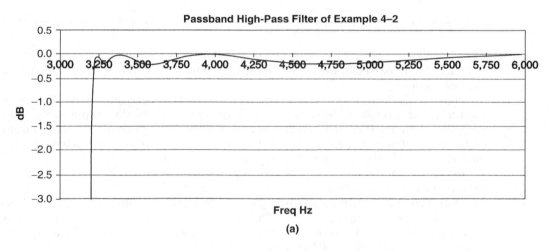

(a)

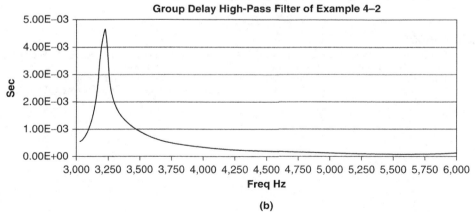

(b)

FIGURE **4-14** High-pass filter of Example 4-2: (*a*) passband response; (*b*) group delay.

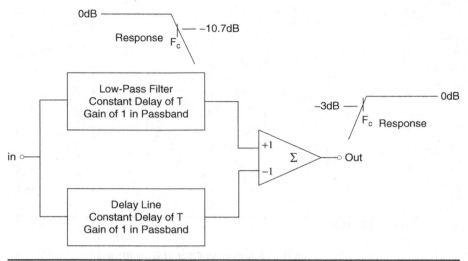

FIGURE 4-15 Constant-delay high-pass filter.

If a constant passband delay (i.e., linear phase) for a high-pass filter is needed, this can become a major problem.

Figure 4-15 illustrates an architecture where a constant-delay low-pass filter can be used to create a constant-delay high-pass filter. By combining the output of the low-pass filter with a delay line in the manner shown, the two signals are subtracted from each other, so that in the passband of the low-pass filter its output is subtracted from the output of the delay line, creating the stopband for the high-pass filter.

In the stopband of the low-pass filter, this subtraction does not occur, so the passband of the high-pass filter begins.

Note that the 10.7-dB attenuation point of the low-pass filter becomes the 3-dB point of the resulting high-pass filter. This is because –10.7 dB corresponds to a multiplication factor of 0.293, which when combined with a 0-dB signal from the delay line, winds up with 0.293 – 1, or –0.707, which is –3 dB.

References

Bruton, L. T. "Active Filter Design Using Generalized Impedance Converters." *EDN*, February, 1973.
Geffe, P. (1963). *Simplified Modern Filter Design*. New York: John F. Rider.
Williams, A. B. (1975). *Active Filter Design*. Dedham, Massachusetts: Artech House.
Williams, A. B. "Design Active Elliptic Filters Easily from Tables." *Electronic Design* 19, no. 21 (October 14, 1971): 76–79.

CHAPTER 5

Band-Pass Filters

5.1   LC Band-Pass Filters

Band-pass filters were classified in Sec. 2.1 as either narrowband or wideband. If the ratio of upper cutoff frequency to lower cutoff frequency is over an octave, the filter is considered a wideband type. The specification is separated into individual low-pass and high-pass requirements and is simply treated as a cascade of low-pass and high-pass filters.

The design of narrowband filters becomes somewhat more difficult. The circuit configuration must be appropriately chosen, and suitable transformations may have to be applied to avoid impractical element values. In addition, as the filter becomes narrower, the element Q requirements increase, and component tolerances and stability become more critical.

5.1.1   Wideband Filters

Wideband band-pass filters are obtained by cascading a low-pass filter and a high-pass filter. The validity of this approach is based on the assumption that the filters maintain their individual responses even though they are cascaded.

The impedance observed at the input or output terminals of an LC low-pass or high-pass filter approaches the resistive termination at the other end at frequencies in the passband. This is apparent from the equivalent circuit of the low-pass filter at DC and the high-pass filter at infinite frequency. At DC, the inductors become short circuits and capacitors become open circuits; at infinite frequency, the opposite conditions occur. If a low-pass and high-pass filter are cascaded and both filters are designed to have equal source and load terminations and identical impedances, the filters will each be properly terminated in their passbands if the cutoff frequencies are separated by at least one or two octaves.

If the separation between passbands is insufficient, the filters will interact because of impedance variations. This effect can be minimized by isolating the two filters through an attenuator. Usually 3 dB of loss is sufficient. Further attenuation provides increased isolation. Table 5-1 contains values for T and π attenuators ranging from 1 to 10 dB at an impedance level of 500 Ω. These networks can be impedance-scaled to the filter impedance level R if each resistor value is multiplied by $R/500$.

Example 5-1   Design of a Wideband LC Band-Pass Filter

Required:

An LC band-pass filter
3 dB at 500 and 2,000 Hz
40-dB minimum at 100 and 4,000 Hz
$R_s = R_L = 600$ Ω

dB	R_1	R_2	R_a	R_b
1	28.8	4,330	8,700	57.7
2	57.3	2,152	4,362	116
3	85.5	1,419	2,924	176
4	113	1,048	2,210	239
5	140	822	1,785	304
6	166	669	1,505	374
7	191	558	1,307	448
8	215	473	1,161	528
9	238	406	1,050	616
10	260	351	963	712

TABLE 5-1 T and π Attenuators

Result:

(*a*) Since the ratio of upper cutoff frequency to lower cutoff frequency is 4:1, a wideband approach will be used. The requirement is first separated into individual low-pass and high-pass specifications:

High-pass filter:
3 dB at 500 Hz
40-dB minimum at 100 Hz

Low-pass filter:
3 dB at 2,000 Hz
40-dB minimum at 4,000 Hz

(*b*) The low-pass and high-pass filters are designed independently, using the design methods outlined in Secs. 3.1 and 4.1 as follows:

Low-pass filter:
Compute the low-pass steepness factor using Eq. (2-11):

$$A_s = \frac{f_s}{f_c} = \frac{4,000 \text{ Hz}}{2,000 \text{ Hz}} = 2$$

Figure 2-43 indicates that a fifth-order 0.25-dB Chebyshev normalized low-pass filter provides over 40 dB of attenuation at 2 rad/s. The normalized low-pass filter is obtained from Table 10-29 and is shown in Fig. 5-1*a*. The filter is frequency- and impedance-scaled by multiplying all inductors by Z/FSF and dividing all capacitors by Z × FSF, where Z is 600 and the frequency-scaling factor (FSF) is $2\pi f_c$ or 12,560. The denormalized low-pass filter is shown in Fig. 5-1*b*.

High-pass filter:
Compute the high-pass steepness factor using Eq. (2-13):

$$A_s = \frac{f_s}{f_c} = \frac{500 \text{ Hz}}{100 \text{ Hz}} = 5$$

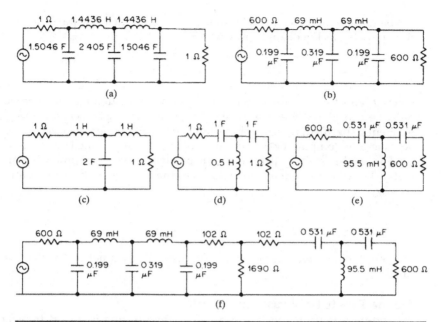

FIGURE 5-1 The *LC* wideband band-pass filter of Example 5-1: (*a*) normalized low-pass filter; (*b*) scaled low-pass filter; (*c*) normalized low-pass filter for high-pass requirement; (*d*) transformed high-pass filter; (*e*) scaled high-pass filter; and (*f*) combined network.

Using Fig. 2-34, an $n = 3$ Butterworth normalized low-pass filter is selected to meet the attenuation requirement. The normalized filter values are found in Table 10-2 and are shown in Fig. 5-1*c*. Since the low-pass filter is to be transformed into a high-pass filter, the dual configuration was selected. By reciprocating element values and replacing inductors with capacitors and vice versa, the normalized high-pass filter of Fig. 5-1*d* is obtained. The network is then denormalized by multiplying all inductors by Z/FSF and dividing all capacitors by $Z \times$ FSF, where Z is 600 Ω and FSF is 3,140. The denormalized high-pass filter is illustrated in Fig. 5-1*e*.

(*c*) The low-pass and high-pass filters can now be combined. A 3-dB T pad will be used to provide some isolation between filters since the separation of cutoffs is only two octaves. The pad values are obtained by multiplying the resistances of Table 5-1, corresponding to 3 dB by 600 Ω/500 Ω or 1.2, and rounding off to standard 1 percent values. The final circuit is shown in Fig. 5-1*f*.

5.1.2 Narrowband Filters

Narrowband band-pass filter terminology was introduced in Sec. 2.1 using the concept of band-pass *Q*, which was defined by Eq. (2-16) as follows:

$$Q_{bp} = \frac{f_0}{BW_{3\,dB}}$$

where f_0 is the geometric center frequency and BW is the 3-dB bandwidth. The geometric center frequency was given by Eq. (2-14) as:

$$f_0 = \sqrt{f_L f_u}$$

where f_L and f_u are the lower and upper 3-dB limits, respectively.

Band-pass filters obtained by transformation from a low-pass filter exhibit geometric symmetry, as Eq. (2-15) shows; that is,

$$f_0 = \sqrt{f_1 f_2}$$

where f_1 and f_2 are any two frequencies with equal attenuation. Geometric symmetry must be considered when normalizing a band-pass specification. For each stopband frequency specified, the corresponding geometric frequency is calculated and a steepness factor is computed based on the more severe requirement.

For band-pass Qs of 10 or more, the passband response approaches arithmetic symmetry. The center frequency then becomes the average of the 3-dB points, as Eq. (2-17) shows:

$$f_0 = \frac{f_L + f_u}{2}$$

The stopband will also become arithmetically symmetrical as the Q increases even further.

The Low-Pass to Band-Pass Transformation

A band-pass transfer function can be obtained from a low-pass transfer function by replacing the frequency variable with a new variable, which is given by

$$f_{bp} = f_0 \left(\frac{f}{f_0} - \frac{f_0}{f} \right) \tag{5-1}$$

When f is equal to f_0, the band-pass center frequency, the response corresponds to that at DC for the low-pass filter.

If the low-pass filter has a 3-dB cutoff of f_c, the corresponding band-pass frequency f can be found by solving

$$\pm f_c = f_0 \left(\frac{f}{f_0} - \frac{f_0}{f} \right) \tag{5-2}$$

The $\pm$ signs occur because a low-pass filter has a mirrored response at negative frequencies in addition to the normal response. Solving Eq. (5-2) for f, we obtain

$$f = \pm \frac{f_c}{2} \pm \sqrt{\left(\frac{f_c}{2} \right)^2 + f_0^2} \tag{5-3}$$

Equation (5-3) implies that the band-pass response has two positive frequencies corresponding to the low-pass response at $\pm f_c$, as well as two negative frequencies with identical responses. These frequencies can be obtained from Eq. (5-3) and are given by

$$f_L = f_0 \left[\sqrt{1 + \left(\frac{f_c}{2f_0} \right)^2} - \frac{f_c}{2f_0} \right] \tag{5-4}$$

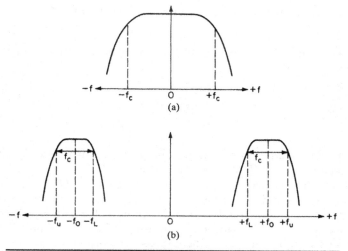

Figure 5-2 The low-pass to band-pass transformation:
(a) low-pass filter; and (b) transformed band-pass filter.

and

$$f_u = f_0 \left[\sqrt{1 + \left(\frac{f_c}{2f_0} \right)^2} + \frac{f_c}{2f_0} \right] \qquad (5\text{-}5)$$

The band-pass 3-dB bandwidth is

$$\text{BW}_{3\text{dB}} = f_u - f_L = f_c \qquad (5\text{-}6)$$

The correspondence between a low-pass filter and the transformed band-pass filter is shown in Fig. 5-2. The response of a low-pass filter to positive frequencies is transformed into the response of the band-pass filter at an equivalent bandwidth. Therefore, a band-pass filter can be obtained by first designing a low-pass filter that has the required response corresponding to the desired bandwidth characteristics of the band-pass filter. The low-pass filter is then transformed into the band-pass filter.

The reactance of a capacitor in the low-pass filter is given by

$$X_c = \frac{1}{j\omega C} \qquad (5\text{-}7)$$

where $\omega = 2\pi f$. If we replace the frequency variable f by the expression of Eq. (5-1), the impedance expression becomes

$$Z = \frac{1}{j\omega C + \dfrac{1}{\dfrac{j\omega}{\omega_0^2 C}}} \qquad (5\text{-}8)$$

where $\omega = 2\pi f_0$. This is the impedance of a parallel resonant LC circuit where the capacitance is still C and the inductance is $1/\omega_0^2 C$. The resonant frequency is ω_0.

The reactance of an inductor in the low-pass filter is

$$X_L = j\omega L \tag{5-9}$$

If we again replace the frequency variable using Eq. (5-1), the resulting impedance expression becomes

$$Z = j\omega L + \cfrac{1}{\cfrac{j\omega}{\omega_0^2 L}} \tag{5-10}$$

This corresponds to a series-resonant LC circuit where the inductance L is unchanged and C is $1/\omega_0^2 L$. The resonant frequency is ω_0.

We can summarize these results by stating that an LC low-pass filter can be transformed into a band-pass filter with the equivalent bandwidth by resonating each capacitor with a parallel inductor and each inductor with a series capacitor. The resonant frequency is f_0, the band-pass filter center frequency. Table 5-2 shows the circuits that result from the low-pass to band-pass transformation.

The Transformation of All-Pole Low-Pass Filters

The LC band-pass filters discussed in this section are probably the most important type of filter. These networks are directly obtained by the band-pass transformation from the LC low-pass values tabulated in Chap. 10. Each normalized low-pass filter defines an infinitely large family of band-pass filters with a geometrically symmetrical response predetermined by the low-pass characteristics.

Low-Pass Branch	Band-Pass Configuration	Circuit Values
Type I		$L = \dfrac{1}{\omega_0^2 C}$ (5-11)
		$C = \dfrac{1}{\omega_0^2 L}$ (5-12)
Type II		$C_a = \dfrac{1}{\omega_0^2 L_a}$ (5-13)
		$L_b = \dfrac{1}{\omega_0^2 C_b}$ (5-14)
Type III		$C_1 = \dfrac{1}{\omega_0^2 L_1}$ (5-15)
		$L_2 = \dfrac{1}{\omega_0^2 C_2}$ (5-16)
Type IV		

TABLE 5-2   The Low-Pass to Band-Pass Transformation

A low-pass transfer function can be transformed to a band-pass type by substituting the frequency variable using Eq. (5-1). This transformation can also be made directly to the circuit elements by first scaling the low-pass filter to the required bandwidth and impedance level. Each coil is then resonated with a series capacitor to the center frequency f_0, and an inductor is introduced across each capacitor to form a parallel tuned circuit that's also resonant at f_0. Every low-pass branch is replaced by the associated band-pass branch, as illustrated by Table 5-2.

Some of the effects of dissipation in low-pass filters were discussed in Sec. 3.1. These effects are even more severe in band-pass filters. The minimum Q requirement for the low-pass elements can be obtained from Fig. 3-8 for a variety of filter types. These minimum values are based on the assumption that the filter elements are predistorted so that the theoretical response is obtained. Since this is not always the case, the branch Qs should be several times higher than the values indicated. When the network undergoes a low-pass to band-pass transformation, the Q requirement is increased by the band-pass Q of the filter. This can be stated as

$$Q_{min} \text{ (band-pass)} = Q_{min} \text{ (low-pass)} \times Q_{bp} \tag{5-17}$$

where $Q_{bp} = f_0 / BW_{3dB}$. As in the low-pass case, the branch Qs should be several times higher than Q_{min}. Since capacitor losses are usually negligible, the branch Q is determined strictly by the inductor losses.

The spread of values in band-pass filters is usually wider than with low-pass filters. For some combinations of impedance and bandwidth, the element values may be impossible or impractical to realize because of their magnitude or the effects of parasitics. When this situation occurs, the designer can use a variety of circuit transformations to obtain a more practical circuit. These techniques are covered in Chap. 8.

The design method can be summarized as follows:

1. Convert the response requirement into a geometrically symmetrical specification.

2. Compute the band-pass steepness factor A_s. Select a normalized low-pass filter from the frequency-response curves of Chap. 2 that makes the passband to stopband transition within a frequency ratio of A_s.

3. Scale the corresponding normalized low-pass filter from the tables of Chap. 10 to the required bandwidth and impedance level of the band-pass filter.

4. Resonate each L and C to f_0 in accordance with Table 5-2.

5. The final design may require manipulation by various transformations so that the values are more practical. In addition, the branch Qs must be well in excess of Q_{min} (band-pass), as given by Eq. (5-17), to obtain near-theoretical results.

Example 5-2 Design of an All-Pole *LC* Band-Pass Filter

Required:

Band-pass filter
A center frequency of 1,000 Hz
3-dB points at 950 and 1,050 Hz
25-dB minimum at 800 Hz and 1,150 Hz
$R_s = R_L = 600 \ \Omega$
An available inductor Q of 100

Result:

(a) Convert to a geometrically symmetrical band-pass requirement:

First, calculate the geometric center frequency using Eq. (2-14).

$$f_0 = \sqrt{f_L f_u} = \sqrt{950 \times 1,050} = 998.8 \text{ Hz}$$

Compute the corresponding geometric frequency for each stopband frequency given, using Eq. (2-18).

$$f_1 f_2 = f_0^2$$

f_1	f_2	$f_2 - f_1$
800 Hz	1,247 Hz	447 Hz
867 Hz	1,150 Hz	283 Hz

The second pair of frequencies will be retained since they represent the more severe requirement. The resulting geometrically symmetrical requirement can be summarized as

$$f_0 = 998.8 \text{ Hz}$$
$$\text{BW}_{3 \text{ dB}} = 100 \text{ Hz}$$
$$\text{BW}_{25 \text{ dB}} = 283 \text{ Hz}$$

(b) Compute the band-pass steepness factor using Eq. (2-19).

$$A_s = \frac{\text{stopband bandwidth}}{\text{passband bandwidth}} = \frac{283 \text{ Hz}}{100 \text{ Hz}} = 2.83$$

(c) Select a normalized low-pass filter that makes the transition from 3 dB to more than 25 dB within a frequency ratio of 2.83:1. Figure 2-34 indicates that an $n = 3$ Butterworth type will satisfy the response requirement. The normalized low-pass filter is found in Table 10-2 and is shown in Fig. 5-3a.

(d) Using Eq. (2-10) and (2-9), denormalize the low-pass filter using a Z of 600 and an FSF of $2\pi f_c$ or 628, where $f_c = 100$ Hz.

$$C_1' = C_3' = \frac{C}{\text{FSF} \times Z} = \frac{1}{628 \times 600} = 2.653 \ \mu\text{F}$$

$$L_2' = \frac{L \times Z}{\text{FSF}} = \frac{2 \times 600}{628} = 1.91 \text{ H}$$

The denormalized low-pass filter is illustrated in Fig. 5-3b.

(e) To make the low-pass to band-pass transformation, resonate each capacitor with a parallel inductor and each inductor with a series capacitor using a resonance frequency of $f_0 = 998.8$ Hz.

$$L_1' = \frac{1}{\omega_0^2 C_1'} = \frac{1}{(6,275)^2 \times 2.653 \times 10^{-6}} = 9.573 \text{ mH} \qquad (5\text{-}11)$$

$$L_3' = L_1' = 9.573 \text{ mH}$$

$$C_2' = \frac{1}{\omega_0^2 L_2'} = \frac{1}{(6,275)^2 \times 1.91} = 0.01329 \ \mu\text{F} \qquad (5\text{-}12)$$

where $\Omega_0 = 2\pi f_0$. The resulting band-pass filter is given in Fig. 5-3c.

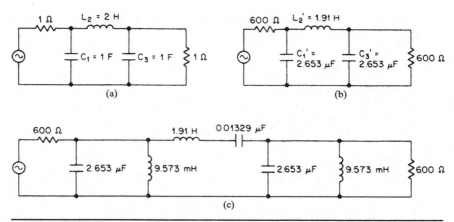

FIGURE 5-3 The band-pass filter of Example 5-2: (a) normalized $n = 3$ Butterworth low-pass filter; (b) low-pass filter scaled to 600 Ω and an f_c of 100 Hz; and (c) transformed band-pass filter.

(f) Estimate if the available inductor Q of 100 is sufficient using Eq. (5-17).

$$Q_{\min} \text{ (band-pass)} = Q_{\min} \text{ (low-pass)} \times Q_{bp} = 2 \times 10 = 20$$

where $Q_{\min}$ (low-pass) was obtained from Fig. 3-8 and Q_{bp} is $f_0/\mathrm{BW}_{3\,dB}$. Since the available Q is well in excess of $Q_{\min}$ (band-pass), the filter response will closely agree with the theoretical predictions.

The response requirement of Example 5-2 was converted to a geometrically symmetrical specification by calculating the corresponding frequency for each stopband frequency specified at a particular attenuation level using the relationship $f_1 f_2 = f_0^2$. The pair of frequencies with the lesser separation was chosen, since this would represent the steeper filter requirement. This technique represents a general method for obtaining the geometrically related frequencies that determine the response requirements of the normalized low-pass filter.

Stopband requirements are frequently specified in an arithmetically symmetrical manner where the deviation on both sides of the center frequency is the same for a given attenuation. Because of the geometric symmetry of band-pass filters, the attenuation for a particular deviation below the center frequency will be greater than for the same deviation above the center frequency. The response curve would then appear compressed on the low side of the passband if plotted on a linear frequency axis. On a logarithmic scale, the curve would be symmetrical.

When the specification is stated in arithmetic terms, the stopband bandwidth on a geometric basis can be computed directly by

$$\mathrm{BW} = f_2 - \frac{f_0^2}{f_2} \tag{5-18}$$

where f_2 is the upper stopband frequency and f_0 is the geometric center frequency as determined from the passband limits. This approach is demonstrated in the following example.

Example 5-3   Design of All-Pole *LC* Band-Pass Filter from an Arithmetically Symmetrical Requirement

Required:

Band-pass filter
A center frequency of 50 kHz
3-dB points at ± 3 kHz (47 kHz, 53 kHz)
30-dB minimum at ± 7.5 kHz (42.5 kHz, 57.5 kHz)
40-dB minimum at ± 10.5 kHz (39.5 kHz, 60.5 kHz)
$Rs = 150 \, \Omega$ $R_L = 300 \, \Omega$

Result:

(*a*) Convert to the geometrically symmetrical band-pass requirement using Eq. (2-14).

$$f_0 = \sqrt{f_L f_u} = \sqrt{47 \times 53 \times 10^6} = 49.91 \text{ kHz}$$

Since the stopband requirement is arithmetically symmetrical, compute the stopband bandwidth using Eq. (5-18).

$$\text{BW}_{30 \text{ dB}} = f_2 - \frac{f_0^2}{f_2} = 57.5 \times 10^3 - \frac{(49.91 \times 10^3)^2}{57.5 \times 10^3} = 14.18 \text{ kHz}$$

$$\text{BW}_{40 \text{ dB}} = 19.33 \text{ kHz}$$

Requirement:

$$f_0 = 49.91 \text{ kHz}$$

$$\text{BW}_{3 \text{ dB}} = 6 \text{ kHz}$$

$$\text{BW}_{30 \text{ dB}} = 14.18 \text{ kHz}$$

$$\text{BW}_{40 \text{ dB}} = 19.33 \text{ kHz}$$

(*b*) Since two stopband bandwidth requirements are given, they must both be converted into band-pass steepness factors using Eq. (2-19).

$$A_s(30 \text{ dB}) = \frac{\text{stopband bandwidth}}{\text{passband bandwidth}} = \frac{14.18 \text{ kHz}}{6 \text{ kHz}} = 2.36$$

$$A_s(40 \text{ dB}) = 3.22$$

(*c*) A normalized low-pass filter must be chosen that provides over 30 dB of rejection at 2.36 rad/s and more than 40 dB at 3.22 rad/s. Figure 2-41 indicates that a fourth-order 0.01-dB Chebyshev filter will meet this requirement. The corresponding low-pass filter can be found in Table 10-27. Since a 2:1 ratio of R_L to R_s is required, the design for a normalized R_s of 2 Ω is chosen and is turned end for end. The circuit is shown in Fig. 5-4*a*.

(*d*) The circuit is now scaled to an impedance level *Z* of 150 and a cutoff of $f_c = 6$ kHz. All inductors are multiplied by *Z*/FSF, and the capacitors are divided by *Z* × FSF, where FSF is $2\pi f_c$. The 1-Ω source and 2-Ω load become 150 and 300 Ω, respectively. The denormalized network is illustrated in Fig. 5-4*b*.

(*e*) The scaled low-pass filter is transformed to a band-pass filter at $f_0 = 49.91$ kHz by resonating each capacitor with a parallel inductor and each inductor with a series capacitor using the general relationship $\omega_0^2 LC = 1$. The resulting band-pass filter is shown in Fig. 5-4*c*.

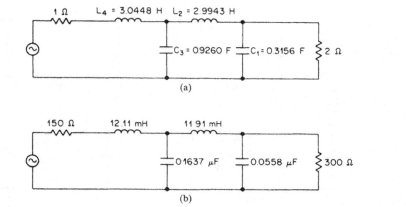

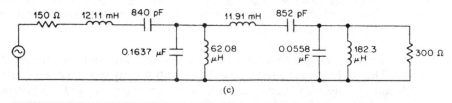

FIGURE 5-4 The band-pass filter of Example 5-3: (a) normalized low-pass filter; (b) scaled low-pass filter; and (c) transformed band-pass filter.

5.1.3   The Design of Parallel Tuned Circuits

The simple RC low-pass circuit of Fig. 5-5a has a 3-dB cutoff corresponding to

$$f_c = \frac{1}{2\pi RC} \tag{5-19}$$

If a band-pass transformation is performed, the circuit of Fig. 5-5b results, where Eq. (5-11) shows us that

$$L = \frac{1}{\omega_0^2 C}$$

The center frequency is f_0, and the 3-dB bandwidth is equal to f_c. The band-pass Q is given by

$$Q_{bp} = \frac{f_0}{BW_{3 \text{ dB}}} = \frac{f_0}{f_c} = \omega_0 RC \tag{5-20}$$

Since the magnitudes of the capacitive and inductive susceptances are equal at resonance by definition, we can substitute $1/\omega_0 L$ for $\omega_0 C$ in Eq. (5-20) and obtain

$$Q_{bp} = \frac{R}{\omega_0 L} \tag{5-21}$$

The element R may be a single resistor, as in Fig. 5-5b, or the parallel combination of both the input and output terminations if an output load resistor is also present.

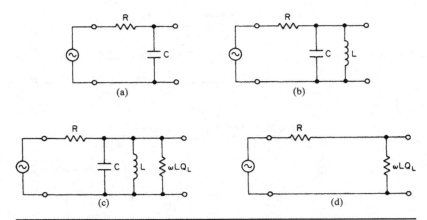

Figure 5-5   The single tuned circuit: (a) *RC* low-pass circuit; (b) result of band-pass transformation; (c) representation of coil losses; (d) equivalent circuit at resonance.

The circuit of Fig. 5-5b is somewhat ideal, since inductor losses are usually unavoidable. (The slight losses usually associated with the capacitor will be neglected.) If the inductor Q is given as Q_L, the inductor losses can be represented as a parallel resistor of $\omega L Q_L$, as shown in Fig. 5-5c. The effective Q of the circuit thus becomes

$$Q_{\text{eff}} = \frac{\dfrac{R}{\omega_0 L} Q_L}{\dfrac{R}{\omega_0 L} + Q_L} \tag{5-22}$$

As a result, the effective circuit Q is somewhat less than the values computed by Eq. (5-20) or (5-21). To compensate for the effect of finite inductor Q, the design Q should be somewhat higher. This value can be found from

$$Q_d = \frac{Q_{\text{eff}} Q_L}{Q_L - Q_{\text{eff}}} \tag{5-23}$$

At a resonance, the equivalent circuit is represented by the resistive voltage divider of Fig. 5-5d, since the reactive elements cancel. The insertion loss at f_0 can be determined by the expression

$$IL_{\text{dB}} = 20 \log \left(1 + \frac{1}{k-1} \right) \tag{5-24}$$

where $k = Q_L / Q_{\text{eff}}$. The insertion loss can be obtained directly from the curve of Fig. 5-6. Clearly, the insertion loss increases dramatically as the inductor Q approaches the required effective Q of the circuit.

The frequency response of a single tuned circuit is expressed by

$$A_{\text{dB}} = 10 \log \left[1 + \left(\frac{BW_x}{BW_{3\,\text{dB}}} \right)^2 \right] \tag{5-25}$$

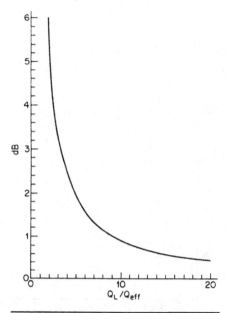

Figure 5-6 Insertion loss versus Q_L/Q_{eff}.

where BW_x is the bandwidth of interest, and BW_{3dB} is the 3-dB bandwidth. The response characteristics are identical to an $n = 1$ Butterworth, so the attenuation curves of Fig. 2-34 can be applied using $BW_x/BW_{3\,dB}$ as the normalized frequency in radians per second.

The phase shift is given by

$$\theta = \tan^{-1}\left(\frac{2\Delta f}{BW_{3dB}}\right) \tag{5-26}$$

where Δf is the frequency deviation from f_0. The output phase shift lags by 45° at the upper 3-dB frequency, and leads by 45° at the lower 3-dB frequency. At DC and infinity, the phase shift reaches +90° and −90°, respectively. Equation (5-26) is plotted in Fig. 5-7.

The group delay can be estimated by the slope of the phase shift at f_0 and results in the approximation

$$T_{gd} = \frac{318}{BW_{3\,dB}} \tag{5-27}$$

where $BW_{3\,dB}$ is the 3-dB bandwidth in hertz and T_{gd} is the resulting group delay in milliseconds.

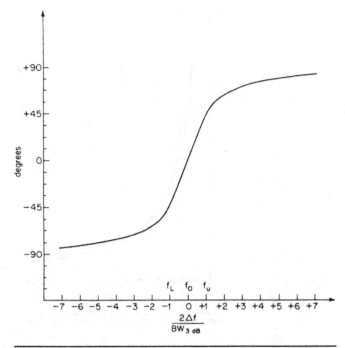

FIGURE 5-7 Phase shift versus frequency.

Example 5-4 Design of a Band-Pass Filter Using a Parallel Tuned Circuit

Required:

An *LC* band-pass filter
A center frequency of 10 kHz
3 dB at ±100 Hz (9.9 kHz, 10.1 kHz)
15-dB minimum at 0±1 kHz (9 kHz, 11 kHz)
Inductor $Q_L = 200$
$R_s = R_L = 6$ kΩ

Result:

(*a*) Convert to the geometrically symmetrical band-pass specification. Since the band-pass Q is much greater than 10, the specified arithmetically symmetrical frequencies are used to determine the following design requirements:

$$f_0 = 10 \text{ kHz}$$

$$\text{BW}_{3 \text{ dB}} = 200 \text{ Hz}$$

$$\text{BW}_{15 \text{ dB}} = 2{,}000 \text{ Hz}$$

(*b*) Compute the band-pass steepness factor using Eq. (2-19).

$$A_s = \frac{\text{stopband bandwidth}}{\text{passband bandwidth}} = \frac{2{,}000}{200} = 10$$

Figure 2-34 indicates that a single tuned circuit ($n = 1$) provides more than 15 dB of attenuation within a bandwidth ratio of 10:1.

(c) Using Eq. (5-23), calculate the design Q to obtain a Q_{eff} equal to $f_0/BW_{3dB} = 50$, considering the inductor Q_L of 200.

$$Q_d = \frac{Q_{\text{eff}}Q_L}{Q_L - Q_{\text{eff}}} = \frac{50 \times 200}{200 - 50} = 66.7$$

(d) Since the source and load are both 6 kΩ, the total resistive loading on the tuned circuit is the parallel combination of both terminations—thus, $R = 3$ kΩ. The design Q can now be used in Eq. (5-20) to compute C.

$$C = \frac{Q_{\text{bp}}}{\omega_0 R} = \frac{66.7}{6.28 \times 10 \times 10^3 \times 3{,}000} = 0.354 \ \mu F$$

The inductance is given by Eq. (5-11).

$$L = \frac{1}{\omega_0^2 C} = \frac{1}{(2\pi \times 10 \text{ kHz})^2 \times 3.54 \times 10^{-7}} = 716 \ \mu H$$

The resulting circuit is shown in Fig. 5-8a, which has the frequency response of Fig. 5-8b. See Sec. 8.1 for a more practical implementation using a tapped inductor.

(e) The circuit insertion loss can be calculated from Eq. (5-24) as follows:

$$IL_{dB} = 20 \log\left(1 + \frac{1}{k-1}\right) = 20 \log 1.333 = 2.5 \text{ dB}$$

where

$$k = \frac{Q_L}{Q_{\text{eff}}} = \frac{200}{50} = 4$$

The low-pass to band-pass transformation illustrated in Fig. 5-5 can also be examined from a pole-zero perspective. The RC low-pass filter has a single real pole at $1/RC$, as shown in Fig. 5-9a, and a zero at infinity. The band-pass transformation results in a pair of complex poles and zeros at the origin and infinity, as illustrated in Fig. 5-9b.

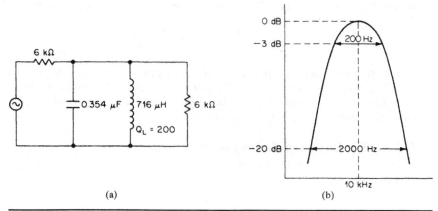

(a) (b)

FIGURE 5-8 The tuned circuit of Example 5-4: (a) circuit; and (b) frequency response.

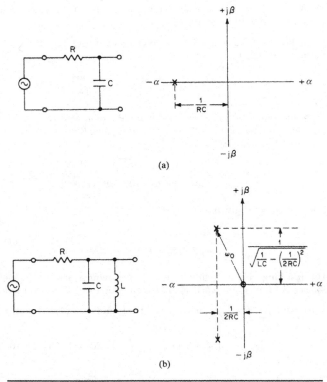

Figure 5-9 The band-pass transformation: (a) the low-pass circuit; and (b) the band-pass circuit.

The radial distance from the origin to the pole is $1/(LC)^{1/2}$, corresponding to ω_0, the resonant frequency. The Q can be expressed by

$$Q = \frac{\omega_0}{2\alpha} \tag{5-28}$$

where α, the real part, is $1/2RC$. The transfer function of the circuit of Fig. 5-9b becomes

$$T(s) = \frac{s}{s^2 + \dfrac{\omega_0}{Q}s + \omega_0^2} \tag{5-29}$$

At ω_0, the impedance of the parallel resonant circuit is at a maximum and is purely resistive, resulting in zero phase shift. If the Q is much less than 10, these effects do not both occur at precisely the same frequency. Series losses of the inductor will also displace the zero from the origin onto the negative real axis.

5.1.4 The Design of Series Tuned Circuits

The losses of an inductor can be conveniently represented by a series resistor determined by

$$R_{\text{coil}} = \frac{\omega L}{Q_L} \tag{5-30}$$

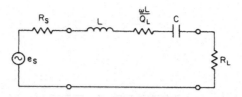

FIGURE 5-10   The series resonant circuit.

If we form a series resonant circuit and include the source and load resistors, we obtain the circuit of Fig. 5-10. Equations (5-24) to (5-27) for insertion loss, frequency response, phase shift, and the group delay of the parallel-tuned circuit apply since the two circuits are duals of each other. The inductance is calculated from

$$L = \frac{R_s + R_L}{\omega_0 \left(\dfrac{1}{Q_{bp}} - \dfrac{1}{Q_L} \right)} \tag{5-31}$$

where Q_{bp} is the required Q and Q_L is the inductor Q. The capacitance is given by Eq. (5-12) as follows:

$$C = \frac{1}{\omega_0^2 L}$$

Example 5-5   Design of a Band-Pass Filter Using a Series Tuned Circuit

Required:

A series tuned circuit
A center frequency of 100 kHz
A 3-dB bandwidth of 2 kHz
$R_s = R_L = 100\ \Omega$
Inductor Q of 400

Result:

(*a*) Compute the band-pass Q using Eq. (2-16).

$$Q_{bp} = \frac{f_0}{BW_{3\,dB}} = \frac{100\ kHz}{2\ kHz} = 50$$

(*b*) Calculate the element values using Eqs. (5-31) and (5-12).

$$L = \frac{R_s + R_L}{\omega_0 \left(\dfrac{1}{Q_{bp}} - \dfrac{1}{Q_L} \right)} = \frac{200}{2\pi \times 10^5 \left(\dfrac{1}{50} - \dfrac{1}{400} \right)} = 18.2\ mH$$

$$C = \frac{1}{\omega_0^2 L} = 139\ pF$$

The circuit is shown in Fig. 5-11.

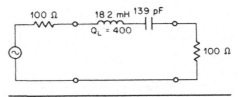

FIGURE 5-11   The series tuned circuit of Example 5-5.

The reader may recall from AC circuit theory that one of the effects of series resonance is a buildup of voltage across both reactive elements. The voltage across either reactive element at resonance is equal to Q times the input voltage, and may be excessively high, causing inductor saturation or capacitor breakdown. In addition, the L/C ratio becomes large as the bandwidth is reduced and will result in impractical element values where high Qs are required. As a result, series-resonant circuits are less desirable than parallel-tuned circuits.

5.1.5   Synchronously Tuned Filters

Tuned circuits can be cascaded to obtain band-pass filters of a higher complexity. Each stage must be isolated from the previous section. If all circuits are tuned to the same frequency, a synchronously tuned filter is obtained. The characteristics of synchronously tuned band-pass filters are discussed in Sec. 2.8, and the normalized frequency response is illustrated by the curves of Fig. 2-77. The design Q of each section was given by Eq. (2-45) as shown:

$$Q_{\text{section}} = Q_{\text{overall}}\sqrt{2^{1/n} - 1}$$

where Q_{overall} is defined by the ratio $f_0/BW_{3\,dB}$ of the composite filter. The individual circuits may be of either the series or the parallel resonant type.

Synchronously tuned filters offer the simplest approximation to a band-pass response. Since all stages are identical and tuned to the same frequency, they are simple to construct and easy to align. The Q requirement of each individual section is less than the overall Q, whereas the opposite is true for conventional band-pass filters. The transient behavior exhibits no overshoot or ringing. On the other hand, the selectivity is extremely poor. To obtain a particular attenuation for a given steepness factor A_s, many more stages are required than for the other filter types. In addition, each section must be isolated from the previous section, so interstage amplifiers are required. The disadvantages generally outweigh the advantages, thus synchronously tuned filters are usually restricted to special applications, such as IF and RF amplifiers.

Example 5-6   Design of a Synchronously Tuned Band-Pass Filter

Required:

A synchronously tuned band-pass filter
A center frequency of 455 kHz
3 dB at ± 5 kHz
30-dB minimum at ± 35 kHz
An inductor Q of 400

Result:

(*a*) Compute the band-pass steepness factor using Eq. (2-19).

$$A_s = \frac{\text{stopband bandwidth}}{\text{passband bandwidth}} = \frac{70 \text{ kHz}}{10 \text{ kHz}} = 7$$

The curves of Fig. 2-77 indicate that a third-order ($n = 3$) synchronously tuned filter satisfies the attenuation requirement.

(*b*) Three sections are required, which are all tuned to 455 kHz and have identical Q. To compute the Q of the individual sections, first calculate the overall Q, which is given by Eq. (2-16) as follows:

$$Q_{bp} = \frac{f_0}{\text{BW}_{3\,\text{dB}}} = \frac{455 \text{ kHz}}{10 \text{ kHz}} = 45.5$$

The section Qs can be found from Eq. (2-48) as shown:

$$Q_{\text{section}} = Q_{\text{overall}} \sqrt{2^{1/n} - 1} = 45.5 \sqrt{2^{1/3} - 1} = 23.2$$

(*c*) The tuned circuits can now be designed using either a series or parallel realization. Let us choose a parallel-tuned circuit configuration using a single-source resistor of 10 kΩ and a high-impedance termination. Since an effective circuit Q of 23.2 is desired and the inductor Q is 400, the design Q is calculated from Eq. (5-23) as shown:

$$Q_d = \frac{Q_{\text{eff}} Q_L}{Q_L - Q_{\text{eff}}} = \frac{23.2 \times 400}{400 - 23.2} = 24.6$$

The inductance is then given by Eq. (5-21):

$$L = \frac{R}{\omega_0 Q_{bp}} = \frac{10 \times 10^3}{2\pi 455 \times 10^3 \times 24.6} = 142 \ \mu\text{H}$$

The resonating capacitor can be obtained from Eq. (5-12):

$$C = \frac{1}{\omega_0^2 L} = 862 \text{ pF}$$

The final circuit is shown in Fig. 5-12 utilizing buffer amplifiers to isolate the three sections.

5.1.6 Narrowband Coupled Resonators

Narrowband band-pass filters can be designed by using coupling techniques where parallel-tuned circuits are interconnected by coupling elements such as inductors or capacitors. Figure 5-13 illustrates some typical configurations.

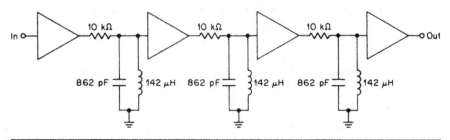

FIGURE 5-12   The synchronously tuned filter of Example 5-6.

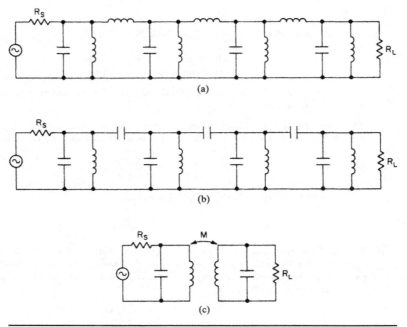

FIGURE 5-13 Coupled resonators: (*a*) inductive coupling; (*b*) capacitive coupling; and (*c*) magnetic coupling.

Coupled resonator configurations are desirable for narrowband filters with band-pass Qs of 10 or more. The values are generally more practical than the elements obtained by the low-pass to band-pass transformation, especially for very high Qs. The tuning is also simpler since it turns out that all nodes are resonated to the same frequency. Of the three configurations shown in Fig. 5-13, the capacitive-coupled configuration is the most desirable from the standpoint of economy and ease of manufacture.

The theoretical justification for the design method is based on the assumption that the coupling elements have a constant impedance with frequency. This assumption is approximately accurate over narrow bandwidths. At DC, the coupling capacitors will introduce additional response zeros. This causes the frequency response to be increasingly unsymmetrical, both geometrically and arithmetically, as we deviate from the center frequency. The response shape will be somewhat steeper on the low-frequency side of the passband, however.

The general form of a capacitive-coupled resonator filter is shown in Fig. 5-14. An nth-order filter requires n parallel-tuned circuits and contains n nodes. Tables 5-3 to 5-12

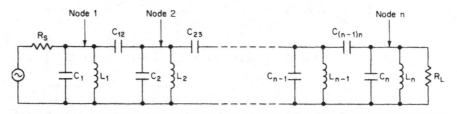

FIGURE 5-14 A general form of a capacitive-coupled resonator filter.

n	q_1	q_n	k_{12}	k_{23}	k_{34}	k_{45}	k_{56}	k_{67}	k_{78}
2	1.414	1.414	0.707						
3	1.000	1.000	0.707	0.707					
4	0.765	0.765	0.841	0.541	0.841				
5	0.618	0.618	1.000	0.556	0.556	1.000			
6	0.518	0.518	1.169	0.605	0.518	0.605	1.169		
7	0.445	0.445	1.342	0.667	0.527	0.527	0.667	1.342	
8	0.390	0.390	1.519	0.736	0.554	0.510	0.554	0.736	1.519

TABLE 5-3   Butterworth Capacitive-Coupled Resonators

n	q_1	q_n	k_{12}	k_{23}	k_{34}	k_{45}	k_{56}	k_{67}	k_{78}
2	1.483	1.483	0.708						
3	1.181	1.181	0.682	0.682					
4	1.046	1.046	0.737	0.541	0.737				
5	0.977	0.977	0.780	0.540	0.540	0.780			
6	0.937	0.937	0.809	0.550	0.518	0.550	0.809		
7	0.913	0.913	0.829	0.560	0.517	0.517	0.560	0.829	
8	0.897	0.897	0.843	0.567	0.520	0.510	0.520	0.567	0.843

TABLE 5-4   0.01-dB Chebyshev Capacitive-Coupled Resonators

n	q_1	q_n	k_{12}	k_{23}	k_{34}	k_{45}	k_{56}	k_{67}	k_{78}
2	1.638	1.638	0.711						
3	1.433	1.433	0.662	0.662					
4	1.345	1.345	0.685	0.542	0.685				
5	1.301	1.301	0.703	0.536	0.536	0.703			
6	1.277	1.277	0.715	0.539	0.518	0.539	0.715		
7	1.262	1.262	0.722	0.542	0.516	0.516	0.542	0.722	
8	1.251	1.251	0.728	0.545	0.516	0.510	0.516	0.545	0.728

TABLE 5-5   0.1-dB Chebyshev Capacitive-Coupled Resonators

n	q_1	q_n	k_{12}	k_{23}	k_{34}	k_{45}	k_{56}	k_{67}	k_{78}
2	1.950	1.950	0.723						
3	1.864	1.864	0.647	0.647					
4	1.826	1.826	0.648	0.545	0.648				
5	1.807	1.807	0.652	0.534	0.534	0.652			
6	1.796	1.796	0.655	0.533	0.519	0.533	0.655		
7	1.790	1.790	0.657	0.533	0.516	0.516	0.533	0.657	
8	1.785	1.785	0.658	0.533	0.515	0.511	0.515	0.533	0.658

TABLE 5-6   0.5-dB Chebyshev Capacitive-Coupled Resonators

n	q_1	q_n	k_{12}	k_{23}	k_{34}	k_{45}	k_{56}	k_{67}
2	2.210	2.210	0.739					
3	2.210	2.210	0.645	0.645				
4	2.210	2.210	0.638	0.546	0.638			
5	2.210	2.210	0.633	0.535	0.538	0.633		
6	2.250	2.250	0.631	0.531	0.510	0.531	0.531	
7	2.250	2.250	0.631	0.530	0.517	0.517	0.530	0.631

TABLE 5-7   1-dB Chebyshev Capacitive-Coupled Resonators

n	q_1	q_n	k_{12}	k_{23}	k_{34}	k_{45}	k_{56}	k_{67}	k_{78}
2	0.5755	0.148	0.900						
3	0.337	2.203	1.748	0.684					
4	0.233	2.240	2.530	1.175	0.644				
5	0.394	0.275	1.910	0.750	0.650	1.987			
6	0.415	0.187	2.000	0.811	0.601	1.253	3.038		
7	0.187	0.242	3.325	1.660	1.293	0.695	0.674	2.203	
8	0.139	0.242	4.284	2.079	1.484	1.246	0.678	0.697	2.286

TABLE 5-8   Bessel Capacitive-Coupled Resonators

n	q_1	q_n	k_{12}	k_{23}	k_{34}	k_{45}	k_{56}	k_{67}	k_{78}
2	0.648	2.109	0.856						
3	0.433	2.254	1.489	0.652					
4	0.493	0.718	1.632	0.718	0.739				
5	0.547	0.446	1.800	0.848	0.584	1.372			
6	0.397	0.468	1.993	1.379	0.683	0.661	1.553		
7	0.316	0.484	2.490	1.442	1.446	0.927	0.579	1.260	
8	0.335	0.363	2.585	1.484	1.602	1.160	0.596	0.868	1.733

TABLE 5-9   Linear Phase with Equiripple Error of 0.05° Capacitive-Coupled Resonators

n	q_1	q_n	k_{12}	k_{23}	k_{34}	k_{45}	k_{56}	k_{67}	k_{78}
2	0.825	1.980	0.783						
3	0.553	2.425	1.330	0.635					
4	0.581	1.026	1.575	0.797	0.656				
5	0.664	0.611	1.779	0.919	0.576	1.162			
6	0.552	0.586	1.874	1.355	0.641	0.721	1.429		
7	0.401	0.688	2.324	1.394	1.500	1.079	0.590	1.045	
8	0.415	0.563	2.410	1.470	1.527	1.409	0.659	0.755	1.335

TABLE 5-10   Linear Phase with Equiripple Error of 0.5° Capacitive-Coupled Resonators

n	q_1	q_n	k_{12}	k_{23}	k_{34}	k_{45}	k_{56}	k_{67}	k_{78}
3	0.404	2.338	1.662	0.691					
4	0.570	0.914	1.623	0.798	0.682				
5	0.891	0.670	1.418	0.864	0.553	1.046			
6	0.883	0.752	1.172	1.029	0.595	0.605	1.094		
7	0.736	0.930	1.130	0.955	0.884	0.534	0.633	1.104	
8	0.738	0.948	1.124	0.866	0.922	0.708	0.501	0.752	1.089

TABLE 5-11   Transitional Gaussian to 6-dB Capacitive-Coupled Resonators

n	q_1	q_n	k_{12}	k_{23}	k_{34}	k_{45}	k_{56}	k_{67}	k_{78}
3	0.415	2.345	1.631	0.686					
4	0.419	0.766	1.989	0.833	0.740				
5	0.534	0.503	2.085	0.976	0.605	1.333			
6	0.543	0.558	1.839	1.442	0.686	0.707	1.468		
7	0.492	0.665	1.708	1.440	1.181	0.611	0.781	1.541	
8	0.549	0.640	1.586	1.262	1.296	0.808	0.569	1.023	1.504

TABLE 5-12   Transitional Gaussian to 12-dB Capacitive-Coupled Resonators

present in tabular form q and k parameters for all-pole filters. These parameters are used to generate the component values for filters that have the form shown in Fig. 5-14. For each network, a q_1 and q_n are given that correspond to the first and last resonant circuit, respectively. The k parameters are given in terms of k_{12}, k_{23}, and so on, and are related to the coupling capacitor shown in Fig. 5-14. The design method proceeds as follows:

1. Compute the desired filter's passband Q, which was given by Eq. (2-16).

$$Q_{bp} = \frac{f_0}{BW_{3\,dB}}$$

2. Determine the q's and k's from the tables corresponding to the chosen filter type and the order of complexity n. Denormalize these coefficients as follows:

$$Q_1 = Q_{bp} \times q_1 \tag{5-32}$$

$$Q_n = Q_{bp} \times q_n \tag{5-33}$$

$$K_{xy} = \frac{k_{xy}}{Q_{bp}} \tag{5-34}$$

3. Choose a convenient inductance value L. The source and load terminations are found from

$$R_s = \omega_0 L Q_1 \tag{5-35}$$

and $$R_L = \omega_0 L Q_n \tag{5-36}$$

4. The total nodal capacitance is determined by

$$C_{\text{node}} = \frac{1}{\omega_0^2 L} \tag{5-37}$$

The coupling capacitors are then computed from

$$C_{xy} = K_{xy}\, C_{\text{node}} \tag{5-38}$$

5. The total capacity connected to each node must be equal to C_{node}. Therefore, the shunt capacitors of the parallel-tuned circuits are equal to the total nodal capacitance C_{node}, minus the values of the coupling capacitors connected to that node. For example

$$C_1 = C_{\text{node}} - C_{12}$$

$$C_2 = C_{\text{node}} - C_{12} - C_{23}$$

$$C_7 = C_{\text{node}} - C_{67} - C_{78}$$

Each node is tuned to f_0, with the adjacent nodes shorted to ground so that the coupling capacitors connected to that node are placed in parallel across the tuned circuit.

The completed filter may require impedance scaling so that the source and load terminating requirements are met. In addition, some of the impedance transformations discussed in Chap. 8 may have to be applied.

The k and q values tabulated in Tables 5-3 to 5-12 are based on infinite inductor Q. In reality, satisfactory results will be obtained for inductor Qs several times higher than Q_{min} (band-pass), determined by Eq. (5-17) in conjunction with Fig. 3-8, which shows the minimum theoretical low-pass Qs.

Example 5-7   Design of a Capacitive-Coupled Resonator Band-Pass Filter

Required:

A band-pass filter
Center frequency of 100 kHz
3 dB at ± 2.5 kHz
35-dB minimum at ±12.5 kHz
Constant delay over the passband

Result:

(a) Since a constant delay is required, a Bessel filter type will be chosen. The low-pass constant-delay properties will undergo a minimum of distortion for the band-pass case since the bandwidth is relatively narrow—that is, the band-pass Q is high. Because the bandwidth is narrow, we can treat the requirements on an arithmetically symmetrical basis.

The band-pass steepness factor is given by Eq. (2-19) as follows:

$$A_s = \frac{\text{stopband bandwidth}}{\text{passband bandwidth}} = \frac{25\ \text{kHz}}{5\ \text{kHz}} = 5$$

The frequency-response curves of Fig. 2-56 indicate that an $n = 4$ Bessel filter provides over 35 dB of attenuation at 5 rad/s. A capacitive-coupled resonator configuration will be used for the implementation.

(b) The q and k parameters for a Bessel filter corresponding to $n = 4$ are found in Table 5-8 and are as follows:

$$q_1 = 0.233$$

$$q_4 = 2.240$$

$$k_{12} = 2.530$$

$$k_{23} = 1.175$$

$$k_{34} = 0.644$$

To denormalize these values, divide each k by the band-pass Q and multiply each q by the same factor as shown in Eq. (2-16):

$$Q_{bp} = \frac{f_0}{BW_{3\,dB}} = \frac{100\ kHz}{5\ kHz} = 20$$

The resulting values are shown using Eqs. (5-32) and (5-34)

$$Q_1 = Q_{bp} \times q_1 = 20 \times 0.233 = 4.66$$

$$Q_4 = 44.8$$

$$K_{12} = \frac{k_{12}}{Q_{bp}} = \frac{2.530}{20} = 0.1265$$

$$K_{23} = 0.05875$$

$$K_{34} = 0.0322$$

(c) Let's choose an inductance of $L = 2.5$ mH. The source and load terminations are calculated using Eqs. (5-35) and (5-36) as shown:

$$R_s = \omega_0 L Q_1 = 6.28 \times 10^5 \times 2.5 \times 10^{-3} \times 4.66 = 7.32\ k\Omega$$

and

$$R_L = \omega_0 L Q_4 = 70.37\ k\Omega$$

where

$$\omega_0 = 2\pi f_0$$

(d) The total nodal capacitance is determined by Eq. (5-37) as shown:

$$C_{node} = \frac{1}{\omega_0^2 L} = 1{,}013\ pF$$

The coupling capacitors can now be calculated using Eq. (5-38):

$$C_{12} = K_{12}\, C_{node} = 0.1265 \times 1.013 \times 10^{-9} = 128.1\ pF$$

$$C_{23} = K_{23}\, C_{node} = 59.5\ pF$$

$$C_{34} = K_{34}\, C_{node} = 32.6\ pF$$

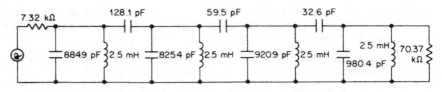

FIGURE 5-15   The capacitive coupled resonator filter of Example 5-7.

The shunt capacitors are determined from

$$C_1 = C_{\text{node}} - C_{12} = 884.9 \text{ pF}$$

$$C_2 = C_{\text{node}} - C_{12} - C_{23} = 825.4 \text{ pF}$$

$$C_3 = C_{\text{node}} - C_{23} - C_{34} = 920.9 \text{ pF}$$

$$C_4 = C_{\text{node}} - C_{34} = 980.4 \text{ pF}$$

The final circuit is shown in Fig. 5-15.

5.1.7   Predistorted Band-Pass Filters

The inductor Q requirements of the band-pass filters are higher than those of low-pass filters since the minimum theoretical branch Q is given by Eq. (5-17) as

$$Q_{\min} \text{ (band-pass)} = Q_{\min} \text{ (low-pass)} \times Q_{\text{bp}}$$

where $Q_{\text{bp}} = f_0/\text{BW}_{3\,\text{dB}}$. In cases where the required filter is extremely narrow, a branch Q many times higher than the minimum theoretical Q may be difficult to obtain. Predistorted band-pass filters can then be used so that exact theoretical results can be obtained with reasonable branch Qs.

Predistorted band-pass filters can be obtained from the normalized predistorted low-pass filters given in Chap. 10 by the conventional band-pass transformation. The low-pass filters must be of the uniform dissipation type, since the lossy-L networks would be transformed to a band-pass filter with losses in the series branches only.

The uniform dissipation networks are tabulated for different values of dissipation factor d. These values relate to the required inductor Q by the relationship

$$Q_L = \frac{Q_{\text{bp}}}{d} \tag{5-39}$$

where $Q_{\text{bp}} = f_0/\text{BW}_{3\,\text{dB}}$.

The losses of a predistorted low-pass filter with uniform dissipation are evenly distributed and occur as both series losses in the inductors and shunt losses across the capacitors. The equivalent circuit of the filter is shown in Fig. 5-16a. The inductor losses were previously given by Eq. (3-2) as

$$R_L = \frac{\omega L}{Q}$$

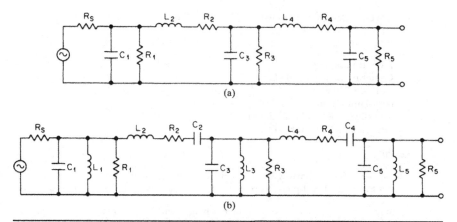

FIGURE 5-16 The location of losses in uniformly predistorted filters: (a) a low-pass filter; and (b) a band-pass filter.

and the capacitor losses were defined by Eq. (3-3) as

$$R_c = \frac{Q}{\omega C}$$

When the circuit is transformed to a band-pass filter, the losses still need to be distributed in series with the series branches and in parallel with the shunt branches, as shown in Fig. 5-16b. In reality, the capacitor losses are minimal and the inductor losses occur in series with the inductive elements in both the series and shunt branches. Therefore, as a narrowband approximation, the losses may be distributed between the capacitors and inductors in an arbitrary manner. The only restriction is that the combination of inductor and capacitor losses in each branch results in a total branch Q equal to the value computed by Eq. (5-39). The combined Q of a lossy inductor and a lossy capacitor in a resonant circuit is given by

$$Q_T = \frac{Q_L Q_C}{Q_L + Q_C} \tag{5-40}$$

where Q_T is the total branch Q, Q_L is the inductor Q, and Q_C is the Q of the capacitor.

The predistorted networks tabulated in Chap. 10 require an infinite termination on one side. In practice, if the resistance used to approximate the infinite termination is large compared with the source termination, satisfactory results will be obtained. If the dual configuration is used, which ideally requires a zero impedance source, the source impedance should be much less than the load termination.

It is usually difficult to obtain inductor Qs precisely equal to the values computed from Eq. (5-39). A Q accuracy within 5 or 10 percent at f_0 is usually sufficient. If greater accuracy is required, an inductor Q higher than the calculated value is used. The Q is then degraded to the exact required value by adding resistors.

Example 5-8   Design of a Predistorted *LC* Band-Pass Filter

Required:

A band-pass filter
Center frequency of 10 kHz
3 dB at ± 250 Hz
60-dB minimum at ± 750 Hz
$R_s = 100\ \Omega$ $R_L = 10\ k\Omega$ minimum
An available inductor Q of 225

Result:

(a) Since the filter is narrow in bandwidth, the requirement is treated in its arithmetically symmetrical form. The band-pass steepness factor is obtained from Eq. (2-19) as follows:

$$A_s = \frac{\text{stopband bandwidth}}{\text{passband bandwidth}} = \frac{1,500\ \text{Hz}}{500\ \text{Hz}} = 3$$

The curves of Fig. 2-43 indicate that a fifth-order ($n = 5$) 0.25-dB Chebyshev filter will meet these requirements. A predistorted design will be used. The corresponding normalized low-pass filters are found in Table 10-33.

(b) The specified inductor Q can be used to compute the required d of the low-pass filter as shown in Eq. (5-39):

$$d = \frac{Q_{bp}}{Q_L} = \frac{20}{225} = 0.0889$$

where $Q_{bp} = f_0 / BW_{3\ dB}$. The circuit corresponding to $n = 5$ and $d = 0.0919$ will be selected since this d is sufficiently close to the computed value. The schematic is shown in Fig. 5-17a.

(c) Using Eqs. (2-10) and (2-9), denormalize the low-pass filter using a frequency-scaling factor (FSF) of $2\pi f_c$ or 3,140, where $f_c = 500$ Hz, the required bandwidth of the band-pass filter, and an impedance-scaling factor Z of 100.

$$C_1' = \frac{C}{FSF \times Z} = \frac{1.0397}{3,140 \times 100} = 3.309\ \mu F$$

$$C_3' = 7.014\ \mu F$$

$$C_5' = 3.660\ \mu F$$

and

$$L_2' = \frac{L \times Z}{FSF} = \frac{1.8181 \times 100}{3,140} = 57.87\ \text{mH}$$

$$L_4' = 55.79\ \text{mH}$$

The denormalized low-pass filter is shown in Fig. 5-17b, where the termination has been scaled to 100 Ω. The filter has also been turned end for end since the high-impedance termination is required at the output and the 100-Ω source at the input.

(d) To transform the circuit into a band-pass filter, resonate each capacitor with an inductor in parallel and each inductor with a series capacitor using a resonant frequency of $f_0 = 10$ kHz. The parallel inductor is computed from Eq. (5-11) as shown:

$$L = \frac{1}{\omega_0^2 C}$$

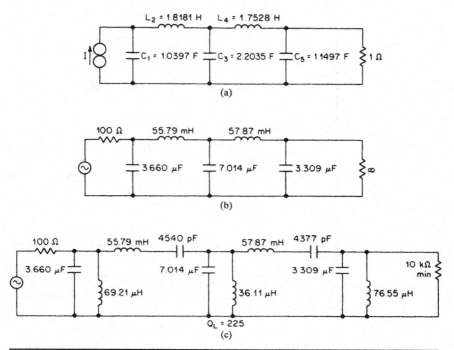

FIGURE 5-17 The predistorted band-pass filter of Example 5-8: (a) normalized low-pass filter; (b) frequency- and impedance-scaled network; and (c) resulting band-pass filter.

and the series capacitor is calculated by Eq. (5-12) as shown:

$$C = \frac{1}{\omega_0^2 L}$$

where both formulas are forms of the general relationship for resonance: $\omega_0^2 LC = 1$. The resulting band-pass filter is given in Fig. 5-17c. The large spread of values can be reduced by applying some of the techniques later discussed in Chap. 8.

5.1.8 Elliptic-Function Band-Pass Filters

Elliptic-function low-pass filters were clearly shown to be far superior to the other filter types in terms of achieving a required attenuation within a given frequency ratio. This superiority is mainly the result of the presence of transmission zeros beginning just outside the passband.

Elliptic-function LC low-pass filters have been extensively tabulated by Saal and Ulbrich and by Zverev (see References). A program called *Filter Solutions Book Version* can be downloaded from the companion website (see App. A) and allows the design of elliptic-function LC filters up to $n = 10$. A second program, ELI 1.0, can be used for odd-order elliptic-function filters up to the 31st order. These networks can be transformed into band-pass filters in the same manner as the all-pole filter types. The elliptic-function band-pass filters will then exhibit the same superiority over the all-pole types as their low-pass counterparts.

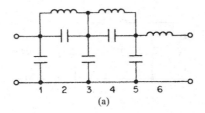

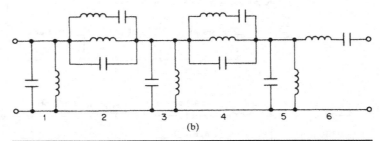

FIGURE 5-18   The low-pass to band-pass transformation of an elliptic-function filter: (a) $n = 6$ low-pass filter; and (b) transformed band-pass configuration.

When an elliptic-function low-pass filter is transformed into a band-pass filter, each low-pass transmission zero is converted into a pair of zeros, one above and one below the passband, and are geometrically related to the center frequency. (For the purposes of this discussion, negative zeros will be disregarded.) The low-pass zeros are directly determined by the resonances of the parallel-tuned circuits in the series branches. When each series branch containing a parallel-tuned circuit is modified by the band-pass transformation, two parallel branch resonances are introduced corresponding to the upper and lower zeros.

A sixth-order elliptic-function low-pass filter structure is shown in Fig. 5-18a. After frequency- and impedance-scaling the low-pass values, we can make a band-pass transformation by resonating each inductor with a series capacitor and each capacitor with a parallel inductor, where the resonant frequency is f_0, the filter center frequency. The circuit in Fig. 5-18b results. The configuration obtained in branches 2 and 4 corresponds to a type III network from Table 5-2.

The type III network realizes two parallel resonances corresponding to a geometrically related pair of transmission zeros above and below the passband. The circuit configuration itself is not very desirable. The elements corresponding to both parallel resonances are not distinctly isolated. Each resonance is determined by the interaction of a number of elements, so tuning is difficult. Also, for very narrow filters, the values may become unreasonable. Fortunately, an alternate circuit exists that provides a more practical relationship between the coils and capacitors. The two equivalent configurations are shown in Fig. 5-19. The alternative configuration utilizes two parallel-tuned circuits where each condition of parallel resonance directly corresponds to a transmission zero.

The type III network in Fig. 5-19 is shown with reciprocal element values. These result when we normalize the band-pass filter to a center frequency of 1 rad/s. Since the

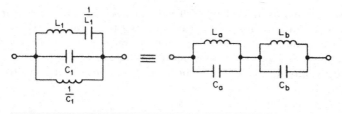

FIGURE 5-19   Equivalent circuit of a type III network.

general equation for resonance $\omega_0^2 LC = 1$ reduces to $LC = 1$ at $\omega_0 = 1$, the resonant elements become reciprocals of each other.

The reason for this normalization is to greatly simplify the equations for the transformation shown in Fig. 5-19. Otherwise, the equations relating the two circuits would be significantly more complex. Therefore, elliptic-function band-pass filters are first designed, normalized, and then scaled to the required center frequency and impedance.

To obtain the normalized band-pass filter, first multiply all L and C values of the normalized low-pass filter by Q_{bp}, which is equal to f_0/BW, where BW is the passband bandwidth. The network can then be transformed directly into a normalized band-pass filter by resonating each inductor with a series capacitor and each capacitor with a parallel inductor. The resonant elements are merely reciprocals of each other since $\omega_0 = 1$.

The transformation of Fig. 5-19 can now be performed. First calculate

$$\beta = 1 + \frac{1}{2 L_1 C_1} + \sqrt{\frac{1}{4 L_1^2 C_1^2} + \frac{1}{L_1 C_1}} \tag{5-41}$$

The values are then obtained from

$$L_a = \frac{1}{C_1 (\beta + 1)} \tag{5-42}$$

$$L_b = \beta L_a \tag{5-43}$$

$$C_a = \frac{1}{L_b} \tag{5-44}$$

and

$$C_b = \frac{1}{L_a} \tag{5-45}$$

The resonant frequencies are given by

$$\Omega_{\infty,a} = \sqrt{\beta} \tag{5-46}$$

and

$$\Omega_{\infty,b} = \frac{1}{\Omega_{\infty,a}} \tag{5-47}$$

After the transformation of Fig. 5-19 is made wherever applicable, the normalized band-pass filter is scaled to the required center frequency and impedance level by multiplying all inductors by Z/FSF and dividing all capacitors by $Z \times FSF$. The frequency-scaling factor in this case is equal to ω_0 ($\omega_0 = 2\pi f_0$), where f_0 is the desired center frequency of the filter. The resonant frequencies in hertz can be found by multiplying all normalized radian resonant frequencies by f_0.

The design of an elliptic-function band-pass filter is demonstrated by the following example.

Example 5-9   Designing an LC Elliptic-Function Band-Pass Filter

Required:

A band-pass filter
1-dB maximum variation from 15 to 20 kHz
50-dB minimum below 14.06 kHz and above 23 kHz
$R_s = R_L = 10\text{ k}\Omega$

Result:

(a) Convert to a geometrically symmetrical band-pass requirement:

First, calculate the geometric center frequency using Eq. (2-14).

$$f_0 = \sqrt{f_L f_u} = \sqrt{15 \times 20 \times 10^6} = 17.32\text{ kHz}$$

Compute the corresponding geometric frequency for each stopband frequency given, using the relationship shown in Eq. (2-18).

$$f_1 f_2 = f_0^2$$

f_1	f_2	$f_2 - f_1$
14.6 kHz	21.34 kHz	7.28 kHz
13.04 kHz	23.00 kHz	9.96 kHz

The first pair of frequencies has the lesser separation and, therefore, represents the more severe requirement. Thus, it will be retained. The geometrically symmetrical requirements can be summarized as

$$f_0 = 17.32\text{ kHz}$$

$$\text{BW}_{1\text{ dB}} = 5\text{ kHz}$$

$$\text{BW}_{50\text{ dB}} = 7.28\text{ kHz}$$

(b) Compute the band-pass steepness factor using Eq. (2-19).

$$A_s = \frac{\text{stopband bandwidth}}{\text{passband bandwidth}} = \frac{7.28\text{ kHz}}{5\text{ kHz}} = 1.456$$

(c) Open *Filter Solutions*.
   Check the *Stop Band Freq* box.
   Enter **0.18** in the *Pass Band Ripple (dB)* box.
   Enter **1** in *Pass Band Freq* box.
   Enter **1.456** in the *Stop Band Freq* box.
   Check the *Frequency Scale Rad/Sec* box.
   Enter **1** for *Source Res* and *Load Res*.

(d) Click the *Set Order* control button to open the second panel.
Enter **50** for the *Stop Band Attenuation (dB)*.
Click the *Set Minimum Order* button and then click *Close*.
6 Order is displayed on the main control panel.
Check the *Even Order Mod* box.

(e) Click the *Synthesize Filter* button.

Two schematics are presented. Select **Lumped Filter 1**, shown in Fig. 5-20a.

(f) The filter must now be converted to a normalized band-pass filter with a center frequency of $\omega_0 = 1$. The band-pass Q is first computed from

$$Q_{bp} = \frac{f_0}{BW} = \frac{17.32 \text{ kHz}}{5 \text{ kHz}} = 3.464$$

Multiply all inductance and capacitance values by Q_{bp}. Then, transform the network into a band-pass filter centered at $\omega_0 = 1$ by resonating each capacitor with a parallel inductor and each inductor with a series capacitor. The resonating elements introduced are simply the reciprocal values, as shown in Fig. 5-20b.

(g) The type III branches will now be transformed in accordance with Fig. 5-19.

For the third branch

$$L_1 = 4.451 \text{ H} \quad C_1 = 1.199 \text{ F}$$

First, compute using Eq. (5-41).

$$\beta = 1 + \frac{1}{2L_1C_1} + \sqrt{\frac{1}{4L_1^2C_1^2} + \frac{1}{L_1C_1}} = 1.5366$$

then use Eqs. (5-42) to (5-45).

$$L_a = \frac{1}{C_1(\beta + 1)} = 0.3288 \text{ H}$$

$$L_b = \beta L_a = 0.5052 \text{ H}$$

$$C_a = \frac{1}{L_b} = 1.9793 \text{ F}$$

$$C_b = \frac{1}{L_a} = 3.0414 \text{ F}$$

The resonant frequencies are calculated as shown in Eqs. (5-46) and (5-47).

$$\Omega_{\infty,a} = \sqrt{\beta} = 1.2396$$

$$\Omega_{\infty,b} = \frac{1}{\Omega_{\infty,b}} = 0.8067$$

For the fifth branch:

$$L_1 = 4.43 \text{ H} \quad C_1 = 0.684 \text{ F}$$

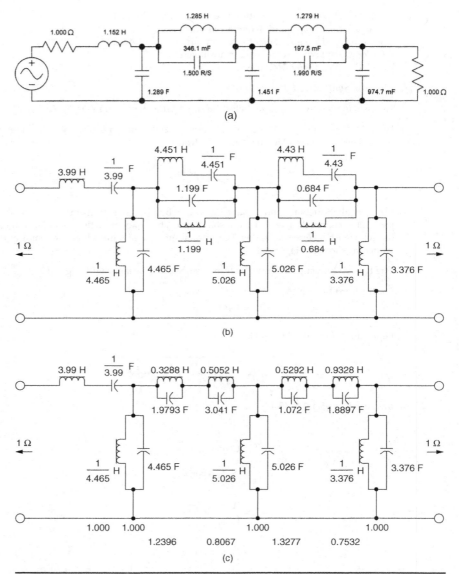

FIGURE 5-20 The elliptic-function band-pass filter: (*a*) normalized low-pass filter; and (*b*) band-pass filter normalized to $\omega_0 = 1$. An elliptic-function band-pass filter; (*c*) transformed type III branches; (*d*) final scaled circuit; and (*e*) frequency response.

then

$$\beta = 1.7627$$

$$L_a = 0.5292 \text{ H}$$

$$L_b = 0.9328 \text{ H}$$

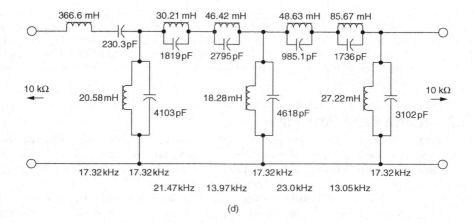

(d)

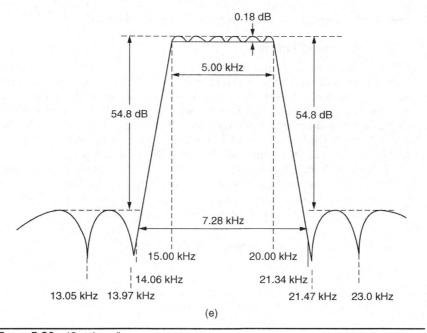

(e)

Figure 5-20   (*Continued*).

$$C_a = 1.072 \text{ F}$$

$$C_b = 1.8897 \text{ F}$$

$$\Omega_{\infty, a} = 1.3277$$

$$\Omega_{\infty, a} = 0.7532$$

The transformed filter is shown in Fig. 5-20c. The resonant frequencies in radians per second are indicated below the schematic.

(*h*) To complete the design, denormalize the filter to a center frequency (f_0) of 17.32 kHz and an impedance level of 10 kΩ. Multiply all inductors by Z/FSF and divide all capacitors by $Z \times \text{FSF}$, where $Z = 10^4$ and $\text{FSF} = 2\pi f_0$ or 1.0882×10^5. The final filter is shown in Fig. 5-20*d*. The resonant frequencies were obtained by directly multiplying the normalized resonant frequencies of Fig. 5-20*c* by the geometric center frequency f_0 of 17.32 kHz. The frequency response is shown in Fig. 5-20*e*.

5.2   Active Band-Pass Filters

When the separation between the upper and lower cutoff frequencies exceeds a ratio of approximately 2, the band-pass filter is considered a wideband type. The specifications are then separated into individual low-pass and high-pass requirements and met by a cascade of active low-pass and high-pass filters.

Narrowband *LC* band-pass filters are usually designed by transforming a low-pass configuration directly into the band-pass circuit. Unfortunately, no such circuit transformation exists for active networks. The general approach involves transforming the low-pass transfer function into a band-pass type. The band-pass poles and zeros are then implemented by a cascade of band-pass filter sections.

5.2.1   Wideband Filters

When LC low-pass and high-pass filters were cascaded, care had to be taken to minimize terminal impedance variations so that each filter maintained its individual response in the cascaded form. Active filters can be interconnected with no interaction because of the inherent buffering of the operational amplifiers. The only exception occurs in the case of the elliptic-function voltage-controlled voltage source (VCVS) filters of Secs. 3.2 and 4.2, where the last section of each filter is followed by an *RC* network to provide the real poles. An amplifier must then be introduced for isolation.

Figure 5-21 shows two simple amplifier configurations that can be used after an active elliptic-function VCVS filter. The gain of the voltage follower is unity. The noninverting amplifier has a gain equal to $R_2/R_1 + 1$. The resistors R_1 and R_2 can have any convenient values since only their ratio is of significance.

Example 5-10   Designing a Wideband Active Band-Pass Filter

Required:

An active band-pass filter
1-dB maximum variation from 3,000 to 9,000 Hz

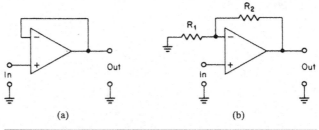

(a)   (b)

Figure 5-21   Isolation amplifiers: (*a*) voltage follower; and (*b*) noninverting amplifier.

35-dB minimum below 1,000 Hz and above 18,000 Hz
Gain of +20 dB

Result:

(*a*) Since the ratio of upper cutoff frequency to lower cutoff frequency is well in excess of an octave, the design will be treated as a cascade of low-pass and high-pass filters. The frequency-response requirement can be restated as the following set of individual low-pass and high-pass specifications:

High-pass filter:	*Low-pass filter:*
1-dB maximum at 3,000 Hz	1-dB maximum at 9,000 Hz
35-dB minimum below 1,000 Hz	35-dB minimum above 18,000 Hz

(*b*) To design the high-pass filter, first compute the high-pass steepness factor using Eq. (2-13).

$$A_s = \frac{f_c}{f_s} = \frac{3,600 \text{ Hz}}{1,000 \text{ Hz}} = 3$$

A normalized low-pass filter must now be chosen that makes the transition from less than 1 dB to more than 35 dB within a frequency ratio of 3:1. An elliptic-function type will be selected. The high-pass filter designed in Example 4-5 illustrates this process and will meet this requirement. The circuit is shown in Fig. 5-22*a*.

(*c*) The low-pass filter is now designed. The low-pass steepness factor is computed by Eq. (2-11) as follows:

$$A_s = \frac{f_s}{f_c} = \frac{18,000 \text{ Hz}}{9,000 \text{ Hz}} = 2$$

A low-pass filter must be selected that makes the transition from less than 1 dB to more than 35 dB within a frequency ratio of 2:1. The curves of Fig. 2-44 indicate that the attenuation of a normalized 0.5-dB Chebyshev filter of a complexity of $n = 5$ is less than 1 dB at 0.9 rad/s and more than 35 dB at 1.8 rad/s, which satisfies the requirements. The corresponding active filter is found in Table 10-39 and is shown in Fig. 5-22*b*.

To denormalize the low-pass circuit, first compute the FSF, which is given by Eq. (2-1) as shown:

$$\text{FSF} = \frac{\text{desired reference frequency}}{\text{existing reference frequency}}$$

$$= \frac{2\pi \times 9,000 \text{ rad/s}}{0.9 \text{ rad/s}} = 62,830$$

The filter is then denormalized by dividing all capacitors by $Z \times \text{FSF}$ and multiplying all resistors by Z, where Z is arbitrarily chosen at 10^4. The denormalized circuit is shown in Fig. 5-22*c*.

(*d*) To complete the design, the low-pass and high-pass filters are cascaded. Since the real-pole *RC* network of the elliptic high-pass filter must be buffered, and since a gain of +20 dB is required, the noninverting amplifier of Fig. 5-21*b* will be used.

The finalized design is shown in Fig. 5-22*d*, where the resistors have been rounded off to standard 1 percent values.

5.2.2 The Band-Pass Transformation of Low-Pass Poles and Zeros

Active band-pass filters are designed directly from a band-pass transfer function. To obtain the band-pass poles and zeros from the low-pass transfer function, a low-pass to

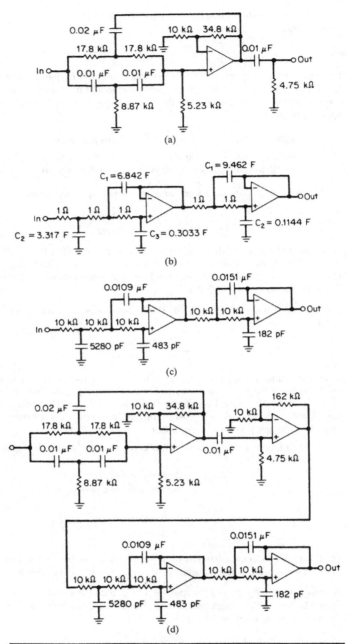

FIGURE 5-22   The wideband band-pass filter of Example 5-10:
(a) the high-pass filter of Example 4-5; (b) normalized low-pass
filter; (c) denormalized low-pass filter; and (d) band-pass filter
configuration.

band-pass transformation must be performed. It was shown in Sec. 5.1 how this transformation can be accomplished by replacing the frequency variable with a new variable, which was given by Eq. (5-1) as

$$f_{bp} = f_0 \left(\frac{f}{f_0} - \frac{f_0}{f} \right)$$

This substitution maps the low-pass frequency response into a band-pass magnitude characteristic.

Two sets of band-pass poles are obtained from each low-pass complex pole pair. If the low-pass pole is real, a single pair of complex poles results for the band-pass case. Also, each pair of imaginary axis zeros is transformed into two pairs of conjugate zeros. This is shown in Fig. 5-23.

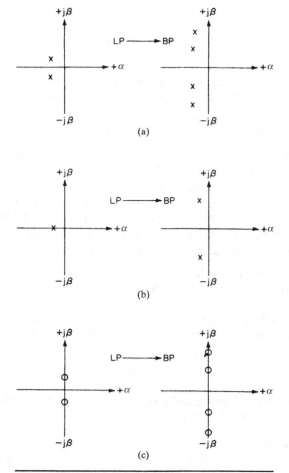

FIGURE 5-23 A low-pass to band-pass transformation: (a) low-pass complex pole pair; (b) low-pass real pole; and (c) low-pass pair of imaginary zeros.

Clearly, the total number of poles and zeros is doubled when the band-pass transformation is performed. However, it is conventional to disregard the conjugate band-pass poles and zeros below the real axis. An n-pole low-pass filter is said to result in an nth-order band-pass filter even though the band-pass transfer function is of the order $2n$. An nth-order active band-pass filter will then consist of n band-pass sections.

Each all-pole band-pass section has a second-order transfer function given by

$$T(s) = \frac{Hs}{s^2 + \dfrac{\omega_r}{Q}s + \omega_r^2} \tag{5-48}$$

where ω_r is equal to $2\pi f_r$, the pole resonant frequency in radians per second, Q is the band-pass section Q, and H is a gain constant.

If transmission zeros are required, the section transfer function will then take the form

$$T(s) = \frac{H\left(s^2 + \omega_\infty^2\right)}{s^2 + \dfrac{\omega_r}{Q}s + \omega_r^2} \tag{5-49}$$

where ω_∞ is equal to $2\pi f_\infty$, the frequency of the transmission zero in radians per second. Active band-pass filters are designed by the following sequence of operations:

1. Convert the band-pass specifications to a geometrically symmetrical requirement as described in Sec. 2.1.

2. Calculate the band-pass steepness factor A_s using Eq. (2-19), and select a normalized filter type from Chap. 2.

3. Look up the corresponding normalized poles (and zeros) from the tables of Chap. 10, or use the *Filter Solutions* program for elliptic-function filters and transform these coordinates into band-pass parameters.

4. Select the appropriate band-pass circuit configuration from the types presented in this chapter, and cascade the required number of sections.

It is convenient to specify each band-pass filter section in terms of its center frequency and Q. Elliptic-function filters will require zeros. These parameters can be directly transformed from the poles (and zeros) of the normalized low-pass transfer function. A numerical procedure will be described for making this transformation.

First make the preliminary calculation using Eq. (2-16).

$$Q_{bp} = \frac{f_0}{BW}$$

where f_0 is the geometric band-pass center frequency and BW is the passband bandwidth. The band-pass transformation is made in the following manner.

Complex Poles
Complex poles can be found in the tables of Chap. 10, having the form

$$-\alpha \pm J\beta$$

where α is the real coordinate and β is the imaginary part.

When using the *Filter Solutions* program to design elliptic-function filters, the program provides the low-pass parameters Q and ω_0. These two parameters can be converted into α and β by using

$$\alpha = \frac{\omega_0}{2Q}$$

$$\beta = \sqrt{\omega_0^2 - \alpha^2}$$

Given α, β, Q_{bp}, and f_0, the following series of calculations results in two sets of values for Q and center frequencies that defines a pair of band-pass filter sections:

$$C = \alpha^2 + \beta^2 \tag{5-50}$$

$$D = \frac{2\alpha}{Q_{bp}} \tag{5-51}$$

$$E = \frac{C}{Q_{bp}^2} + 4 \tag{5-52}$$

$$G = \sqrt{E^2 - 4D^2} \tag{5-53}$$

$$Q = \sqrt{\frac{E + G}{2D^2}} \tag{5-54}$$

$$M = \frac{\alpha Q}{Q_{bp}} \tag{5-55}$$

$$W = M + \sqrt{M^2 - 1} \tag{5-56}$$

$$f_{ra} = \frac{f_0}{W} \tag{5-57}$$

$$f_{rb} = W f_0 \tag{5-58}$$

The two band-pass sections have resonant frequencies of f_{ra} and f_{rb} (in hertz) and identical Qs as given by Eq. (5-54).

Real Poles
A normalized low-pass real pole with a real coordinate of magnitude a_0 is transformed into a single band-pass section with a Q defined by

$$Q = \frac{Q_{bp}}{\alpha_0} \tag{5-59}$$

The section is tuned to f_0, the geometric center frequency of the filter.

Imaginary Zeros. Elliptic-function low-pass filters contain transmission zeros of the form $\pm j\omega_\infty$ as well as poles. These zeros must be transformed along with the poles when a band-pass filter is required. The band-pass zeros can be obtained as follows:

$$H = \frac{\omega_\infty^2}{2Q_{bp}^2} + 1 \tag{5-60}$$

$$Z = \sqrt{H + \sqrt{H^2 - 1}} \tag{5-61}$$

$$f_{\infty,a} = \frac{f_0}{Z} \tag{5-62}$$

$$f_{\infty,b} = Z \times f_0 \tag{5-63}$$

A pair of imaginary band-pass zeros is obtained that occurs at $f_{\infty,a}$ and $f_{\infty,b}$ (in hertz) from each low-pass zero.

Determining Section Gain. The gain of a single band-pass section at the filter geometric center frequency f_0 is given by

$$A_0 = \frac{A_r}{\sqrt{1 + Q^2\left(\dfrac{f_0}{f_r} - \dfrac{f_r}{f_0}\right)^2}} \tag{5-64}$$

where A_r is the section gain at its resonant frequency f_r. The section gain will always be less at f_0 than at f_r since the circuit is peaked to f_r, except for transformed real poles, where $f_r = f_0$. Equation (5-64) will then simplify to $A_0 = A_r$. The composite filter gain is determined by the product of the A_0 values of all the sections.

If the section Q is relatively high ($Q > 10$), Eq. (5-64) can be simplified to

$$A_0 = \frac{A_r}{\sqrt{1 + \left(\dfrac{2Q\Delta f}{f_r}\right)^2}} \tag{5-65}$$

where Δf is the frequency separation between f_0 and f_r.

Example 5-11   Computing Band-Pass Pole Locations and Section Gains

Required:

Determine the pole locations and section gains for a third-order Butterworth band-pass filter with a geometric center frequency of 1,000 Hz, a 3-dB bandwidth of 100 Hz, and a midband gain of + 30 dB.

Result:

(a) The normalized pole locations for an $n = 3$ Butterworth low-pass filter are obtained from Table 10-1 and are

$-0.500 \pm j0.8660$
-1.000

To obtain the band-pass poles, first compute the following using Eq. (2-16)

$$Q_{bp} = \frac{f_0}{BW_{3\,dB}} = \frac{1,000\ Hz}{100\ Hz} = 10$$

The low-pass to band-pass pole transformation is performed as follows, using Eqs. (5-50) to (5-59):

Complex pole:

$$\alpha = 0.500 \qquad \beta = 0.8660$$

$$C = \alpha^2 + \beta^2 = 1.000000 \tag{5-50}$$

$$D = \frac{2\alpha}{Q_{bp}} = 0.1000000 \tag{5-51}$$

$$E = \frac{C}{Q_{bp}^2} + 4 = 4.010000 \tag{5-52}$$

$$G = \sqrt{E^2 - 4D^2} = 4.005010 \tag{5-53}$$

$$Q = \sqrt{\frac{E + G}{2D^2}} = 20.018754 \tag{5-54}$$

$$M = \frac{\alpha Q}{Q_{bp}} = 1.000938 \tag{5-55}$$

$$W = M + \sqrt{M^2 - 1} = 1.044261 \tag{5-56}$$

$$f_{ra} = \frac{f_0}{W} = 957.6\ Hz \tag{5-57}$$

$$f_{rb} = W f_0 = 1,044.3\ Hz \tag{5-58}$$

Real pole:

$$\alpha_0 = 1.0000$$

$$Q = \frac{Q_{bp}}{\alpha_0} = 10 \tag{5-59}$$

$$f_r = f_0 = 1,000\ Hz$$

(*b*) Since a composite midband gain of +30 dB is required, let us distribute the gain uniformly among the three sections. Therefore, $A_0 = 3.162$ for each section corresponding to 110 dB.

The gain at section resonant frequency f_r is obtained from the following form of Eq. (5-64):

$$A_r = A_0 \sqrt{1 + Q^2 \left(\frac{f_0}{f_r} - \frac{f_r}{f_0} \right)^2}$$

The resulting values are

Section 1:

$$f_r = 957.6 \text{ Hz}$$
$$Q = 20.02$$
$$A_r = 6.333$$

Section 2:

$$f_r = 1,044.3 \text{ Hz}$$
$$Q = 20.02$$
$$A_r = 6.335$$

Section 3:

$$f_r = 1,000.0 \text{ Hz}$$
$$Q = 10.00$$
$$A_r = 3.162$$

The block diagram of the realization is shown in Fig. 5-24.

The calculations required for the band-pass pole transformation should be maintained to more than four significant figures after the decimal point to obtain accurate results, since differences of close numbers are involved. Equations (5-55) and (5-56) are especially critical, so the value of M should be computed to five or six places after the decimal point.

5.2.3 Sensitivity in Active Band-Pass Circuits

Sensitivity defines the amount of change of a dependent variable that results from the variation of an independent variable. Mathematically, the sensitivity of y with respect to x is expressed as

$$S_x^y = \frac{dy/y}{dx/x} \tag{5-66}$$

Sensitivity is used as a figure of merit to measure the change in a particular filter parameter, such as Q, or the resonant frequency for a given change in a component value.

Deviations of components from their nominal values occur because of the effects of temperature, aging, humidity, and other environmental conditions in addition to errors due to tolerances. These variations cause changes in parameters such as Q and the center frequency from their design values.

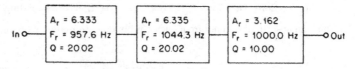

FIGURE 5-24 A block realization of Example 5-11.

For example, let's assume we are given the parameter $S_{R_1}^Q = -3$ for a particular circuit. This means that for a 1 percent increment of R_1, the circuit Q will change 3 percent in the opposite direction.

In addition to component value sensitivity, the operation of a filter is dependent on the active elements. The Q and resonant frequency can be a function of amplifier open-loop gain and phase shift, so the sensitivity to these active parameters is useful in determining an amplifier's suitability for a particular design.

The Q sensitivity of a circuit is a good measure of its stability. With some circuits, the Q can increase to infinity, which implies self-oscillation. Low Q sensitivity of a circuit usually indicates that the configuration will be practical from a stability point of view.

Sometimes the sensitivity is expressed as an equation instead of a numerical value, such as $S_A^Q = 2Q^2$. This expression implies that the sensitivity of Q with respect to amplifier gain A increases with Q^2, so the circuit is not suitable for high Q realizations.

The frequency-sensitivity parameters of a circuit are useful in determining whether the circuit will require resistive trimming and indicate which element should be made variable. It should be mentioned that, in general, only resonant frequency is made adjustable when the band-pass filter is sufficiently narrow. Q variations of 5 or 10 percent are usually tolerable, whereas a comparable frequency error would be disastrous in narrow filters. However, in the case of a state-variable realization, a Q-enhancement effect occurs that's caused by amplifier phase shift. The Q may increase very dramatically, so Q adjustment is usually required in addition to resonant frequency.

5.2.4 All-Pole Band-Pass Configurations
Multiple-Feedback Band-Pass
The circuit in Fig. 5-25a realizes a band-pass pole pair and is commonly referred to as a multiple-feedback band-pass (MFBP) configuration. This circuit features a minimum number of components and a low sensitivity to component tolerances. The transfer function is given by

$$T(s) = \frac{sC/R_1}{s^2C^2 + s2C/R_2 + 1/R_1R_2} \tag{5-67}$$

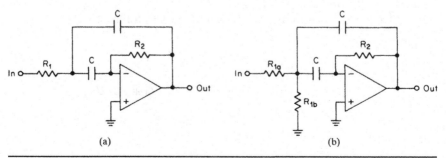

(a)   (b)

FIGURE 5-25 A multiple-feedback band-pass (MFBP) ($Q < 20$): (a) MFBP basic circuit; and (b) modified configuration.

If we equate the coefficients of this transfer function with the general band-pass transfer function of Eq. (5-48), we can derive the following expressions for the element values:

$$R_2 = \frac{Q}{\pi f_r C} \tag{5-68}$$

and

$$R_1 = \frac{R_2}{4Q^2} \tag{5-69}$$

where C is arbitrary.

The circuit gain at resonant frequency f_r is given by

$$A_r = 2Q^2 \tag{5-70}$$

The open-loop gain of the operational amplifier at f_r should be well in excess of $2Q^2$ so that the circuit performance is controlled mainly by the passive elements. This requirement places a restriction on realizable Qs to values typically below 20, depending upon the amplifier type and frequency range.

Extremely high gains occur for moderate Q values because of the Q^2 gain proportionality. Thus, there will be a tendency for clipping at the amplifier output with moderate input levels. Also, the circuit gain is fixed by the Q, which limits flexibility.

An alternate and preferred form of the circuit is shown in Fig. 5-25b. The input resistor R_1 has been split into two resistors, R_{1a} and R_{1b}, to form a voltage divider so that the circuit gain can be controlled. The parallel combination of the two resistors is equal to R_1 in order to retain the resonant frequency. The transfer function of the modified circuit is given by

$$T(s) = -\frac{sR_2C}{s^2R_{1a}R_2C^2 + s2R_{1a}C + (1 + R_{1a}/R_{1b})} \tag{5-71}$$

The values of R_{1a} and R_{1b} are computed from

$$R_{1a} = \frac{R_2}{2A_r} \tag{5-72}$$

and

$$R_{1b} = \frac{R_2/2}{2Q^2 - A_r} \tag{5-73}$$

where A_r is the desired gain at resonant frequency f_r and cannot exceed $2Q^2$. The value of R_2 is still computed from Eq. (5-68).

The circuit sensitivities can be determined as follows:

$$S_{R_{1a}}^{Q} = S_{R_{1a}}^{f_r} = \frac{A_r}{4Q^2} \tag{5-74}$$

$$S_{R_{1b}}^{Q} = S_{R_{1b}}^{f_r} = \frac{1}{2}(1 + A_r/2Q^2) \tag{5-75}$$

$$S_{R_2}^{f_r} = S_C^{f_r} = -\frac{1}{2} \tag{5-76}$$

$$S_{R_2}^{Q} = \frac{1}{2} \tag{5-77}$$

For $Q^2/A \gg 1$, the resonant frequency can be directly controlled by R_{1b} since $S_{R_{1b}}^{fr}$ approaches ½. To use this result, let's assume that the capacitors have 2 percent tolerances and the resistors have a tolerance of 1 percent, which could result in a possible 3 percent frequency error. If frequency adjustment is desired, R_{1b} should be made variable over a minimum resistance range of ±6 percent. This would then permit a frequency adjustment of ±3 percent, since $S_{R_{1b}}^{fr}$ is equal to ½. Resistor R_{1b} should be composed of a fixed resistor in series with a single-turn potentiometer to provide good resolution.

Adjustment of Q can be accomplished by making R_2 adjustable. However, this will affect resonant frequency, and in any event is not necessary for most filters if 1 or 2 percent tolerance parts are used. The section gain can be varied by making R_{1a} adjustable, but again, resonant frequency may be affected.

In conclusion, this circuit is highly recommended for low Q requirements. Although a large spread in resistance values can occur and the Q is limited by amplifier gain, the circuit simplicity, low element sensitivity, and ease of frequency adjustment make it highly desirable.

The following example demonstrates the design of a band-pass filter using the MFBP configuration.

Example 5-12   Design of an Active All-Pole Band-Pass Filter Using the MFBP Configuration

Required:

Design an active band-pass filter with the following specifications:
A center frequency of 300 Hz
3 dB at ±10 Hz
25-dB minimum at ±40 Hz
Essentially zero overshoot to a 300-Hz carrier pulse step
A gain of ±12 dB at 300 Hz

Result:

(*a*) Since the bandwidth is narrow, the requirement can be treated on an arithmetically symmetrical basis. The band-pass steepness factor is given by Eq. (2-19) as shown:

$$A_s = \frac{\text{stopband bandwidth}}{\text{passband bandwidth}} = \frac{80 \text{ Hz}}{20 \text{ Hz}} = 4$$

The curves of Figs. 2-69 and 2-74 indicate that an $n = 3$ transitional gaussian to 6-dB filter will meet the frequency- and step-response requirements.

(*b*) The pole locations for the corresponding normalized low-pass filter are found in Table 10-50 and are as follows:

$-0.9622 \pm j1.2214$
-0.9776

First compute the band-pass Q using Eq. (2-16):

$$Q_{\text{bp}} = \frac{f_0}{\text{BW}_{3\text{ dB}}} = \frac{300 \text{ Hz}}{20 \text{ Hz}} = 15$$

The low-pass poles are transformed to the band-pass form in the following manner using Eqs. (5-50) to (5-59):

Complex pole:

$$\alpha = 0.9622 \quad \beta = 1.2214$$
$$C = 2.417647$$

$$D = 0.128293$$

$$E = 4.010745$$

$$G = 4.002529$$

$$Q = 15.602243$$

$$M = 1.000832$$

$$W = 1.041630$$

$$f_{ra} = 288.0 \text{ Hz}$$

$$f_{rb} = 312.5 \text{ Hz}$$

Real pole:

$$\alpha_0 = 0.9776$$

$$Q = 15.34$$

$$f_r = 300.0 \text{ Hz}$$

(c) A midband gain of +12 dB is required. Let us allocate a gain of +4 dB to each section corresponding to $A_0 = 1.585$. The value of A_r, the resonant frequency gain for each section, is obtained from Eq. (5-64) and is listed in the following table, which summarizes the design parameters of the filters' sections:

	f_r	Q	A_r
Section 1	288.0 Hz	15.60	2.567
Section 2	312.5 Hz	15.60	2.567
Section 3	300.0 Hz	15.34	1.585

(d) Three MFBP band-pass sections will be connected in tandem. The following element values are computed using Eqs. (5-68), (5-72), and (5-73) where C is set equal to 0.1 μF:

Section 1:

$$R_2 = \frac{Q}{\pi f_r C} = \frac{15.6}{\pi \times 288 \times 10^{-7}} = 172.4 \text{ k}\Omega$$

$$R_{1a} = \frac{R_2}{2A_r} = \frac{172.4 \times 10^3}{2 \times 2.567} = 33.6 \text{ k}\Omega$$

$$R_{1b} = \frac{R_2/2}{2Q^2 - A_r} = \frac{86.2 \times 10^3}{2 \times 15.6^2 - 2.567} = 178 \text{ }\Omega$$

Section 2:   **Section 3:**

$R_2 = 158.9 \text{ k}\Omega$ $R_2 = 162.8 \text{ k}\Omega$

$R_{1a} = 30.9 \text{ k}\Omega$ $R_{1a} = 51.3 \text{ k}\Omega$

$R_{1b} = 164 \text{ }\Omega$ $R_{1b} = 174 \text{ }\Omega$

The final circuit is shown in Fig. 5-26. Resistor values have been rounded off to standard 1 percent values, and resistor R_{1b} has been made variable in each section for tuning purposes.

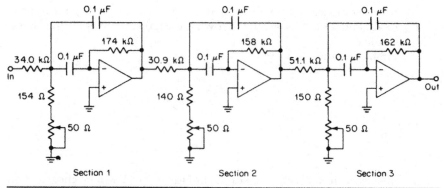

FIGURE 5-26 The MFBP circuit of Example 5-12.

Each filter section can be adjusted by applying a sine wave at the section f_r to the filter input. The phase shift of the section being adjusted is monitored by connecting one channel of an oscilloscope to the section input and the other channel to the section output. A Lissajous pattern is thus obtained. Resistor R_{1b} is then adjusted until the ellipse closes to a straight line.

The Dual-Amplifier Band-Pass (DABP) Structure

The band-pass circuit of Fig. 5-27 was first introduced by Sedra and Espinoza (see References). Truly remarkable performance in terms of available Q, low sensitivity, and flexibility can be obtained in comparison with alternative schemes involving two amplifiers.

The transfer function is given by

$$T(s) = \frac{s2/R_1C}{s^2 + s1/R_1C + 1/R_2R_3C^2} \tag{5-78}$$

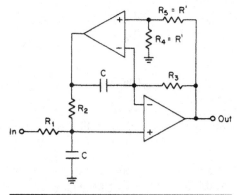

FIGURE 5-27 A dual-amplifier band-pass (DABP) configuration ($Q < 15$).

If we compare this expression with the general band-pass transfer function of Eq. (5-48) and let $R_2 R_3 = R^2$, the following design equations for the element values can be obtained.

First, compute

$$R = \frac{1}{2\pi f_r C} \tag{5-79}$$

then

$$R_1 = QR \tag{5-80}$$

$$R_2 = R_3 = R \tag{5-81}$$

where C is arbitrary. The value of R' in Fig. 5-27 can also be chosen at any convenient value. Circuit gain at f_r is equal to 2.

The following sensitivities can be derived:

$$S_{R_1}^{Q} = 1 \tag{5-82}$$

$$S_{R_2}^{fr} = S_{R_3}^{fr} = S_{R_4}^{fr} = S_{C}^{fr} = -1/2 \tag{5-83}$$

$$S_{R_5}^{fr} = 1/2 \tag{5-84}$$

An interesting result of sensitivity studies is that if the bandwidths of both amplifiers are nearly equivalent, extremely small deviations of Q from the design values will occur. This is especially advantageous at higher frequencies where the amplifier poles have to be taken into account. It is then suggested that a dual-type amplifier be used for each filter section, since both amplifier halves will be closely matched to each other.

A useful feature of this circuit is that resonant frequency and Q can be independently adjusted. Alignment can be accomplished by first adjusting R_2 for resonance at f_r. Resistor R_1 can then be adjusted for the desired Q without affecting the resonant frequency.

Since each section provides a fixed gain of 2 at f_r, a composite filter may require an additional amplification stage if higher gains are needed. If a gain reduction is desired, resistor R_1 can be split into two resistors to form a voltage divider in the same manner as in Fig. 5-25b. The resulting values are

$$R_{1a} = \frac{2R_1}{A_r} \tag{5-85}$$

and

$$R_{1b} = \frac{R_{1a} A_r}{2 - A_r} \tag{5-86}$$

where A_r is the desired gain at resonance.

The spread of element values of the MFBP section previously discussed is equal to $4Q^2$. In comparison, this circuit has a ratio of resistances determined by Q, so the spread is much less.

The DABP configuration has been found to be very useful for designs covering a wide range of Qs and frequencies. Component sensitivity is small, resonant frequency and Q are easily adjustable, and the element spread is low. The following example illustrates the use of this circuit.

Example 5-13   Design of an Active All-Pole Band-Pass Filter Using the DABP Configuration

Required:

Design an active band-pass filter to meet the following specifications:
A center frequency of 3,000 Hz
3 dB at ±30 Hz
20-dB minimum at ±120 Hz

Result:

(*a*) If we consider the requirement as being arithmetically symmetrical, the band-pass steepness factor can be calculated using Eq. (2-19).

$$A_s = \frac{\text{stopband bandwidth}}{\text{passband bandwidth}} = \frac{240\,\text{Hz}}{60\,\text{Hz}} = 4$$

We can determine from the curve of Fig. 2-34 that a second-order Butterworth low-pass filter provides over 20 dB of rejection within a frequency ratio of 4:1. The corresponding poles of the normalized low-pass filter are found in Table 10-1 and are as follows:

$$-0.7071 \pm j0.7071$$

(*b*) To convert these poles to the band-pass form, first compute the following using Eq. (2-16):

$$Q_{bp} = \frac{f_0}{BW_{3\,\text{dB}}} = \frac{3,600\,\text{Hz}}{60\,\text{Hz}} = 50$$

The band-pass poles transformation is performed in the following manner using Eqs. (5-50) to (5-58):

$$\alpha = 0.7071 \quad \beta = 0.7071$$

$$C = 1.000000$$

$$D = 0.028284$$

$$E = 4.000400$$

$$G = 4.000000$$

$$Q = 70.713124$$

$$M = 1.000025$$

$$W = 1.007096$$

$$f_{ra} = 2,978.9\,\text{Hz}$$

$$f_{rb} = 3,021.3\,\text{Hz}$$

(*c*) Two DABP sections will be used. The element values are now computed using Eqs. (5-79) to (5-81), where C is set equal to 0.01 μF and R' is 10 kΩ.

Section 1:

$$f_r = 2,978.9\,\text{Hz}$$

$$Q = 70.7$$

$$R = \frac{1}{2\pi f_r C} = \frac{1}{2\pi \times 2,978.9 \times 10^{-8}} = 5,343\,\Omega$$

$$R_1 = QR = 70.7 \times 5,343 = 377.7\,\text{k}\Omega$$

$$R_2 = R_3 = R = 5,343\,\Omega$$

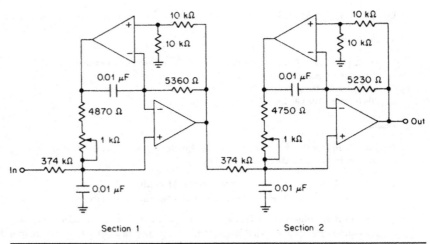

FIGURE 5-28 The DABP filter of Example 5-13.

Section 2:

$$f_r = 3{,}021.3 \text{ Hz}$$

$$Q = 70.7$$

$$R = 5{,}268 \ \Omega$$

$$R_1 = 372.4 \text{ k}\Omega$$

$$R_2 = R_3 = 5{,}268 \ \Omega$$

The circuit is illustrated in Fig. 5-28, where resistors have been rounded off to standard 1 percent values and R_2 is made adjustable for tuning.

Low-Sensitivity Three-Amplifier Configuration

The DABP circuit shown in Fig. 5-27 provides excellent performance and is useful for general-purpose band-pass filtering. A modified version utilizing a total of three amplifiers is shown in Fig. 5-29. This circuit will exhibit performance superior to the DABP configuration, especially at higher frequencies. Using this structure and off-the-shelf op amps, active band-pass filters with moderate percentage bandwidths can be designed to operate in the frequency range approaching 1 to 2 MHz.

The section within the dashed line in Fig. 5-29 realizes a shunt inductance to ground using a configuration called a *gyrator*.* Both op amps within this block should be closely matched to obtain low op-amp sensitivity at high frequencies, so the use of a dual op amp for this location would be highly recommended. The third op amp serves as a voltage follower or buffer to obtain low output impedance.

---

*A gyrator is an impedance-inverting device that converts an impedance Z into a reciprocal impedance $1/G^2 Z$, where G is a constant. Therefore, a capacitor with an impedance $1/SC$ can be converted into an impedance SC/G^2, which corresponds to an inductance of C/G^2. For this circuit, $\sigma = 1/R$, so $L = R^2C$.

FIGURE 5-29 Low-sensitivity three-amplifier band-pass configuration.

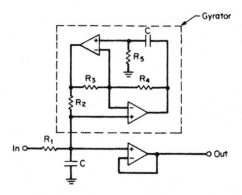

As in the case of the DABP circuit, resonant frequency and Q can be adjusted independently. The design proceeds as follows:

First, select a convenient value for C. Then compute

$$R = \frac{1}{2\pi f_r C} \tag{5-87}$$

Let $R_2 = R_3 = R_4 = R_5 = R$

Then
$$R_1 = QR \tag{5-88}$$

Circuit gain at f_r is unity. For additional gain, the voltage follower can be configured as a noninverting amplifier.

For alignment, first adjust R_2 for resonance at f_r. Resistor R_1 can then be adjusted for the desired Q without affecting the resonant frequency.

Example 5-14 Design of an Active All-Pole Band-Pass Filter Using the Low-Sensitivity Three-Amplifier Configuration

Required:

Design an active band-pass filter to meet the following specifications:

A center frequency of 30 kHz
3 dB at ±300 Hz
20-dB minimum at ±1,200 Hz

Result:

(*a*) Treating the requirement as being arithmetically symmetrical, the band-pass steepness factor is calculated as follows using Eq. (2-19):

$$A_s = \frac{\text{stopband bandwidth}}{\text{passband bandwidth}} = \frac{2,400\,\text{Hz}}{600\,\text{Hz}} = 4$$

From Fig. 2-34, a second-order Butterworth low-pass filter meets the attenuation. The low-pass poles from Table 10-1 are $-0.7071 \pm j0.7071$.

(b) These poles must now be converted to the band-pass form. The procedure is as follows, using Eq. (2-16) and Eqs. (5-50) to (5-58):

$$Q_{bp} = \frac{f_0}{BW_{3\,dB}} = \frac{30,000 \text{ Hz}}{600 \text{ Hz}} = 50$$

$$\alpha = 0.7071 \quad \beta = 0.7071$$

$$C = 1$$

$$D = 0.028284$$

$$E = 4.000400$$

$$G = 4.000000$$

$$Q = 70.7131124$$

$$M = 1.000025$$

$$W = 1.0070996$$

$$f_{ra} = 29.789 \text{ kHz}$$

$$f_{rb} = 30.213 \text{ kHz}$$

(c) Compute the element values for the circuit in Fig. 5-29 using $C = 0.01 \, \mu F$, as shown in Eqs. (5-87) and (5-88).

Section 1:

$$f_r = 29.789 \text{ kHz}$$

$$Q = 70.71$$

$$R = \frac{1}{2\pi f_r C} = \frac{1}{2\pi \times 29.789 \times 10^{-5}}$$

$$= 534.3 \, \Omega$$

$$R_1 = QR = 70.7 \times 534.3 = 37.8 \text{ k}\Omega \, \Omega$$

Section 2:

$$f_r = 30.213 \text{ kHz}$$

$$Q = 70.71$$

$$R = 526.8 \, \Omega$$

$$R_1 = 37.2 \text{ k}\Omega$$

The circuit of this example is illustrated in Fig. 5-30 using standard 1 percent resistor values and a potentiometer for frequency adjustment.

The State-Variable (Biquad) All-Pole Circuit

The state-variable, or biquad, configuration was first introduced in Sec. 3.2 for use as a low-pass filter section. A band-pass output is available as well. The biquad approach features excellent sensitivity properties and the capability to control resonant frequency

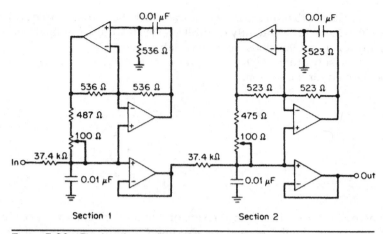

FIGURE 5-30 The band-pass filter of Example 5-14.

and Q independently. It is especially suited for constructing precision-active filters in a standard form.

The circuit in Fig. 5-31 is the all-pole band-pass form of the general biquadratic configuration. The transfer function is given by

$$T(s) = \frac{s/CR_4}{s^2 + s/CR_1 + 1/R_2R_3C^2}$$ (5-89)

If we equate this expression to the general band-pass transfer function of Eq. (5-48), the circuit resonant frequency and 3-dB bandwidth can be expressed as

$$f_r = \frac{1}{2\pi\ C\sqrt{R_2R_3}}$$ (5-90)

and

$$BW_{3\ dB} = \frac{1}{2\pi R_1 C}$$ (5-91)

where $BW_{3\ dB}$ is equal to f_r/Q.

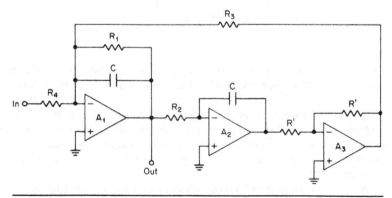

FIGURE 5-31 A biquad all-pole circuit ($Q < 200$).

Equations (5-90) and (5-91) indicate that the resonant frequency and 3-dB bandwidth can be independently controlled. This feature is highly desirable and can lead to many useful applications, such as variable filters.

If we substitute f_r/Q for $BW_{3\,dB}$ and set $R_2 = R_3$, the following design equations can be derived for the section:

$$R_1 = \frac{Q}{2\pi f_r C} \tag{5-92}$$

$$R_2 = R_3 = \frac{R_1}{Q} \tag{5-93}$$

and

$$R_4 = \frac{R_1}{A_r} \tag{5-94}$$

where A_r is the desired gain at resonant frequency f_r. The values of C and R' in Fig. 5-31 can be conveniently selected. By making R_3 and R_1 adjustable, the resonant frequency and Q, respectively, can be adjusted.

The sensitivity factors are

$$S_{R_2}^{f_r} = S_{R_3}^{f_r} = S_C^{f_r} = -1/2 \tag{5-95}$$

$$S_{R_1}^Q = 1 \tag{5-96}$$

and

$$S_\mu^Q = \frac{2Q}{\mu} \tag{5-97}$$

where μ is the open-loop gain of amplifiers A_1 and A_2. The section Q is then limited by the finite gain of the operational amplifier.

Another serious limitation occurs because of finite amplifier bandwidth. Thomas (see References) has shown that as the resonant frequency increases for a fixed design Q, the actual Q remains constant over a broad band and then begins to increase, eventually becoming infinite (oscillatory). This effect is called Q enhancement.

If we assume that the open-loop transfer function of the amplifier has a single pole, the effective Q can be approximated by

$$Q_{\text{eff}} = \frac{Q_{\text{design}}}{1 - \dfrac{2\,Q_{\text{design}}}{\mu_0 \omega_c}(2\omega_r - \omega_c)} \tag{5-98}$$

where ω_r is the resonant frequency, ω_c is the 3-dB breakpoint of the open-loop amplifier gain, and μ_0 is the open-loop gain at DC. As ω_r increases, the denominator approaches zero.

The Q-enhancement effect can be minimized by having a high gain-bandwidth product. If the amplifier requires external frequency compensation, the compensation can be made lighter than the recommended values. The state-variable circuit is well suited for light compensation, since the structure contains two integrators, which have a stabilizing effect.

A solution suggested by Thomas is to introduce a leading phase component in the feedback loop, which compensates for the lagging phase caused by finite amplifier bandwidth. This can be achieved by introducing a capacitor in parallel with resistor R_3 with the value

$$C_p = \frac{4}{\mu_0 \omega_c R_3} \tag{5-99}$$

Probably the most practical solution is to make resistor R_1 variable. The Q may be determined by measuring the 3-dB bandwidth. R_1 is adjusted until the ratio $f_r / BW_{3\,dB}$ is equal to the required Q.

As the Q is enhanced, the section gain is also increased. Empirically, it has been found that correcting for the gain enhancement compensates for the Q enhancement as well. R_1 can be adjusted until the measured gain at f_r is equal to the design value of A_r used in Eq. (5-94). Although this technique is not as accurate as determining the actual Q from the 3-dB bandwidth, it certainly is much more convenient and will usually be sufficient.

The biquad is a low-sensitivity filter configuration suitable for precision applications. Circuit Qs of up to 200 can be realized over a broad frequency range. The following example demonstrates the use of this structure.

Example 5-15 Design of an Active All-Pole Band-Pass Filter Using the Biquad Configuration

Required:

Design an active band-pass filter satisfying the following specifications:
A center frequency of 2,500 Hz
3 dB at ±15 Hz
15-dB minimum at ±45 Hz
A gain of +12 dB at 2,500 Hz

Result:

(*a*) The band-pass steepness factor is determined from Eq. (2-19) as shown:

$$A_s = \frac{\text{stopband bandwidth}}{\text{passband bandwidth}} = \frac{90\,\text{Hz}}{30\,\text{Hz}} = 3$$

Using Fig. 2-42, we find that a second-order 0.1-dB Chebyshev normalized low-pass filter will meet the attenuation requirements. The corresponding poles found in Table 10-23 are as follows:

$$-0.6125 \pm j0.7124$$

(*b*) To transform these low-pass poles to the band-pass form, first compute the following using Eq. (2-16)

$$Q_{bp} = \frac{f_0}{BW_{3\,dB}} = \frac{2,500\,\text{Hz}}{30\,\text{Hz}} = 83.33$$

The band-pass poles are determined using the following series of computations. Since the filter is very narrow, an extended number of significant figures will be used in Eqs. (5-50) through (5-58) to maintain accuracy.

$$\alpha = 0.6125 \quad \beta = 0.7124$$
$$C = 0.882670010$$
$$D = 0.014700588$$

$$E = 4.000127115$$

$$G = 4.000019064$$

$$Q = 136.0502228$$

$$M = 1.000009138$$

$$W = 1.004284182$$

$$f_{ra} = 2,489.3 \text{ Hz}$$

$$f_{rb} = 2,510.7 \text{ Hz}$$

(c) Since a midband gain of +12 dB is required, each section will be allocated a midband gain of +6 dB corresponding to $A_0 = 2.000$. The gain A_r at the resonant frequency of each section is determined from Eq. (5-65) and is listed in the following table.

	f_r	Q	A_r
Section 1	2,489.3 Hz	136	3.069
Section 2	2,510.7 Hz	136	3.069

(d) Two biquad sections in tandem will be used. C is 0.1 μF and R' is 10 kΩ. The element values are computed as follows using Eqs. (5-92) to (5-94):

Section 1:

$$R_1 = \frac{Q}{2\pi f_r C} = \frac{136}{2\pi \times 2,489.3 \times 10^{-7}} = 86.9 \text{ k}\Omega$$

$$R_2 = R_3 = \frac{R_1}{Q} = \frac{86.9 \times 10^3}{136} = 639 \text{ k}\Omega$$

$$R_4 = \frac{R_1}{A_r} = \frac{86.9 \times 10^3}{3.069} = 28.3 \text{ k}\Omega$$

Section 2:

$$R_1 = 86.2 \text{ k}\Omega$$

$$R_2 = R_3 = 634 \text{ }\Omega$$

$$R_4 = 28.1 \text{ k}\Omega$$

The final circuit is shown in Fig. 5-32. Resistors R_3 and R_1 are made variable so that resonant frequency and Q can be adjusted. Standard values of 1 percent resistors have been used.

The Q-Multiplier Approach

Certain active band-pass structures, such as the MFBP configuration of Sec. 5.2, are severely Q-limited because of insufficient amplifier gain or other inadequacies. The technique outlined in this section uses a low-Q-type band-pass circuit within a Q-multiplier structure, which increases the circuit Q to the desired value.

A band-pass transfer function with unity gain at resonance can be expressed as

$$T(s) = \frac{\dfrac{\omega_r}{Q}s}{s^2 + \dfrac{\omega_r}{Q}s + \omega_r^2} \tag{5-100}$$

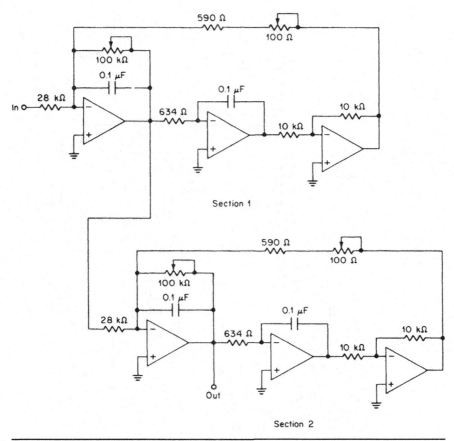

FIGURE 5-32   The biquad circuit of Example 5-15.

If the corresponding circuit is combined with a summing amplifier in the manner shown in Fig. 5-33a, where β is an attenuation factor, the following overall transfer function can be derived:

$$T(s) = \frac{\dfrac{\omega_r}{Q}s}{s^2 + \dfrac{\dfrac{\omega_r}{Q}}{\dfrac{Q}{1-\beta}}s + \omega_r^2} \qquad (5\text{-}101)$$

The middle term of the denominator has been modified so that the circuit Q is given by $Q/(1 - \beta)$, where $0 < \beta < 1$. By selecting a β sufficiently close to unity, the Q can be increased by the factor $1/(1 - \beta)$. The circuit gain is also increased by the same factor.

If we use the MFBP section for the band-pass circuit, the Q-multiplier configuration will take the form shown in Fig. 5-33b. Since the MFBP circuit is inverting, an inverting amplifier can also be used for summing.

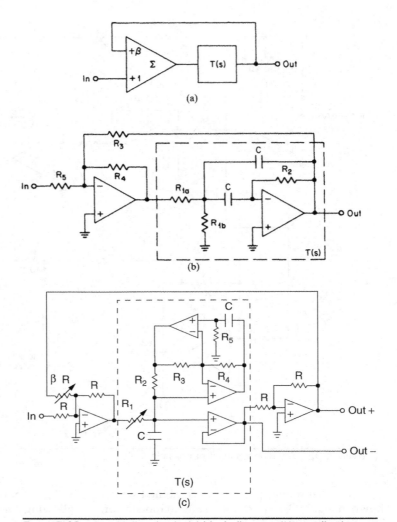

FIGURE 5-33 Q-multiplier circuit: (a) block diagram; (b) a realization using an MFBP section; (c) realization using low-sensitivity band-pass configuration.

The value of β can be found from

$$\beta = 1 - \frac{Q_r}{Q_{eff}} \tag{5-102}$$

where Q_{eff} is the effective overall Q and Q_r is the design Q of the band-pass section. The component values are determined by the following equations:

$$R_3 = \frac{R}{\beta} \tag{5-103}$$

$$R_4 = R \tag{5-104}$$

and
$$R_5 = \frac{R}{(1 - \beta)A_r} \tag{5-105}$$

where R can be conveniently chosen and A_r is the desired gain at resonance.

Design equations for the MFBP section were derived in Sec. 5.2 in Eqs. (5-68), (5-72), and (5-73), and are repeated here corresponding to unity gain.

$$R_2 = \frac{Q_r}{\pi f_r C}$$

$$R_{1a} = \frac{R_2}{2}$$

and
$$R_{1b} = \frac{R_{1a}}{2Q_r^2 - 1}$$

The value of C can be freely chosen.

The configuration of Fig. 5-33b is not restricted to the MFBP section. Figure 5-33c, for example, utilizes the low-sensitivity configuration of Fig. 5-29 combined with the Q-multiplier circuit of Fig. 5-33a. Equations (5-87) and (5-88) are used for the T(s) block. Resistor R_1 is made adjustable so Q can be variable. Further Q enhancement can occur by adjusting βR. As a result, very high Qs can be achieved by combining these two adjustments. Note that the circuit has differential outputs, which can be useful for connecting to a transformer.

The state-variable all-pole band-pass circuit may be used instead of the MFBP section. The only requirements are that the filter section be of an inverting type and that the gain be unity at resonance. This last requirement is especially critical because of the positive feedback nature of the circuit. Small gain errors could result in large overall Q variations when β is close to 1. It may then be desirable to adjust section gain precisely to unity.

Example 5-16   Design of an Active Band-Pass Filter Section Using the Q-Multiplier Configuration

Required:

Design a single band-pass filter section with the following characteristics:
A center frequency of 3,600 Hz
3-dB bandwidth of 60 Hz
A gain of 3

Result:

(a) The band-pass Q is given by Eq. (2-16) as

$$Q_r = \frac{f_0}{\mathrm{BW}_{3\,\mathrm{dB}}} = \frac{3,600\,\mathrm{Hz}}{60\,\mathrm{Hz}} = 60$$

A Q-multiplier implementation using the MFBP section will be employed.

(b) Let us use a Q_r of 10 for the MFBP circuit. The following component values are computed using Eqs. (5-68), (5-72), and (5-73), where C is set equal to 0.01 μF.

$$R_2 = \frac{Q_r}{\pi f_r C} = \frac{10}{\pi 3,600 \times 10^{-8}} = 88.4 \text{ k}\Omega$$

$$R_{1a} = \frac{R_2}{2} = \frac{88.4 \times 10^3}{2} = 44.2 \text{ k}\Omega$$

$$R_{1b} = \frac{R_{1a}}{2Q_r^2 - 1} = \frac{44.2 \times 10^3}{2 \times 10^2 - 1} = 222 \ \Omega$$

The remaining values are given by the following design equations—Eqs. (5-102) to (5-104)—where R is chosen at 10 kΩ and gain A_r is equal to 3:

$$\beta = 1 - \frac{Q_r}{Q_{\text{eff}}} = 1 - \frac{10}{60} = 0.8333$$

$$R_3 = \frac{R}{\beta} = \frac{10^4}{0.8333} = 12.0 \text{ k}\Omega$$

$$R_4 = R = 10 \text{ k}\Omega$$

$$R_5 = \frac{R}{(1 - \beta)A_r} = \frac{10^4}{(1 - 0.8333)3} = 20 \text{ k}\Omega$$

The resulting circuit is shown in Fig. 5-34 using standard resistor values. R_{1b} has been made adjustable for tuning.

5.2.5   Elliptic-Function Band-Pass Filters

An active elliptic-function band-pass filter is designed by first transforming the low-pass poles and zeros to the band-pass form using the formulas of Sec. 5.2. The band-pass poles and zeros are then implemented using active structures.

Normalized low-pass poles and zeros for elliptic-function low-pass filters can be obtained in terms Q, ω_0, ω_∞, and α_0 using the *Filter Solutions* program.

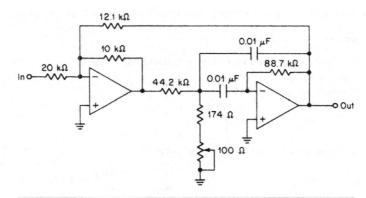

Figure 5-34   The Q-multiplier section of Example 5-16.

The general form of a band-pass transfer function containing zeros was given in Sec. 5.2 in Eq. (5-49) as

$$T(s) = \frac{H\left(s^2 + \omega_\infty^2\right)}{s^2 + \dfrac{\omega_r}{Q}s + \omega_r^2}$$

Elliptic-function band-pass filters are composed of cascaded first-order band-pass sections. When n is odd, $n - 1$ zero-producing sections are required, along with a single all-pole section. When n is even, $n - 2$ zero-producing sections are used, along with two all-pole networks.

This section discusses the VCVS and biquad configurations, which have a transfer function in the form of Eq. (5-49) and are used in the design of active elliptic-function band-pass filters.

VCVS Network

Section 3.2 discussed the design of active elliptic-function low-pass filters using an RC section containing a voltage-controlled voltage source (VCVS). The circuit is repeated in Fig. 5-35a. This structure is not restricted to the design of low-pass filters extensively. Transmission zeros can be obtained at frequencies either above or below the pole locations as required by the band-pass transfer function.

First, calculate

$$a = \frac{1}{Q} \tag{5-106}$$

$$b = \left(\frac{f_\infty}{f_r}\right)^2 \tag{5-107}$$

$$c = 2\pi f_r \tag{5-108}$$

where Q, f_∞, and f_r are the band-pass parameters corresponding to the general-form band-pass transfer function given in Eq. (5-49).

The element values are computed as follows:

Select C.

Then

$$C_1 = C \tag{5-109}$$

$$C_3 = C_4 = \frac{C_1}{2} \tag{5-110}$$

and

$$C_2 \geq \frac{C_1(b - 1)}{4} \tag{5-111}$$

$$R_3 = \frac{1}{cC_1\sqrt{b}} \tag{5-112}$$

$$R_1 = R_2 = 2R_3 \tag{5-113}$$

$$R_4 = \frac{4\sqrt{b}}{cC_1(1 - b) + 4cC_2} \tag{5-114}$$

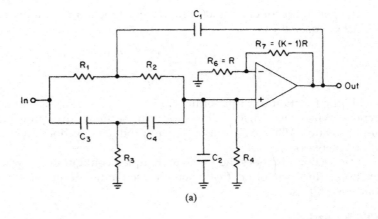

(a)

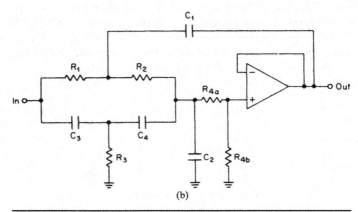

(b)

FIGURE 5-35 The VCVS elliptic-function band-pass section:
(a) circuit for $K > 1$; and (b) circuit for $K < 1$.

$$K = 2 + \frac{2C_2}{C_1} - \frac{a}{2\sqrt{b}} + \frac{2}{C_1\sqrt{b}} \left(\frac{1}{cR_4} - aC_2 \right) \qquad (5\text{-}115)$$

$$R_6 = R \qquad (5\text{-}116)$$

$$R_7 = (K - 1)R \qquad (5\text{-}117)$$

where R can be arbitrarily chosen.

In the event that K is less than 1, the circuit in Fig. 5-35b is used. Resistor R_4 is split into two resistors, R_{4a} and R_{4b}, which are given by

$$R_{4a} = (1 - K) R_4 \qquad (5\text{-}118)$$

and

$$R_{4b} = KR_4 \qquad (5\text{-}119)$$

The section Q can be controlled independently of resonant frequency by making R_6 or R_7 adjustable when $K > 1$. The resonant frequency, however, is not easily adjusted. Experience has shown that with 1 percent resistors and capacitors, section Qs of up to 10 can be realized with little degradation to the overall filter response due to component tolerances.

The actual circuit Q cannot be measured directly since the section's 3-dB bandwidth is determined not only by the design Q but by the transmission zero as well. Nevertheless, the VCVS configuration uses a minimum number of amplifiers and is widely used by low-Q elliptic-function realizations. The design technique is demonstrated in the following example.

Example 5-17 Design of an Active Elliptic-Function Band-Pass Filter Using the VCVS Configuration

Required:

> An active band-pass filter
> A center frequency of 500 Hz
> 1-dB maximum at ±100 Hz (400 Hz, 600 Hz)
> 35-dB minimum at ±363 Hz (137 Hz, 863 Hz)

Result:

(*a*) Convert to geometrically symmetrical band-pass requirements:

First, calculate the geometric center frequency using Eq. (2-14).

$$f_0 = \sqrt{f_L f_u} = \sqrt{400 \times 600} = 490.0 \text{ Hz}$$

Since the stopband requirement is arithmetically symmetrical, compute stopband bandwidth using Eq. (5-18).

$$\text{BW}_{35 \text{ dB}} = f_2 - \frac{f_0^2}{f_2} = 863 - \frac{490^2}{863} = 584.8 \text{ Hz}$$

The band-pass steepness factor is given by Eq. (2-19) as shown:

$$A_s = \frac{\text{stopband bandwidth}}{\text{passband bandwidth}} = \frac{584.8 \text{ Hz}}{200 \text{ Hz}} = 2.924$$

(*b*) Open **Filter Solutions.**
Check the **Stop Band Freq** box.
Enter **0.18** in the **Pass Band Ripple (dB)** box.
Enter **1** in the **Pass Band Freq** box.
Enter **2.924** in the **Stop Band Freq** box.
Check the **Frequency Scale Rad/Sec** box.

(*c*) Click the **Set Order** control button to open the second panel.
Enter **35** for the **Stopband Attenuation (dB)**.
Click the **Set Minimum Order** button and then click **Close.**
3 Order is displayed on the main control panel.

(*d*) Click the **Transfer Function** button.
Check the **Casc** box.

The following is displayed:

Continuous Transfer Function

$$Wn = 3.351$$

$$\frac{.1134 \quad (S^2 + 11.23)}{(S + .883) \quad (S^2 + .7697*S + 1.441)}$$

$$Wo = 1.2$$
$$Q = 1.56$$

3rd Order Low Pass Elliptic

Pass Band Frequency = 1.000 Rad/Sec	Stop Band Ratio = 2.924
Pass Band Ripple = 180.0 mdB	Stop Band Frequency = 2.924 Rad/Sec
	Stop Band Attenuation = 37.51 dB

(e) The normalized low-pass design parameters are summarized as follows:

$$\text{Section } Q = 1.56$$

$$\text{Section } \omega_0 = 1.2$$

$$\text{Section } \omega_\infty = 3.351$$

$$\alpha_0 = 0.883 \text{ (from the denominator)}$$

The pole coordinates in rectangular form are

$$\alpha = \frac{\omega_0}{2Q} = 0.3646$$

$$\beta = \sqrt{\omega_0^2 - \alpha^2} = 1.1367$$

(f) To determine the band-pass parameters, first compute the following using Eq. (2-16):

$$Q_{bp} = \frac{f_0}{BW_{1\,dB}} = \frac{490\,\text{Hz}}{200\,\text{Hz}} = 2.45$$

The poles and zeros are transformed as follows using Eqs. (5-50) to (5-63):

Complex pole:

$$\alpha = 0.3846 \quad \beta = 1.1367$$

$$C = 1.440004$$

$$D = 0.313959$$

$$E = 4.239901$$

$$G = 4.193146$$

$$Q = 6.540396$$

$$M = 1.026709$$

$$W = 1.259369$$

$$f_{ra} = 389 \text{ Hz}$$

$$f_{rb} = 617 \text{ Hz}$$

Real pole:

$$\alpha_0 = 0.883$$

$$Q = 2.7746$$

$$f_r = 490 \text{ Hz}$$

Zero:

$$\omega_\infty = 3.351$$

$$H = 1.935377$$

$$Z = 1.895359$$

$$f_{\infty, a} = 258.5 \text{ Hz}$$

$$f_{\infty, a} = 928.7 \text{ Hz}$$

The band-pass parameters are summarized in the following table, where the zeros are arbitrarily assigned to the first two sections:

Section	f_r	Q	f_∞
1	389 Hz	6.54	258.5 Hz
2	617 Hz	6.54	928.7 Hz
3	490 Hz	2.77	

(g) Sections 1 and 2 are realized using the VCVS configuration of Fig. 5-35. The element values are computed as follows, using Eqs. (5-106) to (5-115), where R' and R are both 10 kΩ.

Section 1:

$$f_r = 389 \text{ Hz}$$

$$Q = 6.54$$

$$f_\infty = 259 \text{ Hz}$$

$$a = 0.15291$$

$$b = 0.4433$$

$$c = 2444$$

Let $\quad\quad\quad\quad C = 0.02 \ \mu\text{F}$

then $\quad\quad\quad\quad C_1 = 0.02 \ \mu\text{F}$

$$C_3 = C_4 = 0.02 \ \mu\text{F}$$

$$C_2 \geq -0.0027835 \ \mu\text{F}$$

Let $\quad\quad\quad\quad C_2 = 0$

$$R_3 = 30.725 \text{ k}\Omega$$

$$R_1 = R_2 = 61.450 \text{ k}\Omega$$

$$R_4 = 97.866 \text{ k}\Omega$$

$$K = 2.5131$$

Section 2:

$$f_r = 617 \text{ Hz}$$

$$Q = 6.54$$

$$f_\infty = 929 \text{ Hz}$$

$$a = 0.15291$$

$$b = 2.2671$$

$$c = 3877$$

Let $\quad\quad\quad\quad C = 0.02 \ \mu\text{F}$

then $\quad\quad\quad\quad C_1 = 0.02 \ \mu\text{F}$

$$C_3 = C_4 = 0.01 \ \mu\text{F}$$

$$C_2 \geq -0.006335 \ \mu\text{F}$$

Let $\quad\quad\quad\quad C_2 = 0.01 \ \mu\text{F}$

$$R_3 = 8566 \ \Omega$$

$$R_1 = R_2 = 17.132 \text{ k}\Omega$$

$$R_4 = 105.98 \text{ k}\Omega$$

$$K = 3.0093$$

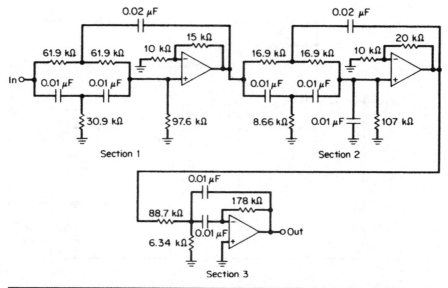

FIGURE 5-36 The circuit of the elliptic-function band-pass filter in Example 5-17.

(*h*) Section 3 must be of the all-pole type, so the MFBP configuration of Fig. 5-25*b* will be used, where C is chosen as 0.01 μF and the section gain A_r is set to unity, using Eqs. (5-68), (5-72), and (5-73) as follows:

Section 3:

$$f_r = 490 \text{ Hz}$$

$$Q = 2.77$$

$$R_2 = 179.9 \text{ k}\Omega$$

$$R_{1a} = 89.97 \text{ k}\Omega$$

$$R_{1b} = 6.27 \text{ k}\Omega$$

The complete circuit is shown in Fig. 5-36 using standard 1 percent resistor values.

State-Variable (Biquad) Circuit

The all-pole band-pass form of the state-variable or biquad section was discussed in Sec. 5.2. With the addition of an operational amplifier, the circuit can be used to realize transmission zeros as well as poles. The configuration is shown in Fig. 5-37. This circuit is identical to the elliptic-function low-pass and high-pass filter configurations of Secs. 3.2 and 4.2. By connecting R_5 either to node 1 or node 2, the zero can be located above or below the resonant frequency.

On the basis of sensitivity and flexibility, the biquad configuration has been found to be the optimum method of constructing precision-active elliptic-function band-pass filters. Section Qs of up to 200 can be obtained, whereas the VCVS section is limited to Qs below 10. Resonant frequency f_r, Q, and notch frequency f_∞ can be independently monitored and adjusted.

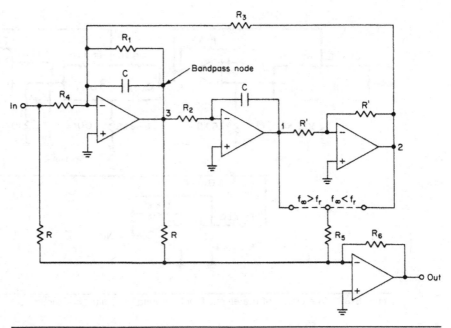

Figure 5-37   A biquad elliptic-function band-pass configuration.

For the case where $f_\infty < f_r$, the transfer function is given by

$$T(s) = -\frac{R_6}{R} \frac{s^2 + \dfrac{1}{R_2R_3C^2}\left(1 - \dfrac{R_3R}{R_4R_5}\right)}{s^2 + \dfrac{1}{R_1C}s + \dfrac{1}{R_2R_3C^2}} \tag{5-120}$$

and when $f_\infty > f_r$, the corresponding transfer function is

$$T(s) = -\frac{R_6}{R} \frac{s^2 + \dfrac{1}{R_2R_3C^2}\left(1 + \dfrac{R_3R}{R_4R_5}\right)}{s^2 + \dfrac{1}{R_1C}s + \dfrac{1}{R_2R_3C^2}} \tag{5-121}$$

If we equate the transfer-function coefficients to those of the general band-pass transfer function (with zeros) of Eq. (5-49), the following series of design equations can be derived:

$$R_1 = R_4 = \frac{Q}{2\pi f_r C} \tag{5-122}$$

$$R_2 = R_3 = \frac{R_1}{Q} \tag{5-123}$$

$$R_5 = \frac{f_r^2 R}{Q\left|f_r^2 - f_\infty^2\right|} \tag{5-124}$$

for $f_\infty > f_r$:
$$R_6 = \frac{f_r^2 R}{f_\infty^2}$$
(5-125)

and when $f_\infty < f_r$:
$$R_6 = R$$
(5-126)

where C and R can be conveniently selected. The value of R_6 is based on unity section gain. The gain can be raised or lowered by proportionally changing R_6.

The section can be tuned by implementing the following steps in the indicated sequence. Both resonant frequency and Q are monitored at the band-pass output occurring at node 3, whereas the notch frequency f_∞ is observed at the section output.

1. *Resonance frequency f_r:* If R_3 is made variable, the section resonant frequency can be adjusted. Resonance is monitored at node 3 (see Fig. 5-37) and can be determined by the 180° phase shift method.

2. *Q adjustment:* The section Q is controlled by R_1 and can be directly measured at node 3. The configuration is subject to the Q-enhancement effect discussed in Sec. 5.2 under "All-Pole Band-Pass Configurations," so a Q adjustment is normally required. The Q can be monitored in terms of the 3-dB bandwidth at node 3, or R_1 can be adjusted until unity gain occurs between the section input and node 3 with f_r applied.

3. *Notch frequency f_∞:* Adjustment of the notch frequency (transmission zero) usually is not required if the circuit is previously tuned to f_r, since f_∞ will usually then fall in. If an adjustment is desired, the notch frequency can be controlled by making R_5 variable.

The biquad approach is a highly stable and flexible implementation for precision-active elliptic-function filters. The ability to independently adjust resonance frequency, Q, and the notch frequency preclude its use when Qs in excess of 10 are required. Stable Qs of up to 200 are obtainable.

Example 5-18   Design of an Active Elliptic-Function Band-Pass Filter Using the Biquad Configuration

Required:

An active band-pass filter
A center frequency 500 Hz
0.2-dB maximum at ±50 Hz (450 Hz, 550 Hz)
30-dB minimum at ±130 Hz (370 Hz, 630 Hz)

Result:

(*a*) Convert to the geometrically symmetrical requirement using Eqs. (2-14), (5-18), and (2-19), as follows:

$$f_0 = \sqrt{f_L f_u} = \sqrt{450 \times 550} = 497.5 \text{ Hz}$$

$$\text{BW}_{30\,\text{dB}} = f_2 - \frac{f_0^2}{f_2} = 630 - \frac{497.5^2}{630} = 237.1 \text{ Hz}$$

$$A_s = \frac{\text{stopband bandwidth}}{\text{passband bandwidth}} = \frac{237.1 \text{ Hz}}{100 \text{ Hz}} = 2.371$$

(b) Open *Filter Solutions*.
Check the *Stop Band Freq* box.
Enter **0.18** in the *Pass Band Ripple (dB)* box.
Enter **1** in the *Pass Band Freq* box.
Enter **2.371** in the *Stop Band Freq* box.
Check the *Frequency Scale Rad/Sec* box.

(c) Click the *Set Order* control button to open the second panel.
Enter **30** for the *Stopband Attenuation (dB)*.
Click the *Set Minimum Order* button and then click *Close*.
3 Order is displayed on the main control panel.

(d) Click the *Transfer Function* button.
Check the *Casc* box.

The following is displayed:

Continuous Transfer Function

$$Wn = 2.705$$

$$\frac{.1784 \quad (S^2 + 7.32)}{(S^2 + .7321*S + 1.435) \quad (S + .9105)}$$

$$Wo = 1.198$$
$$Q = 1.636$$

3rd Order Low Pass Elliptic

Pass Band Frequency = 1.000 Rad/Sec   Stop Band Ratio = 2.371
Pass Band Ripple = 180.0 mdB   Stop Band Frequency = 2.371 Rad/Sec
Stop Band Attenuation = 31.59 dB

(e) The normalized low-pass design parameters are summarized as follows:

$$\text{Section } Q = 1.636$$

$$\text{Section } \omega_0 = 1.198$$

$$\text{Section } \omega_\infty = 2.705$$

$$\alpha_0 = 0.9105 \text{ (from the denominator)}$$

The pole coordinates in rectangular form are

$$\alpha = \frac{\omega_0}{2Q} = 0.3661$$

$$\beta = \sqrt{\omega_0^2 - \alpha^2} = 1.1408$$

(*f*) The band-pass pole-zero transformation is now performed. First compute the following using Eq. (2-16):

$$Q_{bp} = \frac{f_0}{BW_{0.2\,dB}} = \frac{497.5\,Hz}{100\,Hz} = 4.975$$

The transformation proceeds as follows, using Eqs. (5-50) to (5-63):

Complex pole:

$$\alpha = 0.3661 \quad \beta = 1.1408$$
$$C = 1.435454$$
$$D = 0.147176$$
$$E = 4.05800$$
$$G = 4.047307$$
$$Q = 13.678328$$
$$M = 1.006560$$
$$W = 1.121290$$
$$f_{ra} = 443.6\,Hz$$
$$f_{rb} = 557.8\,Hz$$

Real pole:

$$\alpha_0 = 0.9105$$
$$Q = 5.464$$
$$f_r = 497.5\,Hz$$

Zero:

$$\omega_\infty = 2.705$$
$$H = 1.147815$$
$$Z = 1.308154$$
$$f_{\infty,a} = 380.31\,Hz$$
$$f_{\infty,b} = 650.81\,Hz$$

The computed band-pass parameters are summarized in the following table. The zeros are assigned to the first two sections.

Section	f_r	Q	f_∞
1	443.6 Hz	13.7	380.3 Hz
2	557.8 Hz	13.7	650.8 Hz
3	497.5 Hz	5.46	

(g) Sections 1 and 2 will be realized in the form of the biquad configuration of Fig. 5-37 and using Eqs. (5-122) to (5-124) and (5-126), where R' and R are both 10 kΩ and $C = 0.047$ μF.

Section 1:

$$f_r = 443.6 \text{ Hz}$$

$$Q = 13.7$$

$$f_\infty = 380.3 \text{ Hz}$$

$$R_1 = R_4 = \frac{Q}{2\pi f_r C} = 104.58 \text{ k}\Omega$$

$$R_2 = R_3 = \frac{R_1}{Q} = 7.63 \text{ k}\Omega$$

$$R_5 = \frac{f_r^2 R}{Q\left|f_r^2 - f_\infty^2\right|} = 2.76 \text{ k}\Omega$$

$$R_6 = R = 10 \text{ k}\Omega$$

Section 2:

$$f_r = 557.8 \text{ Hz}$$

$$Q = 13.7$$

$$f_\infty = 650.8 \text{ Hz}$$

Using Eqs. (5-122 to (5-125)

$$R_1 = R_4 = 83.17 \text{ k}\Omega$$

$$R_2 = R_3 = 6.07 \text{ k}\Omega$$

$$R_5 = 2.02 \text{ k}\Omega$$

$$R_6 = 7.35 \text{ k}\Omega$$

(h) The MFBP configuration of Fig. 5-25b will be used for the all-pole circuit of section 3. The value of C is 0.047 μF, and A_r is unity, as shown by Eqs. (5-68), (5-72), and (5-73).

Section 3:

$$f_r = 497.5 \text{ Hz}$$

$$Q = 5.46$$

$$R_2 = 74.3 \text{ k}\Omega$$

$$R_{1a} = 37.1 \text{ k}\Omega$$

$$R_{1b} = 634 \text{ }\Omega$$

The resulting filter is shown in Fig. 5-38, where standard 1 percent resistors are used. The resonant frequency and Q of each section have been made adjustable.

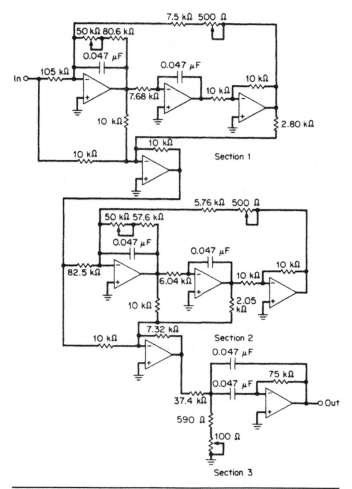

FIGURE 5-38   The biquad elliptic-function band-pass filter of
Example 5-18.

References

Huelsman, L. P. (1968). *Theory and Design of Active RC Circuits*. New York: McGraw-Hill.

Saal, R., and E. Ulbrich. "On the Design of Filters by Synthesis." *IRE Transactions on Circuit Theory*, December 1958.

Sedra, A. S., and J. L. Espinoza. "Sensitivity and Frequency Limitations of Biquadratic Active Filters." *IEEE Transactions on Circuits and Systems CAS-22*, no. 2, February 1975.

Thomas, L. C. "The Biquad: Part I – Some Practical Design Considerations." *IEEE Transactions on Circuit Theory CT–18*, May 1971.

Tow, J. "A Step-by-Step Active Filter Design." *IEEE Spectrum 6*, December 1969.

Williams, A. B. *Active Filter Design*. Dedham, Massachusetts: Artech House, 1975.

Williams, A. B. "Q-Multiplier Techniques Increase Filter Selectivity." *EDN*, October 5, 1975, 74–76.

Zverev, A. I. (1967). *Handbook of Filter Synthesis*, New York: John Wiley and Sons.

CHAPTER 6

Band-Reject Filters

6.1 *LC* Band-Reject Filters

Normalization of a band-reject requirement and the definitions of the response shape parameters were discussed in Sec. 2.1. Like band-pass filters, band-reject networks can also be derived from a normalized low-pass filter by a suitable transformation.

In Sec. 5.1, we discussed the design of wideband band-pass filters by cascading a low-pass filter and a high-pass filter. In a similar manner, wideband band-reject filters can also be obtained by combining low-pass and high-pass filters. Both the input and output terminals are paralleled, and each filter must have a high input and output impedance in the band of the other filter to prevent interaction. Therefore, the order n must be odd and the first and last branches should consist of series elements. These restrictions make the design of band-reject filters by combining low-pass and high-pass filters undesirable. The impedance interaction between filters is a serious problem unless the separation between cutoffs is many octaves, so the design of band-reject filters is best approached by transformation techniques.

6.1.1 The Band-Reject Circuit Transformation

Band-pass filters were obtained by first designing a low-pass filter with a cutoff frequency equivalent to the required bandwidth and then resonating each element to the desired center frequency. The response of the low-pass filter at DC then corresponds to the response of the band-pass filter at the center frequency.

Band-reject filters are designed by initially transforming the normalized low-pass filter into a high-pass network with a cutoff frequency equal to the required bandwidth, and at the desired impedance level. Every high-pass element is then resonated to the center frequency in the same manner as band-pass filters.

This corresponds to replacing the frequency variable in the high-pass transfer function with a new variable, which is given by

$$f_{\mathrm{br}} = f_0 \left(\frac{f}{f_0} - \frac{f_0}{f} \right) \tag{6-1}$$

As a result, the response of the high-pass filter at DC is transformed to the band-reject network at the center frequency. The bandwidth response of the band-reject filter is identical to the frequency response of the high-pass filter. The high-pass to band-reject transformation is shown in Fig. 6-1. Negative frequencies, of course, are strictly of theoretical interest, so only the response shape corresponding to positive frequencies is applicable. As in the case of band-pass filters, the response curve exhibits geometric symmetry.

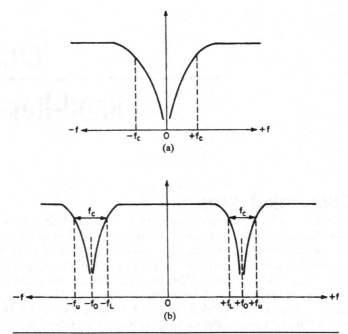

Figure 6-1   The band-reject transformation: (a) high-pass filter response; and (b) transformed band-reject filter response.

The design procedure can be summarized as follows:

1. Normalize the band-reject filter specification and select a normalized low-pass filter that provides the required attenuation within the computed steepness factor.

2. Transform the normalized low-pass filter to a normalized high-pass filter. Then scale the high-pass filter to a cutoff frequency equal to the desired bandwidth and to the preferred impedance level.

3. Resonate each element to the center frequency by introducing a capacitor in series with each inductor and an inductor in parallel with each capacitor to complete the design. The transformed circuit branches are summarized in Table 6-1.

6.1.2   All-Pole Band-Reject Filters

Band-reject filters can be derived from any all-pole or elliptic-function *LC* low-pass network. Although not as efficient as elliptic-function filters, the all-pole approach results in a simpler band-reject structure where all sections are tuned to the center frequency.

The following example demonstrates the design of an all-pole band-reject filter.

Example 6-1   Design an All-Pole *LC* Band-Reject Filter

Required:

Band-reject filter

A center frequency of 10 kHz

3 dB at ±250 Hz (9.75 kHz, 10.25 kHz)

	High-Pass Branch	Band-Reject Configuration	Circuit Values
Type I	C	L C	$L = \dfrac{1}{\omega_0^2 C}$ (6-2)
Type II	L	L C	$C = \dfrac{1}{\omega_0^2 L}$ (6-3)
Type III	C_b L_a	L_b C_b L_a C_a	$C_a = \dfrac{1}{\omega_0^2 L_a}$ (6-4) $L_b = \dfrac{1}{\omega_0^2 C_b}$ (6-5) $C_1 = \dfrac{1}{\omega_0^2 L_1}$ (6-6)
Type IV	L_1 C_2	L_1 C_1 L_2 C_2	$L_2 = \dfrac{1}{\omega_0^2 C_2}$ (6-7)

TABLE 6-1 The High-Pass to Band-Reject Transformation

30-dB minimum at ±100 Hz (9.9 kHz, 10.1 kHz)

A source and load impedance of 600 Ω

Result:

(a) Convert to a geometrically symmetrical requirement. Since the bandwidth is relatively narrow, the specified arithmetically symmetrical frequencies will determine the following design parameters:

$$F_0 = 10 \text{ kHz}$$

$$BW_{3 \text{ dB}} = 500 \text{ Hz}$$

$$BW_{30 \text{ dB}} = 200 \text{ Hz}$$

(b) Compute the band-reject steepness factor using Eq. (2-20).

$$A_s = \frac{\text{passband bandwidth}}{\text{stopband bandwidth}} = \frac{500 \text{ Hz}}{200 \text{ Hz}} = 2.5$$

The response curves of Fig. 2-45 indicate that an $n = 3$ Chebyshev normalized low-pass filter with a 1-dB ripple provides over 30 dB of attenuation within a frequency ratio of 2.5:1. The corresponding circuit is found in Table 10-31 and is shown in Fig. 6-2a.

(c) To transform the normalized low-pass circuit into a normalized high-pass filter, replace inductors with capacitors and vice versa using reciprocal element values. The transformed structure is shown in Fig. 6-2b.

(d) The normalized high-pass filter is scaled to a cutoff frequency of 500 Hz corresponding to the desired bandwidth and to an impedance level of 600 Ω. The capacitors are divided by $Z \times FSF$,

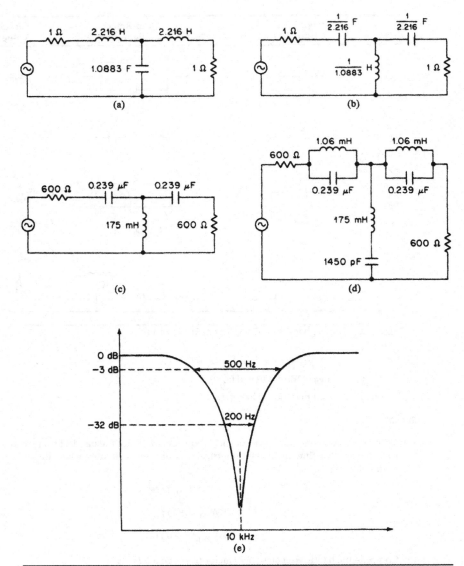

FIGURE 6-2 The band-reject filter of Example 6-1: (*a*) normalized low-pass filter; (*b*) transformed normalized high-pass filter; (*c*) frequency- and impedance-scaled high-pass filter; (*d*) transformed band-reject filter; and (*e*) frequency response.

and the inductors are multiplied by Z/FSF, where Z is 600 and the FSF (frequency-scaling factor) is given by $2\pi f_c$, where f_c is 500 Hz. The scaled high-pass filter is illustrated in Fig. 6-2c.

(*e*) To make the high-pass to band-reject transformation, resonate each capacitor with a parallel inductor and each inductor with a series capacitor. The resonating inductors for the series branches are both given by

$$L = \frac{1}{\omega_0^2 C} = \frac{1}{(2\pi 10 \times 10^3)^2 \times 0.239 \times 10^{-6}} = 1.06 \text{ mH} \tag{6-2}$$

The tuning capacitor for the shunt inductor is determined from

$$C = \frac{1}{\omega_0^2 L} = \frac{1}{\left(2\pi 10 \times 10^3\right)^2 \times 0.175} = 1,450 \text{ pF} \tag{6-3}$$

The final filter is shown in Fig. 6-2d, where all peaks are tuned to the center frequency of 10 kHz. The theoretical frequency response is illustrated in Fig. 6-2e.

When a low-pass filter undergoes a high-pass transformation followed by a band-reject transformation, the minimum Q requirement is increased by a factor equal to the Q of the band-reject filter. This can be expressed as

$$Q_{\min}(\text{band-reject}) = Q_{\min}(\text{low-pass}) \times Q_{br} \tag{6-8}$$

where values for $Q_{\min}$ (low-pass) are given in Fig. 3-8 and $Q_{br} 5 f_0 / BW_{3 dB}$. The branch Q should be several times larger than $Q_{\min}$ (band-reject) to obtain near-theoretical results.

The equivalent circuit of a band-reject filter at the center frequency can be determined by replacing each parallel-tuned circuit by a resistor of $\omega_0 L Q_L$ and each series-tuned circuit by a resistor of $\omega_0 L / Q_L$. These resistors correspond to the branch impedances at resonance, where ω_0 is $2\pi f_0$, L is the branch inductance, and Q_L is the branch Q, which is normally determined only by the inductor losses.

It is then apparent that at the center frequency, the circuit can be replaced by a resistive voltage divider. The amount of attenuation that can be obtained is then directly controlled by the branch Qs. Let's determine the attenuation of the circuit in Example 6-1 for a finite value of inductor Q.

Example 6-2   Estimate Maximum Band-Reject Rejection as a Function of Q

Required:

Estimate the amount of rejection obtainable at the center frequency of 10 kHz for the band-reject filter of Example 6-1. An inductor Q of 100 is available, and the capacitors are assumed to be lossless. Also determine if the Q is sufficient to retain the theoretical passband characteristics.

Result:

(a) Compute the equivalent resistances at resonance for all tuned circuits using Eqs. (5-21) and (5-30).

Parallel-tuned circuits:

$$R = \omega_0 L Q_L = 2\pi \times 10^4 \times 1.06 \times 10^{-3} \times 100 = 6,660 \ \Omega$$

Series-tuned circuits:

$$R = \frac{\omega_0 L}{Q_L} = \frac{2\pi \times 10^4 \times 0.175}{100} = 110 \ \Omega$$

(b) The equivalent circuit at 10 kHz is shown in Fig. 6-3. Using conventional circuit analysis methods, such as mesh equations or approximation techniques, the overall loss is found to be 58 dB. Since the flat loss due to the 600-Ω terminations is 6 dB, the relative attenuation at 10 kHz will be 52 dB.

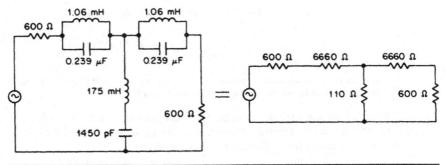

FIGURE 6-3 The equivalent circuit at the center frequency for the filter in Figure 6-2.

(c) The curves of Fig. 3-8 indicate that an $n = 3$ Chebyshev filter with a 1-dB ripple has a minimum theoretical Q requirement of 4.5. The minimum Q of the band-reject filter is given by

$$Q_{min}(\text{band-reject}) = Q_{min}(\text{low-pass}) \times Q_{br} = \frac{10,000}{500} = 90$$

Therefore, the available Q of 100 is barely adequate, and some passband rounding will occur in addition to the reduced stopband attenuation. The resulting effect on frequency response is shown in Fig. 6-4.

6.1.3 Elliptic-Function Band-Reject Filters

The superior properties of the elliptic-function family of filters can also be applied to band-reject requirements. Extremely steep characteristics in the transition region between passband and stopband can be achieved much more efficiently than with all-pole filters.

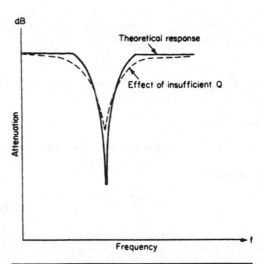

FIGURE 6-4 The effects of insufficient Q upon a band-reject filter.

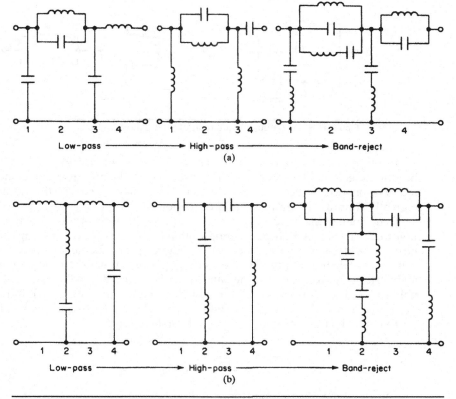

FIGURE 6-5 The band-reject transformation of elliptic-function filters: (a) standard configuration; and (b) dual configuration.

Saal and Ulbrich, as well as Zverev (see References), have extensively tabulated the *LC* values for normalized elliptic-function low-pass networks. Using the *Filter Solutions* program or the ELI 1.0 program, low-pass filters can be directly designed using the filter requirements as the program input rather than engaging normalized tables. These circuits can then be transformed into high-pass filters, and subsequently, to a band-reject filter in the same manner as the all-pole filters.

Since each normalized low-pass filter can be realized in dual forms, the resulting band-reject filters can also take on two different configurations, as illustrated in Fig. 6-5.

Branch 2 of the standard band-reject filter circuit corresponds to the type III network shown in Table 6-1. This branch provides a pair of geometrically related zeros, one above and one below the center frequency. These zeros result from two conditions of parallel resonance. However, the circuit configuration itself is not very desirable. The elements corresponding to the individual parallel resonances are not distinctly isolated, since each resonance is determined by the interaction of a number of elements. This makes tuning somewhat difficult. For very narrow filters, the element values also become somewhat unreasonable.

An identical situation occurred during the band-pass transformation of an elliptic-function low-pass filter discussed in Sec. 5.1. An equivalent configuration was presented as an alternate and is repeated in Fig. 6-6.

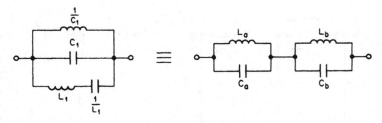

Figure 6-6   The equivalent circuit of a type III network.

The type III network of Fig. 6-6 has reciprocal element values, which occur when the band-reject filter has been normalized to a 1-rad/s center frequency, since the equation of resonance, $\omega_0^2 LC = 1$, then reduces to $LC = 1$. The reason for this normalization is to greatly simplify the transformation equations.

To normalize the band-reject filter circuit, first transform the normalized low-pass filter to a normalized high-pass configuration in the conventional manner by replacing inductors with capacitors and vice versa using reciprocal element values. The high-pass elements are then multiplied by the factor Q_{br}, which is equal to f_0/BW, where f_0 is the geometric center frequency of the band-reject filter and BW is the bandwidth. The normalized band-reject filter can be directly obtained by resonating each inductor with a series capacitor and each capacitor with a parallel inductor using reciprocal values.

To make the transformation of Fig. 6-6, first compute

$$\beta = 1 + \frac{1}{2L_1 C_1} + \sqrt{\frac{1}{4L_1^2 C_1^2} + \frac{1}{L_1 C_1}} \tag{6-9}$$

The values are then found from

$$L_a = \frac{1}{C_1(\beta + 1)} \tag{6-10}$$

$$L_b = \beta L_a \tag{6-11}$$

$$C_a = \frac{1}{L_b} \tag{6-12}$$

$$C_b = \frac{1}{L_a} \tag{6-13}$$

The resonant frequencies for each tuned circuit are given by

$$\Omega_{\infty, a} = \sqrt{\beta} \tag{6-14}$$

and

$$\Omega_{\infty, b} = \frac{1}{\Omega_{\infty, a}} \tag{6-15}$$

After the normalized band-reject filter has undergone the transformation of Fig. 6-6 wherever applicable, the circuit can be scaled to the desired impedance level and frequency. The inductors are multiplied by Z/FSF, and capacitors are divided by $Z \times FSF$.

FIGURE 6-7 The equivalent circuit of the type IV network.

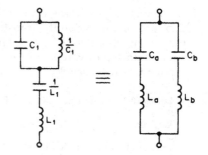

The value of Z is the desired impedance level, and the FSF in this case is equal to ω_0 ($\omega_0 = 2\pi f_0$). The resulting resonant frequencies in hertz are determined by multiplying the normalized radian resonant frequencies by f_0.

Branch 2 of the band-reject filter derived from the dual low-pass structure of Fig. 6-5b corresponds to the type IV network of Table 6-1. This configuration realizes a pair of finite zeros resulting from two conditions of series resonance. However, as in the case of the type III network, the individual resonances are determined by the interaction of all the elements, which makes tuning difficult and can result in unreasonable values for narrow filters. An alternative configuration is shown in Fig. 6-7 consisting of two series-resonant circuits in parallel.

To simplify the transformation equations, the type IV network requires reciprocal values, so the band-reject filter must be normalized to a 1-rad/s center frequency. This is accomplished as previously described, and the filter is subsequently denormalized after the transformations have been made.

The transformation is accomplished as follows:

First, compute

$$\beta = 1 + \frac{1}{2L_1 C_1} + \sqrt{\frac{1}{4L_1^2 C_1^2} + \frac{1}{L_1 C_1}} \qquad (6\text{-}16)$$

then

$$L_a = \frac{(\beta + 1)L_1}{\beta} \qquad (6\text{-}17)$$

$$C_a = \frac{1}{(\beta + 1)L_1} \qquad (6\text{-}18)$$

$$L_b = \frac{1}{C_a} \qquad (6\text{-}19)$$

$$C_b = \frac{1}{L_a} \qquad (6\text{-}20)$$

$$\Omega_{\infty, a} = \sqrt{\beta} \qquad (6\text{-}21)$$

$$\Omega_{\infty, b} = \frac{1}{\Omega_{\infty, a}} \qquad (6\text{-}22)$$

The standard configuration of the elliptic-function filter is usually preferred over the dual circuit so that the transformed low-pass zeros can be realized using the structure of Fig. 6-6. Parallel-tuned circuits are generally more desirable than series-tuned circuits since they can be transformed to alternate L/C ratios to optimize Q and reduce capacitor values (see Sec. 8.2 on tapped inductors).

Example 6-3   Design an LC Elliptic-Function Band-Reject Filter

Required:

Design a band-reject filter to satisfy the following requirements:

   1-dB maximum at 2,200 and 2,800 Hz
   50-dB minimum at 2,300 and 2,700 Hz
   A source and load impedance of 600 Ω

Result:

(*a*) Convert to a geometrically symmetrical requirement. First, calculate the geometric center frequency using Eq. (2-14).

$$f_0 = \sqrt{f_L f_u} = \sqrt{2,200 \times 2,800} = 2,482 \text{ Hz}$$

Compute the corresponding geometric frequency for each stopband frequency given using Eq. (2-18).

$$f_1 f_2 = f_0^2$$

f_1	f_2	$f_2 - f_1$
2,300 Hz	2,678 Hz	378 Hz
2,282 Hz	2,700 Hz	418 Hz

The second pair of frequencies is retained since they represent the steeper requirement. The complete geometrically symmetrical specification can be stated as

$$f_0 = 2,482 \text{ Hz}$$

$$\text{BW}_{1 \text{ dB}} = 600 \text{ Hz}$$

$$\text{BW}_{50 \text{ dB}} = 418 \text{ Hz}$$

(*b*) Compute the band-reject steepness factor using Eq. (2-20).

$$A_s = \frac{\text{passband bandwidth}}{\text{stopband bandwidth}} = \frac{600 \text{ Hz}}{400 \text{ Hz}} = 1.435$$

A normalized low-pass filter must be chosen that makes the transition from less than 1 dB to more than 50 dB within a frequency ratio of 1.435. An elliptic-function filter will be used.

(*c*) Open *Filter Solutions.*
   Check the *Stop Band Freq* box.
   Enter **0.18** in the *Pass Band Ripple (dB)* box.
   Enter **1** in the *Pass Band Freq* box.
   Enter **1.435** in the *Stop Band Freq* box.
   The *Frequency Scale Rad/Sec* box should be checked.
   Enter **1** for *Source Res* and *Load Res.*

(d) Click the **Set Order** control button to open the second panel.
Enter **50** for the **Stop Band Attenuation (dB)**.
Click the **Set Minimum Order** button and then click **Close**.
6 Order is displayed on the main control panel.
Check the **Even Order Mod** box.

(e) Click the **Synthesize Filter** button.

Two schematics are presented by *Filter Solutions*. Use Lumped Filter 1, which is shown in Fig. 6-8a.

(f) The normalized low-pass filter is now transformed into a normalized high-pass structure by replacing all inductors with capacitors, and vice versa, using reciprocal values. The resulting filter is given in Fig. 6-8b.

(g) To obtain a normalized band-reject filter so that the transformation of Fig. 6-6 can be performed, first multiply all the high-pass elements by Q_{br} ($f_0 / \mathrm{BW}_{x\,\mathrm{dB}}$), resulting in

$$Q_{br} = \frac{f_0}{\mathrm{BW}_{1\,\mathrm{dB}}} = \frac{2{,}482\,\mathrm{Hz}}{600\,\mathrm{Hz}} = 4.137$$

The modified high-pass filter is shown in Fig. 6-8c.

(h) Each high-pass inductor is resonated with a series capacitor and each capacitor is resonated with a parallel inductor to obtain the normalized band-reject filter. Since the center frequency is 1 rad/s, the resonant elements are simply the reciprocal of each other, as illustrated in Fig. 6-8d.

(i) The type III networks of the second and fourth branches are now transformed to the equivalent circuit of Fig. 6-6 as follows using Eqs. (6-9) to (6-15):

The type III network of the third branch:

$$L_1 = 11.428\,\mathrm{H} \qquad C_1 = 3.270\,\mathrm{F}$$

$$\beta = 1 + \frac{1}{2L_1 C_1} + \sqrt{\frac{1}{4L_1^2 C_1^2} + \frac{1}{L_1 C_1}} = 1.1775 \tag{6-9}$$

$$L_a = \frac{1}{C_1(\beta + 1)} = 0.1404\,\mathrm{H} \tag{6-10}$$

$$L_b = \beta L_a = 0.1654\,\mathrm{H} \tag{6-11}$$

$$C_a = \frac{1}{L_b} = 6.047\,\mathrm{F} \tag{6-12}$$

$$C_b = \frac{1}{L_a} = 7.1205\,\mathrm{F} \tag{6-13}$$

$$\Omega_{\infty,\,a} = \sqrt{\beta} = 1.0851 \tag{6-14}$$

$$\Omega_{\infty,\,b} = \frac{1}{\Omega_{\infty,\,a}} = 0.92155 \tag{6-15}$$

The type III network of the fourth branch:

$$L_1 = 20.08\,\mathrm{H} \quad C_1 = 3.260\,\mathrm{F}$$

$$\beta = 1.13147$$

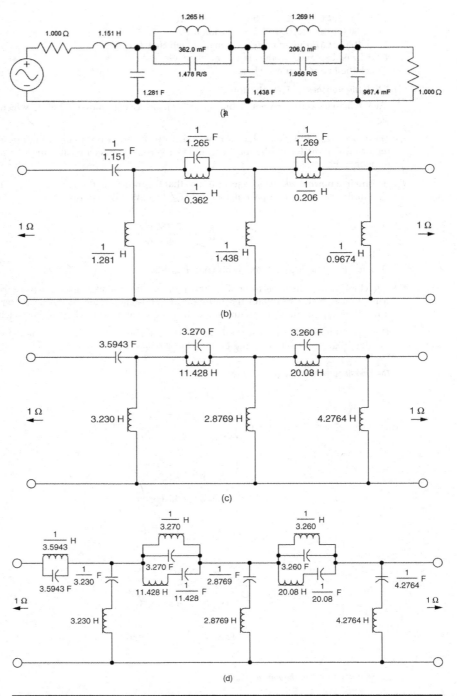

FIGURE 6-8 Elliptic-function band-reject filter: (*a*) normalized low-pass filter; (*b*) transformed high-pass filter; (*c*) high-pass filter with elements multiplied by Q_{br}; (*d*) normalized band-reject filter;

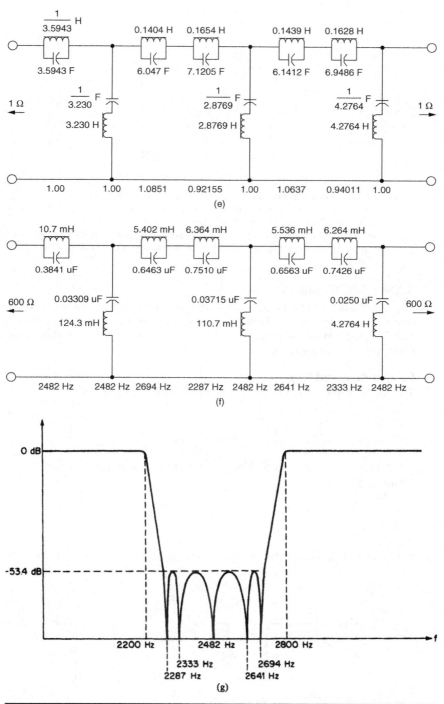

FIGURE 6-8 *(Continued)* Elliptic-function band-reject filter: *(e)* transformed type III network; *(f)* frequency- and impedance-scaled circuit; and *(g)* frequency response.

$$L_a = 0.139 \text{ H}$$

$$L_b = 0.1628 \text{ H}$$

$$C_a = 6.1412 \text{ F}$$

$$C_b = 6.9486 \text{ F}$$

$$C_b = 6.9486 \text{ F}$$

$$\Omega_{\infty,a} = 1.0637$$

$$\Omega_{\infty,b} = 0.94011$$

The resulting normalized band-reject filter is shown in Fig. 6-8e.

(j) The final filter can now be obtained by frequency- and impedance-scaling the normalized band-reject filter to a center frequency of 2,482 Hz and 600 Ω. The inductors are multiplied by Z/FSF, and the capacitors are divided by Z 3 FSF, where Z is 600 and the FSF is $2\pi f_0$, where f_0 is 2,482 Hz. The circuit is given in Fig. 6-8f, where the resonant frequencies of each section were obtained by multiplying the normalized frequencies by f_0. The frequency response is illustrated in Fig. 6-8g.

6.1.4   Null Networks

A null network can be loosely defined as a circuit intended to reject either a single frequency or a very narrow band of frequencies, and is frequently referred to as a trap. Notch depth rather than rate of roll-off is the prime consideration, and the circuit is restricted to a single section.

Parallel-Resonant Trap

The RC high-pass circuit of Fig. 6-9a has a 3-dB cutoff given by

$$f_c = \frac{1}{2\pi RC} \tag{6-23}$$

A band-reject transformation will result in the circuit in Fig. 6-9b. The value of L is computed from

$$L = \frac{1}{\omega_0^2 C} \tag{6-24}$$

where $\omega_0 = 2\pi f_0$. The center frequency is f_0, and the 3-dB bandwidth is f_c.

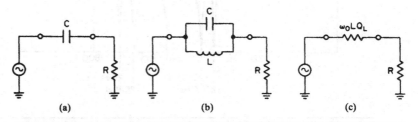

Figure 6-9   A parallel-resonant trap: (a) RC high-pass filter; (b) results of a band-reject transformation; and (c) equivalent circuit at f_0.

The frequency response of a first-order band-reject filter can be expressed as

$$A_{dB} = 10 \log \left[1 + \left(\frac{BW_{3\,dB}}{BW_{x\,dB}} \right)^2 \right]$$

(6-25)

where $BW_{3\,dB}$ is the 3-dB bandwidth corresponding to f_c in Eq. (6-23), and where $BW_{x\,dB}$ is the bandwidth of interest. The response can also be determined from the normalized Butterworth attenuation curves of Fig. 2-34 corresponding to $n = 1$, where $BW_{3\,dB}/BW_{x\,dB}$ is the normalized bandwidth.

The impedance of a parallel-tuned circuit at resonance is equal to $\omega_0 L Q_L$, where Q_L is the inductor Q and the capacitor is assumed to be lossless. We can then represent the band-reject filter at f_0 by the equivalent circuit in Fig. 6-9c. After some algebraic manipulation involving Eqs. (6-23) and (6-24) and the circuit in Fig. 6-9c, we can derive the following expression for the attenuation at resonance of the $n = 1$ band-reject filter of Fig. 6-9:

$$A_{dB} = 20 \log \left(\frac{Q_L}{Q_{br}} + 1 \right)$$

(6-26)

where $Q_{br} 5 f_0 / BW_{3\,dB}$. Equation (6-26) is plotted in Fig. 6-10. When Q_{br} is high, the required inductor Q may become prohibitively large in order to attain sufficient attenuation at f_0.

The effect of insufficient inductor Q will not only reduce relative attenuation, but will also cause some rounding of the response near the cutoff frequencies. Therefore, the ratio Q_L / Q_{br} should be as high as possible.

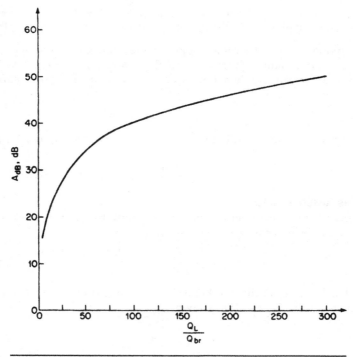

Figure 6-10 Attenuation vs. Q_L/Q_{br}.

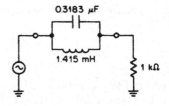

FIGURE 6-11   The parallel-resonant trap of Example 6-4.

Example 6-4   Design a Parallel-Resonant Trap

Required:

Design a parallel-resonant circuit that has a 3-dB bandwidth of 500 Hz and a center frequency of 7,500 Hz. The source resistance is zero, and the load is 1 kΩ. Also determine the minimum inductor Q for a relative attenuation of at least 30 dB at 7,500 Hz.

Result:

(*a*) Compute the value of the capacitor using Eq. (6-23).

$$C = \frac{1}{2\pi f_c R} = \frac{1}{2\pi 500 \times 1,000} = 0.3183 \ \mu F$$

The inductance is calculated using Eq. (6-24).

$$L = \frac{1}{\omega_0^2 C} = \frac{1}{(2\pi 7,500)^2 \times 3.183 \times 10^{27}} - 1.415 \ \text{mH}$$

The resulting circuit is shown in Fig. 6-11.

(*b*) The required ratio of Q_L/Q_{br} for 30-dB attenuation at f_0 can be determined from Fig. 6-10 or Eq. (6-26), and is approximately 30. Therefore, the inductor Q should exceed 30 Q_{br} or 450, where $Q_{br} = f_0/BW_{3\,dB}$.

Frequently, it is desirable to operate the band-reject network between equal source and load terminations instead of a voltage source, as in Fig. 6-9. If a source and load resistor are specified where both are equal to R, Eq. (6-23) is modified to

$$f_c = \frac{1}{4\pi RC} \tag{6-27}$$

When the source and load are unequal, the cutoff frequency is given by

$$f_c = \frac{1}{2\pi(R_S + R_L)C} \tag{6-28}$$

Series-Resonant Trap

An $n = 1$ band-reject filter can be derived from the RL high-pass filter of Fig. 6-12*a*. The 3-dB cutoff is determined from

$$f_c = \frac{R}{2\pi L} \tag{6-29}$$

The band-reject filter of Fig. 6-12*b* is obtained by resonating the coil with a series capacitor using Eq. (6-24) as shown:

$$C = \frac{1}{\omega_0^2 L}$$

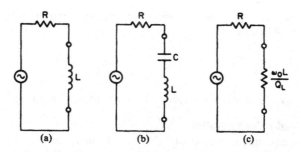

FIGURE **6-12**   A series-resonant trap: (a) RL high-pass filter; (b) result of band-reject transformation; and (c) equivalent circuit at f_0.

The center frequency is f_0, and the 3-dB bandwidth is equal to f_c. The series losses of an inductor can be represented by a resistor of $\omega_0 L / Q_L$. The equivalent circuit of the band-reject network at resonance is given by the circuit in Fig. 6-12c and the attenuation computed from either Eq. (6-26) or Fig. 6-10.

Example 6-5   Design a Series-Resonant Trap

Required:

Design a series-resonant circuit with a 3-dB bandwidth of 500 Hz and a center frequency of 7,500 Hz, as in the previous example. The source impedance is 1 kΩ, and the load is assumed infinite.

Result:

Compute the element values using the relationships in Eqs. (6-29) and (6-24) as shown:

$$L = \frac{R}{2\pi f_c} = \frac{1,000}{2\pi 500} = 0.318 \text{ H}$$

and

$$C = \frac{1}{\omega_0^2 L} = \frac{1}{(2\pi 7500)^2 \, 0.318} = 1420 \text{ pF}$$

The circuit is given in Fig. 6-13.

When a series-resonant trap is to be terminated with a load resistance equal to the source, the high-pass 3-dB cutoff and resulting 3-dB bandwidth of the band-reject filter are given by

$$f_c = \frac{R}{4\pi L} \tag{6-30}$$

FIGURE **6-13**   The series-resonant trap of Example 6-5.

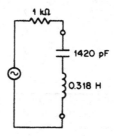

For the more general case where source and load are unequal, the cutoff frequency is determined from

$$f_c = \frac{R_{eq}}{2\pi L} \tag{6-31}$$

where R_{eq} is the equivalent value of the source and load resistors in parallel.

The Bridged-T Configuration

The resonant traps previously discussed suffer severe degradation of notch depth unless an inductor Q is many magnitudes greater than Q_{br}. The bridged-T band-reject structure can easily provide rejection of 60 dB or more with practical values of inductor Q. The configuration is shown in Figure 6-14a.

To understand the operation of the circuit, let us first consider the equivalent circuit of a center-tapped inductor with a coefficient of magnetic coupling equal to unity, which is shown in Fig. 6-14b. The inductance between terminals A and C corresponds to L of Fig. 6-14a. The inductance between A and B or B and C is equal to $L/4$, since, as the reader may recall, the impedance across one-half of a center-tapped autotransformer is one-fourth the overall impedance. This occurs because the impedance is proportional to the turns ratio squared.

The impedance of a parallel-tuned circuit at resonance was previously determined to be equivalent to a resistor of $\omega_0 L Q_L$. Since the circuit in Fig. 6-14a is center-tapped, the equivalent three-terminal network is shown in Fig. 6-14c. The impedance between A and C is still $\omega_0 L Q_L$. A negative resistor must then exist in the middle shunt branch so that the impedance across one-half of the tuned circuit is one-fourth the overall

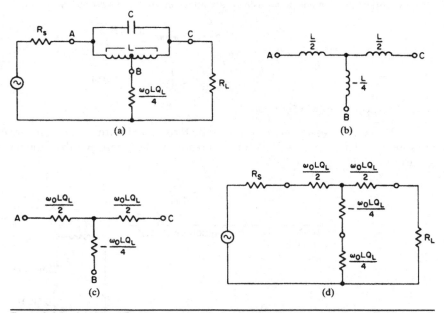

Figure 6-14   A bridged-T null network: (a) circuit configuration; (b) equivalent circuit of center-tapped inductor; (c) tuned circuit equivalent at resonance; and (d) bridged-T equivalent circuit at resonance.

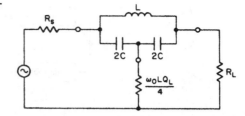

FIGURE 6-15 An alternate form of bridged-T.

impedance, or $\omega_0 LQ_L/4$. Of course, negative resistors or inductors are physically impossible as individual passive two-terminal elements, but they can be embedded within an equivalent circuit.

If we combine the equivalent circuit in Fig. 6-14c with the bridged-T network in Fig. 6-14a, we obtain the circuit in Fig. 6-14d. The positive and negative resistors in the center branch will cancel, resulting in infinite rejection of center frequency. The degree of rejection actually obtained is dependent upon a variety of factors, such as center-tap accuracy, the coefficient of coupling, and the magnitude of Q_L. When the bridged-T configuration is implemented after modifying the parallel trap design in Fig. 6-9b by adding a center tap and a resistor of $\omega_0 LQ_L/4$, a dramatic improvement in notch depth will usually occur.

A center-tapped inductor is not always available or practical. An alternative form of a bridged-T is given in Fig. 6-15. The parallel resonant trap design of Fig. 6-9 is modified by splitting the capacitor into two capacitors of twice the value, and a resistor of $\omega_0 LQ_L/4$ is introduced. The two capacitors should be closely matched.

In conclusion, the bridged-T structure is an economical and effective means of increasing the available notch rejection of a parallel resonant trap without increasing the inductor Q. However, as a final general comment, a single null section can provide high rejection only at a single frequency or relatively narrow band of frequencies for a given 3-dB bandwidth, since $n = 1$. The stability of the circuit then becomes a significant factor. A higher-order band-reject filter design can have a wider stopband and yet maintain the same 3-dB bandwidth.

6.2 Active Band-Reject Filters

This section considers the design of active band-reject filters for both wideband and narrowband applications. Active null networks are covered, and the popular twin-T circuit is discussed in detail.

6.2.1 Wideband Active Band-Reject Filters

Wideband filters can be designed by first separating the specification into individual low-pass and high-pass requirements. Low-pass and high-pass filters are then independently designed and combined by paralleling the inputs and summing both outputs to form the band-reject filter.

A wideband approach is valid when the separation between cutoffs is an octave or more for all-pole filters so that minimum interaction occurs in the stopband when the outputs are summed (see Sec. 2.1 and Fig. 2-13). Elliptic-function networks will require less separation since their characteristics are steeper.

An inverting amplifier is used for summing and can also provide gain. Filters can be combined using the configuration in Fig. 6-16a, where R is arbitrary and A is the

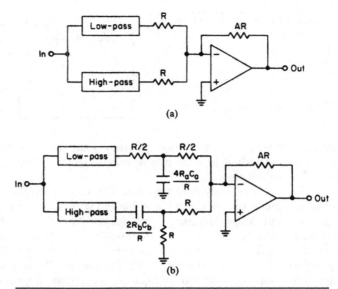

FIGURE 6-16 Wideband band-reject filters: (a) the combining of filters with low output impedance; and (b) combined filters requiring RC real poles.

desired gain. The individual filters should have a low output impedance to avoid loading by the summing resistors.

The VCVS elliptic-function low-pass and high-pass filters of Secs. 3.2 and 4.2 each require an RC termination on the last stage to provide the real pole. These elements can be combined with the summing resistors, resulting in the circuit in Fig. 6-16b. R_a and C_a correspond to the denormalized values of R_5 for the low-pass filter in Fig. 3-20. The denormalized high-pass filter real-pole values are R_b and C_b. If only one filter is of the VCVS type, the summing network of the filter with the low output impedance can be replaced by a single resistor with a value of R.

When one or both filters are of the elliptic-function type, the ultimate attenuation obtainable is determined by the filter with the lesser value of A_{min} since the stopband output is the summation of the contributions of both filters.

Example 6-6 Design a Wideband Band-Reject Filter

Required:

Design an active band-reject filter with 3-dB points at 100 and 400 Hz, and greater than 35 dB of attenuation between 175 and 225 Hz.

Result:

(a) Since the ratio of upper cutoff to lower cutoff is well in excess of an octave, a wideband approach can be used. First, separate the specification into individual low-pass and high-pass requirements.

Low-pass:	High-pass:
3 dB at 100 Hz	3 dB at 400 Hz
35-dB minimum at 175 Hz	35-dB minimum at 225 Hz

(*b*) The low-pass and high-pass filters can now be independently designed as follows:

Low-pass filter:

Compute the steepness factor using Eq. (2-11).

$$A_s = \frac{f_s}{f_c} = \frac{175 \text{ Hz}}{100 \text{ Hz}} = 1.75$$

An $n = 5$ Chebyshev filter with a 0.5-dB ripple is chosen using Fig. 2-44. The normalized active low-pass filter values are given in Table 10-39, and the circuit is shown in Fig. 6-17*a*.

To denormalize the filter, multiply all resistors by Z and divide all capacitors by $Z \times$ FSF, where Z is conveniently selected at 10^5 and the FSF is $2\pi f_c$, where f_c is 100 Hz. The denormalized low-pass filter is given in Fig. 6-17*b*.

High-pass filter:

Compute the steepness factor using Eq. (2-13).

$$A_s = \frac{f_c}{f_s} = \frac{400 \text{ Hz}}{225 \text{ Hz}} = 1.78$$

An $n = 5$ Chebyshev filter with a 0.5-dB ripple will also satisfy the high-pass requirement. A high-pass transformation can be performed on the normalized low-pass filter in Fig. 6-17*a* to obtain the circuit in Fig. 6-17*c*. All resistors have been replaced with capacitors and vice versa using reciprocal element values.

The normalized high-pass filter is then frequency- and impedance-scaled by multiplying all resistors by Z and dividing all capacitors by $Z \times$ FSF, where Z is chosen at 10^5 and FSF is $2\pi f_c$, using an f_c of 400 Hz. The denormalized high-pass filter is shown in Fig. 6-17*d* using standard 1 percent resistor values.

(*c*) The individual low-pass and high-pass filters can now be combined using the configuration in Fig. 6-16*a*. Since no gain is required, A is set equal to unity. The value of R is conveniently selected at 10 kΩ, resulting in the circuit in Fig. 6-17*e*.

6.2.2   Band-Reject Transformation of Low-Pass Poles

The wideband approach to the design of band-reject filters using combined low-pass and high-pass networks is applicable to bandwidths of typically an octave or more. If the separation between cutoffs is insufficient, interaction in the stopband will occur, resulting in inadequate stopband rejection (see Fig. 2-13).

A more general approach involves normalizing the band-reject requirement and selecting a normalized low-pass filter type that meets these specifications. The corresponding normalized low-pass poles are then directly transformed to the band-reject form and realized using active sections.

A band-reject transfer function can be derived from a low-pass transfer function by substituting the frequency variable f with a new variable given by

$$f_{\text{br}} = \cfrac{1}{f_0 \left(\cfrac{f}{f_0} - \cfrac{f_0}{f} \right)} \qquad (6\text{-}32)$$

This transformation combines the low-pass to high-pass and subsequent band-reject transformation discussed in Sec. 6.1 so that a band-reject filter can be obtained directly from the low-pass transfer function.

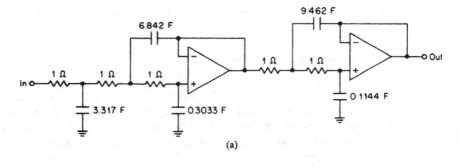

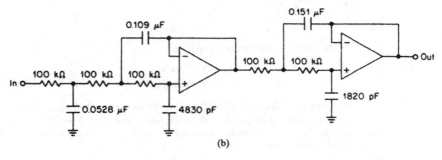

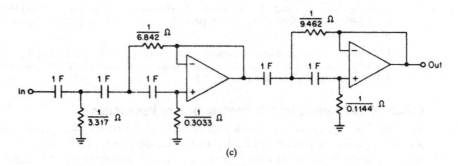

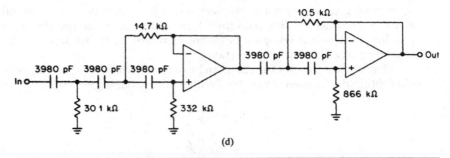

Figure 6-17   The wideband band-reject filter of Example 6-6: (a) normalized low-pass filter; (b) denormalized low-pass filter; (c) transformed normalized high-pass filter; (d) denormalized high-pass filter.

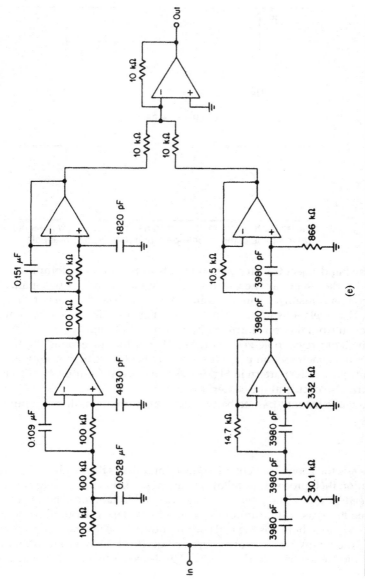

Figure 6-17 (*Continued*) The wideband band-reject filter of Example 6-6: (*e*) combining filters to obtain a band-reject response.

261

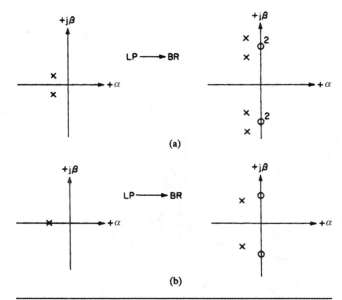

FIGURE 6-18 The band-reject transformation of low-pass poles: (a) low-pass complex pole pair; and (b) low-pass real pole.

The band-reject transformation results in two pairs of complex poles and a pair of second-order imaginary zeros from each low-pass complex pole pair. A single low-pass real pole is transformed into a complex pole pair and a pair of first-order imaginary zeros. These relationships are illustrated in Fig. 6-18. The zeros occur at center frequency and result from the transformed low-pass zeros at infinity.

The band-reject pole-zero pattern in Fig. 6-18a corresponds to two band-reject sections where each section provides a zero at center frequency and also provides one of the pole pairs. The pattern in Fig. 6-18b is realized by a single band-reject section where the zero also occurs at the center frequency.

To make the low-pass to band-reject transformation, first compute

$$Q_{br} = \frac{f_0}{BW} \tag{6-33}$$

where f_0 is the geometric center frequency and BW is the passband bandwidth. The transformation then proceeds as follows in the next section for complex poles and real poles.

Complex Poles. The tables in Chap. **10** contain tabulated poles corresponding to the all-pole low-pass filter families discussed in Chap. 2. Complex poles are given in the form $-\alpha \pm j\beta$, where α is the real coordinate and β is the imaginary part. Given α, β, Q_{br}, and f_0, the following computations [see Eqs. (6-34) to (6-44)] result in two sets of values for Q and frequency that defines two band-reject filter sections. Each section also has a zero at f_0.

$$C = \alpha^2 + \beta^2 \tag{6-34}$$

$$D = \frac{\alpha}{Q_{br}C} \tag{6-35}$$

$$E = \frac{\beta}{Q_{br}C} \tag{6-36}$$

$$F = E^2 - D^2 + 4 \tag{6-37}$$

$$G = \sqrt{\frac{F}{2} + \sqrt{\frac{F^2}{4} + D^2 E^2}} \tag{6-38}$$

$$H = \frac{DE}{G} \tag{6-39}$$

$$K = \frac{1}{2}\sqrt{(D+H)^2 + (E+G)^2} \tag{6-40}$$

$$Q = \frac{K}{D+H} \tag{6-41}$$

$$f_{ra} = \frac{f_0}{K} \tag{6-42}$$

$$f_{rb} = Kf_0 \tag{6-43}$$

$$f_\infty = f_0 \tag{6-44}$$

The two band-reject sections have resonant frequencies of f_{ra} and f_{rb} (in hertz) and identical Qs given by Eq. (6-41). In addition, each section has a zero at f_0, the filter geometric center frequency.

Real Poles. A normalized low-pass real pole with a real coordinate of α_0 is transformed into a single band-reject section with a Q given by

$$Q = Q_{br}\alpha_0 \tag{6-45}$$

This section-resonant frequency is equal to f_0. The section must also have a transmission zero at f_0.

Example 6-7   Calculate Pole and Zero Locations for a Band-Reject Filter

Required:

Determine the pole and zero locations for a band-reject filter with the following specifications:

A center frequency of 3,600 Hz
3 dB at ±150 Hz
40-dB minimum at ±30 Hz

Result:

(*a*) Since the filter is narrow, the requirement can be treated directly in its arithmetically symmetrical form:

$$f_0 = 3600 \text{ Hz}$$

$$BW_{3\text{ dB}} = 300 \text{ Hz}$$

$$BW_{40\text{ dB}} = 60 \text{ Hz}$$

The band-reject steepness factor is given by Eq. (2-20) as

$$A_s = \frac{\text{passband bandwidth}}{\text{stopband bandwidth}} = \frac{300 \text{ Hz}}{60 \text{ Hz}} = 5$$

(b) An $n = 3$ Chebyshev normalized low-pass filter with a 0.1-dB ripple is selected using Fig. 2-42. The corresponding pole locations are found in Table 10-23 and are

$-0.3500 \pm j0.8695$
-0.6999

First, make the preliminary computation using Eq. (6-33) as shown:

$$Q_{br} = \frac{f_0}{BW_{3 \text{ dB}}} = \frac{3,600 \text{ Hz}}{300 \text{ Hz}} = 12$$

The low-pass to band-reject pole transformation is performed as follows using Eqs. (6-34) to (6-45):

Complex-pole transformation:

$$\alpha = 0.3500 \qquad \beta = 0.8695$$

$$C = \alpha^2 + \beta^2 = 0.878530$$

$$D = \frac{\alpha}{Q_{br}C} = 0.033199$$

$$E = \frac{\beta}{Q_{br}C} = 0.082477$$

$$F = E^2 - D^2 + 4 = 4.0005700$$

$$G = \sqrt{\frac{F}{2} + \sqrt{\frac{F^2}{4} + D^2 E^2}} = 2.001425$$

$$H = \frac{DE}{G} = 0.001368$$

$$K = \frac{1}{2}\sqrt{(D + H)^2 + (E + G)^2} = 1.042094$$

$$Q = \frac{K}{D + H} = 30.15$$

$$f_{ra} = \frac{f_0}{K} = 3,455 \text{ Hz}$$

$$f_{rb} = Kf_0 = 3,752 \text{ Hz}$$

$$f_\infty = f_0 = 3,600 \text{ Hz}$$

Real-pole transformation:

$$\alpha_0 = 0.699$$

$$Q = Q_{br}\alpha_0 = 8.40$$

$$f_r = f_\infty = f_0 = 3,600 \text{ Hz}$$

The block diagram is shown in Fig. 6-19.

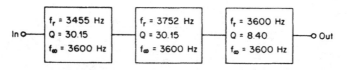

FIGURE 6-19 The block diagram of Example 6-7.

6.2.3 Narrowband Active Band-Reject Filters

Narrowband active band-reject filters are designed by first transforming a set of normalized low-pass poles to the band-reject form. The band-reject poles are computed in terms of resonant frequency f_r, Q, and f_∞ using the results of Sec. 6.2 and are then realized with active band-reject sections.

The VCVS Band-Reject Section

Complex low-pass poles result in a set of band-reject parameters where f_r and f_∞ do not occur at the same frequency. Band-reject sections are then required that permit independent selection of f_r and f_∞ in their design procedure. Both the VCVS and biquad circuits covered in Sec. 5.2 under "Elliptic-Function Band-Pass Filters" have this degree of freedom.

The VCVS realization is shown in Fig. 6-20. The design equations were given in Sec. 5.2 under "Elliptic-Function Band-Pass Filters" and are repeated here for convenience, where f_r, Q, and f_∞ are obtained by the band-reject transformation procedure of Sec. 6.2. The values are computed as follows:

First, calculate

$$a = \frac{1}{Q} \tag{6-46}$$

$$b = \left(\frac{f_\infty}{f_r}\right)^2 \tag{6-47}$$

$$c_1 = 2\pi f_r \tag{6-48}$$

Select C, then
$$C_1 = C \tag{6-49}$$

$$C_3 = C_4 = \frac{C_1}{2} \tag{6-50}$$

$$C_2 \geq \frac{C_1(b-1)}{4} \tag{6-51}$$

$$R_3 = \frac{1}{cC_1\sqrt{b}} \tag{6-52}$$

$$R_1 = R_2 = 2R_3 \tag{6-53}$$

$$R_4 = \frac{4\sqrt{b}}{cC_1(1-b) + 4cC_2} \tag{6-54}$$

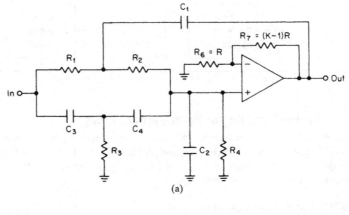

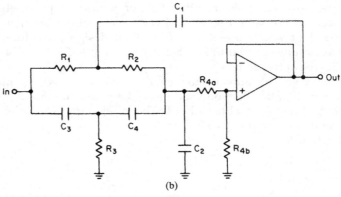

FIGURE 6-20 A VCVS realization for band-reject filters: (a) circuit for $K > 1$; and (b) circuit for $K < 1$.

$$K = 2 + \frac{2C_2}{C_1} - \frac{a}{2\sqrt{b}} + \frac{2}{C_1\sqrt{b}}\left(\frac{1}{cR_4} - aC_2\right) \tag{6-55}$$

$$R_6 = R \qquad \text{and} \qquad R_7 = (K-1)R \tag{6-56}$$

where R can be arbitrarily chosen.

The circuit in Fig. 6-20a is used when $K > 1$. In the cases where $K < 1$, the configuration in Fig. 6-20b is utilized, where

$$R_{4a} = (1 - K)R_4 \tag{6-57}$$

and

$$R_{4b} = KR_4 \tag{6-58}$$

The section gain at DC is given by

$$\text{Section gain} = \frac{bKC_1}{4C_2 + C_1} \tag{6-59}$$

The gain of the composite filter in the passband is the product of the DC gains of all the sections.

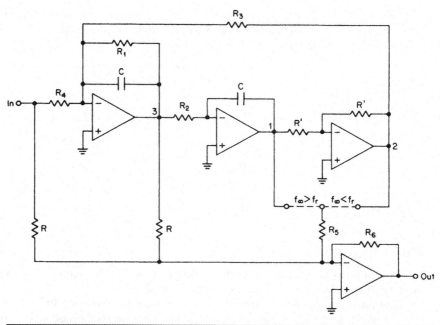

FIGURE 6-21 Biquad band-reject realization.

The VCVS structure has a number of undesirable characteristics. Although the circuit Q can be adjusted by making R_6 or R_7 variable when $K > 1$, the Q cannot be independently measured, since the 3-dB bandwidth at the output is affected by the transmission zero. Resonant frequency f_r or the notch frequency f_∞ cannot be easily adjusted, since these parameters are determined by the interaction of a number of elements. Also, the section gain is fixed by the design parameters. Another disadvantage of the circuit is that a large spread in capacitor values* may occur so that standard values cannot be easily used. Nevertheless, the VCVS realization makes effective use of a minimum number of operational amplifiers in comparison with other implementations, and is widely used. However, because of its lack of adjustment capability, its application is generally restricted to Qs below 10 and with 1 percent component tolerances.

The State-Variable Band-Reject Section

The biquad, or state-variable, elliptic-function band-pass filter section discussed in Sec. 5.2 is highly suitable for implementing band-reject transfer functions. The circuit is given in Fig. 6-21. By connecting resistor R_5 to either node 1 or node 2, the notch frequency f_∞ will be located above or below the pole-resonant frequency f_r.

Section Qs of up to 200 can be obtained. The design parameters f_r, Q, and f_∞, as well as the section gain, can be independently chosen, monitored, and adjusted. From the point of view of low sensitivity and maximum flexibility, the biquad approach is the most desirable method of realization.

---

*The elliptic-function configuration of the VCVS uniform capacitor structure given in Sec. 3.2 can be used at the expense of additional sensitivity.

The design equations were stated in Sec. 5.2 under "Elliptic-Function Band-Pass Filters" and are repeated here for convenience, where f_r, Q, and f_∞ are given and the values of C, R, and R' can be arbitrarily chosen.

$$R_1 = R_4 = \frac{Q}{2\pi f_r C} \tag{6-60}$$

$$R_2 = R_3 = \frac{R_1}{Q} \tag{6-61}$$

$$R_5 = \frac{f_r^2 R}{Q\left|f_r^2 - f_\infty^2\right|} \tag{6-62}$$

for $f_\infty > f_r$:

$$R_6 = \frac{f_r^2 R}{f_\infty^2} \tag{6-63}$$

and when $f_\infty > f_r$:

$$R_6 = R \tag{6-64}$$

The value of R_6 is based on unity section gain at DC. The gain can be raised or lowered by proportionally increasing or decreasing R_6.

Resonance is adjusted by monitoring the phase shift between the section input and node 3 using a Lissajous pattern and adjusting R_3 for 180° phase shift with an input frequency of f_r.

The Q is controlled by R_1 and can be measured at node 3 in terms of section 3-dB bandwidth, or R_1 can be adjusted until unity gain occurs between the input and node 3 with f_r applied. Because of the Q-enhancement effect discussed in Sec. 5.2 under "All-Pole Band-Pass Configuration," a Q adjustment is usually necessary.

The notch frequency is then determined by monitoring the section output for a null. Adjustment is normally not required since the tuning of f_r will usually bring in f_∞ with acceptable accuracy. If an adjustment is desired, R_5 can be made variable.

Sections for Transformed Real Poles. When a real pole undergoes a band-reject transformation, the result is a single pole pair and a single set of imaginary zeros. Complex poles resulted in two sets of pole pairs and two sets of zeros. The resonant frequency f_r of the transformed real pole is exactly equal to the notch frequency f_∞; thus, the design flexibility of the VCVS and biquad structures is not required.

A general second-order band-pass transfer function can be expressed as

$$T(s) = \frac{\dfrac{\omega_r}{Q}s}{s^2 + \dfrac{\omega_r}{Q}s + \omega_r^2} \tag{6-65}$$

where the gain is unity at ω_r. If we realize the circuit in Fig. 6-22 where $T(s)$ corresponds to the preceding transfer function, the composite transfer function at the output is given by

$$T(s) = \frac{s^2 + \omega_r^2}{s^2 + \dfrac{\omega_r}{Q}s + \omega_r^2} \tag{6-66}$$

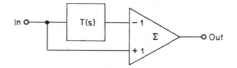

FIGURE **6-22** The band-reject configuration for $f_r = f_\infty$.

This corresponds to a band-reject transfer function with a transmission zero at f_r (that is, $f_\infty = f_r$). The occurrence of this zero can also be explained intuitively from the structure in Fig. 6-22. Since $T(s)$ is unity at f_r, both input signals to the summing amplifier will then cancel, resulting in no output signal.

These results indicate that band-reject sections for transformed real poles can be obtained by combining any of the all-pole band-pass circuits of Sec. 5.2 in the configuration of Fig. 6-22. The basic design parameters are the required f_r and Q of the band-reject section, which are directly used in the design equations for the band-pass circuits.

By combining these band-pass sections with summing amplifiers, the three band-reject structures in Fig. 6-23 can be derived. The design equations for the band-pass sections were given in Sec. 5.2 and are repeated here, where C, R, and R' can be arbitrarily chosen.

The MFBP band-reject section of fig. 6-23a ($f_r = f_\infty$) is given by

$$R_2 = \frac{Q}{\pi f_r C} \tag{6-67}$$

$$R_{1a} = \frac{R_2}{2} \tag{6-68}$$

$$R_{1b} = \frac{R_{1a}}{2Q^2 - 1} \tag{6-69}$$

The DABP band-reject section of fig. 6-23b ($f_r = f_\infty$) is given by

$$R_1 = \frac{Q}{2\pi f_r C} \tag{6-70}$$

$$R_2 = R_3 = \frac{R_1}{Q} \tag{6-71}$$

The biquad band-reject section of fig. 6-23c ($f_r = f_\infty$) is given by

$$R_1 = R_4 = \frac{Q}{2\pi f_r C} \tag{6-72}$$

$$R_2 = R_3 = \frac{R_1}{Q} \tag{6-73}$$

These equations correspond to unity band-pass gain for the MFBP and biquad circuits so that cancellation at f_r will occur when the section input and band-pass output signals are equally combined by the summing amplifiers. Since the DABP section has a gain of 2 and has a noninverting output, the circuit in Fig. 6-23b has been modified accordingly so that cancellation occurs.

Tuning can be accomplished by making R_{1b}, R_2, and R_3 variable in the MFBP, DABP, and biquad circuits, respectively. In addition, the biquad circuit will usually require R_1

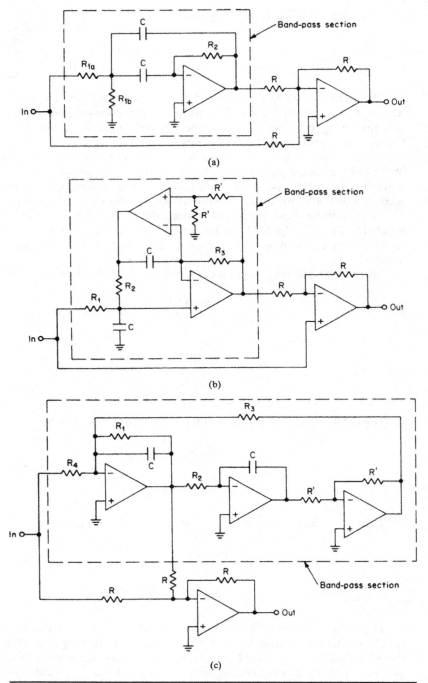

FIGURE 6-23   The band-reject circuits for $f_r = f_\infty$: (a) MFBP band-reject section ($Q < 20$); (b) DABP band-reject section ($Q < 150$); and (c) biquad band-reject section ($Q\ 4 < 200$).

to be made adjustable to compensate for the Q-enhancement effect (see Sec. 5.2 under "All-Pole Band-Pass Configurations"). The circuit can be tuned by adjusting the indicated elements for either a null at f_r measured at the circuit output or for $0°$ or $180°$ phase shift at f_r observed between the input and the output of the band-pass section. If the band-pass section gain is not sufficiently close to unity for the MFBP and biquad case, and 2 for the DABP circuit, the null depth may be inadequate.

Example 6-8   Design an Active Band-Reject Filter

Required:

Design an active band-reject filter from the band-reject parameters determined in Example 6-7 with a gain of +6 dB.

Result:

(a) The band-reject transformation in Example 6-7 resulted in the following set of requirements for a three-section filter:

Section	f_r	Q	f_∞
1	3,455 Hz	30.15	3,600 Hz
2	3,752 Hz	30.15	3,600 Hz
3	3,600 Hz	8.40	3,600 Hz

(b) Two biquad circuits in tandem will be used for sections 1 and 2 followed by a DABP band-reject circuit for section 3. The value of C is chosen at 0.01 μF and R, as well as R', at 10 kΩ. Since the DABP section has a gain of 2 at DC, which satisfies the 6-dB gain requirement, both biquad sections should then have unity gain. The element values are determined as follows using Eqs. (6-60) to (6-64) and Eqs. (6-70) and (6-71):

Section 1 (biquad of Fig. 6-21):

$$f_r = 3,455 \text{ Hz} \qquad Q = 30.15 \qquad f_\infty = 3,600 \text{ Hz}$$

$$R_1 = R_4 = \frac{Q}{2\pi f_r C} = \frac{30.15}{2\pi \times 3,455 \times 10^{-8}} = 138.9 \text{ k}\Omega$$

$$R_2 = R_3 = \frac{R_1}{Q} = \frac{138.9 \times 10^3}{30.15} = 4,610 \ \Omega$$

$$R_5 = \frac{f_r^2 R}{Q\left|f_r^2 - f_\infty^2\right|} = \frac{3,455^2 \times 10^4}{30.15 \left|3,455^2 - 3,600^2\right|} = 3,870 \ \Omega$$

$$R_6 = \frac{f_r^2 R}{f_\infty^2} = \frac{3,455^2 \times 10^4}{3,600^2} = 9,210 \ \Omega$$

Section 2 (biquad of Fig. 6-21):

$$f_r = 3,752 \text{ Hz} \qquad Q = 30.15 \qquad f_\infty = 3,600 \text{ Hz}$$

$$R_1 = R_4 = 127.9 \text{ k}\Omega$$

$$R_2 = R_3 = 4,240 \ \Omega$$

$$R_5 = 4,180\ \Omega$$

$$R_6 = 10\ \text{k}\Omega$$

Section 3 (DABP of Fig. 6-23):

$$f_r = f_\infty = 3,600\ \text{Hz} \qquad Q = 8.40$$

$$R_1 = \frac{Q}{2\pi f_r C} = \frac{8.40}{2\pi \times 3,600 \times 10^{-8}} = 37.1\ \text{k}\Omega$$

$$R_2 = R_3 = \frac{R_1}{Q} = \frac{37.1 \times 10^3}{8.40} = 4,420\ \Omega$$

The final circuit is shown in Fig. 6-24 with standard 1 percent resistor values. The required resistors have been made variable so that the resonant frequencies can be adjusted for all sections and, in addition, the Q is variable for the biquad circuits.

6.2.4 Active Null Networks

Active null networks are single sections used to provide attenuation at either a single frequency or over a narrow band of frequencies. The most popular sections are of the twin-T form, so this circuit will be discussed in detail along with some other structures.

The Twin-T

The twin-T was first discovered by H. W. Augustadt in 1934. Although this circuit is passive by nature, it is also used in many active configurations to obtain a variety of different characteristics.

The circuit in Fig. 6-25a is an RC bridge structure where balance or an output null occurs at 1 rad/s when all arms have an equal impedance ($0.5-j0.5\ \Omega$). The circuit is redrawn in the form of a symmetrical lattice in Fig. 6-25b (refer to Guillemin and Stewart in the References for detailed discussions of the lattice). The lattice in Fig. 6-25b can be redrawn again in the form of two parallel lattices, as shown in Fig. 6-25c.

If identical series elements are present in both the series and shunt branches of a lattice, the element may be extracted and symmetrically placed outside the lattice structure. A 1-Ω resistor satisfies the requirement for the upper lattice, and a 1-F capacitor satisfies the lower lattice. Removal of these components to outside the lattice results in the twin-T in Fig. 6-25d.

The general form of a twin-T is shown in Fig. 6-25e. The value of R_1 is computed from

$$R_1 = \frac{1}{2\pi f_0 C} \tag{6-74}$$

where C is arbitrary. This denormalizes the circuit in Fig. 6-25d so that the null now occurs at f_0 instead of at 1 rad/s.

When a twin-T is driven from a voltage source and terminated in an infinite load,[†] the transfer function is given by

$$T(s) = \frac{s^2 + \omega_0^2}{s^2 + 4\omega_0 s + \omega_0^2} \tag{6-75}$$

[†]Since the source and load are always finite, the value of R_1 should be in the vicinity of $\sqrt{R_s R_L}$, provided that the ratio R_L/R_S is in excess of 10.

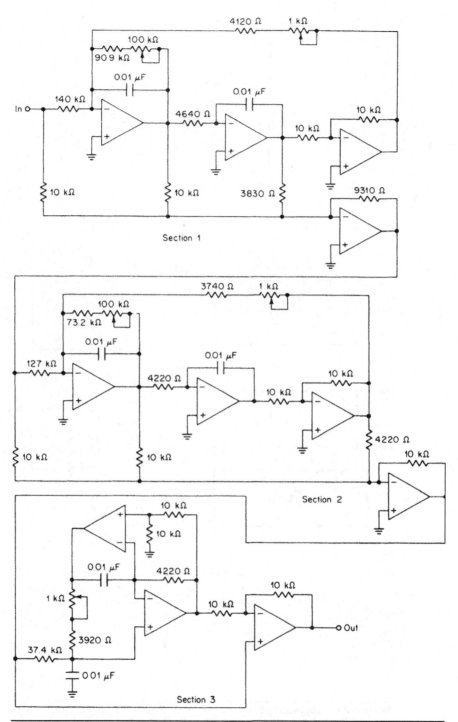

FIGURE 6-24   The band-reject filter of Example 6-8.

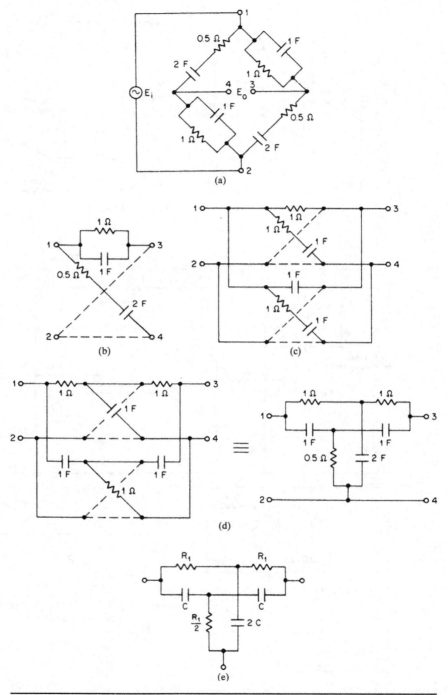

Figure 6-25 A derivation of the twin-T: (a) *RC* bridge; (b) lattice circuit; (c) parallel lattice; (d) twin-T equivalent; and (e) general form of the twin-T.

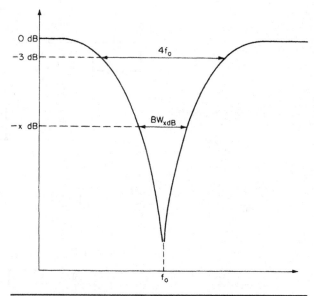

Figure 6-26   The frequency response of a twin-T.

If we compare this expression with the general transfer function of a second-order pole-zero section, as given by Eq. (6-66), we can determine that a twin-T provides a notch at f_0 with a Q of ¼. The attenuation at any bandwidth can be computed by

$$A_{\mathrm{dB}} = 10 \log\left[1 + \left(\frac{4f_0}{\mathrm{BW}_{x\,\mathrm{dB}}}\right)^2\right]$$ (6-76)

The frequency response is shown in Fig. 6-26, where the requirement for geometric symmetry applies.

Twin-T with Positive Feedback

The twin-T has gained widespread usage as a general-purpose null network. However, a major shortcoming is a fixed Q of ¼. This limitation can be overcome by introducing positive feedback.

The transfer function of the circuit in Fig. 6-27a can be derived as

$$T(s) = \frac{\beta}{1 + K(\beta - 1)}$$ (6-77)

If β is replaced by Eq. (6-75), the transfer function of a twin-T, the resulting circuit transfer function expression becomes

$$T(s) = \frac{s^2 + \omega_0^2}{s^2 + 4\omega_0(1 - K)s + \omega_0^2}$$ (6-78)

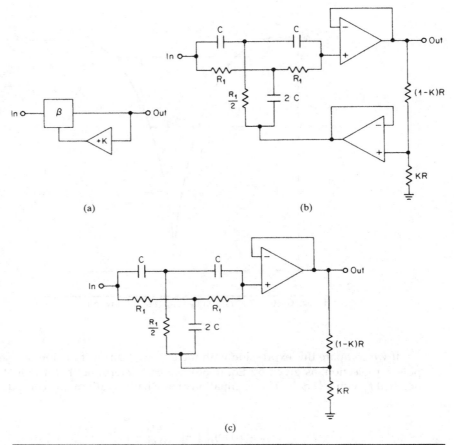

(a)

(b)

(c)

FIGURE 6-27 Twin-T with positive feedback: (a) block diagram; (b) circuit realization; and (c) simplified configuration $R_1 \gg (1 - K) R$.

The corresponding Q is then

$$Q = \frac{1}{4(1 - K)} \tag{6-79}$$

By selecting a positive K of < 1 and sufficiently close to unity, the circuit Q can be dramatically increased. The required value of K can be determined by

$$K = 1 - \frac{1}{4Q} \tag{6-80}$$

The block diagram in Fig. 6-27a can be implemented using the circuit in Fig. 6-27b, where R is arbitrary. By choosing C and R so that $R_1 \gg (1 - K)R$, the circuit may be simplified to the configuration in Fig. 6-27c, which uses only one amplifier.

The attenuation at any bandwidth is given by

$$A_{dB} = 10 \log \left[1 + \left(\frac{f_0}{Q \times BW_{x\,dB}} \right)^2 \right] \tag{6-81}$$

Equation (6-81) is the general expression for the attenuation of a single band-reject section where the resonant frequency and notch frequency are identical (that is, $f_r = f_\infty$). The attenuation formula can be expressed in terms of the 3-dB bandwidth as follows:

$$A_{dB} = 10 \log \left[1 + \left(\frac{BW_{3\,dB}}{BW_{x\,dB}} \right)^2 \right] \tag{6-82}$$

The attenuation characteristics can also be determined from the frequency-response curve of a normalized $n = 1$ Butterworth low-pass filter (see Fig. 2-34) by using the ratio $BW_{3\,dB}/BW_{x\,dB}$ for the normalized frequency.

The twin-T in its basic form or in the positive-feedback configuration is widely used for single-section band-reject sections. However, it suffers from the fact that tuning cannot be easily accomplished. Tight component tolerances may then be required to ensure sufficient accuracy of tuning and adequate notch depth. About a 40- to 60-dB rejection at the notch could be expected using 1 percent components.

Example 6-9 Design a Twin-T Band-Reject Filter Using Positive Feedback

Required:

Design a single null network with a center frequency of 1,000 Hz and a 3-dB bandwidth of 100 Hz. Also determine the attenuation at the 30-Hz bandwidth.

Result:

(*a*) A twin-T structure with positive feedback will be used. To design the twin-T, first choose a capacitance C of 0.01 μF. The value of R_1 is given by Eq. (6-74) as

$$R_1 = \frac{1}{2\pi f_0 C} = \frac{1}{2\pi \times 10^3 \times 10^{-8}} = 15.9 \text{ k}\Omega$$

(*b*) The required value of K for the feedback network is calculated from Eq. (6-80) as shown:

$$K = 1 - \frac{1}{4Q} = 1 - \frac{1}{4 \times 10} = 0.975$$

where $Q\,5f_0/BW_{3\,dB}$.

(*c*) The single amplifier circuit in Fig. 6-27c will be used. If R is chosen at 1 kΩ, the circuit requirement for $R_1 \gg (1-K)R$ is satisfied. The resulting section is shown in Fig. 6-28.

(*d*) To determine the attenuation at a bandwidth of 30 Hz, calculate the following using Eq. (6-82):

$$A_{dB} = 10 \log \left[1 + \left(\frac{BW_{3\,dB}}{BW_{x\,dB}} \right)^2 \right] = 10 \log \left[1 + \left(\frac{100 \text{ Hz}}{30 \text{ Hz}} \right)^2 \right] = 10.8 \text{ dB}$$

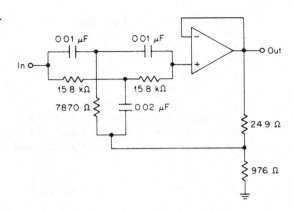

FIGURE **6-28**   The twin-T network of Example 6-9.

Band-Pass Structure Null Networks

Section 6.2 under "Narrowband Active Band-Reject Filters" showed how a first-order band-pass section can be combined with a summing amplifier to obtain a band-reject circuit for transformed real poles, where $f_r = f_\infty$. Three types of sections were illustrated in Fig. 6-23, corresponding to different Q ranges of operation. These same sections can be used as null networks. They offer more flexibility than the twin-T since the null frequency can be adjusted to compensate for component tolerances. In addition, the DABP and biquad circuits permit Q adjustment as well.

The design formulas were given by Eqs. (6-67) to (6-73). The values of f_r and Q in the equations correspond to the section center frequency and Q, respectively.

Frequently, a band-pass and band-reject output are simultaneously required. A typical application might involve the separation of signals for comparison of in-band and out-of-band spectral energy. The band-reject sections in Fig. 6-23 can each provide a band-pass output from the band-pass section along with the null output signal. An additional feature of this technique is that the band-pass and band-reject outputs will track.

References

Guillemin, E. A. (1935). *Communication Networks*, Vol. 2. New York: John Wiley and Sons.

Saal, R., and E. Ulbrich. "On the Design of Filters by Synthesis." *IRE Transactions on Circuit Theory*, December 1958.

Stewart, J. L. (1956). *Circuit Theory and Design*. New York: John Wiley and Son.

Tow, J. "A Step-by-Step Active Filter Design." *IEEE Spectrum* 6, December 1969, 64–68.

Williams, A. B. (1975). *Active Filter Design*. Dedham, Massachusetts: Artech House.

Zverev, A. I. (1967). *Handbook of Filer Synthesis*. New York: John Wiley and Sons.

CHAPTER 7

Networks for the Time Domain

7.1 All-Pass Transfer Functions

Up until now, the networks we've discussed were used to obtain a desired amplitude versus frequency characteristic. No less important is the all-pass family of filters. This class of networks exhibits a flat frequency response but introduces a prescribed phase shift versus frequency. All-pass filters are frequently called *delay equalizers.*

If a network is to be an all-pass type, the absolute magnitudes of the numerator and denominator of the transfer function must be related by a fixed constant at all frequencies. This condition will be satisfied if the zeros are the images of the poles. Since poles are restricted to the left-half quadrants of the complex frequency plane to maintain stability, the zeros must occur in the right-half plane as the mirror image of the poles about the $j\omega$ axis. Figure 7-1 illustrates the all-pass pole-zero representations in the complex frequency plane for first-order and second-order all-pass transfer functions.

7.1.1 First-Order All-Pass Transfer Functions

The real pole-zero pair of Fig. 7-1*a* has a separation of $2\alpha_0$ between the pole and zero and corresponds to the following first-order all-pass transfer function:

$$T(s) = \frac{s - \alpha_0}{s + \alpha_0} \tag{7-1}$$

To determine the absolute magnitude of $T(s)$, compute

$$|T(s)| = \frac{|s - \alpha_0|}{|s + \alpha_0|} = \frac{\sqrt{\alpha_0^2 + \omega^2}}{\sqrt{\alpha_0^2 + \omega^2}} = 1 \tag{7-2}$$

where $s = j\omega$. For any value of frequency, the numerator and denominator of Eq. (7-2) are equal, so the transfer function is clearly all-pass and has an absolute magnitude of unity at all frequencies.

The phase shift is given by

$$\beta(\omega) = -2 \tan^{-1} \frac{\omega}{\alpha_0} \tag{7-3}$$

where $\beta(\omega)$ is in radians.

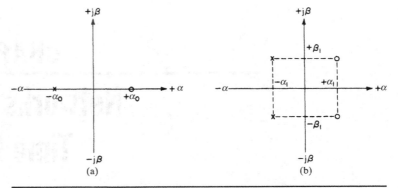

Figure 7-1 All-pass pole-zero patterns: (a) first-order all-pass transfer function; and (b) second-order all-pass transfer function.

The phase shift versus the ratio ω/α_0, as defined by Eq. (7-3), is plotted in Fig. 7-2. The phase angle is $0°$ at DC and $-90°$ at $\omega = \alpha_0$. The phase shift asymptotically approaches $-180°$ with increasing frequency.

The group delay was defined in Sec. 2.2, under "Effect of Nonuniform Time Delay," as the derivative of the phase shift, which results in

$$T_{gd} = -\frac{d\beta(\omega)}{d\omega} = \frac{2\alpha_0}{\alpha_0^2 + \omega^2} \tag{7-4}$$

If Eq. (7-4) is plotted with respect to ω for different values of α_0, a family of curves is obtained, as shown in Fig. 7-3. First-order all-pass sections exhibit maximum delay at DC and decreasing delay with increasing frequency. For small values of α_0, the delay becomes large at low frequencies and decreases quite rapidly above this range. The delay at DC is found by setting ω equal to zero in Eq. (7-4), which results in

$$T_{gd}(\text{DC}) = \frac{2}{a_0} \tag{7-5}$$

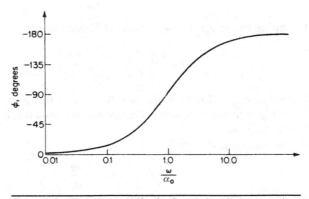

Figure 7-2 The phase shift of a first-order all-pass section.

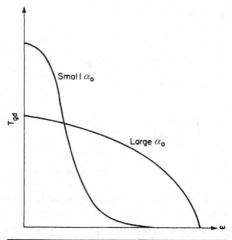

FIGURE 7-3 Group delay of first-order all-pass transfer functions.

7.1.2 Second-Order All-Pass Transfer Functions

The second-order all-pass transfer function represented by the pole-zero pattern of Fig. 7-1b is given by

$$T(s) = \frac{s^2 - \dfrac{\omega_r}{Q}s + \omega_r^2}{s^2 + \dfrac{\omega_r}{Q}s + \omega_r^2} \tag{7-6}$$

where ω_r and Q are the pole-resonant frequency (in radians per second) and the pole Q. These terms may also be computed from the real and imaginary pole-zero coordinates of Fig. 7-1b by

$$\omega_r = \sqrt{\alpha_1^2 + \beta_1^2} \tag{7-7}$$

and

$$Q = \frac{\omega_r}{2\alpha_1} \tag{7-8}$$

The absolute magnitude of $T(s)$ is found to be

$$|T(s)| = \frac{\sqrt{\left(\omega_r^2 - \omega^2\right)^2 + \dfrac{\omega^2 \omega_r^2}{Q^2}}}{\sqrt{\left(\omega_r^2 - \omega^2\right)^2 + \dfrac{\omega^2 \omega_r^2}{Q^2}}} = 1 \tag{7-9}$$

which is all-pass.

The phase shift in radians is

$$\beta(\omega) = -2\tan^{-1}\left(\dfrac{\dfrac{\omega\omega_r}{Q}}{\omega_r^2 - \omega^2}\right) \tag{7-10}$$

and the group delay is given by

$$T_{gd} = \dfrac{2Q\omega_r\left(\omega^2 + \omega_r^2\right)}{Q^2\left(\omega^2 - \omega_r^2\right) + \omega^2 - \omega_r^2} \tag{7-11}$$

The phase and delay parameters of first-order transfer functions are relatively simple to manipulate since they are a function of a single design parameter α_0. A second-order type, however, has two design parameters, Q and ω_r.

The phase shift of a second-order transfer function is $-180°$ at $\omega = \omega_r$. At DC, the phase shift is zero, and at frequencies well above ω_r, the phase asymptotically approaches $-360°$.

The group delay reaches a peak that occurs very close to ω_r. As the Q is made larger, the peak delay increases, the delay response becomes sharper, and the delay at DC decreases, as shown in Fig. 7-4.

The frequency of maximum delay is slightly below ω_r and is expressed in radians per second by

$$\omega(T_{gd,\,\text{max}}) = \omega_r\sqrt{\sqrt{4 - \dfrac{1}{Q^2}} - 1} \tag{7-12}$$

For all practical purposes, the maximum delay occurs at ω_r for Qs in excess of 2. By setting $\omega = \omega_r$ in Eq. (7-11), the delay at ω_r is given by

$$T_{gd,\text{max}} = \dfrac{4Q}{\omega_r} = \dfrac{2Q}{\pi f_r} \tag{7-13}$$

FIGURE 7-4 Group delay of second-order all-pass transfer functions.

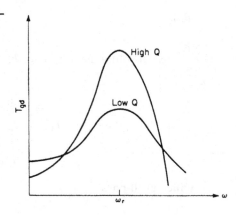

If we set $\omega = 0$, the delay at DC is found from

$$T_{gd}(\text{DC}) = \frac{2}{Q\omega_r} = \frac{1}{Q\pi f_r} \tag{7-14}$$

7.2   Delay Equalizer Sections

Passive or active networks that realize first- or second-order all-pass transfer functions are called *delay equalizers,* since they are normally used to provide a required delay characteristic without disturbing the amplitude response. All-pass networks can be realized in a variety of configurations, both passive and active. Equalizers with adjustable characteristics can also be designed, and are discussed in Chap. 12.

7.2.1   *LC* All-Pass Structures

First-Order Constant-Resistance Circuit

The lattice of Fig. 7-5*a* realizes a first-order all-pass transfer function. The network is also a constant-resistance type, which means that the input impedance has a constant value of R over the entire frequency range. Constant-resistance networks can be cascaded with no interaction so that composite delay curves can be built up by accumulating the individual delay contributions. The lattice has an equivalent unbalanced form, shown in Fig. 7-5*b*. The design formulas are given by

$$L = \frac{2R}{\alpha_0} \tag{7-15}$$

$$C = \frac{2}{\alpha_0 R} \tag{7-16}$$

where R is the desired impedance level and α_0 is the real pole-zero coordinate. The phase shift and delay properties were defined by Eqs. (7-3) to (7-5).

The circuit of Fig. 7-5*b* requires a center-tapped inductor with a coefficient of magnetic coupling K equal to unity.

Second-Order Constant-Resistance Sections

A second-order all-pass lattice with constant-resistance properties is shown in Fig. 7-6*a*. The circuit may be transformed into the unbalanced bridged-T form of Fig. 7-6*b*. The elements are given by

$$L_a = \frac{2R}{\omega_r Q} \tag{7-17}$$

$$C_a = \frac{Q}{\omega_r R} \tag{7-18}$$

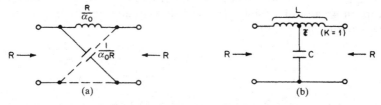

Figure 7-5   First-order *LC* equalizer section: (*a*) lattice form; and (*b*) unbalanced form.

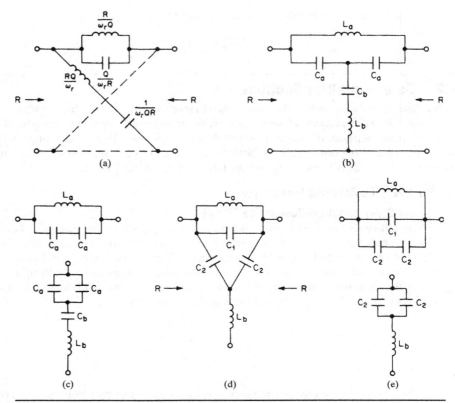

FIGURE 7-6 Second-order section $Q > 1$: (a) lattice form; (b) unbalanced form; (c) circuit for measuring branch resonances; (d) circuit modified by a T-to-pi transformation; and (e) resonant branches of modified circuit.

$$L_b = \frac{QR}{2\omega_r} \tag{7-19}$$

$$C_b = \frac{2Q}{\omega_r(Q^2 - 1)R} \tag{7-20}$$

For tuning and test purposes, the section can be split into parallel- and series-resonant branches by opening the shunt branch and shorting the bridging or series branch, as shown in Fig. 7-6c. Both circuits will resonate at ω_r.

The T-to-pi transformation was first introduced in Sec. 4.1. This transformation may be applied to the T of capacitors that are embedded in the section of Fig. 7-6b to reduce capacitor values if desired. The resulting circuit is given in Fig. 7-6d. Capacitors C_1 and C_2 are computed as follows:

$$C_1 = \frac{C_a^2}{2C_a + C_b} \tag{7-21}$$

$$C_2 = \frac{C_a C_b}{2C_a + C_b} \tag{7-22}$$

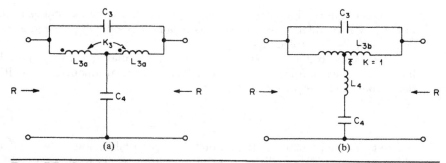

Figure 7-7 Second-order section $Q < 1$: (a) circuit with a controlled coefficient of coupling; and (b) circuit with unity coefficient of coupling.

The branch resonances are obtained by opening the shunt branch and then shorting the bridging branch, which results in the parallel- and series-resonant circuits of Fig. 7-6e. Both resonances occur at ω_r.

Close examination of Eq. (7-20) indicates that C_b will be negative if the Q is less than 1. (If $Q = 1$, C_b can be replaced by a short.) This restricts the circuits of Fig. 7-6 to those cases where the Q is in excess of unity. Fortunately, this is true in most instances.

In those cases where the Q is below 1, the configurations of Fig. 7-7 are used. The circuit of Fig. 7-7a uses a single inductor with a controlled coefficient of coupling given by

$$K_3 = \frac{1 - Q^2}{1 + Q^2} \tag{7-23}$$

The element values are given by

$$L_{3a} = \frac{(Q^2 + 1)R}{2Q\omega_r} \tag{7-24}$$

$$C_3 = \frac{Q}{2\omega_r R} \tag{7-25}$$

$$C_4 = \frac{2}{Q\omega_r R} \tag{7-26}$$

It is not always convenient to control the coefficient of coupling of a coil to obtain the specific value required by Eq. (7-23). A more practical approach uses the circuit of Fig. 7-7b. The inductor L_{3b} is center-tapped and requires a unity coefficient of coupling (typical values of 0.95 or greater can usually be obtained and are acceptable). The values of L_{3b} and L_4 are computed from

$$L_{3b} = 2(1 + K_3)L_{3a} \tag{7-27}$$

$$L_4 = \frac{(1 - K_3)L_{3a}}{2} \tag{7-28}$$

The sections of Fig. 7-7 may be tuned to ω_r in the same manner as in the equalizers in Fig. 7-6. A parallel-resonant circuit is obtained by opening C_4, and a series-resonant circuit will result by placing a short across C_3.

The second-order section of Figs. 7-6 and 7-7 may not always be all-pass. If the inductors have insufficient Q, a notch will occur at the resonances and will have a notch depth that can be approximated by

$$A_{dB} = 20 \log \frac{Q_L + 4Q}{Q_L - 4Q} \qquad (7\text{-}29)$$

where Q_L is the inductor Q. If the notch is unacceptable, adequate coil Q must be provided, or amplitude-equalization techniques used, as discussed in Sec. 8.4.

Minimum Inductor All-Pass Sections

The bridged-T circuit of Fig. 7-8 realizes a second-order all-pass transfer function with a single inductor. The section is not a constant-resistance type, and operates between a zero impedance source and an infinite load. If the ratio of load to source is well in excess of 10, satisfactory results will be obtained. The elements are computed by

$$C = \frac{Q}{4\omega_r R} \qquad (7\text{-}30)$$

and

$$L = \frac{1}{\omega_r^2 C} \qquad (7\text{-}31)$$

The value of R can be chosen as the geometric mean of the source and load impedance ($\sqrt{R_S R_L}$). The LC circuit is parallel resonant at ω_r.

The reader is reminded that the design parameters Q and ω_r are determined from the delay parameters defined in Sec. 7.1, which covers all-pass transfer functions.

A notch will occur at resonance due to a finite inductor Q and can be calculated from

$$A_{dB} = 20 \log \frac{4R + \omega_r L Q_L}{4R - \omega_r L Q_L} \qquad (7\text{-}32)$$

If R is set equal to $\omega_r L Q_L / 4$, the notch attenuation becomes infinite, and the circuit is then identical to the bridged-T null network of Sec. 6.1.

Two sets of all-pass poles and zeros corresponding to a fourth-order transfer function can also be obtained by using a minimum-inductance-type structure. The circuit configuration is shown in Fig. 7-9.

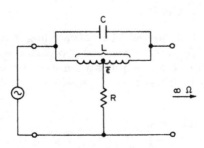

FIGURE 7-8 Minimum inductor type, second-order section.

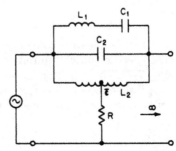

FIGURE 7-9 A fourth-order minimum inductance, all-pass structure.

Upon being given two sets of equalizer parameters Q_1, ω_{r1} and Q_2, ω_{r2} as defined in Sec. 7.1, the following equations are used to determine the element values:

First, compute

$$A = \omega_1^2 \omega_{r2}^2 \tag{7-33}$$

$$B = A\left(\frac{1}{\omega_{r2}Q_2} + \frac{1}{\omega_{r1}Q_1}\right) \tag{7-34}$$

$$C = \frac{Q_1 Q_2}{A(Q_2\omega_{r1} + Q_1\omega_{r2})} \tag{7-35}$$

$$D = \frac{Q_1 Q_2 \left(\omega_{r1}^2 + \omega_{r2}^2\right) + \omega_{r1}\omega_{r2}}{ABQ_1Q_2} - C - \frac{1}{AB^2C} \tag{7-36}$$

$$E = \frac{1}{ABCD} \tag{7-37}$$

The element values are then given by

$$L_1 = \frac{4ER}{A} \tag{7-38}$$

$$C_1 = \frac{AD}{4R} \tag{7-39}$$

$$L_2 = \frac{4BR}{A} \tag{7-40}$$

$$C_2 = \frac{AC}{4R} \tag{7-41}$$

The value of R is generally chosen as the geometric mean of the source and load terminations, as with the second-order minimum-inductance section. The series and parallel branch-resonant frequencies are found from

$$\omega_{L1C1} = \frac{1}{\sqrt{ED}} \tag{7-42}$$

and

$$\omega_{L2C2} = \frac{1}{\sqrt{BC}} \tag{7-43}$$

7.2.2 Active All-Pass Structures

First- and second-order all-pass transfer functions can be obtained by using an active approach. The general form of the active all-pass section is represented by the block diagram of Fig. 7-10, where $T(s)$ is a first- or second-order transfer function with a gain of unity.

First-Order Sections

The transfer function of the circuit of Fig. 7-10 is given by

$$\frac{E_{\text{out}}}{E_{\text{in}}} = 2T(s) - 1 \tag{7-44}$$

FIGURE 7-10 The general form of an active all-pass section.

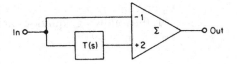

If $T(s)$ is a first-order RC high-pass network with the transfer function $sCR/(sCR+1)$, the composite transfer function becomes

$$\frac{E_{out}}{E_{in}} = \frac{s - 1/RC}{s + 1/RC} \tag{7-45}$$

This expression corresponds to the first-order all-pass transfer function of Eq. (7-1), where

$$\alpha_0 = \frac{1}{RC} \tag{7-46}$$

The circuit can be directly implemented by the configuration of Fig. 7-11a, where R' is arbitrary. The phase shift is then given by

$$\beta(\omega) = -2\tan^{-1}\omega RC \tag{7-47}$$

and the delay is found from

$$T_{gd} = \frac{2RC}{(\omega RC)^2 + 1} \tag{7-48}$$

At DC, the delay is a maximum and is computed from

$$T_{gd}(DC) = 2RC \tag{7-49}$$

The corresponding phase shift is shown in Fig. 7-2. A phase shift of $-90°$ occurs at $\omega = 1/RC$ and approaches $-180°$ and $0°$ at DC and infinity, respectively. By making the element R variable, an all-pass network can be obtained with a phase shift adjustable between 0 and $-180°$.

A sign inversion of the phase will occur if the circuit of Fig. 7-11b is used. The circuit will remain all-pass and first-order, and the group delay is still defined by Eqs. (7-48) and (7-49).

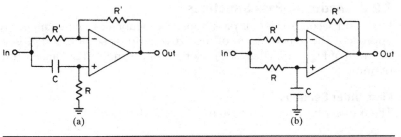

FIGURE 7-11 First-order all-pass sections: (a) circuit with lagging phase shift; and (b) circuit with leading phase shift.

Second-Order Section

If $T(s)$ in Fig. 7-10 is a second-order band-pass network with the general band-pass transfer function

$$T(s) = \frac{\dfrac{\omega_r}{Q} s}{s^2 + \dfrac{\omega_r}{Q} s + \omega_r^2} \tag{7-50}$$

the composite transfer function then becomes

$$\frac{E_{\text{out}}}{E_{\text{in}}} = 2T(s) - 1 = -\frac{s^2 - \dfrac{\omega_r}{Q} s + \omega_r^2}{s^2 + \dfrac{\omega_r}{Q} s + \omega_r^2} \tag{7-51}$$

which corresponds to the second-order all-pass expression given by Eq. (7-6) (except for a simple sign inversion). Therefore, a second-order all-pass equalizer can be obtained by implementing the structure of Fig. 7-10 using a single active band-pass section for $T(s)$.

Section 5-2 discussed the multiple-feedback band-pass (MFBP), dual-amplifier band-pass (DABP), and biquad all-pole band-pass sections. Each circuit can be combined with a summing amplifier to generate a delay equalizer.

The MFBP equalizer is shown in Fig. 7-12a. The element values are given by

$$R_2 = \frac{2Q}{\omega_r C} = \frac{Q}{\pi f_r C} \tag{7-52}$$

$$R_{1a} = \frac{R_2}{2} \tag{7-53}$$

$$R_{1b} = \frac{R_{1a}}{2Q^2 - 1} \tag{7-54}$$

The values of C and R can be arbitrarily chosen, and A in Fig. 7-12a corresponds to the desired gain.

The maximum delay, which occurs at f_r, was given by Eq. (7-13). This expression can be combined with Eqs. (7-52) and (7-54), so the element values can be expressed alternatively in terms of $T_{gd'\text{max}}$ as follows for $Q > 2$:

$$R_2 = \frac{T_{gd,\text{max}}}{2C} \tag{7-55}$$

$$R_{1b} = \frac{R_2}{(\pi f_r T_{gd,\text{max}})^2 - 2} \tag{7-56}$$

where R_{1a} remains $R_2/2$.

The MFBP section can be tuned by making R_{1b} variable. R_{1b} can then be adjusted until $180°$ of phase shift occurs between the input and output of the band-pass section at f_r. In order for the response to be all-pass, the band-pass section gain must be exactly unity at resonance. Otherwise, an amplitude ripple will occur in the frequency-response characteristic in the vicinity of f_r.

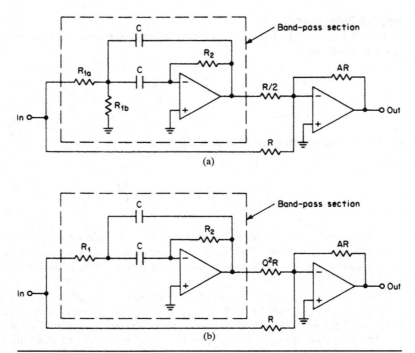

Figure 7-12 The MFBP delay equalizer $Q < 20$: (a) circuit for $0.707 < Q < 20$; and (b) circuit for $Q < 0.707$.

The section Q is limited to values below 20, or is expressed in terms of delay, as in the following:

$$T_{gd,\max} < \frac{40}{\pi f_r} \tag{7-57}$$

Experience has indicated that required Qs are usually well under 20, so this circuit will suffice in most cases. However, if the Q is below 0.707, the value of R_{1b}, as given by Eq. (7-54), is negative, so the circuit of Fig. 7-12b is used. The value of R_1 is given by

$$R_1 = \frac{R_2}{4Q^2} \tag{7-58}$$

In the event that higher Qs are required, the DABP section can be applied to the block diagram of Fig. 7-10. Since the DABP circuit has a gain of 2 and is noninverting, the implementation shown in Fig. 7-13 is used. The element values are given by

$$R_1 = \frac{Q}{\omega_r C} = \frac{Q}{2\pi f_r C} \tag{7-59}$$

and

$$R_2 = R_3 = \frac{R_1}{Q} \tag{7-60}$$

where C, R, and R' can be conveniently chosen. Resistor R_2 may be made variable if tuning is desired. The Q, and therefore the delay, can also be trimmed by making R_1 adjustable.

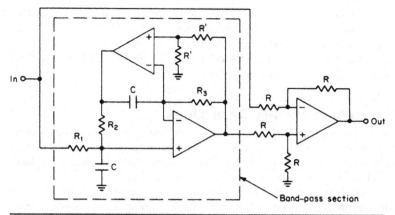

FIGURE 7-13 The DABP delay equalizer $Q < 150$.

The biquad structure can be configured in the form of a delay equalizer. The circuit is shown in Fig. 7-14, and the element values are computed from

$$R_1 = R_4 = \frac{Q}{\omega_r C} = \frac{Q}{2\pi f_r C} \tag{7-61}$$

and

$$R_2 = R_3 = \frac{R_1}{Q} \tag{7-62}$$

where C, R, and R' are arbitrary.

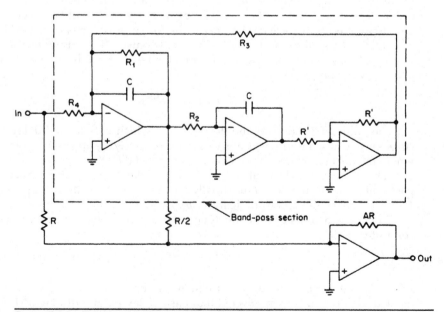

FIGURE 7-14 The biquad delay equalizer $Q < 200$.

Resistor R_3 can be made variable for tuning. The Q is adjusted for the nominal value by making R_1 variable and monitoring the 3-dB bandwidth at the output of the band-pass section, or by adjusting for unity band-pass gain at f_r. The biquad is subject to the Q-enhancement effect discussed in Sec. 5.2, under "All-Pole Band-Pass Configurations," so a Q adjustment is usually required.

7.3   Design of All-Pass Delay Lines

The classical approach to the design of delay lines involves a cascade of identical LC sections (except for the end sections) and uses image-parameter theory [see Wallis (1952)]. This technique is an approximation at best, but does have some advantages such as repetitive values and is covered in Sec. 7.6.

Modern network theory permits us to predict the delay of networks accurately and to obtain a required delay in a much more efficient manner than with the classical approach. The Bessel linear phase with equiripple error and transitional filters all feature a constant delay. The curves in Chap. 2 indicate that for $n > 3$, the flat delay region is extended well into the stopband. If a delay line is desired, a low-pass filter implementation is not the best approach from a delay-bandwidth perspective. A significant portion of the constant-delay region would be attenuated.

All the low-pass transfer functions covered can be implemented by using an all-pass realization to overcome the bandwidth limitations. This results in a precise and efficient means of designing delay lines.

7.3.1   The Low-Pass to All-Pass Transformation

A low-pass transfer function can be transformed to an all-pass transfer function simply by introducing zeros in the right-half plane of the $j\omega$ axis corresponding to each pole. If the real and complex poles tabulated in Chap. 11 are realized using the first- and second-order all-pass structures of Sec. 7.2, complementary zeros will also occur. When a low-pass to all-pass transformation is made, the low-pass delay is increased by a factor of exactly 2 because of the additional phase shift contributions of the zeros.

An all-pass delay-bandwidth factor can be derived from the delay curves of Chap. 2, which is given by

$$TU = \omega_u T_{gd}(\text{DC}) \qquad (7\text{-}63)$$

The value of T_{gd} (DC) is the delay at DC, which is twice the delay shown in the curves because of the all-pass transformation, and ω_u is the upper-limit radian frequency, where the delay deviates a specified amount from the DC value.

Table 7-1 lists the delay at DC, ω_u, and the delay-bandwidth product TU for an all-pass realization of the Bessel maximally flat delay family. Values are provided for both 1 and 10 percent deviations of delay at ω_u.

To choose a transfer-function type and determine the complexity required, first compute

$$TU_{\text{req}} = 2\pi f_{gd} T_{gd} \qquad (7\text{-}64)$$

where f_{gd} is the maximum desired frequency of operation and T_{gd} is the nominal delay needed. A network is then selected that has a delay-bandwidth factor TU that exceeds TU_{req}.

N	T_{gd} (DC)	1 Percent Deviation		10 Percent Deviation	
		ω_u	TU	ω_u	TU
2	2.72	0.412	1.121	0.801	2.179
3	3.50	0.691	2.419	1.109	3.882
4	4.26	0.906	3.860	1.333	5.679
5	4.84	1.120	5.421	1.554	7.521
6	5.40	1.304	7.042	1.737	9.380
7	5.90	1.478	8.720	1.912	11.280
8	6.34	1.647	10.440	2.079	13.180
9	6.78	1.794	12.160	2.227	15.100

TABLE 7-1   All-Pass Bessel Delay Characteristics

Compute the delay-scaling factor (DSF), which is the ratio of the normalized delay at DC to the required nominal delay. For example:

$$\mathrm{DSF} = \frac{T_{gd}(\mathrm{DC})}{T_{gd}} \tag{7-65}$$

The corresponding poles of the filter selected are denormalized by the DSF and can then be realized by the all-pass circuits of Sec. 7.2.

A real pole α_0 is denormalized by the formula

$$\alpha'_0 = \alpha_0 \times \mathrm{DSF} \tag{7-66}$$

Complex poles tabulated in the form $\alpha + j\beta$ are denormalized and transformed into the all-pass section design parameters ω_r and Q by the relationships

$$\omega_r = \mathrm{DSF}\sqrt{\alpha^2 + \beta^2} \tag{7-67}$$

$$Q = \frac{\omega_r}{2\alpha\mathrm{DSF}} \tag{7-68}$$

The parameters α'_0, ω_r, and Q are then directly used in the design equations for the circuits of Sec. 7.2.

Sometimes the required delay-bandwidth factor TU_{req}, as computed by Eq. (7-64), is in excess of the TU factors available from the standard filter families tabulated. The total delay required can then be subdivided into N smaller increments and realized by N delay lines in cascade, since the delays will add algebraically.

7.3.2   *LC* Delay Lines

LC delay lines are designed by first selecting a normalized filter type and then denormalizing the corresponding real and complex poles, all in accordance with Sec. 7.3, under "The Low-Pass to All-Pass Transformation."

The resulting poles and associated zeros are then realized using the *LC* all-pass circuit of Sec. 7.2. This procedure is best illustrated by the following design example.

Example 7-1   Design of a 1mS *LC* Delay Line

Required:

Design a passive delay line to provide 1 ms of delay constant within 10 percent from DC to 3,200 Hz. The source and load impedances are both 10 kΩ.

Result:

(*a*) Compute the required delay-bandwidth factor using Eq. (7-64).

$$TU_{\text{req}} = 2\pi f_{gd} T_{gd} = 2\pi 3{,}200 \times 0.001 = 20.1$$

A linear phase design with an equiripple error of 0.5° will be chosen. The delay characteristics for the corresponding low-pass filters are shown in Fig. 2-64. The delay at DC of a normalized all-pass network for $n = 9$ is equal to 7.5 s, which is twice the value obtained from the curves. Since the delay remains relatively flat to 3 rad/s, the delay-bandwidth factor is given by Eq. (7-63) as shown:

$$TU = \omega_u T_{gd}(\text{DC}) = 3 \times 7.5 = 22.5$$

Since *TU* is in excess of TU_{req}, the $n = 9$ design will be satisfactory.

(*b*) The low-pass poles are found in Table 10-45 and are as follows:

$$-0.5688 \pm j0.7595$$
$$-0.5545 \pm j1.5089$$
$$-0.5179 \pm j2.2329$$
$$-0.4080 \pm j2.9028$$
$$-0.5728$$

Four second-order all-pass sections and a single first-order section will be required. The delay-scaling factor is given by Eq. (7-65) as shown:

$$\text{DSF} = \frac{T_{gd}(\text{DC})}{T_{gd}} = \frac{7.5}{10^{-3}} = 7{,}500$$

The denormalized design parameters ω_r and Q for the second-order sections are computed by Eqs. (7-67) and (7-68), respectively, and are tabulated as follows:

Section	α	β	ω_r	Q
1	0.5688	0.7595	7,117	0.8341
2	0.5545	1.5089	12,057	1.450
3	0.5179	2.2329	17,191	2.213
4	0.4080	2.9028	21,985	3.592

The design parameter α_0' for section 5 corresponding to the real pole is found from Eq. (7-66) as shown:

$$\alpha_0' = \alpha_0 \times \text{DSF} = 4{,}296$$

where α_0 is 0.5728.

(c) The element values can now be computed as follows:

Section 1:

Since the Q is less than unity, the circuit of Fig. 7-7b will be used. The element values are found from Eqs. (7-23) to (7-28) as shown:

$$K_3 = \frac{1 - Q^2}{1 + Q^2} = \frac{1 - 0.8341^2}{1 + 0.8341^2} = 0.1794$$

$$L_{3a} = \frac{(Q^2 + 1)R}{2Q\omega_r} = \frac{(0.8341^2 + 1)10^4}{2 \times 0.8341 \times 7,117} = 1.428 \text{ H}$$

$$C_3 = \frac{Q}{2\omega_r R} = \frac{0.8341}{2 \times 7,117 \times 10^4} = 5,860 \text{ pF}$$

$$C_4 = \frac{2}{Q\omega_r R} = \frac{2}{0.8341 \times 7,117 \times 10^4} = 0.0337 \text{ } \mu\text{F}$$

$$L_{3b} = 2(1 + K_3)L_{3a} = 3.368 \text{ H}$$

$$L_4 = \frac{(1 - K_3)L_{3a}}{2} = 0.586 \text{ H}$$

Sections 2 through 4:

Since the Qs are in excess of unity, the circuit of Fig. 7-6b will be used. The values for section 2 are found from Eqs. (7-17) to (7-20) as shown:

$$L_a = \frac{2R}{\omega_r Q} = \frac{2 \times 10^4}{12,057 \times 1.450} = 1.144 \text{ H}$$

$$C_a = \frac{Q}{\omega_r R} = \frac{1.450}{12,057 \times 10^4} = 0.012 \text{ } \mu\text{F}$$

$$L_b = \frac{QR}{2\omega_r} = \frac{1.450 \times 10^4}{2 \times 12,057} = 0.601 \text{ H}$$

$$C_b = \frac{2Q}{\omega_r(Q^2 - 1)R} = \frac{2 \times 1.450}{12,057(1.45^2 - 1)10^4} = 0.0218 \text{ } \mu\text{F}$$

In the same manner, the remaining element values can be computed, which results in

Section 3:

$$L_a = 0.526 \text{ H}$$

$$C_a = 0.0129 \text{ } \mu\text{F}$$

$$L_b = 0.644 \text{ H}$$

$$C_b = 6,606 \text{ pF}$$

Section 4:

$$L_a = 0.253 \text{ H}$$

$$C_a = 0.0163 \text{ } \mu\text{F}$$

$$L_b = 0.817 \text{ H}$$

$$C_b = 2,745 \text{ pF}$$

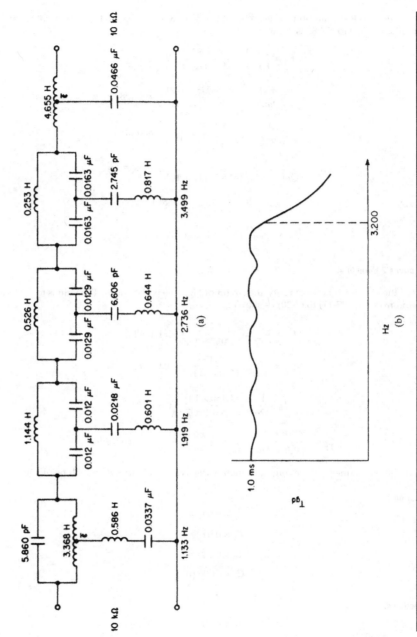

FIGURE 7-15 The 1-ms delay line of Example 7-1: (a) delay-line circuit; and (b) frequency response.

Section 5:

The remaining first-order all-pass section is realized using the circuit of Fig. 7-5b. The element values are given by Eqs. (7-15) and (7-16) as shown:

$$L = \frac{2R}{\alpha_0'} = \frac{2 \times 10^4}{4,296} = 4.655 \text{ H}$$

$$C = \frac{2}{\alpha_0' R} = \frac{2}{4,296 \times 10^4} = 0.0466 \ \mu\text{F}$$

(d) The resulting delay line is illustrated in Fig. 7-15a. The resonant frequencies shown are in hertz and correspond to $\omega_r/2\pi$ for each section. The center-tapped inductors require a unity coefficient of coupling. The delay characteristics as a function of frequency are also shown in Fig. 7-15b.

The delay line of Example 7-1 requires a total of nine inductors. If the classical design approach [see Wallis (1952)], which is based on image-parameter theory, were used, the resulting delay line would use about twice as many coils. Although the inductors would all be uniform in value (except for the end sections), this feature is certainly not justified by the added cost and complexity.

7.3.3 Active Delay Lines

An active delay line is designed by initially choosing a normalized filter and then denormalizing the associated poles in the same manner as in the case of LC delay lines. The resulting all-pass design parameters are implemented using the first- and second-order active structures of Sec. 7.2.

Active delay lines do not suffer from the Q limitations of LC delay lines and are especially suited for low-frequency applications where inductor values may become impractical. The following example illustrates the design of an active delay line.

Example 7-2   Design of a 100 μS Active Delay Line

Required:

Design an active delay line with a delay of 100 μs constant within 3 percent to 3 kHz. A gain of 10 is also required.

Result:

(a) Compute the required delay-bandwidth factor using Eq. (7-64).

$$TU_{req} = 2\pi f_{gd} T_{gd} = 2\pi 3{,}000 \times 10^{-4} = 1.885$$

A Bessel-type all-pass network will be chosen. Table 7-1 indicates that for a delay deviation of 1 percent, a complexity of $n = 3$ has a delay-bandwidth factor of 2.419, which is in excess of the required value.

(b) The Bessel low-pass poles are given in Table 10-41, and the corresponding values for $n = 3$ are

$$-1.0509 \pm j1.0025$$

$$-1.3270$$

Two sections are required consisting of a first-order and second-order type. The delay-scaling factor is computed using Eq. (7-65) as shown:

$$\text{DSF} = \frac{T_{gd}(\text{DC})}{T_{gd}} = \frac{3.5}{10^{-4}} = 3.5 \times 10^4$$

where T_{gd} (DC) is obtained from Table 7-1 and T_{gd} is 100 ms, the design value.
The second-order section design parameters are calculated using Eqs. (7-67) and (7-68).

$$\omega_r = \text{DSF}\sqrt{\alpha^2 + \beta^2} = 3.5 \times 10^4 \sqrt{1.0509^2 + 1.0025^2} = 50,833$$

and

$$Q = \frac{\omega_r}{2\alpha \text{DSF}} = \frac{50,833}{2 \times 1.509 \times 3.5 \times 10^4} = 6.691$$

The first-order section design parameter is given by Eq. (7-66) as shown:

$$\alpha'_0 = \alpha_0 \times \text{DSF} = 1.325 \times 3.5 \times 10^4 = 46,450$$

(c) The element values are computed as follows:

The second-order section:

The MFBP equalizer section of Fig. 7-12b will be used corresponding to $Q < 0.707$, where $R = 10$ kΩ, $C = 0.01$ μF, and $A = 10$. The element values are found from Eqs. (7-52) and (7-58) as shown:

$$R_2 = \frac{2Q}{\omega_r C} = \frac{2 \times 0.691}{50,833 \times 10^{-8}} = 2,719 \ \Omega$$

$$R_1 = \frac{R_2}{4Q^2} = \frac{2,719}{4 \times 0.691^2} = 1,424 \ \Omega$$

The first-order section:

The first-order section of Fig. 7-11a will be used, where R' is chosen at 10 kΩ, C at 0.01 μF, and α'_0 is 46,450. The value of R is given by Eq. (7-46) as follows:

$$R = \frac{1}{\alpha'_0 C} = \frac{1}{46,450 \times 10^{-8}} = 2,153 \ \Omega$$

(d) The resulting 100-μs active delay line is shown in Fig. 7-16 using standard 1 percent resistor values.

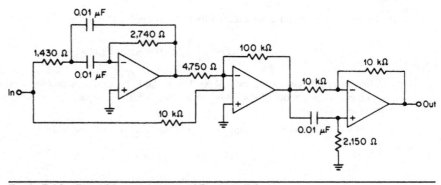

FIGURE 7-16   The 100-μs delay line of Example 7-2.

7.4 Delay Equalization of Filters

The primary emphasis in previous chapters has been the attenuation characteristics of filters. However, if the signal consists of a modulated waveform, the delay characteristics are also of importance. To minimize distortion of the input signal, a constant delay over the frequency range of interest is desirable. Typically, this would be 6 dB or so. The greater the attenuation, the less significant the impact of delay variation (delay distortion), since the spectral contributions of the attenuated signals are reduced.

The Bessel linear phase with equiripple error and transitional filter families all exhibit a flat delay. However, the amplitude response is less selective than that of other families. Frequently, the only solution to an attenuation requirement is a Butterworth, Chebyshev, or elliptic-function filter type. To also maintain the delay constant, delay equalizers would be required.

It is important to recognize that there are trade-offs between steep attenuation requirements and flatness of delay. For example, the higher the ripple of a Chebyshev filter, the steeper the rate of attenuation, but also the larger the delay deviation from flatness, especially around the corner frequency. Delay distortion also grows larger with increasing order n and steepness of elliptic-function filters, as well as ripple. Steep elliptic-function filters and high-order Chebyshev filters (see Sec. 2.4) are especially difficult to equalize since their delay characteristics near cutoff exhibit sharp delay peaks (hornlike in appearance).

Delay equalizer networks are frequently at least as complex as the filter being equalized. The number of sections required is dependent on the initial delay curve, the portion of the curve to be equalized, and the degree of equalization necessary. A very crude approximation to the number of equalizer sections required is given by

$$n = 2\Delta_{BW}\Delta_T + 1 \qquad (7\text{-}69)$$

where Δ_{BW} is the bandwidth of interest in hertz, and Δ_T is the delay distortion over Δ_{BW} in seconds.

The approach to delay equalization discussed in this section is graphical rather than analytical. A closed-form solution to the delay equalization of filters is not available. However, computer programs can be obtained that achieve a least-squares approximation to the required delay specifications, and are preferred to trial-and-error techniques. (See Note 1.)

Simply stated, delay equalization of a filter involves designing a network that has a delay shape that complements the delay curve of the filter being equalized. The composite delay will then exhibit the required flatness. Although the absolute delay increases as well, this result is usually of little significance, since it is the delay variation over the band of interest that disperses the spectral components of the signal. Typical delay curves of a band-pass filter, the delay equalizer network, and the composite characteristics are shown in Fig. 7-17.

To equalize the delay of a low-pass filter graphically, the highest frequency of interest and corresponding delay should be scaled to 1 rad/s so that the lower portion of the curve falls within the frequency region between DC and $\omega = 1$. This is accomplished by multiplying the delay axis by $2\pi f_h$, where f_h is the highest frequency to be equalized. The frequency axis is also divided by f_h and interpreted in radians per second so that f_h is transformed to 1 rad/s and all other frequencies are normalized to this point.

Note 1: The full version of *Filter Solutions* available from Nuhertz Technologies (www .nuhrtz.com) uses a proprietary approach to automatically perform equalization of low-pass and band-pass filters. The final results can then be manually "tweaked" by adjusting the pole-zero locations while observing the changes in group delay in real time.

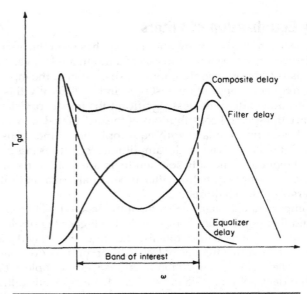

FIGURE 7-17   The delay equalization of a band-pass filter.

The normalized low-pass filter delay curves shown in Chapter 2 for the various filter families may also be used directly. In either case, the required equalizer delay characteristic is obtained by subtracting the delay curve from a constant equal to the maximum delay that occurs over the band. The resulting curve is then approximated by adding the delay contributions of equalizer sections. A sufficient number of sections is used to obtain the required composite delay flatness.

When a suitable match to the required curve is found, the equalizer parameters may be directly used to design the equalizer, and the circuit is then denormalized to the required frequency range and impedance level. Alternatively, the equalizer parameters can first be denormalized and the equalizer designed directly.

Band-pass filters are equalized in a manner similar to low-pass filters. The delay curve is first normalized by multiplying the delay axis by $2\pi f_0$, where f_0 is the filter center frequency. The frequency axis is divided by f_0 and interpreted in radians per second so that the center frequency is 1 rad/s and all other frequencies are normalized to the center frequency. A complementary curve is found, and appropriate equalizer sections are used until a suitable fit occurs. The equalizer is then denormalized.

7.4.1   First-Order Equalizers

First-order all-pass transfer functions were first introduced in Sec. 7.1. The delay of a first-order all-pass section is characterized by a maximum delay at low frequencies and decreasing delay with increasing frequency. As the value of α_0 is reduced, the delay tends to peak at DC and will roll off more rapidly with increasing frequencies.

The delay of a first-order all-pass section was given in Sec. 7.1 by Eq. (7-4) as shown:

$$T_{gd} = \frac{2\alpha_0}{\alpha_0^2 + \omega^2}$$

	ω, rad/s										
α_0	0	0.1	0.2	0.3	0.4	0.5	0.6	0.7	0.8	0.9	1.0
0.05	40.00	8.00	2.35	1.08	0.62	0.40	0.28	0.20	0.16	0.12	0.10
0.10	20.00	10.00	4.00	2.00	1.18	0.77	0.54	0.40	0.31	0.24	0.20
0.15	13.33	9.23	4.80	2.67	1.64	1.10	0.78	0.59	0.45	0.36	0.29
0.20	10.00	8.00	5.00	3.08	2.00	1.38	1.00	0.75	0.59	0.47	0.38
0.25	8.00	6.90	4.88	3.28	2.25	1.60	1.18	0.91	0.71	0.57	0.47
0.30	6.67	6.00	4.62	3.33	2.40	1.76	1.33	1.03	0.82	0.67	0.55
0.35	5.71	5.28	4.31	3.29	2.48	1.88	1.45	1.14	0.92	0.75	0.62
0.40	5.00	4.71	4.00	3.20	2.50	1.95	1.54	1.23	1.00	0.82	0.69
0.45	4.44	4.24	3.71	3.08	2.48	1.99	1.60	1.30	1.07	0.89	0.75
0.50	4.00	3.85	3.45	2.94	2.44	2.00	1.64	1.35	1.12	0.94	0.80
0.55	3.64	3.52	3.21	2.80	2.38	1.99	1.66	1.39	1.17	0.99	0.84
0.60	3.33	3.24	3.00	2.67	2.31	1.97	1.67	1.41	1.20	1.03	0.88
0.65	3.08	3.01	2.81	2.54	2.23	1.93	1.66	1.42	1.22	1.05	0.91
0.70	2.86	2.80	2.64	2.41	2.15	1.89	1.65	1.43	1.24	1.08	0.94
0.75	2.67	2.62	2.49	2.30	2.08	1.85	1.63	1.43	1.25	1.09	0.96
0.80	2.50	2.46	2.35	2.19	2.00	1.80	1.60	1.42	1.25	1.10	0.98
0.85	2.35	2.32	2.23	2.09	1.93	1.75	1.57	1.40	1.25	1.11	0.99
0.90	2.22	2.20	2.12	2.00	1.86	1.70	1.54	1.38	1.24	1.11	0.99
0.95	2.11	2.08	2.02	1.91	1.79	1.65	1.50	1.36	1.23	1.11	1.00
1.00	2.00	1.98	1.92	1.83	1.72	1.60	1.47	1.34	1.22	1.10	1.00
1.25	1.60	1.59	1.56	1.51	1.45	1.38	1.30	1.22	1.14	1.05	0.98
1.50	1.33	1.33	1.31	1.28	1.24	1.20	1.15	1.09	1.04	0.98	0.92
1.75	1.14	1.14	1.13	1.11	1.09	1.06	1.02	0.99	0.95	0.90	0.86
2.00	1.00	1.00	0.99	0.98	0.97	0.94	0.92	0.89	0.86	0.83	0.80

TABLE 7-2   The First-Order Equalizer Delay in Seconds

Working directly with Eq. (7-4) is somewhat tedious, so a table of delay values for α_0 ranging between 0.05 and 2.00 at frequencies from $\omega = 0$ to $\omega = 1$ is provided in Table 7-2. This table can be directly used to determine the approximate α_0 necessary to equalize the normalized filter delay. A more exact value of α_0 can then be determined from Eq. (7-4) if desired.

Use of Table 7-2 is best illustrated by an example, as follows.

Example 7-3   Design of an *LC* and Active Delay Equalizer for a Low-Pass Filter

Required:

Design a delay equalizer for an $n = 5$ Butterworth low-pass filter with a 3-dB cutoff of 1,600 Hz. The delay variation should not exceed 75 μs from DC to 1,600 Hz.

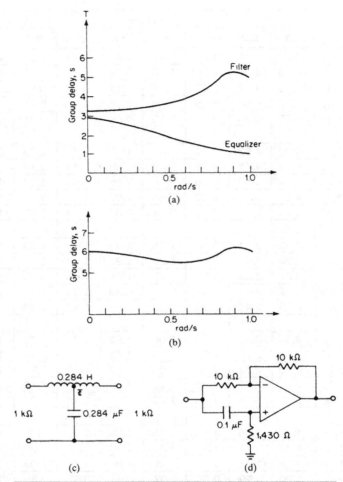

Figure 7-18   The delay equalization of Example 7-3: (*a*) filter and equalizer delay curves; (*b*) composite delay curve; (*c*) *LC* equalizer; and (*d*) active equalizer.

Result:

(*a*) The Butterworth normalized delay curves of Fig. 2-35 can be used directly since the region between DC and 1 rad/s corresponds to the frequency range of interest. The curve for $n = 5$ indicates that the peak delay occurs near 0.9 rad/s and is approximately 1.9 s greater than the value at DC. This corresponds to a denormalized variation of $1.9/2\pi f_h$, or 190 μs, where f_h is 1,600 Hz, so an equalizer is required.

(*b*) Examination of Table 7-2 indicates that a first-order equalizer with an α_0 of 0.7 has a delay at DC that is approximately 1.8 s greater than the delay at 0.9 rad/s, so a reasonable fit to the required shape should occur.

The delay of the normalized filter and the first-order equalizer for $\alpha_0 = 0.7$ is shown in Fig. 7.18*a*. The combined delay is given in Fig. 7-18*b*. The peak-to-peak delay variation is about 0.7 s, which corresponds to a denormalized delay variation of $0.7/2\pi f_h$, or 70 μs.

(*c*) The first-order equalizer parameter $\alpha_0 = 0.7$ is denormalized by the factor $2\pi f_h$, resulting in $\alpha_0' = 7037$. The corresponding passive equalizer is designed as shown in Eqs. (7-15) and (7-16), where the impedance level R is chosen to be 1 kΩ:

$$L = \frac{2R}{\alpha'_0} = \frac{2 \times 10^3}{7,037} = 0.284 \text{ H}$$

$$C = \frac{2}{\alpha'_0 R} = \frac{2}{7,037 \times 10^3} = 0.284 \ \mu\text{F}$$

The first-order LC equalizer section is shown in Fig. 7-18c.

(d) An active first-order equalizer section can also be designed using the circuit of Fig. 7-11a. If we select a C of 0.1 μF, where R' = kΩ, the value of R is given by Eq. (7-46) as shown:

$$R = \frac{1}{\alpha'_0 C} = \frac{1}{7,037 \times 10^{-7}} = 1,421 \ \Omega$$

The active equalizer circuit is illustrated in Fig. 7-18d.

Highly selective low-pass filters, such as the elliptic-function type, have a corresponding delay characteristic that increases very dramatically near cutoff. First-order all-pass sections cannot then provide a complementary delay shape, so they are limited to applications involving low-pass filters of moderate selectivity.

7.4.2 Second-Order Equalizers

First-order equalizers have a maximum delay at DC and a single design parameter α_0, which limits their use. Second-order sections have two design parameters, ω_r and Q. The delay shape is band-pass in nature and can be made broad or sharp by changing the Q. The peak delay frequency is determined by the design parameter ω_r. As a result of this flexibility, second-order sections can be used to equalize virtually any type of delay curve. The only limitations are in the number of sections the designer is willing to use and the effort required to achieve a given degree of equalization.

The group delay of a second-order all-pass section was given by Eq. (7-11) as

$$T_{gd} = \frac{2Q\omega_r \left(\omega^2 + \omega_r^2\right)}{Q^2 \left(\omega^2 + \omega_r^2\right)^2 + \omega^2 \omega_r^2}$$

If we normalize this expression by setting ω_r equal to 1, we obtain

$$T_{gd} = \frac{2Q(\omega^2 + 1)}{Q^2(\omega^2 - 1)^2 + \omega^2} \tag{7-70}$$

To determine the delay at DC, we can set ω_r equal to zero, which results in

$$T_{gd}(\text{DC}) = \frac{2}{Q} \tag{7-71}$$

For Qs below 0.577, the maximum delay occurs at DC. As the Q is increased, the frequency where maximum delay occurs approaches 1 rad/s and is given by

$$\omega(T_{gd,\,\text{max}}) = \sqrt{\sqrt{4 - \frac{1}{Q^2}} - 1} \tag{7-72}$$

For Qs of 2 or more, the maximum delay can be assumed to occur at 1 rad/s and may be determined from

$$T_{gd,\,\text{max}} = 4Q \tag{7-73}$$

Equations (7-70) to (7-72) are evaluated in Table 7-6 for Qs ranging from 0.25 to 10.

To use Table 7-6 directly, first normalize the curve to be equalized so that the minimum delay occurs at 1 rad/s. Then, select an equalizer from the table that provides the best fit for a complementary curve.

A composite curve is then plotted. If the delay ripple is excessive, additional equalizer sections are required to fill in the delay gaps. The data of Table 7-6 can again be used by scaling the region to be equalized to a 1-rad/s center and selecting a complementary equalizer shape from the table. The equalizer parameters can then be shifted to the region of interest by scaling.

The procedure described is an oversimplification of the design process. The equalizer responses will interact with each other, so each delay region to be filled in cannot be treated independently. Every time a section is added, the previous sections may require an adjustment of their design parameters.

Delay equalization generally requires considerably more skill than the actual design of filters. Standard pole-zero patterns are defined for the different filter families, whereas the design of equalizers involves the approximation problem where a pole-zero pattern must be determined for a suitable fit to a curve. The following example illustrates the use of second-order equalizer sections to equalize delay.

Example 7-4 Design of a Delay Equalizer for a Band-Pass Filter

Required:

A band-pass filter with the delay measurements of Table 7-3 must be equalized to within 700 μs.
The corresponding delay curve is plotted in Fig. 7-19a.

Result:

(a) Since the minimum delay occurs at 1,000 Hz, normalize the curve by dividing the frequency axis by 1,000 Hz and multiplying the delay axis by 2π 1,000. The results are shown in Table 7-4 and plotted in Fig. 7-19b.

(b) An equalizer is required that has a nominal delay peak of 10 s at 1 rad/s relative to the delay at 0.5 and 1.5 rad/s. Examination of Table 7-6 indicates that the delay corresponding to a Q of 2.75 will meet this requirement.

If we add this delay, point by point, to the normalized delay of Table 7-4, the values of Table 7-5 will be obtained.

The corresponding curve is plotted in Fig. 7-19c. This curve can be denormalized by dividing the delay by 2π1,000 and multiplying the frequency axis by 1,000, resulting in the final curve of Fig. 7-19d. The differential delay variation over the band is about 675 μs.

Frequency, Hz	Delay, μs
500	1,600
600	960
700	640
800	320
900	50
1,000	0
1,100	160
1,200	480
1,300	800
1,400	1,120
1,500	1,500

TABLE 7-3 Specified Delay

Frequency, rad/s	Delay, s
0.5	10.1
0.6	6.03
0.7	4.02
0.8	2.01
0.9	0.31
1.0	0
1.1	1.01
1.2	3.02
1.3	5.03
1.4	7.04
1.5	9.42

TABLE 7-4 Normalized Delay

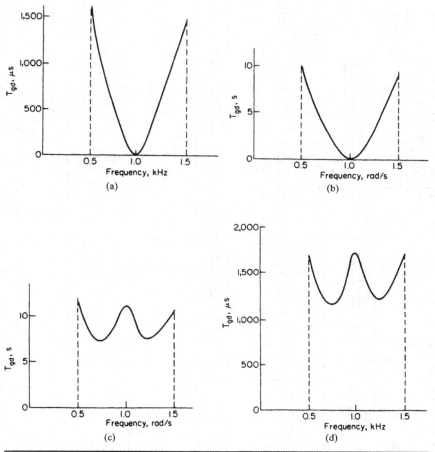

FIGURE 7-19 The delay equalization of Example 7-4: (*a*) unequalized delay; (*b*) normalized delay; (*c*) normalized equalized delay; and (*d*) denormalized equalized delay.

Frequency, rad/s	Delay, s
0.5	11.6
0.6	8.19
0.7	7.36
0.8	7.58
0.9	9.50
1.0	11.0
1.1	8.89
1.2	7.64
1.3	7.83
1.4	8.86
1.5	10.7

TABLE 7-5 Equalized Delay

Table 7-6   The Delay of Normalized Second-Order Section ($\omega_r = 1$ rad/s)

Q	T_{gd}(DC)	$\omega(T_{gd,max})$	$T_{gd,max}$	0.1	0.2	0.3	0.4	0.5	0.6	0.7	0.8	0.9	1.0	1.1	1.2	1.3	1.4	1.5	1.6	1.7	1.8	1.9	2.0	3.0	4.0	5.0
0.25	8.00	DC	8.00	7.09	5.33	3.85	2.84	2.19	1.76	1.42	1.27	1.11	1.00	0.91	0.84	0.78	0.73	0.69	0.66	0.62	0.60	0.57	0.55	0.38	0.28	0.21
0.50	4.00	DC	4.00	3.96	3.85	3.67	3.45	3.20	2.94	2.69	2.44	2.21	2.00	1.81	1.64	1.49	1.35	1.23	1.12	1.03	0.94	0.87	0.80	0.40	0.24	0.15
0.75	2.67	0.700	3.51	2.70	2.79	2.94	3.12	3.31	3.46	3.51	3.45	3.27	3.00	2.69	2.36	2.06	1.79	1.56	1.36	1.19	1.05	0.93	0.83	0.33	0.18	0.11
1.00	2.00	0.856	4.31	2.04	2.16	2.37	2.68	3.08	3.53	3.97	4.26	4.28	4.00	3.52	2.99	2.48	2.05	1.71	1.43	1.20	1.03	0.88	0.77	0.27	0.14	0.09
1.25	1.60	0.913	5.23	1.64	1.76	1.97	2.30	2.77	3.40	4.16	4.87	5.22	5.00	4.32	3.50	2.76	2.18	1.73	1.40	1.15	0.96	0.81	0.69	0.23	0.11	0.07
1.50	1.33	0.941	6.18	1.37	1.47	1.67	1.99	2.47	3.18	4.16	5.28	6.09	6.00	5.06	3.90	2.92	2.20	1.69	1.33	1.07	0.88	0.73	0.62	0.20	0.10	0.06
1.75	1.14	0.957	7.15	1.17	1.27	1.45	1.75	2.22	2.95	4.05	5.54	6.88	7.00	5.75	4.20	2.99	2.17	1.62	1.24	0.98	0.80	0.66	0.55	0.17	0.08	0.05
2.00	1.00	0.968	8.13	1.03	1.12	1.28	1.56	2.00	2.72	3.89	5.66	7.59	8.00	6.38	4.41	2.99	2.10	1.53	1.16	0.91	0.73	0.60	0.50	0.15	0.07	0.04
2.25	0.89	0.975	9.12	0.91	0.99	1.15	1.40	1.82	2.52	3.71	5.69	8.20	9.00	6.94	4.54	2.95	2.01	1.44	1.08	0.83	0.66	0.54	0.45	0.14	0.06	0.04
2.50	0.80	0.980	10.1	0.82	0.90	1.04	1.27	1.66	2.33	3.52	5.66	8.74	10.0	7.44	4.60	2.88	1.92	1.35	1.00	0.77	0.61	0.50	0.41	0.12	0.06	0.04
2.75	0.73	0.983	11.1	0.75	0.82	0.94	1.16	1.53	2.16	3.34	5.57	9.19	11.0	7.88	4.62	2.80	1.82	1.27	0.93	0.72	0.57	0.46	0.38	0.11	0.05	0.03
3.00	0.67	0.986	12.1	0.69	0.75	0.87	1.07	1.41	2.02	3.16	5.45	9.57	12.0	8.25	4.60	2.70	1.73	1.20	0.87	0.67	0.53	0.43	0.35	0.10	0.05	0.03
3.25	0.61	0.988	13.1	0.63	0.69	0.80	0.99	1.31	1.89	2.99	5.31	9.88	13.0	8.57	4.55	2.60	1.65	1.13	0.82	0.62	0.49	0.40	0.33	0.09	0.05	0.03
3.50	0.57	0.990	14.1	0.59	0.64	0.75	0.92	1.23	1.77	2.84	5.15	10.1	14.0	8.84	4.48	2.50	1.56	1.06	0.77	0.58	0.46	0.37	0.31	0.09	0.04	0.03
3.75	0.53	0.991	15.1	0.55	0.60	0.70	0.86	1.15	1.67	2.69	5.00	10.3	15.0	9.06	4.40	2.41	1.49	1.01	0.73	0.55	0.43	0.35	0.29	0.08	0.04	0.02
4.00	0.50	0.992	16.1	0.51	0.56	0.65	0.81	1.08	1.57	2.56	4.84	10.4	16.0	9.23	4.30	2.31	1.42	0.95	0.69	0.52	0.41	0.33	0.27	0.08	0.04	0.02
4.25	0.47	0.993	17.1	0.48	0.53	0.62	0.76	1.02	1.49	2.44	4.68	10.5	17.0	9.36	4.20	2.22	1.35	0.91	0.65	0.49	0.38	0.31	0.26	0.07	0.04	0.02
4.50	0.44	0.994	18.1	0.46	0.50	0.58	0.72	0.97	1.41	2.33	4.52	10.6	18.0	9.46	4.10	2.14	1.29	0.86	0.62	0.47	0.36	0.29	0.24	0.07	0.03	0.02
4.75	0.42	0.994	19.1	0.43	0.47	0.55	0.69	0.92	1.35	2.22	4.37	10.6	19.0	9.52	3.99	2.06	1.24	0.82	0.59	0.44	0.35	0.29	0.23	0.07	0.03	0.02
5.00	0.40	0.995	20.1	0.41	0.45	0.52	0.65	0.87	1.28	2.13	4.23	10.6	20.0	9.56	3.89	1.98	1.18	0.79	0.56	0.42	0.33	0.27	0.22	0.06	0.03	0.02
6.00	0.33	0.997	24.0	0.34	0.38	0.44	0.54	0.73	1.08	1.82	3.71	10.3	24.0	9.48	3.48	1.71	1.01	0.67	0.47	0.36	0.28	0.22	0.18	0.05	0.03	0.02
7.00	0.29	0.997	28.0	0.29	0.32	0.38	0.47	0.63	0.93	1.58	3.28	9.83	28.0	9.18	3.13	1.51	0.88	0.58	0.41	0.31	0.24	0.19	0.16	0.04	0.02	0.01
8.00	0.25	0.998	32.0	0.26	0.28	0.33	0.41	0.55	0.82	1.39	2.94	9.28	32.0	8.77	2.82	1.34	0.78	0.51	0.36	0.27	0.21	0.17	0.14	0.04	0.02	0.01
9.00	0.22	0.999	36.0	0.23	0.25	0.29	0.36	0.49	0.73	1.24	2.65	8.73	36.0	8.32	2.57	1.20	0.70	0.45	0.32	0.24	0.19	0.15	0.12	0.03	0.02	0.01
10.00	0.20	0.999	40.0	0.21	0.23	0.26	0.33	0.44	0.66	1.13	2.41	8.19	40.0	7.87	2.35	1.09	0.69	0.41	0.30	0.22	0.17	0.13	0.11	0.03	0.02	0.01

The equalizer of Example 7-4 provides over a 2:1 reduction in the differential delay. Further equalization can be obtained with two addition equalizers to fill in the concave regions around 750 and 1,250 Hz.

7.5   Wideband 90° Phase-Shift Networks

Wideband 90° phase-shift networks have a single input and two output ports. Both outputs maintain a constant phase difference of 90° within a prescribed error over a wide range of frequencies. The overall transfer function is all-pass. These networks are widely used in the design of single-sideband systems and in other applications requiring 90° phase splitting.

Bedrosian (see References) solved the approximation problem for this family of networks on a computer. The general structure is shown in Fig. 7-20a and consists of N and P networks. Each network provides real-axis pole-zero pairs and is all-pass. The transfer function is of the form

$$T(s) = \frac{(s - \alpha_1)(s - \alpha_2)\cdots(s - \alpha_{n/2})}{(s + \alpha_1)(s + \alpha_2)\cdots(s + \alpha_{n/2})}$$

(7-74)

where $n/2$ is the order of the numerator and denominator polynomials. The total complexity of both networks is then n.

Real-axis all-pass transfer functions can be realized using a cascade of passive or active first-order sections. Both versions are shown in Figs. 7-20b and c.

The transfer functions tabulated in Table 7-7 approximate a 90° phase difference in an equiripple manner. This approximation occurs within the bandwidth limits ω_L and ω_u, as shown in Fig. 7-21. These frequencies are normalized so that $\sqrt{\omega_L \omega_u} = 1$. For a specified bandwidth ratio ω_u/ω_L, the individual band limits can be found from

$$\omega_L = \sqrt{\frac{\omega_L}{\omega_u}}$$

(7-75)

and

$$\omega_u = \sqrt{\frac{\omega_u}{\omega_L}}$$

(7-76)

As the total complexity n is made larger, the phase error decreases for a mixed bandwidth ratio, or, for a fixed phase error, the bandwidth ratio will increase.

To use Table 7-7, first determine the required bandwidth ratio from the frequencies given. A network is then selected that has a bandwidth ratio ω_u/ω_L that exceeds the requirements and a phase error $\pm\Delta\phi$ that is acceptable.

A frequency-scaling factor (FSF) is determined from

$$\text{FSF} = 2\pi f_0$$

(7-77)

where f_0 is the geometric mean of the specified band limits, or $\sqrt{f_L f_u}$. The tabulated α's are then multiplied by the FSF for denormalization. The resulting pole-zero pairs can be realized by a cascade of active or passive first-order sections for each network.

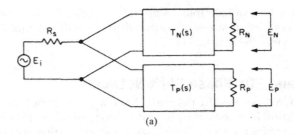

(a)

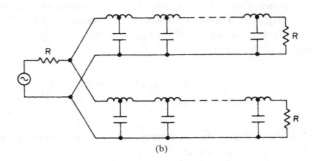

(b)

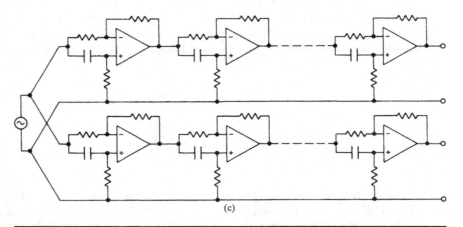

(c)

FIGURE 7-20 Wideband 90° phase shift networks: (a) the general structure; (b) a passive realization; and (c) an active realization.

n	$\Delta\phi$	α_N	α_P
\multicolumn{4}{c}{$\omega_u/\omega_L = 1{,}146$}			
6	6.84°	43.3862	8.3350
		2.0264	0.4935
		0.1200	0.0231
8	2.12°	59.7833	14.4159
		4.8947	1.6986
		0.5887	0.2043
		0.0694	0.0167
10	0.66°	75.8845	20.4679
		8.3350	3.5631
		1.5279	0.6545
		0.2807	0.1200
		0.0489	0.0132
\multicolumn{4}{c}{$\omega_u/\omega_L = 573.0$}			
6	4.99°	34.3132	7.0607
		1.9111	0.5233
		0.1416	0.0291
8	1.39°	47.0857	11.8249
		4.3052	1.6253
		0.6153	0.2323
		0.0846	0.0212
10	0.39°	59.6517	16.5238
		7.0607	3.2112
		1.4749	0.6780
		0.3114	0.1416
		0.0605	0.0168
\multicolumn{4}{c}{$\omega_u/\omega_L = 286.5$}			
4	13.9°	16.8937	2.4258
		0.4122	0.0592
6	3.43°	27.1337	5.9933
		1.8043	0.5542
		0.1669	0.0369
8	0.84°	37.0697	9.7136
		3.7944	1.5566
		0.6424	0.2636
		0.1030	0.0270

TABLE 7-7   Pole-Zero Locations for 90° Phase-Shift Networks* (*Continued*)

n	$\Delta\phi$	α_N	α_P
		$\omega_u/\omega_L = $ **286.5**	
10	0.21°	46.8657	13.3518
		5.9933	2.8993
		1.4247	0.7019
		0.3449	0.1669
		0.0749	0.0213
		$\omega_u/\omega_L = $ **143.2**	
4	10.2°	13.5875	2.2308
		0.4483	0.0736
8	0.46°	29.3327	8.0126
		3.3531	1.4921
		0.6702	0.2982
		0.1248	0.0341
10	0.10°	37.0091	10.8375
		5.1050	2.6233
		1.3772	0.7261
		0.3812	0.1959
		0.0923	0.0270
		$\omega_u/\omega_L = $ **81.85**	
4	7.58°	11.4648	2.0883
		0.4789	0.0918
6	1.38°	18.0294	4.5017
		1.6316	0.6129
		0.2221	0.0555
8	0.25°	24.4451	6.8929
		3.0427	1.4432
		0.6929	0.3287
		0.1451	0.0409
10	0.046°	30.7953	9.2085
		4.5017	2.4248
		1.3409	0.7458
		0.4124	0.2221
		0.1086	0.0325
		$\omega_u/\omega_L = $ **57.30**	
4	6.06°	10.3270	2.0044
		0.4989	0.0968
6	0.99°	16.1516	4.1648
		1.5873	0.6300
		0.2401	0.0619

TABLE 7-7   Pole-Zero Locations for 90° Phase-Shift Networks* (*Continued*)

n	$\Delta\phi$	α_N	α_P
		$\omega_u/\omega_L = 57.30$	
8	0.16°	21.8562	6.2817
		2.8648	1.4136
		0.7074	0.3491
		0.1592	0.0458
10	0.026°	27.5087	8.3296
		4.1648	2.3092
		1.3189	0.7582
		0.4331	0.2401
		0.1201	0.0364
		$\omega_u/\omega_L = 28.65$	
4	3.57°	8.5203	1.6157
		0.5387	0.1177
6	0.44°	13.1967	3.6059
		1.5077	0.6633
		0.2773	0.0758
8	0.056°	17.7957	5.2924
		2.5614	1.3599
		0.7354	0.3904
		0.1890	0.0562
10	0.0069°	22.3618	6.9242
		3.6059	2.1085
		1.2786	0.7821
		0.4743	0.2773
		0.1444	0.0447
		$\omega_u/\omega_L = 11.47$	
4	1.31°	5.9339	1.5027
		0.5055	0.1280
6	0.10°	10.4285	3.0425
		1.4180	0.7052
		0.3287	0.0959
8	0.0075°	14.0087	4.3286
		2.2432	1.2985
		0.7701	0.4458
		0.2310	0.0714

*Numerical values for this table are obtained from S. D. Bedrosian, "Normalized Design of 90° Phase-Difference Networks," *IRE Transactions on Circuit Theory,* June 1960.

TABLE 7-7 Pole-Zero Locations for 90° Phase-Shift Networks* (*Continued*)

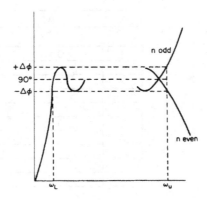

FIGURE 7-21 Wideband 90° phase shift approximation.

The following example illustrates the design of a 90° phase shift network.

Example 7-5   Design of an Active Wideband 90° Phase Splitter

Required:

Design a network with dual outputs that maintain a phase difference of 90° within ±0.2° over the frequency range of 300 to 3,000 Hz. The circuit should be all-pass and active.

Result:

(a) Since a 10:1 bandwidth ratio is required (3,000 Hz/300 Hz), the design corresponding to $n = 6$ and $\omega_u/\omega_L = 11.47$ is chosen. The phase shift error will be ±0.1°.

(b) The normalized real pole-zero coordinates for both networks are given as follows:

P Network	N Network
$\alpha_1 = 10.4285$	$\alpha_4 = 3.0425$
$\alpha_2 = 1.4180$	$\alpha_5 = 0.7052$
$\alpha_3 = 0.3287$	$\alpha_6 = 0.0959$

The frequency-scaling factor is obtained using Eq. (7-77).

$$\text{FSF} = 2\pi f_0 = 2\pi \times 948.7 = 5{,}961$$

where f_0 is $\sqrt{300 \times 3{,}000}$. The pole-zero coordinates are multiplied by the FSF, resulting in the following set of denormalized values for α.

P Network	N Network
$\alpha_1' = 62{,}164$	$\alpha_4' = 18{,}136$
$\alpha_2' = 8{,}453$	$\alpha_5' = 4{,}204$
$\alpha_3' = 1{,}959$	$\alpha_6' = 517.7$

(c) The P and N networks can now be realized using the active first-order all-pass circuit of Sec. 7.2 and Fig. 7-11a.

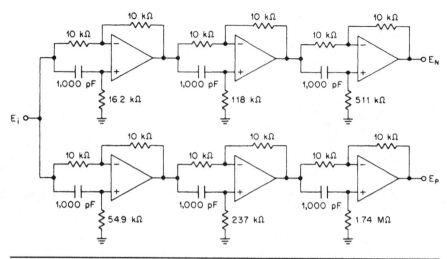

FIGURE 7-22   The wideband 90° phase shift network of Example 7-5.

If we let $R' = 10\ k\Omega$ and $C = 1,000\ pF$, the value of R is given by Eq. (7-46) as shown:

$$R = \frac{1}{\alpha_0 C}$$

Using the denormalized α's for the P and N networks, the following values are obtained:

Section	P Network	N Network
1	R = 16.09 kΩ	R = 55.14 kΩ
2	R = 118.3 kΩ	R = 237.9 kΩ
3	R = 510.5 kΩ	R = 1.749 MΩ

The final circuit is shown in Fig. 7-22 using standard 1 percent resistor values.

7.6   Design of Passive Delay Lines with Repetitious Elements

7.6.1   An All-Pass Delay Line

The LC delay lines of Sec. 7.3 all require a variety of values for the inductors and capacitors. This becomes somewhat of an issue during manufacturing, as multiple inductor values have to be procured from coil manufacturers, so there is no advantage of economy of quantity. The delay lines in this section have repetitious inductor values (as well as capacitors), except for the image parameter design, which uses just three different inductor values no matter what the complexity.

The two critical parameters of a delay line are the bandwidth, usually expressed as the 3-dB point, and the flatness of the delay over that bandwidth. However, by using an all-pass approach, in theory, frequency response should always be flat. That is not always the case, as the components used are not ideal. Inductors do not have infinite Q, zero stray capacitance, and do not have a unity coefficient of magnetic coupling if a tap is involved.

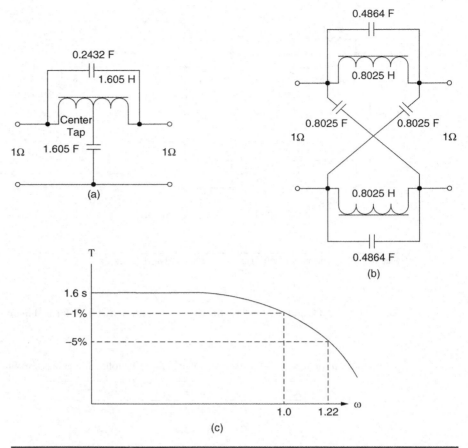

FIGURE 7-23 All-pass delay line section: (a) bridged-T configuration; (b) lattice structure; and (c) normalized delay.

The best results using the tapped inductor approach shown in this section are generally achieved below 20 KHz, where long delays may be needed. Insertion loss will be a function mainly of the DC resistance of the inductors, so it can be easily estimated to ensure it is acceptable if a large number of sections is involved. If not, inductors with lower losses can be used.

Figure 7-23 shows a single normalized one-stage delay section in both bridged-T and lattice form. The lattice configuration does not require a center-tapped inductor, but does use more elements. The inductor Q corresponding to the normalized frequency of 1.6 rad/s should be as high as possible to avoid a notch in the amplitude response.

The design procedure is quite simple. First determine whether you are seeking 1 percent or 5 percent flatness. Then determine the FSF by

$$FSF = \frac{2\pi F_d}{\omega_A} \tag{7-78}$$

where F_d is the highest frequency of interest for the delay and ω_A is either 1 or 1.22, depending on whether the flatness requirement is 1 percent or 5 percent, as shown in Fig. 7-23.

Compute:
$$T_s = \frac{1.6}{FSF} \tag{7-79}$$

where T_s is the delay per denormalized section. The number of sections required is determined by

$$N = \frac{T_d}{T_s} \tag{7-80}$$

where T_d is the total delay required. Round off N to the next highest number.
Then compute a new FSF by

$$T_{snew} = \frac{T_d}{N_{new}} \tag{7-81}$$

then

$$FSF_{new} = \frac{1.6}{T_{snew}} \tag{7-82}$$

With a new FSF and a desired impedance level, the circuit can be scaled (denormalized). This completes the design procedure. If necessary, the total delay can be trimmed by adding or removing one section at a time.

7.6.2 Image Parameter Unsymmetrical Delay Line

Figure 7-24 illustrates a delay line based on image parameter. Although the image parameter design method is not as precise as the other methods used in this book, it is convenient and can result in convenient element values. Typically, the delay will remain within 2 percent over the band, except for the effect of component tolerance variations. The terminating impedance is Z_o, which is normally the image impedance, but in this design we will take the liberty of using R_o rather than a complex image impedance.

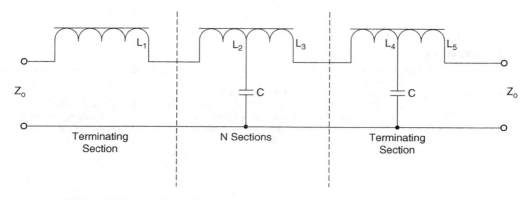

FIGURE 7-24   Image parameter unsymmetrical delay line.

As in the case of the all-pass delay line circuit discussed earlier, the coefficient of coupling of the tapped inductor is quite critical, so this approach is restricted to relatively low frequencies. Also, DC resistance of the inductors should be minimized to avoid a high insertion loss.

The element values are computed as follows:

$$N = \frac{T_d BW_{3\,dB}}{0.23} \tag{7-83}$$

where T_d is the total delay and $BW_{3\,dB}$ is the 3-dB bandwidth in Hz.

Compute the delay per section by

$$T_s = \frac{T_d}{N + 1} \tag{7-84}$$

The values are computed by

$$L = R_o T_s \tag{7-85}$$

$$C = \frac{T_s}{R_o} \tag{7-86}$$

$$L_2 = \frac{L}{7,225} \tag{7-87}$$

$$L' = 56.25 L_2 \tag{7-88}$$

$$L_3 = L' + L_2 + 2\sqrt{L'L_2} \tag{7-89}$$

$$L_1 = \frac{L' - L_2}{2} \tag{7-90}$$

$$L_4 = 2.1 L_2 \tag{7-91}$$

$$L_5 = \frac{L'}{2.1} + L_4 + 2\sqrt{L'L_2} \tag{7-92}$$

References

Bedrosian, S. D. "Normalized Design of 90° Phase-Difference Networks." *IRE Transactions on Circuit Theory* CT-7, June 1960.

Geffe, P. R. (1963). *Simplified Modern Filter Design*. New York: John F. Rider.

Lindquist, C. S. (1977). *Active Network Design*. Long Beach, California: Steward and Sons.

Wallis, C. M. "Design of Low-Frequency Constant Time Delay Lines." *AIEE Proceedings* 71, 1952.

Williams, A. B. "An Active Equalizer with Adjustable Amplitude and Delay." *IEEE Transactions on Circuit Theory* CT-16, November 1969.

CHAPTER 8

Refinements in *LC* Filter Design and the Use of Resistive Networks

8.1 Introduction

The straightforward application of the design techniques outlined for *LC* filters will not always result in practical element values or desirable circuit configurations. Extreme cases of impedance or bandwidth can produce designs that may be extremely difficult or even impossible to realize. This chapter is concerned mainly with circuit transformations that enable impractical designs to be transformed into alternative configurations with the identical response and using more practical elements. Also, the use of resistive networks to supplement *LC* filters or to function independently is covered.

8.2 Tapped Inductors

An extremely useful tool for eliminating impractical element values is the transformer. As the reader may recall from introductory AC circuit analysis, a transformer with a turns ratio N will transform an impedance by a factor of N^2. A parallel element can be shifted between the primary and secondary at will, provided that its impedance is modified by N^2.

Figure 8-1 illustrates how a tapped inductor is used to reduce the value of a resonating capacitor. The tuned circuit of Fig. 8-1*a* is first modified by introducing an impedance step-up transformer, as shown in Fig. 8-1*b*, so that capacitor C can be moved to the secondary and reduced by a factor of N^2. This can be carried a step further, resulting in the circuit of Fig. 8-1*c*. The transformer has been absorbed as a continuation of the inductor, resulting in an autotransformer. The ratio of the overall inductance to the tap inductance becomes N^2.

As an example, let's modify the tuned circuit of Example 5-4, shown in Fig. 8-2*a*. To reduce the capacitor from 0.354 μF to 0.027 μF, the overall inductance is increased by the impedance ratio 0.354 μF/0.027 μF, resulting in the circuit of Fig. 8-2*b*. The resonant frequency remains unchanged since the overall *LC* product is still the same.

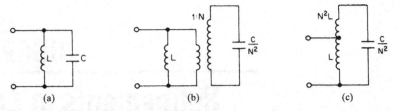

FIGURE 8-1 The tapped inductor: (*a*) basic tuned circuit; (*b*) the introduction of a transformer; and (*c*) absorbed transformer.

As a further example, let's consider *LC* elliptic-function low-pass filters. The parallel-resonant circuits may also contain high-capacity values, which can be reduced by this method. Figure 8-3 shows a section of a low-pass filter. To reduce the resonating capacitor to 0.354 μF, the overall inductance is increased by the factor 1.055 μF/0.1 μF and a tap is provided at the original inductance value.

The tapped coil is useful not only for reducing resonating capacitors, but also for transforming entire sections of a filter, including terminations. The usefulness of the tapped inductor is limited only by the ingenuity and resourcefulness of the designer. Figure 8-4 illustrates some applications of this technique using designs from previous examples. In the case of Fig. 8-4*a*, where a tapped coil enables operation from unequal terminations, the same result could have been achieved using Bartlett's bisection theorem or other methods (see Sec. 3.1). However, the transformer approach results in maximum power transfer (minimum insertion loss). The circuits of Figs. 8-4*b* and *c* demonstrate how element values can be manipulated by taps. The tapped

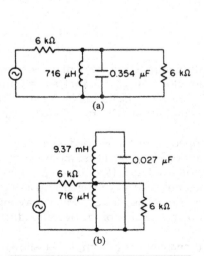

FIGURE 8-2 Reducing the resonant capacitor value: (*a*) tuned circuit; and (*b*) modified circuit.

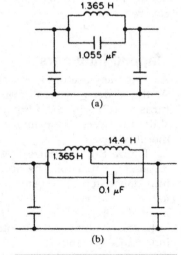

FIGURE 8-3 The application of a tapped inductor in elliptic-function low-pass filters: (*a*) a filter section; and (*b*) tapped inductor.

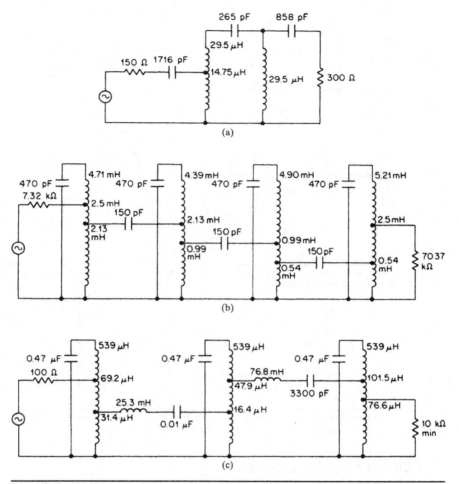

FIGURE 8-4 Applications of tapped inductors: (*a*) the high-pass filter of Example 4-1 modified for unequal terminations; (*b*) the filter of Example 5-7 modified for standard capacitor values; and (*c*) the filter of Example 5-8 modified for standard capacitor values.

inductance values shown are all measured from the grounded end of the shunt inductors. Series branches can be manipulated up or down in their impedance level by multiplying the shunt inductance taps on both sides of the branch by the desired impedance-scaling factor.

Transformers or autotransformers are by no means ideal. Imperfect coupling within the magnetic structure will result in a leakage inductance that can cause spurious responses at higher frequencies, as shown in Fig. 8-5. These effects can be minimized by using near-unity turns ratios. Another solution is to leave a portion of the original capacity at the tap for high-frequency bypassing. This method is shown in Fig. 8-6.

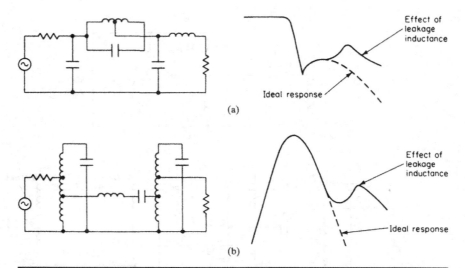

FIGURE 8-5 Spurious responses from leakage inductance: (a) in a low-pass filter; and (b) in a band-pass filter.

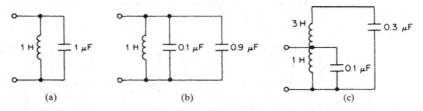

FIGURE 8-6 Preventing spurious response from leakage inductance: (a) initial circuit; (b) split capacity; and (c) transformed circuit.

8.3 Circuit Transformations

Circuit transformations fall into two categories: equivalent circuits or narrowband approximations. The impedance of a circuit branch can be expressed as a ratio of two polynomials in s, similar to a transfer function. If two branches are equivalent, their impedance expressions are identical. A narrowband approximation to a particular filter branch is valid only over a small frequency range. Outside of this region, the impedances depart considerably. Thus, the filter response is affected. As a result, narrowband approximations are essentially limited to band-pass filters having a small percentage bandwidth (narrow band).

8.3.1 Norton's Capacitance Transformer

Let's consider the circuit of Fig. 8-7a, consisting of impedance Z interconnected between impedances Z_1 and Z_2. If we want to raise impedance Z_2 by a factor of N^2 without disturbing an overall transfer function (except for possibly a constant multiplier), a transformer can be introduced, as shown in Fig. 8-7b.

Determinant manipulation can provide us with an alternative approach. The nodal determinant of a two-port network is given by

$$\begin{vmatrix} Y_{11} & -Y_{12} \\ -Y_{21} & Y_{22} \end{vmatrix}$$

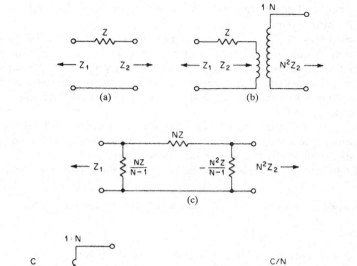

FIGURE 8-7 Norton's capacitance transformer: (*a*) a general two-port network; (*b*) transformer step-up of output impedance; (*c*) the Norton impedance transformation; and (*d*) the Norton capacitance transformation.

where Y_{11} and Y_{22} are the input and output nodal admittance, respectively, and Y_{12} and Y_{21} are the transfer admittances, which are normally equal to each other.

If we consider the two-port network of Fig. 8-7*a*, the nodal determinant becomes

$$\begin{vmatrix} \dfrac{1}{Z_1}+\dfrac{1}{Z} & -\dfrac{1}{Z} \\ -\dfrac{1}{Z} & \dfrac{1}{Z_2}+\dfrac{1}{Z} \end{vmatrix}$$

To raise the impedance of the output or Y_{22} node by N^2, the second row and second column are multiplied by $1/N$, resulting in

$$\begin{vmatrix} \dfrac{1}{Z_1}+\dfrac{1}{Z} & -\dfrac{1}{NZ} \\ -\dfrac{1}{NZ} & \dfrac{1}{N^2Z_2}+\dfrac{1}{N^2Z} \end{vmatrix}$$

This determinant corresponds to the circuit of Fig. 8-7*c*. The Y_{11} total nodal admittance is unchanged, and the Y_{22} total nodal admittance has been reduced by N^2, or the impedance has been increased by N^2. This result was originated by Norton and is called *Norton's transformation.*

If the element Z is a capacitor C, this transformation can be applied to obtain the equivalent circuit of Fig. 8-7d. This transformation is important since it can be used to modify the impedance on one side of a capacitor by a factor of N^2 without a transformer. However, the output shunt capacitor introduced is negative. A positive capacitor must then be present external to the network so that the negative capacitance can be absorbed.

If an N^2 of less than unity is used, the impedance at the output node will be reduced. The shunt capacitor at the input node will then become negative and must be absorbed by an external positive capacitor across the input.

The following example illustrates the use of the capacitance transformer.

Example 8-1   Using the Capacitance Transformation to Lower Inductor Values

Required:

Using the capacitance transformation, modify the band-pass filter circuit of Fig. 5-3c so that the 1.91-H inductor is reduced to 100 mH. The source and load impedances should remain 600 Ω.

Result:

(a) The circuit to be transformed is shown in Fig. 8-8a. To facilitate the capacitance transformation, the 0.01329 μF series capacitor is split into two equal capacitors of twice the value and redrawn in Fig. 8-8b.

To reduce the 1.91-H inductor to 100 mH, first lower the impedance of the network to the right of the dashed line in Fig. 8-8b by a factor of 100 mH/1.91 H, or 0.05236. Using the capacitance transformation of Fig. 8-7d, where $N^2 = 0.05236$, the circuit of Fig. 8-8c is obtained, where the input negative capacitor has been absorbed.

(b) To complete the transformation, the output node must be transformed back up in impedance to restore the 600-Ω termination. Again using the capacitance transformation with an N^2 of 600 Ω/ 31.42 Ω or 19.1, the final circuit of Fig. 8-8d is obtained. Because of the symmetrical nature of the circuit of Fig. 8-8b, both capacitor transformations are also symmetrical.

(c) Each parallel-resonant circuit is tuned by opening the inductors of the adjacent series-resonant circuits, and each series-resonant circuit is resonated by shorting the inductors of the adjacent parallel-tuned circuits, as shown in Fig. 8-8e.

8.3.2   Narrowband Approximations

A narrowband approximation to a circuit branch consists of an alternate network that is theoretically equivalent only at a single frequency. Nevertheless, good results can be obtained with band-pass filters that have small percentage bandwidths, typically of up to 20 percent.

The series and parallel RL and RC circuits of Table 8-1 are narrowband approximations that are equivalent at ω_0. This frequency is generally set equal to the band-pass center frequency in Eqs. (8-1) to (8-8). These equations were derived simply by determining the expressions for the network impedances and then equating the real parts and the imaginary parts to solve for the resistive and reactive components, respectively.

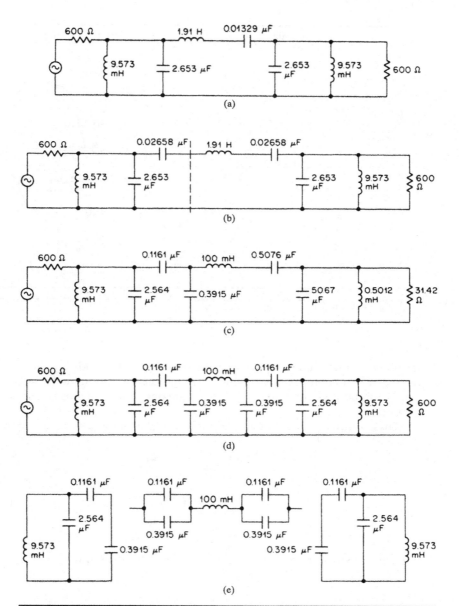

FIGURE 8-8   Capacitance transformation applied to the filter of Example 5-2: (a) the band-pass filter of Example 5-2; (b) split series capacitors; (c) the reduction of the 1.91-H inductor using capacitance transformation; (d) the restoration of the 600-Ω output impedance using capacitance transformation; and (e) equivalent circuits for tuning.

Narrowband approximations can be used to manipulate the source and load terminations of band-pass filters. If a parallel *RC* network is converted to a series *RC* circuit, it is apparent from Eq. (8-8) that the resistor value decreases. When we apply this approximation to a band-pass filter containing a parallel-resonant circuit as the

Circuit	Design Equations
	$L_1 = L_a + \dfrac{R_a^2}{\omega_0^2 L_a}$ (8-1)
	$R_1 = R_a + \dfrac{\omega_0^2 L_a^2}{R_a}$ (8-2)
	$L_a = \dfrac{L_1 R_1^2}{R_1^2 + \omega_0^2 L_1^2}$ (8-3)
	$R_a = \dfrac{\omega_0^2 L_1^2 R_1}{R_1^2 + \omega_0^2 L_1^2}$ (8-4)
	$C_2 = \dfrac{C_b}{1 + \omega_0^2 C_b^2 R_b^2}$ (8-5)
	$R_2 = R_b + \dfrac{1}{\omega_0^2 C_b^2 R_b}$ (8-6)
	$C_b = C_2 + \dfrac{1}{\omega_0^2 R_2^2 C_2}$ (8-7)
	$R_b = \dfrac{R_2}{1 + \omega_0^2 C_2^2 R_2^2}$ (8-8)

TABLE 8-1 Narrowband Approximations

terminating branch, the source or load resistor can be made smaller. To control the degree of reduction so that a desired termination can be obtained, the shunt capacitor is first subdivided into two capacitors where only one capacitor is associated with the termination.

These results are illustrated in Fig. 8-9. The element values are given by

$$C_2 = \frac{1}{\omega_0 \sqrt{R_1 R_2 - R_2^2}} \tag{8-9}$$

and

$$C_1 = C_T - \frac{1}{\omega_0} \sqrt{\frac{R_1 - R_2}{R_1^2 R_2}} \tag{8-10}$$

where the restrictions $R_2 < R_1$ and $(R_1 - R_2)/(R_1^2 R_2) < \omega_0^2 C_T^2$ apply.

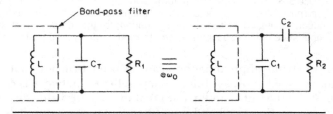

FIGURE 8-9 A narrowband transformation of terminations.

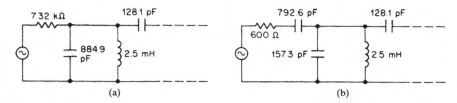

FIGURE 8-10   The narrowband source transformation of Example 8-2: (a) source input to the filter of Example 5-7; and (b) transformed source.

Example 8-2   Using a Narrowband Transformation to Lower Source Impedance

Required:

Modify the 100-kHz band-pass filter of Example 5.7 for a source impedance of 600 Ω.

Result:

The filter is shown in Fig. 8-10a. If we use the narrowband source transformation of Fig. 8-9, the values are given by Eqs. (8-9) and (8-10) as shown:

$$C_2 = \frac{1}{\omega_0\sqrt{R_1R_2 - R_2^2}} = \frac{1}{2\pi \times 10^5\sqrt{7.32 \times 6 \times 10^5 - 600^2}}$$

$$= 792.6 \text{ pF}$$

$$C_1 = C_T - \frac{1}{\omega_0}\sqrt{\frac{R_1 - R_2}{R_1^2 R_2}} = 884.9 \times 10^{-12}$$

$$- \frac{1}{2\pi \times 10^5}\sqrt{\frac{7.32 \times 10^3 - 600}{7,320^2 \times 600}} = 157.3 \text{ pF}$$

The resulting filter is illustrated in Fig. 8-10b.

8.4   Designing with Parasitic Capacitance

As a first approximation, inductors and capacitors are considered pure lumped reactive elements. Most physical capacitors are nearly perfect reactances. Inductors, on the other hand, have impurities that can be detrimental in many cases. In addition to the highly critical resistive losses, distributed capacity across the coil will occur because of inter-turn capacitance of the coil winding and other stray capacities involving the core. The equivalent circuit of an inductor is shown in Fig. 8-11.

The result of this distributed capacitance is to create the effect of a parallel-resonant circuit instead of an inductor. If the coil is to be located in shunt with an external capacitance, the external capacitor value can be decreased accordingly, thus absorbing the distributed capacitance.

FIGURE 8-11   Equivalent circuit of an inductor.

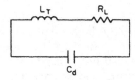

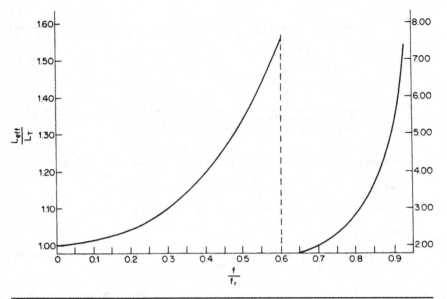

FIGURE 8-12   The effective inductance with frequency.

The distributed capacity across the inductor in a series-resonant circuit causes parallel resonances resulting in nulls in the frequency response. If the self-resonant frequency is too low, the null may even occur in the passband, thus severely distorting the expected response.

To determine the effective inductance of a practical inductor, the coil is resonated to the frequency of interest with an external capacitor, and the effective inductance is calculated using the standard formula for resonance. The effective inductance can also be found from

$$L_{\text{eff}} = \frac{L_T}{1 - \left(\dfrac{f}{f_r}\right)^2} \tag{8-11}$$

where L_T is the true (low-frequency) inductance, f is the frequency of interest, and f_r is the inductor's self-resonant frequency. As f approaches f_r, the value of L_{eff} will increase quite dramatically and will become infinite at self-resonance. Equation (8-11) is plotted in Fig. 8-12.

To compensate for the effect of distributed capacity in a series-resonant circuit, the true inductance L_T can be appropriately decreased so that the effective inductance given by Eq. (8-11) is the required value. However, the Q of a practical series-resonant circuit is given by

$$Q_{\text{eff}} = Q_L \left| 1 - \left(\frac{f}{f_r}\right)^2 \right| \tag{8-12}$$

where Q_L is the Q of the inductor as determined by the series losses (that is, $Q_L = \Omega L_T / R_L$). The effective Q is therefore reduced by the distributed capacity.

Distributed capacity is determined by the mechanical parameters of the core and winding, and as a result is subject to change due to mechanical stresses and so on. Therefore, for maximum stability, the distributed capacity should be kept as small as possible.

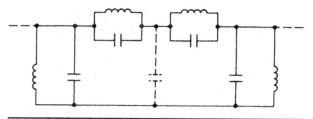

FIGURE 8-13   An elliptic-function band-pass filter.

Another form of parasitic capacity is stray capacitance between the circuit nodes and ground. These strays may be especially harmful at high frequencies and with high-impedance nodes. In the case of low-pass filters where the circuit nodes already have shunt capacitors to ground, these strays can usually be neglected, especially when the impedance levels are low.

A portion of an elliptic-function band-pass filter is shown in Fig. 8-13. The stray capacity at nodes not connected to ground by a design capacitor may cause problems since these nodes have high impedances.

Geffe (see References) has derived a transformation to introduce a design capacitor from the junction of the parallel-tuned circuits to the ground. The stray capacity can then be absorbed. The design of elliptic-function band-pass filters was discussed in Sec. 5.1. This transformation is performed upon the filter while it is normalized to a 1-rad/s center frequency and a 1-Ω impedance level. A section of the normalized network is shown in Fig. 8-14a. The transformation proceeds as follows:

Choose an arbitrary value of $m < 1$, then

$$n = 1 - \frac{L_b}{L_c}\frac{1-m}{m^2} \tag{8-13}$$

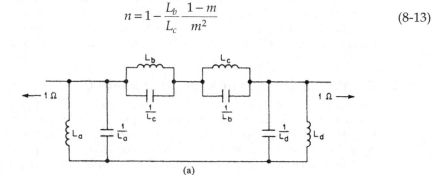

(a)

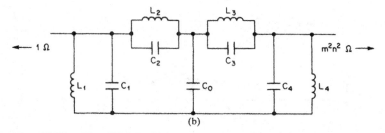

(b)

FIGURE 8-14   A transformation to absorb stray capacitance:
(a) normalized filter section; and (b) transformed circuit.

$$C_0 = \frac{1-n}{n^2 L_c} - \frac{1-m}{mn^2 L_b} \tag{8-14}$$

$$C_1 = \frac{1}{L_a} - \frac{1-n}{nL_c} \tag{8-15}$$

$$C_2 = \frac{1}{nL_C} \tag{8-16}$$

$$C_3 = \frac{1}{mn^2 L_b} \tag{8-17}$$

$$C_4 = \frac{1-m}{m^2 n^2 L_b} + \frac{1}{n^2 L_d} \tag{8-18}$$

$$L_1 = \frac{L_a}{1 - \dfrac{L_a}{L_b}\dfrac{1-n}{n}} \tag{8-19}$$

$$L_2 = nL_b \tag{8-20}$$

$$L_3 = mn^2 L_c \tag{8-21}$$

$$L_4 = \frac{m^2 n^2 L_c}{1 + \dfrac{L_c}{L_d}\dfrac{m^2}{1-m}} \tag{8-22}$$

The resulting network is given in Fig. 8-14*b*. The output node has been transformed to an impedance level of $m^2 n^2\ \Omega$. Therefore, all the circuitry to the right of this node, up to and including the termination, must be impedance-scaled by this same factor. The filter is subsequently denormalized by scaling to the desired center frequency and impedance level.

8.5   Amplitude Equalization for Inadequate *Q*

Insufficient element *Q* will cause a sagging or rounding of the frequency response in the region of cutoff. Some typical cases are shown in Fig. 8-15, where the solid curve represents the theoretical response. Finite *Q* will also result in less rejection in the vicinity of any stopband zeros and increased filter insertion loss.

Amplitude-equalization techniques can be applied to compensate for the sagging response near cutoff. A passive amplitude equalizer will not actually "boost" the corner

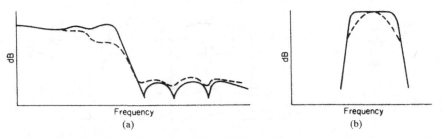

FIGURE 8-15   The effects of insufficient *Q*: (*a*) low-pass response; and (*b*) band-pass response.

response, since a gain cannot be achieved as with active equalizer circuits. However, the equalizer will introduce attenuation, except in the region of interest, therefore resulting in a boost in terms of the relative response.

Amplitude equalizers used for low Q compensation are of the band-pass type. They have either constant-impedance or nonconstant-impedance characteristics. The constant-impedance types can be cascaded with each other and with the filter with no interaction. The nonconstant-impedance equalizer sections are less complex but will result in some interaction when cascaded with other networks. However, for a boost of 1 or 2 dB, these effects are usually minimal and can be neglected.

Both types of equalizers are shown in Fig. 8-16. The nonconstant-impedance type can be used in either the series or shunt form. In general, the shunt form is preferred since the resonating capacitor may be reduced by tapping the inductor.

To design a band-pass equalizer, the following characteristics must be determined from the curve to be equalized:

A_{dB} = total amount of equalization required in decibels
f_r = frequency corresponding to A_{dB}
f_b = frequency corresponding to $A_{dB}/2$

These parameters are illustrated in Fig. 8-17, where the corner response and corresponding equalizer are shown for both upper and lower cutoff frequencies.

To design the equalizer, first compute K from

$$A_{dB} = 20 \log K \tag{8-23}$$

Then calculate b where

$$b = \frac{f_r}{f_b} \tag{8-24}$$

or

$$b = \frac{f_b}{f_r} \tag{8-25}$$

selecting whichever b is greater than unity.

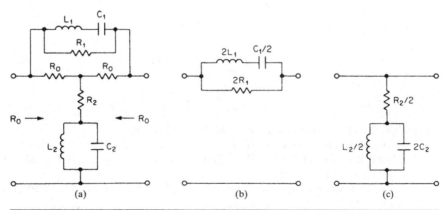

FIGURE 8-16 Band-pass-type amplitude equalizers: (*a*) constant-impedance type; (*b*) series nonconstant-impedance type; and (*c*) shunt nonconstant-impedance type.

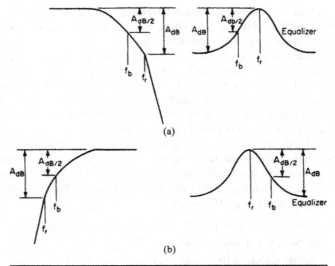

FIGURE 8-17 Band-pass equalization of the corner response:
(a) equalization of the upper cutoff; and (b) equalization of the
lower cutoff.

The element values corresponding to the sections of Fig. 8-16 are found as
follows:

$$L_1 = \frac{R_0(K-1)}{2\pi f_b(b^2-1)\sqrt{K}} \tag{8-26}$$

$$C_1 = \frac{1}{(2\pi f_r)^2 L_1} \tag{8-27}$$

$$L_2 = \frac{R_0(b^2-1)\sqrt{K}}{2\pi f_b b^2(K-1)} \tag{8-28}$$

$$C_2 = \frac{1}{(2\pi f_r)^2 L_2} \tag{8-29}$$

$$R_1 = R_0(K-1) \tag{8-30}$$

$$R_2 = \frac{R_0}{K-1} \tag{8-31}$$

where R_0 is the terminating impedance of the filter.

To equalize a low-pass or high-pass filter, a single equalizer is required at the cutoff.
For band-pass or band-reject filters, a pair of equalizer sections is needed for the upper
and lower cutoff frequencies.

The following example illustrates the design of an equalizer to compensate for
low Q.

Example 8-3   Using an Amplitude Equalizer to Compensate for Low Q

Required:

A low-pass filter should have a theoretical roll-off of 0.1 dB at 2,975 Hz, but has instead the following response in the vicinity of cutoff due to insufficient Q:

2,850 Hz   −0.5 dB
2,975 Hz   −1.0 dB

Design a shunt nonconstant-impedance equalizer to restore the sagging response. The filter impedance level is 1,000 Ω.

Result:

First, make the following preliminary computations using Eqs. (8-23), (8-24), (8-28), (8-29), and (8-31):

$$K = 10^{A_{dB}/20} = 10^{1/20} = 1.122$$

$$b = \frac{f_r}{f_b} = \frac{2,975 \text{ Hz}}{2,850 \text{ Hz}} = 1.0439$$

then

$$L_2 = \frac{R_0(b^2 - 1)\sqrt{K}}{2\pi f_b b^2(K - 1)} = \frac{10^3(1.0439^2 - 1)\sqrt{1.122}}{2\pi 2,850 \times 1.0439^2(1.122 - 1)} = 40.0 \text{ mH}$$

$$C_2 = \frac{1}{(2\pi f_r)^2 L_2} = \frac{1}{(2\pi 2,975)^2 \times 0.04} = 0.0715 \ \mu F$$

$$R_2 = \frac{R_0}{K - 1} = \frac{1,000}{1.122 - 1} = 8,197 \ \Omega$$

The resulting equalizer is shown in Fig. 8-18 using the circuit of Fig. 8-16c.

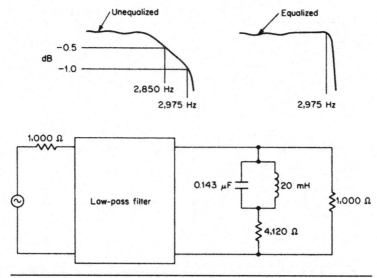

FIGURE **8-18**   The equalizer of Example 8-3.

8.6   Coil-Saving Elliptic-Function Band-Pass Filters

If an even-order elliptic-function low-pass filter (as shown in Fig. 8-19a) is transformed into a band-pass filter using the methods of Sec. 5.1, the band-pass circuit of Fig. 8-19b is obtained.

A method has been developed to transform the low-pass filter into the configuration of Fig. 8-19c. The transfer function is unchanged, except for a constant multiplier, and $1/2\ (n-2)$ coils are saved in comparison with the conventional transformation. These structures are called minimum-inductance or zigzag band-pass filters. However, this transformation requires a very large number of calculations (see Saal and Ulbrich in References) and is therefore considered impractical without a computer.

Geffe (see References) has presented a series of formulas so that this transformation can be performed on an $n = 4$ low-pass network. The normalized low-pass element values for $n = 4$ can be determined with the *Filter Solutions* program or found in either Zverev's *Handbook of Filter Synthesis* or Saal's "Der Entwurf von Filtern mit Hilfe des Kataloges Normierter Tiefpasse" (see References). Although these calculations are laborious, the downloadable Excel spreadsheet of formulas from the companion website (see App. A) performs these computations.

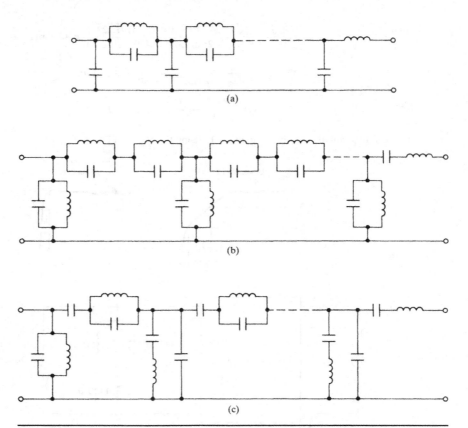

Figure 8-19 Coil-saving band-pass transformation: (*a*) an even-order elliptic-function low-pass filter; (*b*) conventional band-pass transformation; and (*c*) minimum-inductance band-pass transformation.

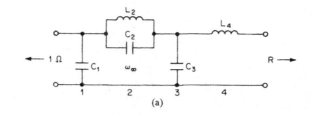

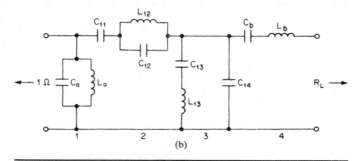

Figure 8-20 The minimum-inductance transformation for $n = 4$: (a) an $n = 4$ low-pass filter; and (b) the transformed band-pass filter.

The low-pass filter and corresponding band-pass network are shown in Fig. 8-20. The following preliminary computations are required:

$$Q_{bp} = \frac{f_0}{BW} \tag{8-32}$$

$$a = \frac{\omega_\infty}{2Q_{bp}} \tag{8-33}$$

$$x = a + \sqrt{a^2 + 1} \tag{8-34}$$

$$t_1 = 1 + \frac{c_3}{c_2} \tag{8-35}$$

$$T = \frac{1 + t_1 x^2}{t_1 + x^2} \tag{8-36}$$

$$k = \frac{Q_{bp} T}{t_1} \tag{8-37}$$

$$t_2 = \frac{x^2}{x^2 + t_1} \tag{8-38}$$

$$t_3 = \frac{t_1 t_2}{T} \tag{8-39}$$

$$\alpha = 1 - \frac{1}{x^2} \tag{8-40}$$

$$\beta = x^2 - 1 \tag{8-41}$$

$$A = \frac{C_3 k \alpha}{T} \tag{8-42}$$

$$B = \frac{t_2 t_3}{C_3 k \beta} \tag{8-43}$$

The band-pass element values can now be computed as follows:

$$R_L = t_3^2 R \tag{8-44}$$

$$C_{11} = \frac{C_3 k \beta}{t_1 t_2} \tag{8-45}$$

$$C_{12} = \frac{C_3 k \alpha}{T - 1} \tag{8-46}$$

$$L_{12} = \frac{1}{x^2 C_{12}} \tag{8-47}$$

$$C_{13} = \frac{C_{11}(T - 1)}{t_2} \tag{8-48}$$

$$L_{13} = \frac{x^2}{C_{13}} \tag{8-49}$$

$$C_{14} = \frac{C_3 k \alpha}{t_2} \tag{8-50}$$

$$L_a = \frac{1}{Q_{bp}\left(C_1 + \dfrac{C_3}{t_1}\right)} \tag{8-51}$$

$$C_a = \frac{1}{L_a} - A \tag{8-52}$$

$$L_b = t_3^2 Q_{bp} L_4 \tag{8-53}$$

$$C_b = \frac{1}{L_b - B} \tag{8-54}$$

The band-pass filter of Fig. 8-20b must be denormalized to the required impedance level and center frequency f_0. Since the source and load impedance levels are unequal, either the tapped inductor or the capacitance transformation can be used to obtain equal terminations if required.

The transmission zero above the passband is provided by the parallel resonance of $L_{12}C_{12}$ in branch 2, and the lower zero corresponds to the series resonance of $L_{13}C_{13}$ in branch 3. The circuits of branches 2 and 3 each have conditions of both series and

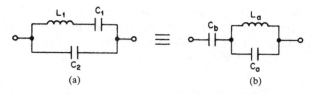

Figure 8-21 Equivalent branches: (a) type 1 network; and (b) type 2 network.

parallel resonance and can be transformed from one form to the other. The following equations relate the type 1 and 2 networks shown in Fig. 8-21.

For a type 1 network:

$$L_1 = L_a \left(1 + \frac{C_a}{C_b} \right)^2 \tag{8-55}$$

$$C_1 = C_b \, \frac{1}{1 + \dfrac{C_a}{C_b}} \tag{8-56}$$

$$C_2 = C_a \, \frac{1}{1 + \dfrac{C_a}{C_b}} \tag{8-57}$$

$$f_{\text{series}} = \frac{1}{2\pi\sqrt{L_1 C_1}} \tag{8-58}$$

$$f_{\text{par}} = \frac{1}{2\pi\sqrt{\dfrac{L_1 C_1 C_2}{C_1 + C_2}}} \tag{8-59}$$

For a type 2 network:

$$L_a = L_1 \, \frac{1}{\left(1 + \dfrac{C_2}{C_1} \right)^2} \tag{8-60}$$

$$C_a = C_2 \left(1 + \frac{C_2}{C_1} \right) \tag{8-61}$$

$$C_b = C_1 + C_2 \tag{8-62}$$

$$f_{\text{series}} = \frac{1}{2\pi\sqrt{L_a(C_a + C_b)}} \tag{8-63}$$

$$f_{\text{par}} = \frac{1}{2\pi\sqrt{L_a C_a}} \tag{8-64}$$

In general, the band-pass series arms are of the type 2 form and the shunt branches are of the type 1 form, as in Fig. 8-19c. The tuning usually consists of adjusting the parallel resonances of the series branches and the series resonances of the shunt branches—in other words, the transmission zeros.

8.7   Filter Tuning Methods

LC filters are typically assembled using elements with 1 or 2 percent tolerances. For many applications, the deviation in the desired response caused by component variations may be unacceptable, so the adjustment of elements will be required. It has been found that wherever resonances occur, the LC product of the resonant circuit is significantly more critical than the L/C ratio. As a result, filter adjustment normally involves adjusting each tuned circuit for resonance at the specified frequency.

Adjustment techniques are based on the impedance extremes that occur at resonance. In the circuit of Fig. 8-22a, an output null will occur at parallel resonance because of voltage-divider action. Series LC circuits are tuned using the circuit of Fig. 8-22b, where an output null will also occur at resonance.

The adjustment method in both cases involves setting the oscillator for the required frequency and adjusting the variable element, usually the inductor, for an output null. Resistors R_L and R_s are chosen so that an approximately 20- to 30-dB drop occurs between the oscillator and the output at resonance. These values can be estimated from

$$R_L \approx \frac{2\pi f_r L_p Q_L}{20} \qquad (8\text{-}65)$$

and

$$R_s \approx \frac{40\pi f_r L_s}{Q_L} \qquad (8\text{-}66)$$

where Q_L is the inductor Q. A feature of this technique is that no tuning errors result from stray capacity across the VTVM. (Note that the term "VTVM" is used to represent a high-input impedance voltage (dB) measuring meter and not necessarily a *vacuum tube volt meter*.) Care should be taken that the oscillator does not have excessive distortion, since a sharp null may then be difficult to obtain. Also, excessive levels should be avoided, as detuning can occur from inductor saturation effects. The oscillator and VTVM can be replaced by the tracking generator output and the signal input of a network analyzer. A swept measurement can be made to determine resonant frequency.

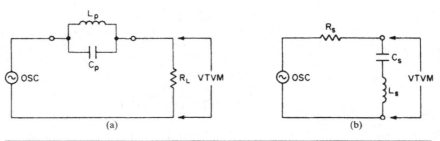

Figure 8-22   Test circuits for adjusting resonant frequencies: (a) adjusting parallel resonance; and (b) adjusting series resonance.

When inductor Qs are below 10, sharp nulls cannot be obtained. A more desirable tuning method is to adjust for the condition of zero phase shift at resonance, which will be more distinct than the null. The circuits of Fig. 8-22 can still be used in conjunction with an oscilloscope that has both vertical and horizontal inputs. One channel monitors the oscillator and the other channel is connected to the output instead of using the VTVM. A Lissajous pattern is obtained, and the tuned circuit is then adjusted for a closed ellipse. Some network analyzers can display phase shift and can be used as well.

Certain construction practices must be used so that the assembled filter can be tuned. There must be provision for access to each tuned circuit on an individual basis. This is usually accomplished by leaving all the grounds disconnected until after tuning so that each branch can be individually inserted into the tuning configurations of Fig. 8-22 with all the other branches present.

8.8   Measurement Methods

This section discusses some major filter parameters and describes techniques for their measurement. Also, some misconceptions associated with these characteristics are clarified. All measurements should be made using rated operating levels so that the results are meaningful. After fabrication, filters should be subjected to insertion-loss and frequency-response measurements as a minimum production test.

8.8.1   Insertion Loss and Frequency Response

The frequency response of filters is always considered relative to the attenuation occurring at a particular reference frequency. The actual attenuation at this reference is called *insertion loss.*

The classical definition of insertion loss is the decrease in power delivered to the load when a filter is inserted between the source and the load. Using Fig. 8-23, the insertion loss is given by

$$\text{IL}_{\text{dB}} = 10 \ \log \frac{P_{L1}}{P_{L2}} \tag{8-67}$$

where P_{L1} is the power delivered to the load with both switches in position 1 (filter bypassed) and P_{L2} is the output power with both switches in position 2. Equation (8-67) can also be expressed in terms of a voltage ratio as

$$\text{IL}_{\text{dB}} = 10 \ \log \frac{E_{L1}^2/R_L}{E_{L2}^2/R_L} = 20 \ \log \frac{E_{L1}}{E_{L2}} \tag{8-68}$$

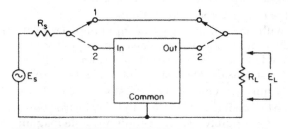

FIGURE 8-23   The test circuit for insertion loss.

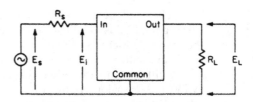

Figure 8-24 The test circuit for frequency response.

so a decibel meter can be used at the output to measure insertion loss directly in terms of output voltage

The classical definition of insertion loss may be somewhat inapplicable when the source and load terminations are unequal. In reality, if the filter were not used, the source and load would probably be connected through an impedance-matching transformer instead of a direct connection. Therefore, the comparison to Fig. 8-23 would be invalid.

An alternate definition is *transducer loss*, which is defined as the decrease in power delivered to the load when an ideal impedance-matching transformer is replaced by the filter. The test circuit of Fig. 8-23 can still be used if a correction factor is added to Eq. (8-68). The resulting expression then becomes

$$\text{IL}_{\text{dB}} = 20 \ \log \frac{E_{L1}}{E_{L2}} + 20 \ \log \frac{R_s + R_L}{2\sqrt{R_s R_L}} \tag{8-69}$$

Frequency response, or relative attenuation, is measured using the test circuit of Fig. 8-24. The input source E_s is set to the reference frequency, and the level is arbitrarily set for a 0-dB reference at the output. As the input frequency is changed, the variation in output level is the relative attenuation.

It must be understood that the variation of the ratio E_L/E_s is the frequency response. The ratio E_L/E_i is of no significance since it reflects the frequency response of the filter when driven by a voltage source. As a source frequency is varied, the voltage E_s must be kept constant. However, any attempt to keep E_i constant will distort the response shape since the voltage-divider action between R_s and the filter input impedance must normally occur to satisfy the transfer function.

The oscillator source itself, E_s, may contain some internal impedance. Nevertheless, the value of R_s should correspond to the design source impedance since the internal impedance of E_s is allowed for by maintaining the terminal voltage of E_s at a constant value.

A network analyzer is commonly used in the industry to measure frequency response and can generate test results in various electronic forms, which can then be saved and manipulated.

8.8.2 Input Impedance of Filter Networks

The input or output impedance of filters must frequently be determined to ensure compatibility with external circuitry. The input impedance of a filter is the impedance measured at the input terminals with the output appropriately terminated. Conversely, the output impedance can be measured by terminating the input.

Let us first consider the test circuit of Fig. 8-25*a*. A common fallacy is to adjust the value of R until $|E_2|$ is equal to $1/2 |E_1|$—that is, a 6-dB drop. The input impedance Z_{11} is then said to be equal to R. However, this will be true only if Z_{11} is purely resistive. As an example, if Z_{11} is purely reactive and its absolute magnitude is equal to R, the value of E_2 will be 0.707 dB or 3 dB below E_1 and not 6 dB.

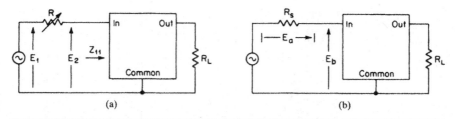

FIGURE 8-25   The measurement of input impedance: (a) an indirect method; and (b) a direct method.

Using the circuit of Fig. 8-25a, an alternative approach will result in greater accuracy. If R is adjusted until a 20-dB drop occurs between $|E_2|$ and $|E_1|$, then $|E_1|$ is determined by $R/10$. The accuracy will be within 10 percent. For even more accurate results, the 40-dB method can be used where R is adjusted for a 40-dB drop. The magnitude of Z_{11} is then given by $R/100$.

If a more precise measurement is required, a floating meter can be used in the configuration of Fig. 8-25b. The input impedance is then directly given by

$$|Z_{11}| = \frac{|E_b|}{|E_a|} R_s \tag{8-70}$$

Most network analyzers can also measure impedance.

Return Loss

Return loss is a figure of merit that indicates how closely a measured impedance matches a standard impedance, both in magnitude and in phase angle. Return loss is expressed as

$$A_\rho = 20 \log\left|\frac{Z_s + Z_x}{Z_s - Z_x}\right| \tag{8-71}$$

where Z_s is the standard impedance and Z_x is the measured impedance. For a perfect match, the return loss would be infinite.

Return loss can be directly measured using the bridge arrangement of Fig. 8-26. The return loss is given by

$$A_\rho = 20 \log\left|\frac{V_{01}}{V_{02}}\right| \tag{8-72}$$

FIGURE 8-26   The measurement of return loss.

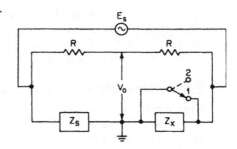

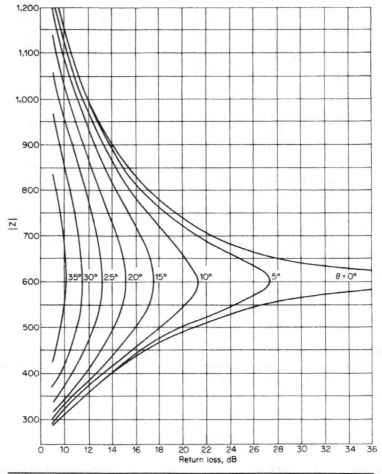

Figure 8-27 Return loss versus $|Z|$.

where V_{01} is the output voltage with the switch closed and V_{02} is the output voltage with the switch open. The return loss can then be read directly using a decibel meter or by using a network analyzer. The value of R is arbitrary, but both resistors must be closely matched to each other.

The family of curves in Fig. 8-27 represents the return loss using a standard impedance of 600 Ω with the phase angle of impedance as a parameter. Clearly, the return loss is very sensitive to the phase angle. If the impedance were exactly 600 Ω at an angle of only 10°, the return loss would be 21 dB. If there was no phase shift, an impedance error of as much as 100 Ω would correspond to 21 dB of return loss.

8.8.3 Time-Domain Characteristics

Step Response

The step response of a filter network is a useful criterion since low transient distortion is a necessary requirement for good transmission of modulated signals. To determine

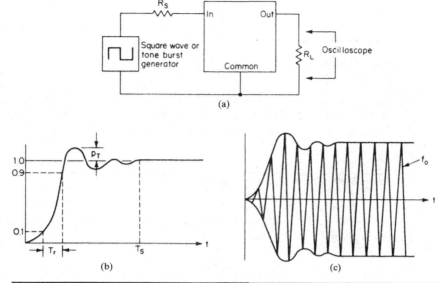

FIGURE 8-28 Step response of networks: (*a*) the test circuit; (*b*) the step response to a DC step; and (*c*) the step response to a tone burst.

the step response of a low-pass filter, an input DC step is applied. For band-pass filters, a carrier step is used where the carrier frequency is equal to the filter center frequency f_0. Since it is difficult to view a single transient on an oscilloscope unless it's of the storage type, a square-wave generator is used instead of the DC step, and a tone-burst generator is substituted for the carrier step. However, a repetition rate must be chosen that is slow enough so that the transient behavior has stabilized prior to the next pulse in order to obtain meaningful results.

The test configuration is shown in Fig. 8-28*a*. The output waveforms are depicted in Figs. 8-28*b* and *c* for a DC step and tone burst, respectively. The following definitions are applicable:

Percent Overshoot P$_T$
The difference between the peak response and the final steady-state value expressed as a percentage.

Rise Time T$_r$
The interval between 10 and 90 percent of the final value.

Settling Time T$_s$
The time required for the response to settle within a specified percent of its final value.

The waveform definitions shown in Fig. 8-28*b* also apply to Fig. 8-28*c* if we consider the envelope of the carrier waveform instead of the instantaneous values.

8.8.4 Group Delay

The phase shift of a filter can be measured by using the Lissajous pattern method. By connecting the vertical channel of an oscilloscope to the input source and the horizontal

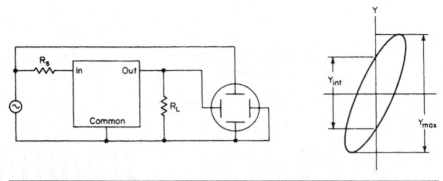

FIGURE 8-29   The measurement of phase shift.

channel to the load, an ellipse is obtained, as shown in Fig. 8-29. The phase angle in degrees is given by

$$\phi = \sin^{-1} \frac{Y_{\text{int}}}{Y_{\text{max}}} \tag{8-73}$$

Since group delay is the derivative with respect to frequency of the phase shift, we can measure the phase shift at two closely spaced frequencies and approximate the delay as follows:

$$T_{\text{gd}} \approx \frac{\Delta\phi}{360\Delta f} \tag{8-74}$$

where $\Delta\phi$ is $\phi_2 - \phi_1$ in degrees, Δf is $f_2 - f_1$ in hertz, and T_{gd} is the group delay at the midfrequency—in other words, $(f_1 + f_2)/2$.

A less accurate method involves determining the 180° phase-shift points. As the frequency is varied throughout the passband of a high-order filter, the phase shift will go through many integer multiples of 180° where the Lissajous pattern adopts a straight line at either 45° or 225°. If we record the separation between adjacent 180° points, the nominal group delay at the midfrequency can be approximated by

$$T_{\text{gd}} \approx \frac{1}{2\Delta f} \tag{8-75}$$

The classical approach to measuring group delay is shown in Fig. 8-30. A sine-wave source, typically 25 Hz, is applied to an amplitude modulator along with a carrier signal. The output consists of an amplitude-modulated signal comprising the carrier and two sidebands at ±25 Hz on either side of the carrier. The signal is then applied to the network under test.

The output signal from the network is of the same form as the input, but the 25-Hz envelope has been shifted in time by an amount equal to the group delay at the carrier frequency. The output envelope is recovered by an AM detector and applied to a phase detector along with a reference 25-Hz signal.

The phase detector output is a DC signal proportional to the phase shift between the 25-Hz reference and the demodulated 25-Hz carrier envelope. As the carrier is varied in frequency, the DC signal will vary in accordance with the change in group delay

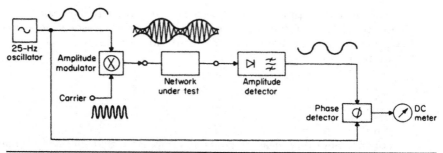

FIGURE 8-30   Direct measurement of group delay.

(differential delay distortion), and the delay can be displayed on a DC meter having the proper calibration. If an adjustable phase shift network is interposed between the AM detector and phase detector, the meter indication can be adjusted to establish a reference level at a desired reference frequency.

The theoretical justification for this scheme is based on the fact that the delay of the envelope is determined by the slope of a line segment interconnecting the phase shift of the two sidebands at ±25 Hz about the carrier—that is, $(\phi_2 - \phi_1)/(\omega_2 - \omega_1)$. This definition is sometimes called the *envelope delay*, for obvious reasons. As the separation between sidebands is decreased, the envelope approaches the theoretical group delay at the carrier frequency since group delay is defined as the derivative of the phase shift. For most measurements, a modulation rate of 25 Hz is adequate. Some network analyzers have a feature where group delay can be measured.

8.8.5   Measuring the *Q* of Inductors

A device called a Q *meter* is frequently used to measure the Q of inductors at a specified frequency. The principle of the Q meter is based on the fact that in a series-resonant circuit, the voltage across each reactive element is Q times the voltage applied to the resonant circuit. The Q can then be directly determined by the ratio of two voltages.

A test circuit is shown in Fig. 8-31*a*. Care should be taken that the applied voltage does not result in an excessive voltage developed across the inductor during the measurement. Also, the resonating capacitor should have a much higher Q than the inductor, and the meter used for the voltage measurements should be a high-impedance type to avoid loading errors.

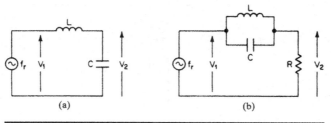

FIGURE 8-31   Measuring coil Q: (*a*) the Q meter method; and (*b*) the parallel-resonant circuit method.

An alternative approach involves measuring the impedance of a parallel-resonant circuit consisting of the inductor and the required resonating capacitor for the frequency of interest. Using the voltage divider of Fig. 8-31b, the Q at resonance is found from

$$Q = \frac{R}{2\pi f_r L}\left(\frac{V_1}{V_2} - 1\right) \qquad (8\text{-}76)$$

A null will occur in V_2 at resonance.

For meaningful Q measurements, it's important that the measurement frequency corresponds to the frequency of interest of the filter since coil Q can decrease quite dramatically outside a particular range. In low-pass and high-pass filters, the Q should be measured at the cutoff, and for band-pass and band-reject filters, the center frequency is the frequency of interest.

8.9   Designing For Unequal Impedances

8.9.1   Exponentially Tapered Impedance Scaling

Bartlett's bisection theorem, discussed in Sec. 3.1, allows us to modify a design for unequal impedances if the initial schematic (including values) is totally symmetrical around a center line (mirror image). However, that is not always the case. In addition, the tables in Chap. 10 for all-pole filters with unequal impedances are limited to finite ratios of termination impedances.

The method shown in Fig. 8-32, although not theoretically precise, allows us to gradually taper the impedance of each element of the ladder from source to load. The more branches of the ladder there are, the more gradual becomes the tapering, and the less the change of the original transfer function. Experience has indicated that this works quite well with at least six branches of the ladder (excluding source and load resistors). Note that the load can be higher or lower than the source.

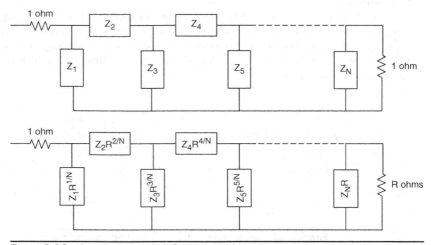

Figure 8-32   A tapered network for unequal impedances.

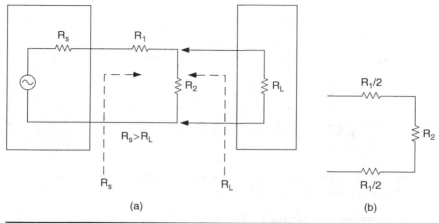

FIGURE 8-33   Minimum loss pads: (*a*) single ended; and (*b*) the balanced version.

8.9.2   Minimum-Loss Resistive Pad for Impedance Matching

The circuit of Fig. 8-33 matches resistive source impedance R_s to resistive load impedance R_L with minimum loss. It can be useful when two different filters designed for different impedances are cascaded, or to match a filter to a termination impedance it is not designed for. The design equations are

$$R_1 = R_s \sqrt{1 - \frac{R_L}{R_s}} \qquad (8\text{-}77)$$

$$R_2 = \frac{R_L}{\sqrt{1 - \dfrac{R_L}{R_s}}} \qquad (8\text{-}78)$$

Note that R_s must be higher than R_L. To form a balanced circuit, split R_1 into two $R_1/2$ resistors as shown. The voltage loss in dB is given by

$$\text{Voltage Loss dB} = 20 \, Log_{10}\left(\frac{R_1\left(R_2 + R_L\right)}{R_2 R_L} + 1\right) \qquad (8\text{-}79)$$

In terms of power, the loss in dB is

$$\text{Power Loss dB} = \text{Voltage Loss dB} - 10 \, Log_{10}\frac{R_s}{R_L} \qquad (8\text{-}80)$$

8.9.3   Design of Unsymmetrical Resistive *T* and π Attenuators for Impedance Matching

The circuit of Fig. 8-33 matches two unequal impedances with minimum loss. To obtain a fixed amount of attenuation between two *unequal* resistive impedances, a *T* or π

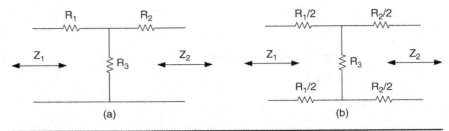

FIGURE 8-34 Unsymmetrical *T* attenuators $Z_2 > Z_1$: (*a*) single ended; and (*b*) the balanced version.

attenuator can be used. The values are computed as follows for the unbalanced and balanced circuits of Figs. 8-34 and 8-35. *Note that Z_1 must be greater than Z_2.*

For an unsymmetrical T attenuator:
First compute

$$K_{min} = \frac{2Z_1}{Z_2} - 1 + 2\sqrt{\frac{Z_1}{Z_2}\left(\frac{Z_1}{Z_2} - 1\right)} \qquad (8\text{-}81)$$

For a required voltage loss in dB:

$$dB_{min\text{-}voltage\ loss} = 10\ Log_{10} K_{min} + 10\ Log_{10}\frac{Z_1}{Z_2} \qquad (8\text{-}82)$$

This is the minimum voltage loss in dB that results in positive resistor values. Select a desired voltage loss ($dB_{voltage\ loss}$) > Minimum Voltage Loss ($dB_{min\text{-}voltage\ loss}$)

$$K = 10^{dB\ voltage\ loss/10 + Log(Z_2/Z_1)} \qquad (8\text{-}83)$$

The resistor values are obtained from

$$R_1 = \frac{Z_1(K+1) - 2\sqrt{KZ_1Z_2}}{K-1} \qquad (8\text{-}84)$$

$$R_2 = \frac{Z_2(K+1) - 2\sqrt{KZ_1Z_2}}{K-1} \qquad (8\text{-}85)$$

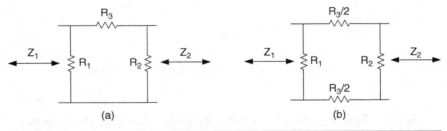

FIGURE 8-35 Unsymmetrical *π* attenuators $Z_2 > Z_1$: (*a*) single ended; and (*b*) the balanced version.

$$R_3 = \frac{2\sqrt{KZ_1Z_2}}{K-1} \tag{8-86}$$

For a required *power loss in dB,* first compute:

$$dB_{min-voltage\ loss} = 10\ Log_{10}K_{min} \tag{8-87}$$

Select a desired power loss ($dB_{power\ loss}$) > Minimum Power Loss ($dB_{min-power\ loss}$)

$$K = 10^{dB/10} \tag{8-88}$$

Resistors R_1, R_2, and R_3 are computed using Eqs. (8-84), (8-85), and (8-86), respectively. The resulting voltage loss is

$$dB_{voltage\ loss} = dB_{power\ loss} + 10\ Log_{10}\frac{Z_1}{Z_2} \tag{8-89}$$

For an unsymmetrical π attenuator:
First, compute using Eq. (8-81):

$$K_{min} = \frac{2Z_1}{Z_2} - 1 + 2\sqrt{\frac{Z_1}{Z_2}\left(\frac{Z_1}{Z_2} - 1\right)}$$

For a required *voltage loss in dB,* use Eq. (8-82):

$$dB_{min-voltage\ loss} = 10\ Log_{10}K_{min} + 10\ Log_{10}\frac{Z_1}{Z_2}$$

This is the minimum voltage loss in dB that results in positive resistor values.
Select a desired voltage loss ($dB_{voltage\ loss}$) > Minimum Voltage Loss ($dB_{min-voltage\ loss}$), as shown in Eq. (8-83):

$$K = 10^{dB\ voltage\ loss/10+Log(Z_2/Z_1)}$$

The resistor values are computed by

$$R_1 = \frac{Z_1(K-1)\sqrt{Z_2}}{(K+1)\sqrt{Z_2} - 2\sqrt{KZ_1}} \tag{8-90}$$

$$R_2 = \frac{Z_2(K-1)\sqrt{Z_1}}{(K+1)\sqrt{Z_1} - 2\sqrt{KZ_2}} \tag{8-91}$$

$$R_3 = \frac{K-1}{2}\sqrt{\frac{Z_1 Z_2}{K}} \tag{8-92}$$

For a required *power loss in dB,* first compute the following using Eq. (8-87):

$$dB_{min-power\ loss} = 10\ Log_{10}K_{min}$$

Select a desired power loss ($\mathrm{dB_{power\ loss}}$) > Minimum Power Loss ($\mathrm{dB_{min\text{-}power\ loss}}$) using Eq. (8-88):

$$K = 10^{\mathrm{dB}/10}$$

Resistors R_1, R_2, and R_3 are computed using Eqs. (8-90), (8-91), and (8-92), respectively. The resulting voltage loss is computed by Eq. (8-89) as follows.

$$\mathrm{dB_{voltage\ loss}} = \mathrm{dB_{power\ loss}} + 10\ \mathrm{Log}_{10}\ \frac{Z_1}{Z_2}$$

8.10   Symmetrical Attenuators

8.10.1   Symmetrical *T* and *π* Attenuators

When source- and load-resistive impedances are equal, a symmetrical *T* or *π* attenuator can be used to symmetrically (bidirectionally) introduce fixed loss where needed. Figures 8-36 and 8-37 illustrate symmetrical *T* and *π* attenuators, respectively.

Section 8.8 discusses return loss, which is a figure of merit that indicates how closely a measured impedance matches a standard impedance, both in magnitude and in phase angle. Return loss is expressed as shown in Eq. (8-71):

$$A_\rho = 20\ \log\left|\frac{Z_s + Z_x}{Z_s - Z_x}\right|$$

where Z_s is the standard impedance and Z_x is the measured impedance. For a perfect match, the return loss would be infinite. If the standard impedance is resistive Z_s and the network is preceded with a symmetrical attenuator of X dB at an impedance level of Z_s, *a minimum return loss of 2X dB is guaranteed* even if the network has impedance extremes of zero or infinity. The attenuator will smooth any impedance gyrations.

For a given loss in dB (the power loss in dB is equal to the voltage loss in dB since the impedances are equal on both sides), first compute

$$K = 10^{\mathrm{dB}/20} \tag{8-93}$$

For a symmetrical T attenuator:

$$R_1 = Z\ \frac{K-1}{K+1} \tag{8-94}$$

$$R_3 = \frac{2\ ZK}{K^2 - 1} \tag{8-95}$$

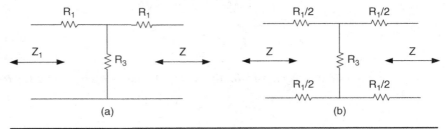

(a)   (b)

Figure 8-36   Symmetrical *T* attenuators: (*a*) single ended; and (*b*) the balanced version.

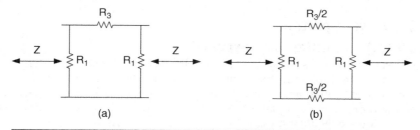

FIGURE 8-37   Symmetrical π attenuators: (*a*) single ended; and (*b*) the balanced version.

For a symmetrical π attenuator:

$$R_1 = Z\,\frac{K+1}{K-1} \tag{8-96}$$

$$R_3 = Z\,\frac{K^2-1}{2K} \tag{8-97}$$

Bridged-*T* Attenuator

Figure 8-38 shows a symmetrical attenuator in a bridged-*T* form. One advantage of this configuration is that only two resistor values have to be changed to vary attenuation. Two of the resistors always remain at the source and load impedances R_0.

The resistors R_1 and R_2 are calculated from

$$R_1 = \frac{R_0}{K-1} \tag{8-98}$$

and

$$R_2 = R_0\,(K-1) \tag{8-99}$$

where, using Eq. (8-93) we have $K = 10^{\mathrm{dB}/20}$.

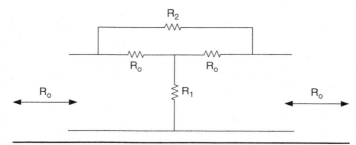

FIGURE 8-38   A bridged-*T* attenuator.

8.11 Power Splitters

8.11.1 Resistive Power Splitters

Resistive power splitters are essentially resistive voltage dividers that distribute a signal in multiple directions while maintaining the same impedance at all ports. They provide no isolation between ports. In other words, even under ideal, perfectly terminated conditions, a signal arriving at any one port appears at all other ports.

Figure 8-39 illustrates an "N"-way splitter, where

$$N = \text{total number of ports} - 1 \tag{8-100}$$

All resistors are equal to R, which is calculated by

$$R = R_0 \frac{N-1}{N+1} \tag{8-101}$$

where R_0 is the impedance at all ports.

Resistive power splitters by their nature are very inefficient. The loss in dB is

$$\text{Power Loss dB} = 10 \, \text{Log}_{10} \frac{1}{N^2} \tag{8-102}$$

So for a two-way splitter (three ports, $N = 2$), the power loss is 6 dB; a four-way splitter has 12-dB loss; and so forth. Since all impedances are matched, the voltage loss in dB is exactly equal to the power loss computed by Eq. (8-102). This loss in dB is also the isolation between ports.

8.11.2 A Magic-*T* Splitter

The circuit of Fig. 8-40 is commonly called a magic-*T* splitter or two-way splitter/combiner. To some extent, it has the functionality of the resistive power splitter, but it differs in two major ways. First, it has *a power loss of 3 dB rather than 6 dB* for the equivalent function of the resistive power splitter. Second, it can have theoretical *infinite isolation between ports A and B*. Thus, a signal entering port A will be prevented from arriving at port B, and vice versa. However, any signals entering either port A or port B will arrive at port S with 3 dB of loss, and any signal applied to port S will arrive at ports A and B with 3 dB of loss.

Observe that port S has an impedance of $R_0/2$. A 2:1 impedance ratio transformer can step this up to R_0.

FIGURE 8-39 A resistive power splitter.

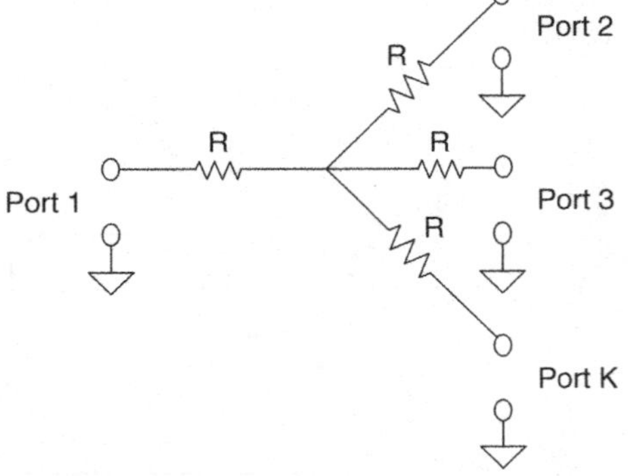

FIGURE 8-40   The magic-*T* splitter/
combiner.

The output signal at port *B* will not change when port *A* is terminated with an impedance other than R_0, even on a short or open (ports *A* and *B* are interchangeable). An impedance mismatch at port *S* will cause a reflection of the signal applied to port *A* onto port *B*. The amount of this reflection is the return loss of the impedance mismatch at port *S* plus 6 dB.

$$\text{Reflection Loss dB} = 20 \, \text{Log}_{10} \left| \frac{R_0 + R_x}{R_0 - R_x} \right| + 6 \text{ dB} \tag{8-103}$$

where R_x is the value of the termination of port *S*.

Operation of the magic-T splitter is not very intuitive from the schematic. Let's examine Fig. 8-41, where the circuit has been redrawn showing a signal applied to port *S*. In the circuit of Fig. 8-41, V_s results in equal currents, I_1 and I_2, in opposite directions through T_1. The resulting voltages, E_1 and E_2, are equal in magnitude but opposite in polarity, so they cancel. From the symmetry of the circuit, the signal at port *S* appears equally at ports *A* and *B*. The impedance seen at port *S* is $R_0/2$ since the terminations of both ports *A* and *B* are reflected to port *S* and are in parallel.

The isolation mechanism between ports *A* and *B* is illustrated in Fig. 8-42. A signal V_A is applied to port *A*. The impedance measured at the input of T_1 between the port *A*

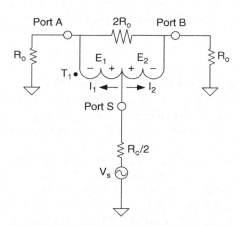

FIGURE 8-41   A magic-*T* with a signal at the
S port.

FIGURE 8-42   The magic-*T* isolation between
ports *A* and *B*.

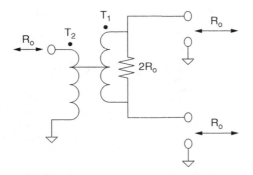

FIGURE 8-43 The final circuit with equal impedances at all ports.

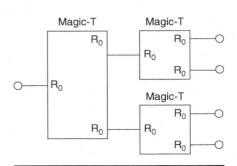

FIGURE 8-44 A four-way splitter-combiner (with a loss of 6 dB).

and port S terminals is $R_0/2$ since T_1 is terminated with $2R_0$ and a four-to-one imped-ance step-down occurs between the port A and port S terminals (center-tapped T_1). As a result, the applied voltage divides evenly, so E_1 and E_S are equal.

Because of transformer action, E_1 results in E_2 which is equal in magnitude. As mea-sured from the input of port B, E_2 and E_S are equal and in series but *opposite* in polarity, so they cancel. Therefore, under these conditions, V_A will not appear at port B, so total isolation will occur. V_A will appear at port S attenuated 3 dB in power.

The final circuit is shown in Fig. 8-43. An additional transformer (autotransformer) with a turns ratio of 1.414:1 has been added to step up $R_0/2$ to R_0. The impedance ratio is 2:1.

Splitter-combiners can be cascaded in a treelike fashion to create additional ports. The circuit of Fig. 8-44 illustrates how a four-way splitter-combiner can be made from three two-way splitter-combiners. All ports must have equal impedances, however. The insertion loss for Fig. 8-44 is 6 dB.

Care should be taken when cascading magic-T circuits that transformers T_1 and T_2 have sufficient bandwidth since the roll-off will be cumulative. As a good rule of thumb, the inductance of T_1 and T_2 should be a minimum of $R_0/(\pi F_L)$, where F_L is the lowest frequency of interest. In addition, the center-tap accuracy should be in the vicinity of 1 percent.

8.12   Introduction of Transmission Zeros to an Existing Design

At times, a complex filter reaches the final stage of testing and shortcomings are found. Rather than go back to the drawing board and start a new design, scrapping the old one, it is better to make a simple modification to the existing design to correct the fre-quency response deficiency.

Section 8.5 discusses amplitude equalization to compensate for inadequate Q effects in the passband. However, this would not resolve any stopband issues. Figure 8-45a shows the response of an all-pole band-pass filter superimposed on a mask that defines the attenuation limits. Clearly the attenuation falls outside the mask on both the low-frequency end and the high-frequency end. Introducing transmission zeros on both the low side and the high side would accelerate the descent into the stopband. The effect of this introduction is shown in Fig. 8-45b.

FIGURE 8-45 Effect of adding transmission zeros: (a) original response; and (b) adding zeros on low and high sides to increase steepness.

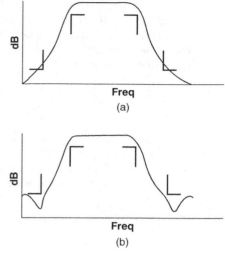

Note that adding these nulls results in a decrease of attenuation beyond the nulls (an amplitude lobe) rather than continued monotonic behavior. The closer the null is to the passband, the higher in amplitude will be the lobe. Also, best results are achieved if the band-pass filter bandwidth is less than 10 percent, since this is all based on narrow-band approximations.

Figure 8-46 shows a conventional series- and parallel-tuned circuit extracted from a band-pass filter schematic and how they are transformed in a manner to add transmission zeros. Note that a high side notch, a low side notch, or both can be introduced as needed.

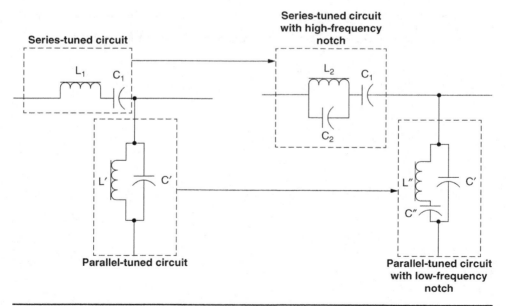

FIGURE 8-46 Introduction of transmission zeros.

The design equations are as follows:

Define F_o as the band-pass filter geometric center frequency.

F_1 is the low-frequency notch (below the passband). F_2 is the high-frequency notch (above the passband).

For a low-frequency notch

$$C'' = \left(\frac{F_o^2}{F_1^2} - 1 \right) C' \tag{8-104}$$

$$L'' = \frac{1}{C''(2\pi F_1)^2} \tag{8-105}$$

where C'' and L'' form a series-resonant circuit at F_1.

To add a high-frequency notch at F_2

$$L_2 = L_1 \left(1 - \frac{F_o^2}{F_2^2} \right) \tag{8-106}$$

$$C_2 = \frac{1}{L_2(2\pi F_2)^2} \tag{8-107}$$

where L_2 and C_2 form a parallel-resonant circuit at F_2.

Both tuned circuits can be transformed into alternate forms. The design equations are given in Sec. 8.6.

References

Geffe, P. R.(1963). *Simplified Modern Filter Design*. New York: John F. Rider.

Saal, R. "Der Entwurf von Filtern mit Hilfe des Kataloges Normierter Tiefpasse," *Telefunken GMBH*, 1963.

Saal, R., and E. Ulbrich. "On the Design of Filters by Synthesis." *IRE Transactions on Circuit Theory* CT-5, December 1958.

Zverev, A. I. (1967). *Handbook of Filter Synthesis*. New York: John Wiley and Sons.

CHAPTER 9

Component Selection for
LC and Active Filters

9.1 Review of Basic Magnetic Principles

9.1.1 Units of Measurement

Magnetic permeability is represented by the symbol μ and is defined by

$$\mu = \frac{B}{H} \tag{9-1}$$

B is the magnetic flux density in lines per square centimeter and is measured in gauss, while H is the magnetizing force in oersteds that produced the flux. Permeability is dimensionless and can be considered a figure of merit of a particular magnetic material since it represents the ease of producing a magnetic flux for a given input. The permeability of air, or that of a vacuum, is 1. As a magnetic material's permeability increases, it becomes more suitable for lower frequencies.

Magnetizing force is caused by current flowing through turns of wire. Thus, H can be determined from ampere-turns by

$$H = \frac{4\pi NI}{10 \text{ mL}} \tag{9-2}$$

where N is the number of turns, I is the current in amperes, and mL is the mean length of the magnetic path in centimeters.

The inductance of a coil is directly proportional to the number of flux linkages per unit current. The total flux is found from

$$\phi = BA = \mu HA = \frac{4\pi NI\mu A}{10 \text{ mL}} \tag{9-3}$$

where A is the cross-sectional area in square centimeters.

The inductance proportionality may then be expressed as

$$L \propto \frac{4\pi NI\mu A}{10 \text{ mL}} \frac{N}{I} \tag{9-4}$$

or directly in henrys by

$$L = \frac{4\pi N^2 \mu A}{\text{mL}} 10^{-9} \tag{9-5}$$

A number of things should be apparent from Eq. (9-5). First of all, the inductance of a coil is directly proportional to the permeability of the core material. If an iron core is inserted into an air-core inductor, the inductance will increase in direct proportion to the iron core's permeability. The inductance is also proportional to N^2.

All the previous design equations make the assumption that the magnetic path is uniform and closed with negligible leakage flux in the surrounding air such as would occur with a single-layer toroidal coil structure. However, this assumption is really never completely valid, so some deviations from the theory can be expected.

The induced voltage of an inductor can be related to the flux density by

$$E_{rms} = 4.44BNfA \times 10^{-8} \tag{9-6}$$

where B is the maximum flux density in gauss, N is the number of turns, f is the frequency in hertz, and A is the cross-sectional area of the core in square centimeters. This important equation is derived from Faraday's law.

As a point of information, the units of flux density B in gauss and the magnetizing force H in oersteds are centimeter-gram-second (CGS) units. International system (SI) units are gaining in usage, mainly outside of the United States. Flux density is then expressed in mT, which is a milli-Tesla and is equivalent to 10 gauss. The magnetizing force is expressed in A/m (ampere/meter) and is equivalent to $4\pi/10^3$ oersteds.

9.1.2 Saturation and DC Polarization

A plot of B versus H is shown in Fig. 9-1. Let's start at point A and increase the magnetizing force to point B. A decrease in magnetizing force will pass through point C, and then D, and then E as it becomes negative. An increasing magnetizing force, again in the positive direction, will travel to B through point F. The enclosed area formed by the curve is called a *hysteresis loop* and results from the energy required to reverse the magnetic molecules of the core. The magnitude of H between points D and A is called *coercive force* and is the amount of H necessary to reduce the residual magnetism in the core to zero.

Permeability was defined as the ratio B/H and can be obtained from the slope of the BH curve. Since we normally deal with low-level AC signals, the region of interest is

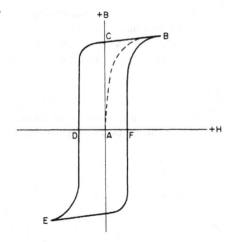

FIGURE 9-1 The hysteresis loop.

restricted to a relatively narrow range. We can then assume that the permeability is determined by the derivative of the curve at the origin. The derivative of a *B/H* curve is sometimes called *incremental permeability*.

If a DC bias is introduced, the quiescent point will move from the origin to a point farther out on the curve. Since the curve tends to flatten out with higher values of *H*, the incremental permeability will decrease, which reduces the inductance. This effect is known as *saturation* and can also occur without a DC bias for large AC signals. Severe waveform distortion usually accompanies saturation. The *B/H* curve for an air core is a straight line at a 45° angle through the origin. The permeability is unity, and no saturation can occur.

9.1.3   Inductor Losses

The *Q* of a coil can be found from

$$Q = \frac{\omega L}{R_{dc} + R_{ac} + R_d} \tag{9-7}$$

where R_{dc} is the copper loss, R_{ac} is the core loss, and R_d is the dielectric loss. Copper loss consists strictly of the DC winding resistance and is determined by the wire size and total length of wire required. The core loss is composed mostly of losses due to eddy currents and hysteresis. Eddy currents are induced in the core material by changing magnetic fields. These circulating currents produce losses that are proportional to the square of the inducing frequency.

When a core is subjected to an AC or pulsating DC magnetic field, the *B* versus *H* characteristics can be represented by the curve of Fig. 9-1. The enclosed area is called a hysteresis loop and results from the energy required to reverse the magnetic domains in the core material. These core losses increase in direct proportion to frequency, since each cycle traverses the hysteresis loop.

The dielectric losses are important at higher frequencies and are determined by the power factor of the distributed capacity. Keeping the distributed capacity small, as well as using wire insulation with good dielectric properties, will minimize dielectric losses.

Above approximately 50 kHz, the current will tend to travel on the surface of a conductor rather than through the cross-section. This phenomenon is called *skin effect*. To reduce this effect, litz wire is commonly used. This wire consists of many braided strands of insulated conductors so that a larger surface area is available in comparison with a single solid conductor of the equivalent cross-section. Above 1 or 2 MHz, solid wire can again be used.

A figure of merit of the efficiency of a coil at low frequencies is the ratio of ohms per henry (Ω/H), where the ohms correspond to R_{dc}—that is, the copper losses. For a given coil structure and permeability, the ratio Ω/H is a constant, independent of the total number of turns, provided that the winding cross-sectional area is kept constant.

9.1.4   Effect of an Air Gap

If an ideal toroidal core has a narrow air gap introduced, the flux will decrease and the permeability will be reduced. The resulting effective permeability can be found from

$$\mu_e = \frac{\mu_i}{1 + \mu_i \left(\dfrac{g}{mL} \right)} \tag{9-8}$$

where μ_i is the initial permeability of the core and g/mL is the ratio of gap to length of the magnetic path. Equation (9-8) applies to closed magnetic structures of any shape if the initial permeability is high and the gap ratio small.

The effect of an air gap is to reduce the permeability and make the coil's characteristics less dependent upon the initial permeability of the core material. A gap will prevent saturation with large AC signals or DC bias and allow tighter control of inductance. However, lower permeability due to the gap requires more turns and, thus, associated copper losses, so a suitable compromise is required.

9.2   Magnetic Materials and Physical Form Factors of Inductors

The actual design of inductors is best left to the specialist since there are many complex factors involved, as well as requirements in terms of appropriate coil-winding machinery. Depending on the frequency range, inductance, desired Q, temperature range, and other operating parameters, the inductor manufacturer is in the best position to recommend the appropriate part. In many cases, this will be a custom job, especially if it has to be adjustable or have high precision. However, it is useful for the filter designer to be aware of the various magnetic materials and form factors inductors come in.

9.2.1   Magnetic Materials

Molybdenum Permalloy Powder Toroidal Cores

Molybdenum permalloy powder (MPP) toroidal cores are manufactured by pulverizing a magnetic alloy consisting of approximately 2 percent molybdenum, 81 percent nickel, and 17 percent iron into a fine powder, insulating the powder with a ceramic binder to form a uniformly distributed air gap, and then compressing it into a toroidal core at extremely high pressures. Finally, the cores are coated with an insulating finish.

MPP cores result in extremely stable inductive components for use below a few hundred kilohertz. Core losses are low over a wide range of available permeabilities. Inductance remains stable with large changes in flux density, frequency, temperature, and DC magnetization due to high resistivity, low hysteresis, and low eddy-current losses.

MPP cores are categorized according to size, permeability, and temperature stability. Generally, the largest core size that physical and economical considerations permit should be chosen. Larger cores offer higher Qs, since flux density is lower due to the larger cross-sectional area, resulting in lower core losses, and the larger window area reduces the copper losses.

Toroidal cores range in size from an outside diameter (OD) of 0.140 to 5.218 in. Available core permeabilities range from 14 to 550. The lower permeabilities are more suitable for use at the higher frequencies, since the core losses are lower.

Ferrite

Ferrites are ceramic structures created by combining iron oxide with oxides or carbonates of other metals, such as manganese, nickel, or magnesium. The mixtures are pressed, fired in a kiln at very high temperatures, and machined into the required shapes.

The major advantage of ferrites over MPP cores is their high resistivity so that core losses are extremely low, even at higher frequencies where eddy-current losses become critical. Additional properties such as high permeability and good stability with time and temperature often make ferrites the best core-material choice for frequencies from 10 kHz to well in the megahertz region.

Powdered Iron

Above 1 or 2 MHz, the core losses of most ferrite materials become prohibitive. Toroidal cores composed of compressed iron powder known as carbonyl iron then become desirable for use up to the VHF range.

These cores consist of finely divided iron particles, which are insulated and then compressed at very high pressures into a toroidal form in a manner similar to MPP cores. The high resistance in conjunction with the very small particles results in good high-frequency performance. Permeability can be controlled to very tight tolerances and with low temperature coefficients. Saturation levels of iron are quite high compared to other materials such as ferrite.

A variety of iron powders within the carbonyl iron family are available and are suitable for use over different frequency bands. The more commonly used materials are listed in Table 9-1 along with the nominal permeability, temperature coefficient, and maximum frequency of operation above which the core losses can become excessive. A major supplier of these cores is Micrometals, Inc. The Micrometals nomenclature for the various materials is provided as well. In general, a material should be selected that has the highest permeability over the frequency range of operation.

Ceramic or Air Core

The material with the lowest permeability ($\mu = 1$) is air. Typically a nonmagnetic core such as ceramic is used as a form for the inductor. Inductors of this form are useful above 10 MHz and up to a few hundred MHz. Air has a m of 1, will not saturate, and has no core losses, so the Q is strictly dependent upon the winding.

Micrometals Designation	Type of Iron Powder	Frequency Range of Operation	Material Permeability μ	Temperature Stability ppm/°C
–1	Carbonyl C	150 KHz to 3 MHz	20	280
–2	Carbonyl E	250 KHz to 10 MHz	10	95
–3	Carbonyl HP	20 KHz to 1 MHz	35	370
–4	Carbonyl J	3 MHz to 40 MHz	9	280
–6	Carbonyl SF	3 MHz to 40 MHz	8.5	35
–7	Carbonyl TH	1 MHz to 25 MHz	9	30
–8	Carbonyl GQ4	20 KHz to 1 MHz	35	255
–10	Carbonyl W	15 MHz to 100 MHz	6	150
–15	Carbonyl GS6	150 KHz to 3 MHz	25	190
–17	Carbonyl	20 MHz to 200 MHz	4	50
–42	Hydrogen Reduced	300 KHz to 80 MHz	40	550
–0	Phenolic	50 MHz to 350 MHz	1	0

TABLE 9-1 Iron-Powder Core Materials

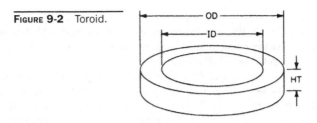

FIGURE 9-2 Toroid.

9.2.2 Magnetic Coil Structures

Toroidal Core Construction

A toroid is the most ideal structure with regard to having a closed magnetic path resulting in almost no magnetic leakage. They are available in many materials, including phenolic, which is essentially an air core. However they can only be adjusted by adding or removing turns and are not extremely efficient, since an open area (the hole) must be left because of their winding method. In addition to the material they are specified by size, as shown in Fig. 9-2.

Pot Core Construction

A typical pot core assembly is shown in Fig. 9-3. A winding supported on a bobbin is mounted in a set of symmetrical ferrite pot core halves. The assembly is held rigid by a metal clamp. An air gap in introduced in the center post of each half, since only the outside surfaces of the pot core halves mate with each other. By introducing an adjustment slug containing a ferrite sleeve, the effect of the gap can be partially neutralized as the slug is inserted into the gap region.

Ferrite cores come in many other shapes besides pot cores. A popular family is the rectangular module (RM) cores. These are square or rectangular in shape so they have a more efficient form factor as far as PC board utilization than round pot cores. Multiple RM cores can be physically packed more tightly together than the round pot cores. The magnetic properties are similar, although RM cores offer less of a closed magnetic path than the round pot cores, so more flux leakage can occur between cores. Figure 9-4 illustrates a cut-away view of an RM core.

9.2.3 Surface-Mount RF Inductors

Surface-mount radio frequency (RF) inductors (sometimes referred to as "chip" inductors) are wound on a solenoid-like core and encapsulated. The two ends of the coil winding are terminated in metal end caps, which are then soldered to the board during assembly. The core itself can consist of phenolic, ceramic, powdered iron, or ferrite, depending on the inductance values and core size. Figure 9-5 shows a cross-sectional view of a wound surface-mount RF inductor.

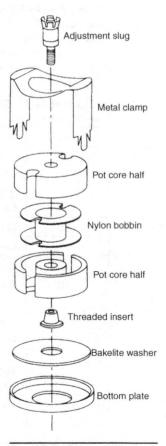

Adjustment slug

Metal clamp

Pot core half

Nylon bobbin

Pot core half

Threaded insert

Bakelite washer

Bottom plate

FIGURE 9-3 Pot core structure.

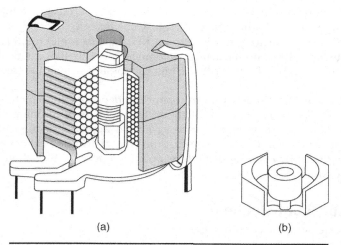

(a)   (b)

FIGURE 9-4   RM core structure.

Surface-mount RF inductors are available in standard Electronic Industries Association (EIA) sizes, such as 0402, 0603, 0805, 1206, and 1210. These numbers describe the approximate body length and width as a multiple of 10 mil inches. For example, 1206 size is 120 mil inches (0.120") long and 60 mil inches (0.060") wide. The height varies with each design. Some manufacturers also offer nonstandard sizes.

Each core material has unique properties. Phenolic or ceramic is essentially an air core with a permeability of 1. An air core never saturates with DC current or large AC signals, and has a zero temperature coefficient of inductance, but because of a μ of only 1, it is best suited for low-inductance high-frequency requirements. Powdered iron, on the other hand, offers higher permeabilities and a very low temperature coefficient. Last comes ferrite, which has the highest permeability but also the highest

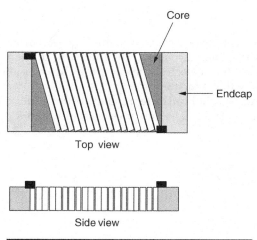

FIGURE 9-5   Cross-sectional view of a wound surface-mount RF inductor.

temperature coefficient. Typically, the higher the permeability, the more the tendency to saturate and the higher the temperature coefficient.

The major factors in selecting an RF inductor are Q (quality factor), self-resonant frequency (SRF), and inductance tolerance. The Q will dramatically affect performance of a filter, so an inductor with the highest possible Q over the frequency range of operation should be chosen.

Section 8-4 discusses self-resonant frequency (SRF), which is the frequency where the distributed capacitance of the inductor forms a resonant circuit with the actual inductance. As the frequency of operation approaches this self-resonant frequency, Q will be dramatically decreased, since the distributed capacitance results from very lossy dielectric material. Therefore, operation near the SRF should be avoided as much as possible.

9.3 Capacitor Selection

An extensive selection of capacitor types is available for the designer to choose from. They differ in terms of construction and electrical characteristics. The abundance of different capacitors often results in a dilemma in choosing the appropriate type for a specific application. Some of the factors to consider are stability, size, losses, voltage rating, tolerances, cost, and construction.

The initial step in the selection process is to determine the capacitor dielectric. These substances include air, glass, ceramic, mica, plastic films, aluminum, and tantalum. The type of mechanical construction must also be chosen. Since miniaturization is usually a prime consideration, the smallest possible capacitors will require thin dielectrics and efficient packaging without degrading performance.

9.3.1 Properties of Dielectrics

Capacitors in their most fundamental form consist of a pair of metallic plates or electrodes separated by an insulating substance called the dielectric. The capacitance in farads is given by

$$C = \frac{kA}{D} \, 8.85 \times 10^{-12} \tag{9-9}$$

where k is the dielectric constant, A is the plate area in square meters, and D is the separation between plates in meters. The dielectric constant of air is 1. Since capacity is proportional to the dielectric constant, the choice of dielectric highly influences the physical size of the capacitor.

A practical capacitor can be represented by the equivalent circuit of Fig. 9-6a, where L_s is the series inductance, R_s is the series resistance, and R_p is the parallel resistance. L_s, R_s,

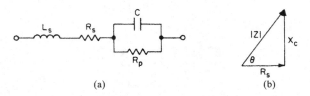

FIGURE 9-6 The equivalent representation of a capacitor: (a) an equivalent circuit; and (b) a simplified vector diagram.

and R_p are all parasitic elements. If we assume L_s negligible and R_p infinite, the equivalent impedance can be found by the vector addition of R_s and the capacitive reactance X_c, as shown in Fig. 9-6b. R_s is often referred to as _ESR_ or _equivalent series resistance._

An important characteristic of a capacitor is the dissipation factor. This parameter is the reciprocal of capacitor Q and is given by

$$d = \frac{1}{Q} = 2\pi f C R_s = \cot \theta \tag{9-10}$$

where f is the frequency of operation. The dissipation factor is an important figure of merit and should be as low as possible. This is especially true for selective _LC_ filters where branch Qs should be high. Capacitor Q must be sufficiently higher than inductor Q so that the effective Q is not degraded.

The power factor is another figure of merit similar to the dissipation factor. Referring to Fig. 9-6 the power factor is defined as

$$\text{PF} = \frac{R_s}{|Z|} = \cos \theta \tag{9-11}$$

For capacitor Q in excess of 10, we can state $d = \text{PF} = 1/Q$. The dissipation factor, or power factor, is sometimes expressed as a percentage.

The shunt resistive element R_p in Fig. 9-6 is often referred to as insulation resistance and results from dielectric leakage currents. A commonly used figure of merit is the $R_p C$ time constant, normally given in $M\Omega \times \mu F$. It is frequently convenient to combine all the losses in terms of a single resistor in parallel with C, which is given by

$$R = \frac{1}{2\pi f C d} \tag{9-12}$$

The temperature coefficient (TC) is the rate of change of capacity with temperature and is usually given in parts per million per degree Celsius (ppm/°C). This parameter is extremely important when filter stability is critical. TC should either be minimized or of a specific nominal value so that cancellation will occur with an inductor's or resistor's TC of the opposite sign. A dielectric material will also have an operating temperature range beyond which the material can undergo permanent molecular changes.

Another important parameter is _retrace_, which is defined as the capacity deviation from the initial value after temperature cycling. To maintain long-term stability in critical filters that are subjected to temperature variation, retrace should be small.

Table 9-2 summarizes some of the properties of the more commonly used dielectric materials when applied to capacitors.

9.3.2 Capacitor Construction

Surface-Mount Capacitors

Surface-mount technology provides circuit densities previously unattainable using older through-hole methods. This technology allows the placement of subminiature passive and active components on both sides of a circuit board. As a result, densities can be nearly quadrupled in practice, since the components are much smaller than their through-hole counterparts and the available board area for their placement is essentially

Capacitor Type	Dielectric Constant	TC, ppm/°C	Dissipation Factor, %	Insulation Resistance, MΩ-μF	Temperature Range, °C
Aluminum	8	±2,500	10	100	−40 to +85
Ceramic NPO/COG	65	±30	0.02	5×10^3	−55 to +125
Ceramic X7R	2,000	±15%[1]	1	5×10^3	−55 to +125
Glass	5	±140	0.001	10^6	−55 to +125
Mica	6	±50	0.001	2.5×10^4	−55 to +150
Paper	3	±800	1.0	5×10^3	−55 to +125
Polycarbonate	3	±50	0.2	5×10^5	−55 to +125
PEN (Polyethylene Napthalate)	3	±200	0.40	10^5	−55 to +125
PET (Polyethylene Terephtalate/ Polyester/Mylar)	3.2	+400	0.50	10^5	−55 to +100
PPS (Polyphenylene Sulphide)	3	+80	0.10	5×10^5	−55 to +150
Polypropylene	2.2	−250	0.05	10^5	−55 to +105
Polystyrene	2.5	−120	0.01	3.5×10^7	−55 to +85
Polysulfone	3.1	+80	0.3	10^5	−55 to +150
Porcelain	5	+120	0.1	5×10^5	−55 to +125
Tantalum	28	+800	4.0	20	−55 to +85
Teflon	2.1	−200	0.04	2.5×10^5	−70 to +250

[1]Over entire temperature range

TABLE 9-2   Properties of Capacitor Dielectrics

doubled. Figure 9-7 illustrates the construction of a typical surface-mount ceramic type capacitor. End terminations allow soldering to pads on the printed circuit board (PCB). Most forms of surface-mount construction do not result in hermetically sealed capacitors. Surface-mount components are sometimes called *chip* components, such as chip capacitors, chip resistors, chip inductors, and so on.

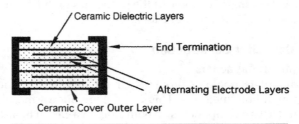

FIGURE 9-7   The typical construction of surface-mount ceramic capacitors.

EIA Size Code	Metric Size Code	Length L	Width W
0201	0603	0.02" (0.6 mm)	0.01" (0.3 mm)
0402	1005	0.04" (1.0 mm)	0.02" (0.5 mm)
0603	1608	0.06" (1.6 mm)	0.03" (0.8 mm)
0805	2012	0.08" (2.0 mm)	0.05" (1.2 mm)
1206	3216	0.12" (3.2 mm)	0.06" (1.6 mm)
1210	3225	0.12" (3.2 mm)	0.10" (2.5 mm)
1812	4532	0.18" (4.5 mm)	0.12" (3.2 mm)
2225	5764	0.22" (5.7 mm)	0.25" (6.4 mm)

TABLE 9-3   Preferred Sizes

Surface-mount capacitors are categorized in terms of size by length and width, as follows in Table 9-3.

The first two significant figures for standard values are given in Table 9-4. The bold values are more readily available. Frequently, a design may require nonstandard values. It is nearly always far more economical to parallel a few off-the-shelf capacitors of standard values when possible than to order a custom-manufactured component unless the quantities involved are substantial.

Axial and Radial Lead Construction

Another method of packaging capacitors is axial lead construction (illustrated in Fig. 9-8).

For film type capacitors such as Mylar, polypropylene, or polystyrene, the electrodes are either a conductive foil such as aluminum or a layer of metallization placed directly on the dielectric film. The electrode and dielectric combination is tightly rolled into a core. The alternate layers of foil or metallization are slightly offset so that leads can be attached.

The assembly is heat-shrunk to form a tight package. It can then be sealed by coating it with epoxy or by encapsulating it in a molded package. A more economical form of sealing is to wrap the capacitor in a plastic film and fill the ends with epoxy.

The most economical form of construction involves inserting the wound capacitor in a polystyrene sleeve and heat-shrinking the entire assembly to obtain a rigid package. This is called *wrap-and-fill*. However, the capacitor is not truly sealed and can be affected by humidity and cleaning solvents entering through the porous end seals.

10	**18**	**33**	**56**
11	20	36	62
12	**22**	**39**	**68**
13	24	43	75
15	**27**	**47**	**82**
16	30	51	91

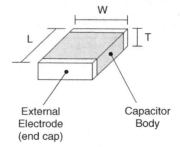

TABLE 9-4   Standard Capacitor Values

FIGURE 9-8 Axial lead construction.

Increased protection against humidity or chemical effects is obtained by sealing the ends with epoxy. If a true hermetic seal is required for extremely harsh environments, the capacitor can be encased in a metal can with glass-to-metal end seals.

An alternative to axial leads (where the leads are in the same plane as the capacitor axis) is *radial leads*. An example of radial leads is shown in Fig. 9-9. The leads are *radial* to the body of the capacitor. The critical dimensions are lead spacing *S*, height *H*, length *L*, and thickness *T*. Axial and radial leaded components are known as "through-hole" since they mount through holes in a printed circuit board rather than on the surface like surface-mount components.

Capacitors such as ceramic, mica, and porcelain contain a more rigid dielectric substance and cannot be rolled like the film capacitors. They are constructed in stacks or layers. Leads are attached to the end electrodes, and the entire assembly is molded or dipped in epoxy. More commonly, they are packaged using surface-mount construction.

Electrolytic Capacitor Construction

Capacitors with the highest possible capacitance per unit volume are the electrolytics made of either aluminum or tantalum. Basically, these capacitors consist of two electrodes immersed in a liquid electrolyte (see Fig. 9-10). One or both electrodes are coated with an extremely thin oxide layer of aluminum or tantalum, forming a film with a high dielectric constant and good electrical characteristics. The electrolyte liquid makes contact between the film and the electrodes. The entire unit is housed in a leakproof metal can. The dielectric film is "formed" by applying a DC voltage between cathode and anode, resulting in a *permanent polarization* of the electrodes. If both plates are "formed," a nonpolar unit will result with half the capacitance of the equivalent polarized type.

A solid-anode tantalum capacitor is composed of a sintered anode pellet on which is a tantalum oxide layer. The pellet is coated with a solid electrolyte of manganese dioxide, which also becomes the cathode. This construction is superior electrically to the other forms of electrolytic construction. Figure 9-11 shows the surface-mount version of a solid tantalum capacitor. Note that the polarity is always indicated on the body of the capacitor.

9.3.3 Selecting Capacitors for Filter Applications

The performance of a capacitor in its application is highly dependent on the choice of dielectric material. Table 9-2 provides some of these properties. The decision is governed

FIGURE 9-9 Radial lead construction.

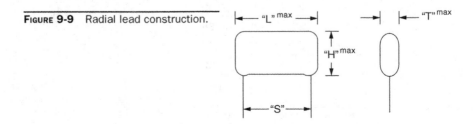

Figure 9-10   An electrolytic capacitor.

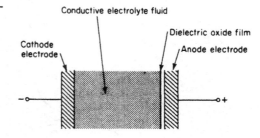

by a variety of factors, such as the value of the capacitor, operating temperature range and stability, frequency range, desired accuracy (tolerance), dissipation factor (*Q*), voltage rating, and so on. The most popular capacitor families are film and ceramic for both active and passive (*LC*) filters.

Film Capacitors

Film capacitors in surface-mount configurations use mainly two construction methods. The most common method of manufacture involves stacking sheets of film dielectric, which are metallized on one side and are called stacked-film chips. The other form of construction involves wound chips rather than stacked, and is called metal electrode leadless face (MELF) chips. Also the through-hole axial and radial type construction methods mentioned previously can be used.

Polyester (Mylar/PET) capacitors are the smallest and most economical of the film types. They should be considered first for general-purpose filters operating below a few hundred kilohertz and at temperatures up to 125°C. The value range is from 1,000 pF up to 10 μF. Values below 1,000 pF are not recommended unless the voltage rating is intentionally increased to a few hundred volts, which forces a thicker and therefore more robust film. PEN (polyethylene napthalate) film capacitors are similar to PET but are available in higher voltage ratings and tighter tolerances. PPS (polyphenylene sulphide) capacitors have the best qualities compared to the PET and PEN types. They can support higher temperatures during both operation and soldering. They can be provided with tighter tolerances, but are generally more expensive than the PET and PEN types.

Polystyrene capacitors probably have the best electric properties of all film capacitors. The temperature coefficient is precisely controlled, almost perfectly linear, and has

Figure 9-11   A tantalum capacitor in a surface-mount package.

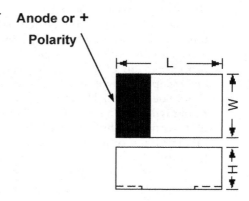

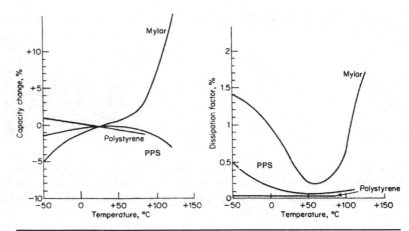

FIGURE 9-12 The capacitance and dissipation factor versus ambient temperature of film capacitors.

a nominal value of −120 ppm/°C. Because of the predictable temperature characteristics, these capacitors are highly suited for *LC* resonant circuits, where the inductors have corresponding positive temperature coefficients. Capacity retrace is typically 0.1 percent. Losses are extremely small, resulting in a dissipation factor of approximately 0.01 percent. The maximum temperature, however, is limited to 85°C. Polypropylene capacitors are comparable in performance to polystyrene, although they have a slightly higher dissipation factor and temperature coefficient. Their maximum temperature rating, however, is 105°C. They are more economical than polystyrene and priced to be nearly competitive with Mylar.

Figure 9-12 compares the capacitance and dissipation factor versus temperature for Mylar (PET), PPS, and polystyrene capacitors. Clearly, polystyrene is superior, although PPS exhibits slightly less capacity variation with temperature over a limited temperature range. The dissipation factor is also a function of frequency, so at higher frequencies, the dielectric losses will increase *d*, thus reducing circuit *Q*s.

Film capacitors are available with standard tolerances of 1, 2½, 5, 10, and 25 percent. For most applications, the 2½ or 5 percent tolerances will be adequate. This is especially true when the resonant frequency is adjustable by tuning the inductor, or when using a potentiometer in active filters. Precision capacitors tend to become more expensive with decreasing tolerance.

If the differential voltage across a capacitor becomes excessive, the dielectric will break down and may become permanently damaged. Most film capacitors are available with DC voltage ratings ranging from 33 to 600 V or more. Since most filters process signals of a few volts, voltage rating is normally not a critical requirement. Since capacitor volume increases with voltage rating, unnecessarily high ratings should be avoided. However, polystyrene capacitors below 0.01 μF with a low voltage rating will have a tendency to change value from the PCB soldering process because of distortion of the dielectric film from heat conducted through the leads. Capacitors rated at 100 V or more will generally be immune, since the dielectric film will be sufficiently thick.

In addition to resistive losses, the equivalent circuit of Fig. 9-6a contains a parasitic inductance L_s. As the frequency of operation is increased, the series resonance of L_s and

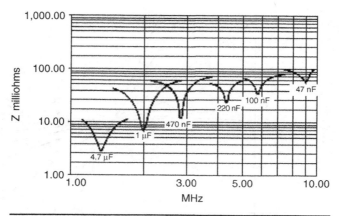

FIGURE 9-13 The self-resonant effect in film capacitors.

C is approached, resulting in a dramatic drop in impedance, as shown in Fig. 9-13. Above self-resonance, the reactance becomes inductive. The self-resonant frequency is highly dependent upon construction and value, but in general, film capacitors are limited to operation below a few megahertz.

Film capacitors do not have piezoelectric properties. This means that any mechanical shock such as vibration will not result in the generation of small voltages.

Ceramic Capacitors

Ceramic capacitors are formed in stacks or layers using the construction illustrated in Fig. 9-7. The ceramic plates are formed by first casting a film slurry of barium titanate and binders, which is then coated with a metallic ink to form electrodes. This results in a high dielectric constant. Another type uses electrically stable paraelectric material, which has lower dielectric constants but highly superior properties. (Barium titanates are not a major part of the composition.) The plates are stacked into layers, fired in ovens, and separated into individual capacitors.

The basic dielectric material may have a dielectric constant *k* as high as 3,000 at room temperature, which can be become as high as 10,000 at 125°C. The temperature at which the maximum *k* occurs is called the *Curie point*. By introducing additives, the Curie point can be lowered, the temperature variation reduced, and negative temperature coefficients obtained.

Ceramic capacitors are manufactured in a variety of shapes, including monolithic chips, which can be directly bonded to a metallized substrate for surface mounting (see Table 9-3 for standard case sizes). They are also available in the traditional disc form.

Ceramic capacitors are available in specific temperature coefficients. Some typical values are shown in Fig. 9-14 , where the P100 has a positive temperature coefficient of 100 ppm/°C and the negative types range from NPO (an approximately zero temperature coefficient) through N4700 (4,700 ppm/°C). These capacitors become very useful for temperature compensation of passive and active filter networks.

There are essentially two classes of ceramic capacitors. The first and most popular class, EIA *Class 1* (and highly recommended for filters), is the temperature compensation type, and more specifically the NPO category (also known as COG), which is extremely stable with temperature.

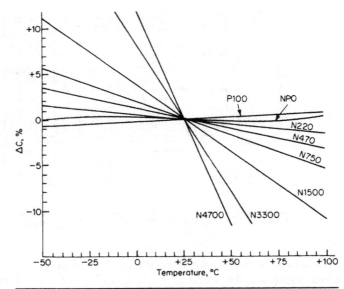

FIGURE 9-14 Temperature characteristics of ceramic capacitors.

The second class, *EIA Class 2,* has much higher dielectric constants to achieve higher volumetric efficiencies for high capacitance values. They are known as the X7R, X5R, Z5U, and Y5V types, which have much higher temperature coefficients, higher dissipation factors, and a piezoelectric characteristic. This means that any shock or vibration will result in a piezoelectric effect where small voltages will be generated. This could wreak havoc where small signals are present, such as in preamplifiers and so on. In addition, low levels of third harmonic distortion can occur due to the capacitors having a high voltage coefficient (change of capacitance with instantaneous voltage). Therefore, this class of ceramic capacitors should be limited to brute-force filtering applications, such as those for power supplies or integrated circuit (IC) decoupling, and should not be used for precision passive or active filters.

Table 9-5 compares the electrical properties of ceramic capacitors.

Parameter	NPO or COG	X7R	X5R	Z5U	Y5V
Operating Temperature Range	−55 to +125°C	−55 to +125°C	−55 to +85°C	10 to +85°C	−30 to +85°C
Temperature Coeff.	0 ±50 ppm/°C	±15%*	±15%*	+22%, −56%*	+22%, −56%*
Dissipation Factor	<0.1%	<2.5%	<3%	<3%	<5%
Insulation Resistance	>1,000 MΩ-μF	>1,000 MΩ-μF	>1,000 MΩ-μF	>100 MΩ-μF	>100 MΩ-μF

*Over the entire temperature range.

TABLE 9-5 Properties of Ceramic Capacitors

Capacitance values range from 0.5 pF to values as high as 2.2 μF. The DC working voltages may be as high as a few kilovolts, or as little as 3 V for the larger capacitance values. Tolerances for the NPO (COG) types can be as close as ± 1 percent.

Ceramic capacitors exhibit minimum parasitic inductance and little variation in the dissipation factor with frequency. These properties make them particularly suited for high frequency use, even into the GHz range. The surface-mount package is best suited for high frequencies, since packaging parasitic effects are minimized.

Mica Capacitors

Mica capacitors, although more costly than the film or ceramic types, have virtually unequaled electrical properties. (Mica itself is a rock that is extremely stable and inert.) Typical temperature coefficients are ± 25 ppm/$°$C. Retrace is better than 0.1 percent, and the dissipation factor is typically 0.01 percent ($Q = 10,000$). Mica capacitors can operate to 150$°$C. Values above 10,000 pF become prohibitively expensive and should be avoided.

Mica capacitors are available in both surface-mount and dipped construction. The dipped type package is similar to that shown in Fig. 9-9.

Dipped mica capacitors are also known as silvered mica. Mica plates are silvered on both sides to form electrodes, stacked, and thermally bonded together. This results in a mechanically stable assembly with characteristics indicative of the mica itself. A coating of epoxy resin is then applied by dipping or other means.

The stacked construction of mica capacitors results in excellent performance well into the gigahertz region. The dissipation factor remains low at high frequencies, and parasitic inductance is small.

Electrolytic Capacitors

Aluminum electrolytic capacitors are intended for low-frequency bypassing or nonprecision timing and are generally unsuitable for active or passive filters. They have unsymmetrical and broad tolerances, such as +80 percent/-20 percent, and require a DC polarization. (If the polarity is reversed, electrolytic capacitors can violently explode.) They also have poor stability and a shelf-life limitation. Large parasitic inductances and series resistances preclude usage at high frequencies.

Tantalum capacitors, on the other hand, can be used in low-frequency passive or active filters that require very large capacity values in a small volume. They have fairly high temperature coefficients (approximately +800 ppm/$°$C) and high dissipation factors (approximately 4 percent). However, these limitations may not be serious when applied to low-selectivity filters. The high-frequency characteristics are superior to those of aluminum electrolytics. Tolerances of 10 or 20 percent are standard.

Tantalum capacitors are formed in a similar manner to aluminum electrolytics and are polarized. However, they have no shelf-life restrictions and can even operate indefinitely without DC polarization. A momentary reverse polarity usually will not damage the capacitor, which is not true in the case of the aluminum electrolytics.

A polarization voltage may not always be present. Nonpolar tantalums can be obtained, but are somewhat more expensive and not always available off the shelf. If two identical tantalum capacitors are series-connected back to back, as shown in Fig. 9-15, a nonpolar type will result. The total value will be $C/2$, corresponding to capacitors in series.

Tantalum capacitors are available in three forms: dry foil, wet, and solid. The dry form consists of foil anodes and cathodes, which are stacked or rolled using a paper

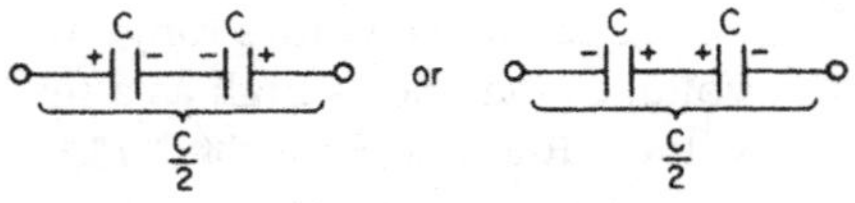

FIGURE 9-15 Nonpolar tantalum capacitors.

spacer and impregnated with electrolyte. The wet forms are constructed using a porous slug of tantalum for the anode electrode and the silver-plated case as the cathode. The unit is filled with sulfuric acid for the electrolyte. Solid tantalums consist of a slab of compressed tantalum powder for the anode with a lead attached. A layer of tantalum pentoxide is formed on the surface for the dielectric, which is then connected to a lead to form the cathode, and the entire assembly is encapsulated in epoxy.

Tantalum capacitors are not recommended where any degree of accuracy is required. Filter designs should be scaled up in impedance so that film or ceramic capacitors can be used.

Trimmer Capacitors

In *LC* filters, particularly those for RF use, it is sometimes more convenient to resonate a tuned circuit by adjusting capacity rather than inductance. A smaller trimmer is then placed in parallel with a fixed resonating capacitor. These trimmers usually consist of air, ceramic, mica, or glass as the dielectric.

Air capacitors consist of two sets of plates: one called the *rotor*, which is mounted on a shaft, and the other the *stator*, which is fixed. As the rotor is revolved, the plates intermesh or overlap without making contact, resulting in increasing capacity. For high-capacity values, the plate size and number must increase dramatically, since the dielectric constant of air is only 1. This becomes a serious limitation when the available room is restricted. Standard air trimmers usually range from maximum values of 15 or 20 pF for small packages and up to 150 pF for larger physical size trimmers.

Ceramic trimmers are smaller than air for comparable values due to an increased dielectric constant. They are usually composed of a single pair of ceramic disks joined at the center in a manner that permits rotation of one of the disks. A silvered region covers part of each disk, forming the plates of the capacitor. As the disk is rotated, the silvered areas begin to overlap, resulting in increasing capacity. Ceramic trimmers are available with maximum capacity values up to 50 pF. Good performance is obtained well into the VHF frequency range this way. Surface-mount packages are also available.

Piston trimmers are composed of a glass or quartz tube with an outside conductive coating corresponding to one electrode. The other plate or electrode is a piston, which by rotation is inserted deeper into the outside tube, resulting in increased capacitance. Multiturn construction results in excellent resolution. Piston trimmers have the best electrical properties of all trimmer types, but are also the most costly. They are suitable for use even at microwave frequencies.

When using a trimmer capacitor, a fixed capacitor is normally placed in parallel with a larger value in order to obtain a finer resolution of adjustment.

9.4 Resistors

A fundamental component of active filters is resistors. Sensitivity studies show that resistors are usually at least as important as capacitors, so their proper selection is crucial to the success of a particular design.

Type	Range, Ω	Standard Tolerances, %	Rating Wattage	Temperature Coefficient, ppm/°C
Carbon composition	1–100 M	5, 10	$\frac{1}{8}, \frac{1}{4}, \frac{1}{2}, 1, 2$	±1,000
Carbon film	1–10 M	2, 5	$\frac{1}{8}, \frac{1}{4}, \frac{1}{2}, 1, 2$	±200
Cermet film	10–22 M	0.5, 1	$\frac{1}{4}, \frac{1}{2}$	±100
Metal film	0.1–1 M	0.1, 0.25, 0.5, 1	$\frac{1}{8}, \frac{1}{4}, \frac{1}{2}, 1, 2$	±25
Wirewound	1–100 K	5, 10, 20	3, 5, 10, 20	±50
Thick film	1–20 M	0.5, 1, 2, 5	$\frac{1}{8}, \frac{1}{4}, \frac{1}{2}, 1$	±100
Thin film	10–100 K	0.05, 1, 2, 5	$\frac{1}{8}, \frac{1}{4}, \frac{1}{2}, 1, 2$	±25

TABLE 9-6   Typical Properties of Fixed Resistors

Resistors are formed by connecting leads across a resistive element. The resistance in ohms is determined by

$$R = \frac{\rho L}{A} \tag{9-13}$$

where ρ is the resistivity of the element in ohm-centimeters, L is the length of the element in centimeters, and A is the cross-sectional area in square centimeters. By using materials of particular resistivities and special geometries, resistors can be manufactured that have the desired properties. Resistors fall into one of two general categories: fixed or variable.

9.4.1   Fixed Resistors

Fixed resistors are normally classified as carbon composition, carbon film, cermet film, metal film, wirewound, thick film, and thin film, according to the resistive element. Of all these types, thick and thin film resistors in chip form are preferred for use in filters and are most readily available. Table 9-6 summarizes some typical properties of fixed resistors.

Thick and Thin Film

Thick film resistors are fabricated by depositing formulated pastes onto a ceramic substrate using a silk screen method. It is essentially an *additive* process where layers of material, including terminations, are added to a substrate. Thin film, on the other hand, is a *subtractive* process. A layer of metallization is sputtered onto a substrate, and material is removed by etching a pattern into the previously applied metallization layer.

Thin film processes are photolithographic, which can result in very precise tolerance resistors with better temperature properties than thick film. However, thick film resistors can handle more power and are more economical.

Thick and thin film resistors for surface-mount applications come in the standard sizes and power ratings, as indicated in Table 9-7 and Fig. 9-16.

Thick and thin film resistors can also be obtained in other package formats besides surface mount (such as axial leaded packages), but for medium- or low-power applications,

EIA Size Code	Metric Size Code	Length L	Width W	Power
0201	0603	0.02" (0.6 mm)	0.01" (0.3 mm)	1/20 watt
0402	1005	0.04" (1.0 mm)	0.02" (0.5 mm)	1/16 watt
0603	1608	0.06" (1.6 mm)	0.03" (0.8 mm)	1/10 watt
0805	2012	0.08" (2.0 mm)	0.05" (1.2 mm)	1/8 watt
1206	3216	0.12" (3.2 mm)	0.06" (1.6 mm)	1/4 watt
1210	3225	0.12" (3.2 mm)	0.10" (2.5 mm)	1/4 watt
1812	4532	0.18" (4.5 mm)	0.12" (3.2 mm)	1/2 watt
2010	5025	0.20" (5.0 mm)	0.10" (2.5 mm)	1/2 watt
2225	5764	0.22" (5.7 mm)	0.25" (6.4 mm)	1 watt

TABLE 9-7 Sizes and Wattage Ratings of Surface-Mount Film Resistors

surface mount is recommended, since the fewest parasitic effects will occur and the packaging is the most economical.

Standard values for precision film resistors are given in Table 9-8. The value is represented by four digits, where the first three digits are significant and are selected from the following table, and the fourth digit (not shown) is the number of zeros. The bold values in this table specifically correspond to standard 1 percent values, and all listed numbers are standard for 0.1, 0.25, and 0.5 percent tolerances.

Metal Film

Metal film resistors are manufactured by depositing nichrome alloys on a rod substrate. Exceptional characteristics can be obtained. Normal tolerances are ±1 percent, but tolerances of ±0.1, ±0.25, and ±0.5 percent are available. Temperature coefficients can be as low as ±15 ppm/°C. Retrace and aging result in changes usually not exceeding 0.25 percent.

Metal film resistors have many highly desirable features. In addition to low temperature coefficients and good long-term stability, they exhibit the lowest noise attainable in resistors. Parasitic effects are minimal and have no significant effect below 10 MHz. Metal film resistors are rugged in design, have excellent immunity to environmental stress, and have high reliability. Although tolerances to 0.1 percent are available, 1 percent values are used almost exclusively because of their lower cost.

The power rating of resistors corresponds to the maximum power that can be continually dissipated with no permanent damage. This rating is normally applicable up to ambient temperatures of 70°C, above which derating is required. In general, a safety margin corresponding to a factor of 2 is desirable to obtain a high level of reliability.

FIGURE 9-16 Surface-mount film resistors.

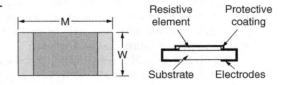

10.0	12.1	14.7	17.8	21.5	26.1	31.6	38.3	46.4	56.2	68.1	82.5
10.1	12.3	14.9	18.0	21.8	26.4	32.0	38.8	47.0	56.9	69.0	83.5
10.2	12.4	15.0	18.2	22.1	26.7	32.4	39.2	47.5	57.6	69.8	84.5
10.4	12.6	15.2	18.4	22.3	27.1	32.8	39.7	48.1	58.3	70.6	85.6
10.5	12.7	15.4	18.7	22.6	27.4	33.2	40.2	48.7	59.0	71.5	86.6
10.6	12.9	15.6	18.9	22.9	27.7	33.6	40.7	49.3	59.7	72.3	87.6
10.7	13.0	15.8	19.1	23.2	28.0	34.0	41.2	49.9	60.4	73.2	88.7
10.9	13.2	16.0	19.3	23.4	28.4	34.4	41.7	50.5	61.2	74.1	89.8
11.0	13.3	16.2	19.6	23.7	28.7	34.8	42.2	51.1	61.9	75.0	90.9
11.1	13.5	16.4	19.8	24.0	29.1	35.2	42.7	51.7	62.6	75.9	92.0
11.3	13.7	16.5	20.0	24.3	29.4	35.7	43.2	52.3	63.4	76.8	93.1
11.4	13.8	16.7	20.3	24.6	29.8	36.1	43.7	53.0	64.2	77.7	94.2
11.5	14.0	16.9	20.5	24.9	30.1	36.5	44.2	53.6	64.9	78.7	95.3
11.7	14.2	17.2	20.8	25.2	30.5	37.0	44.8	54.2	65.7	79.6	96.5
11.8	14.3	17.4	21.0	25.5	30.9	37.4	45.3	54.9	66.5	80.6	97.6
12.0	14.5	17.6	21.3	25.8	31.2	37.9	45.9	55.6	67.3	81.6	98.8

TABLE 9-8   Standard Values for Film Resistors

Cermet Film

Cermet film resistors are manufactured by screening a layer of combined metal and ceramic or glass particles on a ceramic core and firing it at high temperatures. They can provide higher resistance values for a particular size than most other types, up to a few hundred megohms. However, they are somewhat inferior electrically to the metal film type. Their temperature coefficients are higher (typically 200 ppm/°C). Retrace and long-term stability are typically 0.5 percent. Tolerances of 1 percent are standard.

Precision resistors are also available in wirewound form. They consist essentially of resistance wire, such as nichrome, wound on an insulated core. They are costlier than the metal film type and comparable in performance, except for higher parasitics. Wirewound resistors are best suited for applications with higher power requirements.

9.4.2   Variable Resistors

Variable resistors are commonly referred to as potentiometers or trimmers. They are classified by the type of resistance element, such as carbon, cermet, or wirewound, and also by whether they are single- or multiple-turn. The electrical properties of the three basic element types are given in Table 9-9.

Type	Range, Ω	Standard Tolerances, %	Wattage Rating	Temperature Coefficient, ppm/°C
Carbon composition	100–10 M	10, 20	0.5, 1, 2,	±1,000
Cermet	100–1 M	5, 10, 20	0.5, 1	±100
Wirewound	10–100 M	5, 10	0.5, 1, 5, 10	±100

TABLE 9-9   The Typical Properties of Potentiometers

FIGURE 9-17 Potentiometer connections: (a) a potentiometer; and (b) a rheostat configuration.

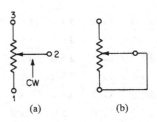

Potentiometers are always three-terminal devices, as depicted in Fig. 9-17a. (CW indicates the direction of travel of the wiper for clockwise rotation.) Since most applications require a two-terminal variable resistor or rheostat rather than a voltage divider, the wiper is normally externally joined to one of the end terminals, as shown in Fig. 9-17b.

Construction

The basic types of construction are the single-turn and multiturn forms. Single-turn potentiometers are rotary devices with a centrally located adjustment hub that contains the movable contact for the wiper. The moveable contact rests upon the circular resistive element. The entire assembly may be exposed or enclosed in a plastic case containing PC pins or surface-mount contacts for the external connections. The adjustable hub is usually slotted for screwdriver access. Some types have a toothed thumbwheel for manual adjustment.

Single-turn trimpots require 270° of rotation to fully traverse the entire resistance element. Increasing the number of turns will improve the operator's ability to make very fine adjustments. Multiturn trimmers contain a threaded shaft. A threaded collar, which functions as a wiper, travels along this shaft, making contact with the resistive element. This general construction is shown in Fig. 9-18.

Trimmer controls, especially the single-turn variety, are subject to movement of the wiper adjustment due to vibration. To assure stability, it is desirable to prevent movement by placing a small amount of a rigid sealer such as Glyptal on the adjuster after circuit alignment.

Types of Resistance Elements

Carbon composition potentiometers are formed by molding a carbon mixture on a nonconductive disk or base containing previously embedded leads. The wiper mechanism is then attached to complete the assembly.

FIGURE 9-18 Multiturn potentiometer construction.

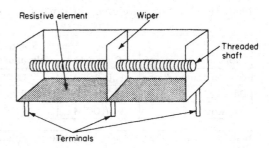

Carbon composition potentiometers are available in both the single- and multiple-turn configurations. They are the most economical of all types and also have the poorest characteristics. TC is about 1000 ppm/°C Retrace and long-term stability are poor, unfortunately. Therefore, carbon composition potentiometers find limited usage in filter circuits.

Cermet film potentiometers are the most commonly used type for filter networks. They are moderate in cost, have a wide range of available values, and have good temperature characteristics and stability. Parasitic effects are minimal. They are manufactured by depositing cermet film on a nonconductive disk or base in thicknesses varying from 0.0005 to 0.005 in.

Wirewound potentiometers are formed by winding resistance wire (usually nichrome) on an insulated base. They have comparable temperature characteristics to cermet, but are higher in cost. Their major attribute is high power capability. However, for most filter requirements, this feature is of little importance.

Wirewound potentiometers have quite different adjustment characteristics than the other types. As the other type wipers are rotated, the resistance varies linearly with degrees of rotation, providing nearly infinite resolution. (Nonlinear tapers are also available, such as logarithmic.) The resistance of wirewound potentiometers, however, changes in discrete steps as the wiper moves from turn to turn. The resolution (or adjustability), therefore, is not as good as the cermet or carbon composition types.

Ratings

Since potentiometers are almost always used as variable elements, the overall tolerances are not critical and are generally 10 or 20 percent. For the same reason, many different standard values are not required. Standard values are given by one significant figure—1, 2, or 5—and range from 10 Ω to 10 MΩ.

9.4.3   Resistor Johnson (Thermal) Noise

At any temperature above absolute zero (−273°C or 0°K), electrons in any substance are in constant motion. Since the direction of motion is random, there is no steady current that can be detected because the randomness results in decorrelation of any short-term current flow. However, a continuous series of random noise pulses occurs, which results in a noise signal known as Johnson noise or thermal noise. The magnitude of noise in a resistor is related to the magnitude of resistance by the following relationship:

$$V_n^2 = 4\,K_b \text{TRB in } V^2/\text{Hz} \tag{9-14}$$

where V_n = noise voltage in volts
K_b = Boltzmann's constant of 1.38×10^{-23} J/°K
T = temperature in °K
R = resistance in ohms
B = bandwidth in Hz

We can greatly simplify this expression if we assume room temperature as follows:

$$V_{\text{noise}} = 4\sqrt{R} \tag{9-15}$$

where R is in KΩ and V_{noise} is in nanovolts per square-root Hz. Therefore, a 1 MΩ resistor would have a Johnson (thermal) noise of 126 nV/$\sqrt{\text{Hz}}$. Figure 9-19 illustrates the Johnson noise generated by a resistor at 25°C in dBm/Hz with a 50-Ω resistive termination.

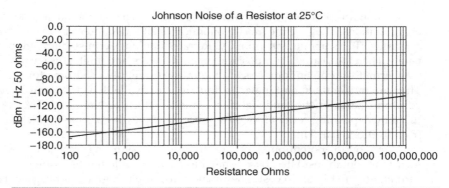

FIGURE 9-19   Johnson noise versus resistance.

Although the noise voltage and resulting power level may be extremely low, if this resistor was part of an active filter with a high gain, the noise could become significant. For low noise requirements, it is best to use as low a resistor value as practical. Noise will increase with temperature and resistance as a square-root proportionality.

The wider the bandwidth, the more the total noise power, so even though the magnitude of the dBm/Hz appears insignificant, the *total* power for a given bandwidth could be much higher. If we convert V_{noise} into watts using V_{noise}^2 / R_{term}, where R_{term} is the termination resistance for the noise, and then multiply the result by the total bandwidth in Hz, we get the *total* noise power over that bandwidth, which could become unacceptable for wide bandwidths and applications with low noise requirements.

References

Ferroxcube Corp. "Linear Ferrite Magnetic Design Manual—Bulletin 550." Saugerties, New York: Ferroxcube Corp.

Ferroxcube Corp. "Linear Ferrite Materials and Components." Saugerties, New York: Ferroxcube Corp.

Magnetics, Inc. *2013 Ferrite Catalog*. Butler, Pennsylvania: Magnetics, Inc.

Ferroxcube Corp. *Powder Cores*. Catalog 2004/2005, Butler, Pennsylvania: Magnetics, Inc.

Ferroxcube Corp. *Molypermalloy Powder Cores*. Catalog MPP-303S. Butler, Pennsylvania: Magnetics, Inc.

Micrometals. *Q Curves for Iron Powder Toroidal Cores*. 2013, Anaheim, California: Micrometals.

Micrometals. Jim Cox, ed. *Iron Powder Cores for High Q Inductors*. 2013 Anaheim, California: Micrometals.

Welsby, V. G. (1950). *The Theory and Design of Inductance Coils*. London: Macdonald and Sons.

CHAPTER 10

Normalized Filter Design Tables

Order n	Real Part $-\alpha$	Imaginary Part $\pm j\beta$
2	0.7071	0.7071
3	0.5000	0.8660
	1.0000	
4	0.9239	0.3827
	0.3827	0.9239
5	0.8090	0.5878
	0.3090	0.9511
	1.0000	
6	0.9659	0.2588
	0.7071	0.7071
	0.2588	0.9659
7	0.9010	0.4339
	0.6235	0.7818
	0.2225	0.9749
	1.0000	
8	0.9808	0.1951
	0.8315	0.5556
	0.5556	0.8315
	0.1951	0.9808
9	0.9397	0.3420
	0.7660	0.6428
	0.5000	0.8660
	0.1737	0.9848
	1.0000	
10	0.9877	0.1564
	0.8910	0.4540
	0.7071	0.7071
	0.4540	0.8910
	0.1564	0.9877

TABLE **10-1** Butterworth Pole Locations

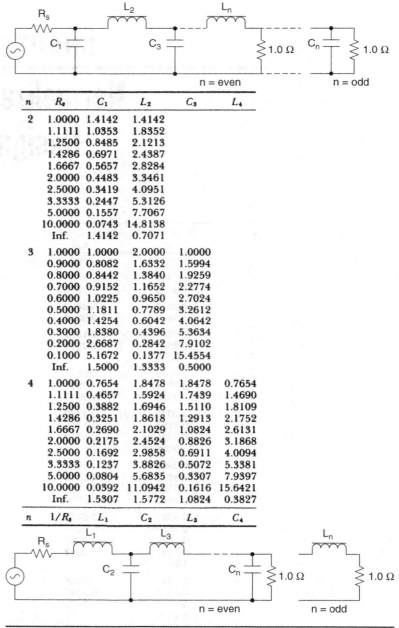

n	R_s	C_1	L_2	C_3	L_4
2	1.0000	1.4142	1.4142		
	1.1111	1.0353	1.8352		
	1.2500	0.8485	2.1213		
	1.4286	0.6971	2.4387		
	1.6667	0.5657	2.8284		
	2.0000	0.4483	3.3461		
	2.5000	0.3419	4.0951		
	3.3333	0.2447	5.3126		
	5.0000	0.1557	7.7067		
	10.0000	0.0743	14.8138		
	Inf.	1.4142	0.7071		
3	1.0000	1.0000	2.0000	1.0000	
	0.9000	0.8082	1.6332	1.5994	
	0.8000	0.8442	1.3840	1.9259	
	0.7000	0.9152	1.1652	2.2774	
	0.6000	1.0225	0.9650	2.7024	
	0.5000	1.1811	0.7789	3.2612	
	0.4000	1.4254	0.6042	4.0642	
	0.3000	1.8380	0.4396	5.3634	
	0.2000	2.6687	0.2842	7.9102	
	0.1000	5.1672	0.1377	15.4554	
	Inf.	1.5000	1.3333	0.5000	
4	1.0000	0.7654	1.8478	1.8478	0.7654
	1.1111	0.4657	1.5924	1.7439	1.4690
	1.2500	0.3882	1.6946	1.5110	1.8109
	1.4286	0.3251	1.8618	1.2913	2.1752
	1.6667	0.2690	2.1029	1.0824	2.6131
	2.0000	0.2175	2.4524	0.8826	3.1868
	2.5000	0.1692	2.9858	0.6911	4.0094
	3.3333	0.1237	3.8826	0.5072	5.3381
	5.0000	0.0804	5.6835	0.3307	7.9397
	10.0000	0.0392	11.0942	0.1616	15.6421
	Inf.	1.5307	1.5772	1.0824	0.3827

n	$1/R_s$	L_1	C_2	L_3	C_4

TABLE 10-2 Butterworth *LC* Element Values

n	R_s	C_1	L_2	C_3	L_4	C_5	L_6	C_7
5	1.0000	0.6180	1.6180	2.0000	1.6180	0.6180		
	0.9000	0.4416	1.0265	1.9095	1.7562	1.3887		
	0.8000	0.4698	0.8660	2.0605	1.5443	1.7380		
	0.7000	0.5173	0.7313	2.2849	1.3326	2.1083		
	0.6000	0.5860	0.6094	2.5998	1.1255	2.5524		
	0.5000	0.6857	0.4955	3.0510	0.9237	3.1331		
	0.4000	0.8378	0.3877	3.7357	0.7274	3.9648		
	0.3000	1.0937	0.2848	4.8835	0.5367	5.3073		
	0.2000	1.6077	0.1861	7.1849	0.3518	7.9345		
	0.1000	3.1522	0.0912	14.0945	0.1727	15.7103		
	Inf.	1.5451	1.6944	1.3820	0.8944	0.3090		
6	1.0000	0.5176	1.4142	1.9319	1.9319	1.4142	0.5176	
	1.1111	0.2890	1.0403	1.3217	2.0539	1.7443	1.3347	
	1.2500	0.2445	1.1163	1.1257	2.2389	1.5498	1.6881	
	1.4286	0.2072	1.2363	0.9567	2.4991	1.3464	2.0618	
	1.6667	0.1732	1.4071	0.8011	2.8580	1.1431	2.5092	
	2.0000	0.1412	1.6531	0.6542	3.3687	0.9423	3.0938	
	2.5000	0.1108	2.0275	0.5139	4.1408	0.7450	3.9305	
	3.3333	0.0816	2.6559	0.3788	5.4325	0.5517	5.2804	
	5.0000	0.0535	3.9170	0.2484	8.0201	0.3628	7.9216	
	10.0000	0.0263	7.7053	0.1222	15.7855	0.1788	15.7375	
	Inf.	1.5529	1.7593	1.5529	1.2016	0.7579	0.2588	
7	1.0000	0.4450	1.2470	1.8019	2.0000	1.8019	1.2470	0.4450
	0.9000	0.2985	0.7111	1.4043	1.4891	2.1249	1.7268	1.2961
	0.8000	0.3215	0.6057	1.5174	1.2777	2.3338	1.5461	1.6520
	0.7000	0.3571	0.5154	1.6883	1.0910	2.6177	1.3498	2.0277
	0.6000	0.4075	0.4322	1.9284	0.9170	3.0050	1.1503	2.4771
	0.5000	0.4799	0.3536	2.2726	0.7512	3.5532	0.9513	3.0640
	0.4000	0.5899	0.2782	2.7950	0.5917	4.3799	0.7542	3.9037
	0.3000	0.7745	0.2055	3.6706	0.4373	5.7612	0.5600	5.2583
	0.2000	1.1448	0.1350	5.4267	0.2874	8.5263	0.3692	7.9079
	0.1000	2.2571	0.0665	10.7004	0.1417	16.8222	0.1823	15.7480
	Inf.	1.5576	1.7988	1.6588	1.3972	1.0550	0.6560	0.2225
n	$1/R_s$	L_1	C_2	L_3	C_4	L_5	C_6	L_7

TABLE 10-2 Butterworth *LC* Element Values (*Continued*)

n	R_s	C_1	L_2	C_3	L_4	C_5	L_6	C_7	L_8	C_9	L_{10}
8	1.0000	0.3902	1.1111	1.6629	1.9616	1.9616	1.6629	1.1111	0.3902		
	1.1111	0.2075	0.7575	0.9925	1.6362	1.5900	2.1612	1.7092	1.2671		
	1.2500	0.1774	0.8199	0.8499	1.7779	1.3721	2.3874	1.5393	1.6246		
	1.4286	0.1513	0.9138	0.7257	1.9852	1.1760	2.6879	1.3490	2.0017		
	1.6667	0.1272	1.0455	0.6102	2.2740	0.9912	3.0945	1.1530	2.4524		
	2.0000	0.1042	1.2341	0.5003	2.6863	0.8139	3.6678	0.9558	3.0408		
	2.5000	0.0822	1.5201	0.3945	3.3106	0.6424	4.5308	0.7594	3.8825		
	3.3333	0.0608	1.9995	0.2919	4.3563	0.4757	5.9714	0.5650	5.2400		
	5.0000	0.0400	2.9608	0.1921	6.4523	0.3133	8.8538	0.3732	7.8952		
	10.0000	0.0198	5.8479	0.0949	12.7455	0.1547	17.4999	0.1846	15.7510		
	Inf.	1.5607	1.8246	1.7287	1.5283	1.2588	0.9371	0.5776	0.1951		
9	1.0000	0.3473	1.0000	1.5321	1.8794	2.0000	1.8794	1.5321	1.0000	0.3473	
	0.9000	0.2242	0.5388	1.0835	1.1859	1.7905	1.6538	2.1796	1.6930	1.2447	
	0.8000	0.2434	0.4623	1.1777	1.0200	1.9542	1.4336	2.4189	1.5318	1.6033	
	0.7000	0.2719	0.3954	1.3162	0.8734	2.1885	1.2323	2.7314	1.3464	1.9812	
	0.6000	0.3117	0.3330	1.5092	0.7361	2.5124	1.0410	3.1516	1.1533	2.4328	
	0.5000	0.3685	0.2735	1.7846	0.6046	2.9734	0.8565	3.7426	0.9579	3.0223	
	0.4000	0.4545	0.2159	2.2019	0.4775	3.6706	0.6771	4.6310	0.7624	3.8654	
	0.3000	0.5987	0.1600	2.9006	0.3539	4.8373	0.5022	6.1128	0.5680	5.2249	
	0.2000	0.8878	0.1054	4.3014	0.2333	7.1750	0.3312	9.0766	0.3757	7.8838	
	0.1000	1.7558	0.0521	8.5074	0.1153	14.1930	0.1638	17.9654	0.1862	15.7504	
	Inf.	1.5628	1.8424	1.7772	1.6202	1.4037	1.1408	0.8414	0.5155	0.1736	
10	1.0000	0.3129	0.9080	1.4142	1.7820	1.9754	1.9754	1.7820	1.4142	0.9080	0.3129
	1.1111	0.1614	0.5924	0.7853	1.3202	1.3230	1.8968	1.6956	2.1883	1.6785	1.2267
	1.2500	0.1388	0.6452	0.6762	1.4400	1.1420	2.0779	1.4754	2.4377	1.5245	1.5861
	1.4286	0.1190	0.7222	0.5797	1.6130	0.9802	2.3324	1.2712	2.7592	1.3431	1.9646
	1.6667	0.1004	0.8292	0.4891	1.8528	0.8275	2.6825	1.0758	3.1895	1.1526	2.4169
	2.0000	0.0825	0.9818	0.4021	2.1943	0.6808	3.1795	0.8864	3.7934	0.9588	3.0072
	2.5000	0.0652	1.2127	0.3179	2.7108	0.5384	3.9302	0.7018	4.7002	0.7641	3.8512
	3.3333	0.0484	1.5992	0.2358	3.5754	0.3995	5.1858	0.5211	6.2118	0.5700	5.2122
	5.0000	0.0319	2.3740	0.1556	5.3082	0.2636	7.7010	0.3440	9.2343	0.3775	7.8738
	10.0000	0.0158	4.7005	0.0770	10.5104	0.1305	15.2505	0.1704	18.2981	0.1872	15.7481
	Inf.	1.5643	1.8552	1.8121	1.6869	1.5100	1.2921	1.0406	0.7626	0.4654	0.1564
n	$1/R_s$	L_1	C_2	L_3	C_4	L_5	C_6	L_7	C_8	L_9	C_{10}

TABLE 10-2 Butterworth *LC* Element Values (*Continued*)

$n = 2$

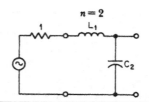

d	L_1	C_2	α_0, dB
0	0.7071	1.414	0
0.05	0.7609	1.410	0.614
0.10	0.8236	1.398	1.22
0.15	0.8974	1.374	1.83
0.20	0.9860	1.340	2.42
0.25	1.094	1.290	2.99
0.30	1.228	1.223	3.53
0.35	1.400	1.138	4.05
0.40	1.628	1.034	4.52
0.45	1.944	0.9083	4.94
0.50	2.414	0.7630	5.30
0.55	3.183	0.5989	5.59
0.60	4.669	0.4188	5.82
0.65	8.756	0.2267	5.96

d	C_2	L_1	α_0, dB

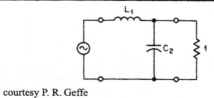

courtesy P. R. Geffe

TABLE 10-3 Butterworth Uniform Dissipation Network

$n = 3$

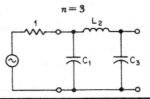

d	C_1	L_2	C_3	α_0, dB
0	0.5000	1.333	1.500	0
0.05	0.5405	1.403	1.457	0.868
0.10	0.5882	1.481	1.402	1.73
0.15	0.6452	1.567	1.334	2.60
0.20	0.7143	1.667	1.250	3.45
0.25	0.8000	1.786	1.149	4.30
0.30	0.9091	1.939	1.026	5.15
0.35	1.053	2.164	0.8743	5.98
0.40	1.250	2.581	0.6798	6.82
0.45	1.538	3.806	0.4126	7.66
d	L_3	C_2	L_1	α_0, dB

courtesy P. R. Geffe

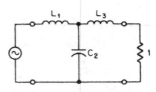

TABLE 10-4 Butterworth Uniform Dissipation Network

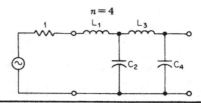

d	L_1	C_2	L_3	C_4	$\alpha_0,$ dB
0	0.3827	1.082	1.577	1.531	0
0.05	0.4144	1.156	1.636	1.454	1.13
0.10	0.4518	1.240	1.701	1.362	2.27
0.15	0.4967	1.339	1.777	1.250	3.39
0.20	0.5515	1.459	1.879	1.113	4.51
0.25	0.6199	1.609	2.039	0.9400	5.63
0.30	0.7077	1.812	2.384	0.7099	6.73
0.35	0.8243	2.124	3.848	0.3651	7.82
d	C_4	L_3	C_2	L_1	$\alpha_0,$ dB

courtesy P. R. Geffe

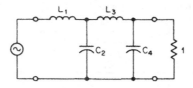

TABLE 10-5 Butterworth Uniform Dissipation Network

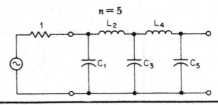

d	C_1	L_2	C_3	L_4	C_5	$a_0,$ dB
0	0.3090	0.8944	1.382	1.694	1.545	0
0.02	0.3189	0.9199	1.412	1.712	1.504	0.562
0.04	0.3294	0.9468	1.443	1.730	1.461	1.12
0.06	0.3406	0.9754	1.476	1.750	1.414	1.69
0.08	0.3526	1.006	1.512	1.771	1.364	2.25
0.10	0.3654	1.038	1.549	1.794	1.309	2.81
0.12	0.3794	1.073	1.589	1.822	1.250	3.37
0.14	0.3943	1.111	1.633	1.854	1.184	3.93
0.16	0.4104	1.151	1.681	1.894	1.113	4.48
0.18	0.4281	1.195	1.734	1.946	1.034	5.04
0.20	0.4472	1.243	1.796	2.018	0.9452	5.59
0.22	0.4681	1.296	1.867	2.124	0.8434	6.15
0.24	0.4911	1.354	1.953	2.300	0.7242	6.70
0.26	0.5165	1.419	2.061	2.631	0.5798	7.25
0.28	0.5446	1.493	2.204	3.453	0.3965	7.79
0.30	0.5760	1.578	2.409	8.084	0.1476	8.34
d	L_5	C_4	L_3	C_2	L_1	$a_0,$ dB

courtesy P. R. Geffe

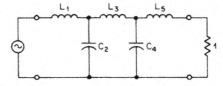

TABLE 10-6 Butterworth Uniform Dissipation Network

$n = 6$

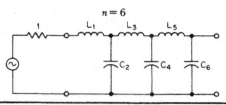

d	L_1	C_2	L_3	C_4	L_5	C_6	α_0, dB
0	0.2588	0.7579	1.202	1.553	1.759	1.533	0
0.02	0.2671	0.7804	1.232	1.581	1.727	1.502	0.671
0.04	0.2760	0.8043	1.264	1.611	1.786	1.446	1.34
0.06	0.2854	0.8297	1.297	1.643	1.802	1.386	2.01
0.08	0.2955	0.8569	1.333	1.679	1.821	1.321	2.68
0.10	0.3064	0.8860	1.372	1.714	1.844	1.250	3.35
0.12	0.3181	0.9172	1.413	1.755	1.874	1.171	4.02
0.14	0.3307	0.9508	1.458	1.802	1.917	1.083	4.69
0.16	0.3443	0.9871	1.508	1.860	1.979	0.9839	5.30
0.18	0.3594	1.027	1.558	1.923	2.080	0.8690	6.00
0.20	0.3754	1.070	1.621	2.008	2.258	0.7313	6.68
0.22	0.3931	1.117	1.690	2.122	2.646	0.5586	7.34
d	C_6	L_5	C_4	L_3	C_2	L_1	α_0, dB

courtesy P. R. Geffe

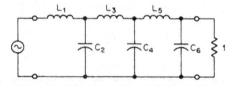

TABLE 10-7 Butterworth Uniform Dissipation Network

$n = 7$

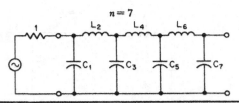

d	C_1	L_2	C_3	L_4	C_5	L_6	C_7	α_0, dB
0	0.2225	0.6560	1.054	1.397	1.659	1.799	1.588	0
0.02	0.2297	0.6759	1.084	1.428	1.684	1.808	1.496	0.781
0.04	0.2373	0.6972	1.114	1.461	1.712	1.818	1.428	1.56
0.06	0.2454	0.7198	1.146	1.496	1.742	1.832	1.354	2.34
0.08	0.2542	0.7440	1.180	1.533	1.775	1.851	1.274	3.12
0.10	0.2636	0.7699	1.217	1.573	1.813	1.878	1.184	3.90
0.12	0.2739	0.7980	1.254	1.614	1.860	1.923	1.085	4.68
0.14	0.2846	0.8281	1.294	1.659	1.910	1.992	0.9701	5.45
0.16	0.2966	0.8608	1.344	1.715	1.979	2.111	0.8350	6.23
0.18	0.3091	0.8960	1.394	1.778	2.073	2.356	0.6679	7.00
0.20	0.3232	0.9243	1.453	1.862	2.233	3.177	0.4220	7.77

d	L_7	C_6	L_5	C_4	L_3	C_2	L_1	α_0, dB

courtesy P. R. Geffe

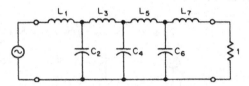

TABLE 10-8 Butterworth Uniform Dissipation Network

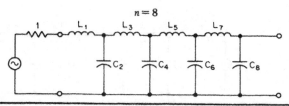

d	L_1	C_2	L_3	C_4	L_5	C_6	L_7	C_8	α_0, dB
0	0.1951	0.5776	0.9371	1.259	1.528	1.729	1.824	1.561	0
0.02	0.2014	0.5954	0.9636	1.290	1.558	1.752	1.830	1.488	0.890
0.04	0.2081	0.6144	0.9918	1.323	1.590	1.777	1.838	1.409	1.78
0.06	0.2152	0.6347	1.022	1.357	1.624	1.806	1.851	1.321	2.67
0.08	0.2229	0.6564	1.054	1.394	1.622	1.839	1.872	1.224	3.56
0.10	0.2312	0.6796	1.088	1.434	1.703	1.880	1.908	1.114	4.45
0.12	0.2400	0.7046	1.124	1.478	1.750	1.932	1.972	0.9856	5.33
0.14	0.2496	0.7316	1.164	1.526	1.804	2.003	2.101	0.8305	6.22
0.16	0.2600	0.7608	1.208	1.579	1.869	2.110	2.414	0.6307	7.10
0.18	0.2713	0.7926	1.255	1.639	1.951	2.294	3.683	0.3439	7.98
d	C_8	L_7	C_6	L_5	C_4	L_3	C_2	L_1	α_0, dB

courtesy P. R. Geffe

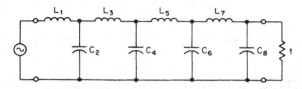

TABLE 10-9 Butterworth Uniform Dissipation Network

$$n = 9$$

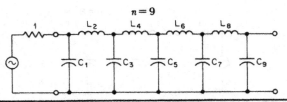

d	C_1	L_2	C_3	L_4	C_5	L_6	C_7	L_8	C_9	$\alpha_0,$ dB
0	0.1736	0.5155	0.8414	1.141	1.404	1.620	1.777	1.842	1.563	0
0.02	0.1793	0.5316	0.8659	1.171	1.435	1.649	1.798	1.845	1.480	1.00
0.04	0.1852	0.5488	0.8921	1.202	1.469	1.680	1.822	1.851	1.388	2.00
0.06	0.1916	0.5671	0.9199	1.236	1.504	1.713	1.850	1.864	1.286	3.00
0.08	0.1984	0.5867	0.9496	1.272	1.543	1.751	1.884	1.891	1.171	4.00
0.10	0.2058	0.6077	0.9814	1.311	1.584	1.794	1.931	1.942	1.036	5.00
0.12	0.2137	0.6303	1.016	1.353	1.630	1.844	1.997	2.054	0.8735	5.99
0.14	0.2223	0.6547	1.053	1.398	1.682	1.907	2.101	2.340	0.6614	6.99
0.16	0.2315	0.6812	1.093	1.448	1.742	1.991	2.293	3.620	0.3486	7.98

d	L_9	C_8	L_7	C_6	L_5	C_4	L_3	C_2	L_1	$\alpha_0,$ dB

courtesy P. R. Geffe

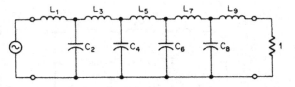

TABLE 10-10 Butterworth Uniform Dissipation Network

$n = 10$

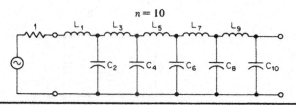

d	L_1	C_2	L_3	C_4	L_5	C_6	L_7	C_8	L_9	C_{10}	$\alpha_0,$ dB
0	0.1564	0.4654	0.7626	1.041	1.292	1.510	1.687	1.812	1.855	1.564	0
0.02	0.1614	0.4800	0.7854	1.069	1.324	1.541	1.714	1.831	1.855	1.471	1.11
0.04	0.1669	0.4956	0.8096	1.099	1.357	1.574	1.744	1.853	1.860	1.367	2.22
0.06	0.1726	0.5123	0.8353	1.132	1.392	1.610	1.777	1.882	1.875	1.249	3.33
0.08	0.1788	0.5301	0.8629	1.166	1.430	1.648	1.814	1.920	1.910	1.114	4.44
0.10	0.1854	0.5493	0.8924	1.203	1.471	1.692	1.860	1.976	1.991	0.9508	5.55
0.12	0.1926	0.5698	0.9242	1.243	1.516	1.741	1.918	2.067	2.201	0.7409	6.65
0.14	0.2003	0.5921	0.9584	1.286	1.566	1.798	1.997	2.239	3.051	0.4349	7.76
d	C_{10}	L_9	C_8	L_7	C_6	L_5	C_4	L_3	C_2	L_1	$\alpha_0,$ dB

courtesy P. R. Geffe

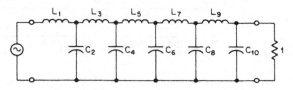

TABLE 10-11   Butterworth Uniform Dissipation Network

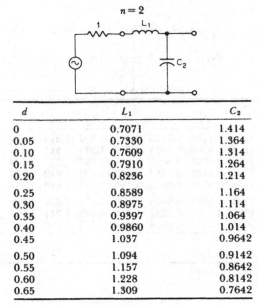

d	L_1	C_2
0	0.7071	1.414
0.05	0.7330	1.364
0.10	0.7609	1.314
0.15	0.7910	1.264
0.20	0.8236	1.214
0.25	0.8589	1.164
0.30	0.8975	1.114
0.35	0.9397	1.064
0.40	0.9860	1.014
0.45	1.037	0.9642
0.50	1.094	0.9142
0.55	1.157	0.8642
0.60	1.228	0.8142
0.65	1.309	0.7642

courtesy P. R. Geffe

TABLE 10-12 Butterworth Lossy-L Network

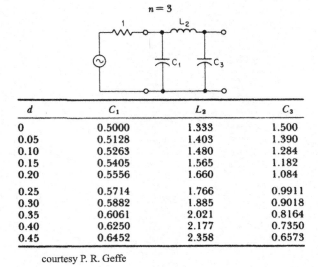

d	C_1	L_2	C_3
0	0.5000	1.333	1.500
0.05	0.5128	1.403	1.390
0.10	0.5263	1.480	1.284
0.15	0.5405	1.565	1.182
0.20	0.5556	1.660	1.084
0.25	0.5714	1.766	0.9911
0.30	0.5882	1.885	0.9018
0.35	0.6061	2.021	0.8164
0.40	0.6250	2.177	0.7350
0.45	0.6452	2.358	0.6573

courtesy P. R. Geffe

TABLE 10-13 Butterworth Lossy-L Network

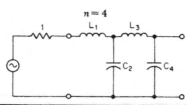

d	L_1	C_2	L_3	C_4
0	0.3827	1.082	1.577	1.531
0.05	0.3979	1.087	1.698	1.362
0.10	0.4144	1.091	1.834	1.205
0.15	0.4323	1.095	1.990	1.061
0.20	0.4518	1.098	2.170	0.9289
0.25	0.4732	1.100	2.380	0.8072
0.30	0.4967	1.102	2.628	0.6955
0.35	0.5227	1.102	2.926	0.5933

courtesy P. R. Geffe

TABLE 10-14 Butterworth Lossy-L Network

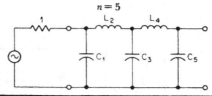

d	C_1	L_2	C_3	L_4	C_5
0	0.3090	0.8944	1.382	1.694	1.545
0.02	0.3129	0.9127	1.369	1.762	1.452
0.04	0.3168	0.9316	1.355	1.834	1.363
0.06	0.3209	0.9514	1.342	1.911	1.278
0.08	0.3251	0.9719	1.327	1.993	1.197
0.10	0.3294	0.9934	1.313	2.080	1.119
0.12	0.3338	1.016	1.298	2.173	1.046
0.14	0.3383	1.039	1.283	2.273	0.9754
0.16	0.3429	1.063	1.268	2.380	0.9086
0.18	0.3477	1.089	1.253	2.494	0.8450
0.20	0.3526	1.116	1.237	2.620	0.7844
0.22	0.3576	1.144	1.221	2.754	0.7269
0.24	0.3628	1.173	1.204	2.901	0.6721
0.26	0.3682	1.204	1.188	3.061	0.6201
0.28	0.3737	1.237	1.171	3.237	0.5076
0.30	0.3794	1.271	1.154	3.431	0.5236

courtesy P. R. Geffe

TABLE 10-15 Butterworth Lossy-L Network

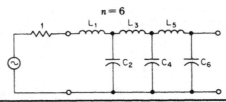

d	L_1	C_2	L_3	C_4	L_5	C_6
0	0.2588	0.7579	1.202	1.553	1.759	1.553
0.02	0.2629	0.7631	1.235	1.519	1.850	1.436
0.04	0.2671	0.7683	1.271	1.485	1.947	1.326
0.06	0.2714	0.7736	1.308	1.451	2.052	1.223
0.08	0.2760	0.7789	1.347	1.417	2.165	1.125
0.10	0.2806	0.7843	1.388	1.383	2.228	1.034
0.12	0.2854	0.7897	1.432	1.349	2.421	0.9487
0.14	0.2904	0.7952	1.478	1.315	2.565	0.8684
0.16	0.2955	0.8007	1.527	1.281	2.723	0.7932
0.18	0.3009	0.8063	1.579	1.248	2.896	0.7227
0.20	0.3064	0.8118	1.634	1.214	3.807	0.6567
0.22	0.3121	0.8174	1.692	1.181	3.298	0.5949

courtesy P. R. Geffe

TABLE 10-16   Butterworth Lossy-L Network

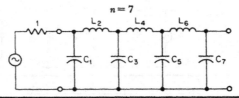

d	C_1	L_2	C_3	L_4	C_5	L_6	C_7
0	0.2225	0.6560	1.054	1.397	1.659	1.799	1.588
0.02	0.2255	0.6688	1.053	1.449	1.602	1.913	1.417
0.04	0.2286	0.6822	1.051	1.504	1.546	2.038	1.288
0.06	0.2318	0.6960	1.048	1.564	1.490	2.173	1.167
0.08	0.2351	0.7104	1.045	1.627	1.436	2.322	1.056
0.10	0.2384	0.7255	1.043	1.694	1.382	2.484	0.9532
0.12	0.2419	0.7412	1.039	1.766	1.330	2.664	0.8581
0.14	0.2454	0.7575	1.036	1.842	1.278	2.862	0.7703
0.16	0.2491	0.7746	1.032	1.924	1.228	3.083	0.6892
0.18	0.2529	0.7924	1.028	2.013	1.178	3.330	0.6144
0.20	0.2568	0.8110	1.024	2.108	1.130	3.609	0.5454

courtesy P. R. Geffe

TABLE 10-17   Butterworth Lossy-L Network

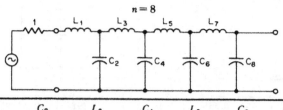

d	L_1	C_2	L_3	C_4	L_5	C_6	L_7	C_8
0	0.1951	0.5776	0.9371	1.259	1.528	1.729	1.824	1.561
0.02	0.1982	0.5829	0.9613	1.243	1.602	1.648	1.963	1.398
0.04	0.2014	0.5884	0.9868	1.227	1.680	1.569	2.116	1.249
0.06	0.2047	0.5939	1.014	1.211	1.764	1.493	2.285	1.113
0.08	0.2081	0.5996	1.042	1.194	1.856	1.419	2.472	0.9894
0.10	0.2116	0.6053	1.071	1.178	1.954	1.347	2.681	0.8768
0.12	0.2152	0.6111	1.102	1.160	2.061	1.278	2.914	0.7743
0.14	0.2190	0.6170	1.134	1.143	2.177	1.211	3.178	0.6810
0.16	0.2229	0.6231	1.169	1.124	2.302	1.147	3.477	0.5962
0.18	0.2270	0.6292	1.206	1.107	2.440	1.084	3.819	0.5191

courtesy P. R. Geffe

TABLE 10-18   Butterworth Lossy-L Network

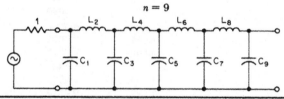

d	C_1	L_2	C_3	L_4	C_5	L_6	C_7	L_8	C_9
0	0.1736	0.5155	0.8414	1.141	1.404	1.620	1.777	1.842	1.563
0.02	0.1761	0.5253	0.8432	1.180	1.371	1.716	1.672	2.006	1.377
0.04	0.1786	0.5354	0.8450	1.221	1.338	1.821	1.571	2.189	1.211
0.06	0.1812	0.5460	0.8467	1.264	1.304	1.934	1.474	2.393	1.061
0.08	0.1839	0.5570	0.8483	1.310	1.271	2.058	1.383	2.623	0.9261
0.10	0.1866	0.5684	0.8497	1.359	1.238	2.193	1.294	2.884	0.8054
0.12	0.1894	0.5802	0.8510	1.412	1.204	2.342	1.211	3.180	0.6971
0.14	0.1924	0.5926	0.8522	1.467	1.171	2.505	1.132	3.521	0.6001
0.16	0.1954	0.6054	0.8533	1.527	1.137	2.686	1.057	3.917	0.5132

courtesy P. R. Geffe

TABLE 10-19   Butterworth Lossy-L Network

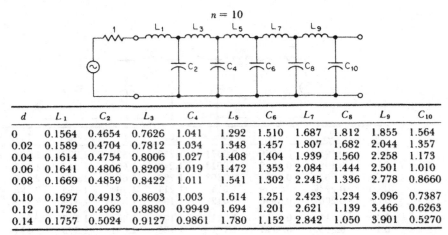

d	L_1	C_2	L_3	C_4	L_5	C_6	L_7	C_8	L_9	C_{10}
0	0.1564	0.4654	0.7626	1.041	1.292	1.510	1.687	1.812	1.855	1.564
0.02	0.1589	0.4704	0.7812	1.034	1.348	1.457	1.807	1.682	2.044	1.357
0.04	0.1614	0.4754	0.8006	1.027	1.408	1.404	1.939	1.560	2.258	1.173
0.06	0.1641	0.4806	0.8209	1.019	1.472	1.353	2.084	1.444	2.501	1.010
0.08	0.1669	0.4859	0.8422	1.011	1.541	1.302	2.245	1.336	2.778	0.8660
0.10	0.1697	0.4913	0.8603	1.003	1.614	1.251	2.423	1.234	3.096	0.7387
0.12	0.1726	0.4969	0.8880	0.9949	1.694	1.201	2.621	1.139	3.466	0.6263
0.14	0.1757	0.5024	0.9127	0.9861	1.780	1.152	2.842	1.050	3.901	0.5270

courtesy P. R. Geffe

TABLE 10-20   Butterworth Lossy-L Network

Order n	C_1	C_2	C_3
2	1.414	0.7071	
3	3.546	1.392	0.2024
4	1.082	0.9241	
	2.613	0.3825	
5	1.753	1.354	0.4214
	3.235	0.3090	
6	1.035	0.9660	
	1.414	0.7071	
	3.863	0.2588	
7	1.531	1.336	0.4885
	1.604	0.6235	
	4.493	0.2225	
8	1.020	0.9809	
	1.202	0.8313	
	1.800	0.5557	
	5.125	0.1950	
9	1.455	1.327	0.5170
	1.305	0.7661	
	2.000	0.5000	
	5.758	0.1736	
10	1.012	0.9874	
	1.122	0.8908	
	1.414	0.7071	
	2.202	0.4540	
	6.390	0.1563	

Reprinted from *Electronics*, McGraw-Hill, Inc., August 18, 1969.

TABLE 10-21   Butterworth Active Low-Pass Values

Order n	Real Part $-\alpha$	Imaginary Part $\pm j\beta$
2	0.6743	0.7075
3	0.4233	0.8663
	0.8467	
4	0.6762	0.3828
	0.2801	0.9241
5	0.5120	0.5879
	0.1956	0.9512
	0.6328	
6	0.5335	0.2588
	0.3906	0.7072
	0.1430	0.9660
7	0.4393	0.4339
	0.3040	0.7819
	0.1085	0.9750
	0.4876	
8	0.4268	0.1951
	0.3618	0.5556
	0.2418	0.8315
	0.08490	0.9808
9	0.3686	0.3420
	0.3005	0.6428
	0.1961	0.8661
	0.06812	0.9848
	0.3923	

TABLE 10-22 0.01-dB Chebyshev Pole Locations

Order n	Real Part $-\alpha$	Imaginary Part $\pm j\beta$
2	0.6104	0.7106
3	0.3490	0.8684
	0.6979	
4	0.2177	0.9254
	0.5257	0.3833
5	0.3842	0.5884
	0.1468	0.9521
	0.4749	
6	0.3916	0.2590
	0.2867	0.7077
	0.1049	0.9667
7	0.3178	0.4341
	0.2200	0.7823
	0.0785	0.9755
	0.3528	
8	0.3058	0.1952
	0.2592	0.5558
	0.1732	0.8319
	0.06082	0.9812
9	0.2622	0.3421
	0.2137	0.6430
	0.1395	0.8663
	0.04845	0.9852
	0.2790	

TABLE 10-23   0.1-dB Chebyshev Pole Locations

Order n	Real Part $-\alpha$	Imaginary Part $\pm j\beta$
2	0.5621	0.7154
3	0.3062	0.8712
	0.6124	
4	0.4501	0.3840
	0.1865	0.9272
5	0.3247	0.5892
	0.1240	0.9533
	0.4013	
6	0.3284	0.2593
	0.2404	0.7083
	0.08799	0.9675
7	0.2652	0.4344
	0.1835	0.7828
	0.06550	0.9761
	0.2944	
8	0.2543	0.1953
	0.2156	0.5561
	0.1441	0.8323
	0.05058	0.9817
9	0.2176	0.3423
	0.1774	0.6433
	0.1158	0.8667
	0.04021	0.9856
	0.2315	

TABLE **10-24** 0.25-dB Chebyshev Pole Locations

Order n	Real Part $-\alpha$	Imaginary Part $\pm j\beta$
2	0.5129	0.7225
3	0.2683	0.8753
	0.5366	
4	0.3872	0.3850
	0.1605	0.9297
5	0.2767	0.5902
	0.1057	0.9550
	0.3420	
6	0.2784	0.2596
	0.2037	0.7091
	0.07459	0.9687
7	0.2241	0.4349
	0.1550	0.7836
	0.05534	0.9771
	0.2487	
8	0.2144	0.1955
	0.1817	0.5565
	0.1214	0.8328
	0.04264	0.9824
9	0.1831	0.3425
	0.1493	0.6436
	0.09743	0.8671
	0.03383	0.9861
	0.1949	

TABLE **10-25**   0.50-dB Chebyshev Pole Locations

Order n	Real Part $-\alpha$	Imaginary Part $\pm j\beta$
2	0.4508	0.7351
3	0.2257	0.8822
	0.4513	
4	0.3199	0.3868
	0.1325	0.9339
5	0.2265	0.5918
	0.08652	0.9575
	0.2800	
6	0.2268	0.2601
	0.1660	0.7106
	0.06076	0.9707
7	0.1819	0.4354
	0.1259	0.7846
	0.04494	0.9785
	0.2019	
8	0.1737	0.1956
	0.1473	0.5571
	0.09840	0.8337
	0.03456	0.9836
9	0.1482	0.3427
	0.1208	0.6442
	0.07884	0.8679
	0.02739	0.9869
	0.1577	

TABLE 10-26   1-dB Chebyshev Pole Locations

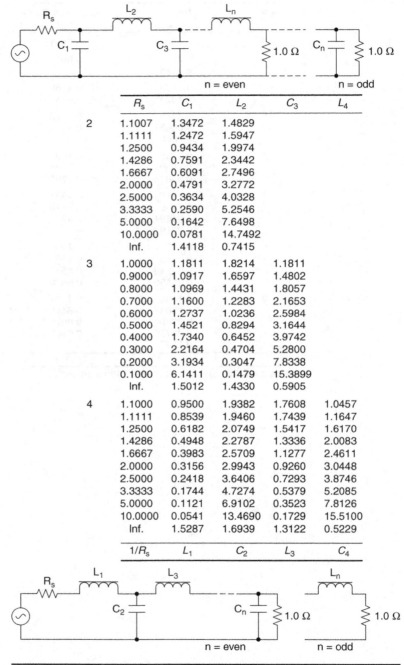

	R_s	C_1	L_2	C_3	L_4
2	1.1007	1.3472	1.4829		
	1.1111	1.2472	1.5947		
	1.2500	0.9434	1.9974		
	1.4286	0.7591	2.3442		
	1.6667	0.6091	2.7496		
	2.0000	0.4791	3.2772		
	2.5000	0.3634	4.0328		
	3.3333	0.2590	5.2546		
	5.0000	0.1642	7.6498		
	10.0000	0.0781	14.7492		
	Inf.	1.4118	0.7415		
3	1.0000	1.1811	1.8214	1.1811	
	0.9000	1.0917	1.6597	1.4802	
	0.8000	1.0969	1.4431	1.8057	
	0.7000	1.1600	1.2283	2.1653	
	0.6000	1.2737	1.0236	2.5984	
	0.5000	1.4521	0.8294	3.1644	
	0.4000	1.7340	0.6452	3.9742	
	0.3000	2.2164	0.4704	5.2800	
	0.2000	3.1934	0.3047	7.8338	
	0.1000	6.1411	0.1479	15.3899	
	Inf.	1.5012	1.4330	0.5905	
4	1.1000	0.9500	1.9382	1.7608	1.0457
	1.1111	0.8539	1.9460	1.7439	1.1647
	1.2500	0.6182	2.0749	1.5417	1.6170
	1.4286	0.4948	2.2787	1.3336	2.0083
	1.6667	0.3983	2.5709	1.1277	2.4611
	2.0000	0.3156	2.9943	0.9260	3.0448
	2.5000	0.2418	3.6406	0.7293	3.8746
	3.3333	0.1744	4.7274	0.5379	5.2085
	5.0000	0.1121	6.9102	0.3523	7.8126
	10.0000	0.0541	13.4690	0.1729	15.5100
	Inf.	1.5287	1.6939	1.3122	0.5229
	$1/R_s$	L_1	C_2	L_3	C_4

TABLE 10-27 0.01-dB Chebyshev Element Values

	R_s	C_1	L_2	C_3	L_4	C_5	L_6	C_7
5	1.0000	0.9766	1.6849	2.0366	1.6849	0.9766		
	0.9000	0.8798	1.4558	2.1738	1.6412	1.2739		
	0.8000	0.8769	1.2350	2.3785	1.4991	1.6066		
	0.7000	0.9263	1.0398	2.6582	1.3228	1.9772		
	0.6000	1.0191	0.8626	3.0408	1.1345	2.4244		
	0.5000	1.1658	0.6985	3.5835	0.9421	3.0092		
	0.4000	1.3983	0.5442	4.4027	0.7491	3.8453		
	0.3000	1.7966	0.3982	5.7721	0.5573	5.1925		
	0.2000	2.6039	0.2592	8.5140	0.3679	7.8257		
	0.1000	5.0406	0.1266	16.7406	0.1819	15.6126		
	Inf.	1.5466	1.7950	1.6449	1.2365	0.4883		
6	1.1007	0.8514	1.7956	1.8411	2.0266	1.6312	0.9372	
	1.1111	0.7597	1.7817	1.7752	2.0941	1.6380	1.0533	
	1.2500	0.5445	1.8637	1.4886	2.4025	1.5067	1.5041	
	1.4286	0.4355	2.0383	1.2655	2.7346	1.3318	1.8987	
	1.6667	0.3509	2.2978	1.0607	3.1671	1.1451	2.3568	
	2.0000	0.2786	2.6781	0.8671	3.7683	0.9536	2.9483	
	2.5000	0.2139	3.2614	0.6816	4.6673	0.7606	3.7899	
	3.3333	0.1547	4.2448	0.5028	6.1631	0.5676	5.1430	
	5.0000	0.0997	6.2227	0.3299	9.1507	0.3760	7.7852	
	10.0000	0.0483	12.1707	0.1623	18.1048	0.1865	15.5950	
	Inf.	1.5510	1.8471	1.7897	1.5976	1.1904	0.4686	
7	1.0000	0.9127	1.5947	2.0021	1.8704	2.0021	1.5947	0.9127
	0.9000	0.8157	1.3619	2.0886	1.7217	2.2017	1.5805	1.2060
	0.8000	0.8111	1.1504	2.2618	1.5252	2.4647	1.4644	1.5380
	0.7000	0.8567	0.9673	2.5158	1.3234	2.8018	1.3066	1.9096
	0.6000	0.9430	0.8025	2.8720	1.1237	3.2496	1.1310	2.3592
	0.5000	1.0799	0.6502	3.3822	0.9276	3.8750	0.9468	2.9478
	0.4000	1.2971	0.5072	4.1563	0.7350	4.8115	0.7584	3.7900
	0.3000	1.6692	0.3716	5.4540	0.5459	6.3703	0.5682	5.1476
	0.2000	2.4235	0.2423	8.0565	0.3604	9.4844	0.3776	7.8019
	0.1000	4.7006	0.1186	15.8718	0.1784	18.8179	0.1879	15.6523
	Inf.	1.5593	1.8671	1.8657	1.7651	1.5633	1.1610	0.4564
	$1/R_s$	L_1	C_2	L_3	C_4	L_5	C_6	L_7

TABLE 10-27   0.01-dB Chebyshev Element Values (*Continued*)

	R_s	C_1	L_2	C_3	L_4	C_5	L_6	C_7	L_8	C_9	L_{10}
8	1.1007	0.8145	1.7275	1.7984	2.0579	1.8695	1.9796	1.5694	0.8966		
	1.1111	0.7248	1.7081	1.7239	2.1019	1.8259	2.0595	1.5827	1.0111		
	1.2500	0.5176	1.7772	1.4315	2.3601	1.5855	2.4101	1.4754	1.4597		
	1.4286	0.4138	1.9422	1.2141	2.6686	1.3723	2.7734	1.3142	1.8544		
	1.6667	0.3336	2.1896	1.0169	3.0808	1.1660	3.2393	1.1369	2.3136		
	2.0000	0.2650	2.5533	0.8313	3.6598	0.9639	3.8820	0.9518	2.9073		
	2.5000	0.2036	3.1118	0.6537	4.5303	0.7653	4.8393	0.7627	3.7524		
	3.3333	0.1474	4.0539	0.4826	5.9828	0.5697	6.4287	0.5718	5.1118		
	5.0000	0.0951	5.9495	0.3170	8.8889	0.3770	9.6002	0.3804	7.7668		
	10.0000	0.0462	11.6509	0.1562	17.6067	0.1870	19.1009	0.1895	15.6158		
	Inf.	1.5588	1.8848	1.8988	1.8556	1.7433	1.5391	1.1412	0.4483		
9	1.0000	0.8854	1.5513	1.9614	1.8616	2.0717	1.8616	1.9614	1.5513	0.8854	
	0.9000	0.7886	1.3192	2.0330	1.6941	2.2249	1.7402	2.1774	1.5478	1.1764	
	0.8000	0.7834	1.1127	2.1959	1.4930	2.4614	1.5603	2.4565	1.4423	1.5076	
	0.7000	0.8273	0.9353	2.4404	1.2924	2.7808	1.3620	2.8093	1.2927	1.8793	
	0.6000	0.9109	0.7761	2.7852	1.0962	3.2140	1.1688	3.2747	1.1233	2.3295	
	0.5000	1.0436	0.6290	3.2805	0.9045	3.8249	0.9710	3.9223	0.9436	2.9193	
	0.4000	1.2542	0.4910	4.0329	0.7167	4.7444	0.7739	4.8900	0.7582	3.7637	
	0.3000	1.6151	0.3599	5.2951	0.5325	6.2792	0.5780	6.4989	0.5697	5.1254	
	0.2000	2.3468	0.2349	7.8274	0.3518	9.3504	0.3835	9.7114	0.3797	7.7882	
	0.1000	4.5556	0.1150	15.4334	0.1743	18.5641	0.1908	19.3382	0.1895	15.6645	
	Inf.	1.5646	1.8884	1.9242	1.8977	1.8425	1.7261	1.5217	1.1273	0.4427	
10	1.1007	0.7970	1.6930	1.7690	2.0395	1.8827	2.0724	1.8529	1.9472	1.5380	0.8773
	1.1111	0.7083	1.6714	1.6921	2.0763	1.8281	2.1308	1.8167	2.0310	1.5541	0.9910
	1.2500	0.5049	1.7353	1.4005	2.3184	1.5706	2.4371	1.5953	2.3952	1.4574	1.4381
	1.4286	0.4037	1.8958	1.1871	2.6178	1.3552	2.7830	1.3895	2.7685	1.3027	1.8327
	1.6667	0.3255	2.1375	0.9942	3.0205	1.1497	3.2370	1.1863	3.2448	1.1300	2.2923
	2.0000	0.2586	2.4932	0.8128	3.5878	0.9497	3.8698	0.9849	3.9004	0.9484	2.8867
	2.5000	0.1988	3.0398	0.6394	4.4418	0.7538	4.8173	0.7849	4.8757	0.7617	3.7333
	3.3333	0.1440	3.9619	0.4723	5.8678	0.5612	6.3951	0.5863	6.4939	0.5722	5.0955
	5.0000	0.0930	5.8175	0.3103	8.7220	0.3715	9.5486	0.3893	9.7217	0.3814	7.7563
	10.0000	0.0451	11.3993	0.1530	17.2866	0.1844	19.0046	0.1938	19.3905	0.1904	15.6234
	Inf.	1.5625	1.8978	1.9323	1.9288	1.8907	1.8309	1.7128	1.5088	1.1173	0.4386
	$1/R_s$	L_1	C_2	L_3	C_4	L_5	C_6	L_7	C_8	L_9	C_{10}

TABLE 10-27 0.01-dB Chebyshev Element Values (*Continued*)

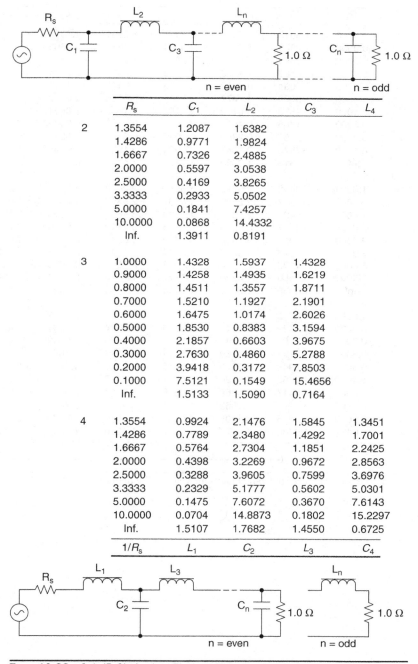

R_s	C_1	L_2	C_3	L_4
2				
1.3554	1.2087	1.6382		
1.4286	0.9771	1.9824		
1.6667	0.7326	2.4885		
2.0000	0.5597	3.0538		
2.5000	0.4169	3.8265		
3.3333	0.2933	5.0502		
5.0000	0.1841	7.4257		
10.0000	0.0868	14.4332		
Inf.	1.3911	0.8191		
3				
1.0000	1.4328	1.5937	1.4328	
0.9000	1.4258	1.4935	1.6219	
0.8000	1.4511	1.3557	1.8711	
0.7000	1.5210	1.1927	2.1901	
0.6000	1.6475	1.0174	2.6026	
0.5000	1.8530	0.8383	3.1594	
0.4000	2.1857	0.6603	3.9675	
0.3000	2.7630	0.4860	5.2788	
0.2000	3.9418	0.3172	7.8503	
0.1000	7.5121	0.1549	15.4656	
Inf.	1.5133	1.5090	0.7164	
4				
1.3554	0.9924	2.1476	1.5845	1.3451
1.4286	0.7789	2.3480	1.4292	1.7001
1.6667	0.5764	2.7304	1.1851	2.2425
2.0000	0.4398	3.2269	0.9672	2.8563
2.5000	0.3288	3.9605	0.7599	3.6976
3.3333	0.2329	5.1777	0.5602	5.0301
5.0000	0.1475	7.6072	0.3670	7.6143
10.0000	0.0704	14.8873	0.1802	15.2297
Inf.	1.5107	1.7682	1.4550	0.6725
$1/R_s$	L_1	C_2	L_3	C_4

TABLE 10-28   0.1-dB Chebyshev Element Values

	R_s	C_1	L_2	C_3	L_4	C_5	L_6	C_7
5	1.0000	1.3013	1.5559	2.2411	1.5559	1.3013		
	0.9000	1.2845	1.4329	2.3794	1.4878	1.4883		
	0.8000	1.2998	1.2824	2.5819	1.3815	1.7384		
	0.7000	1.3580	1.1170	2.8679	1.2437	2.0621		
	0.6000	1.4694	0.9469	3.2688	1.0846	2.4835		
	0.5000	1.6535	0.7777	3.8446	0.9126	3.0548		
	0.4000	1.9538	0.6119	4.7193	0.7333	3.8861		
	0.3000	2.4765	0.4509	6.1861	0.5503	5.2373		
	0.2000	3.5457	0.2950	9.1272	0.3659	7.8890		
	0.1000	6.7870	0.1447	17.9569	0.1820	15.7447		
	Inf.	1.5613	1.8069	1.7659	1.4173	0.6507		
6	1.3554	0.9419	2.0797	1.6581	2.2473	1.5344	1.2767	
	1.4286	0.7347	2.2492	1.4537	2.5437	1.4051	1.6293	
	1.6667	0.5422	2.6003	1.1830	3.0641	1.1850	2.1739	
	2.0000	0.4137	3.0679	0.9575	3.7119	0.9794	2.7936	
	2.5000	0.3095	3.7652	0.7492	4.6512	0.7781	3.6453	
	3.3333	0.2195	4.9266	0.5514	6.1947	0.5795	4.9962	
	5.0000	0.1393	7.2500	0.3613	9.2605	0.3835	7.6184	
	10.0000	0.0666	14.2200	0.1777	18.4267	0.1901	15.3495	
	Inf.	1.5339	1.8838	1.8306	1.7485	1.3937	0.6383	
7	1.0000	1.2615	1.5196	2.2392	1.6804	2.2392	1.5196	1.2615
	0.9000	1.2422	1.3946	2.3613	1.5784	2.3966	1.4593	1.4472
	0.8000	1.2550	1.2449	2.5481	1.4430	2.6242	1.3619	1.6967
	0.6000	1.4170	0.9169	3.2052	1.1092	3.3841	1.0807	2.4437
	0.5000	1.5948	0.7529	3.7642	0.9276	4.0150	0.9142	3.0182
	0.4000	1.8853	0.5926	4.6179	0.7423	4.9702	0.7384	3.8552
	0.3000	2.3917	0.4369	6.0535	0.5557	6.5685	0.5569	5.2167
	0.2000	3.4278	0.2862	8.9371	0.3692	9.7697	0.3723	7.8901
	0.1000	6.5695	0.1405	17.6031	0.1838	19.3760	0.1862	15.8127
	Inf.	1.5748	1.8577	1.9210	1.8270	1.7340	1.3786	0.6307
	$1/R_s$	L_1	C_2	L_3	C_4	L_5	C_6	L_7

TABLE 10-28   0.1-dB Chebyshev Element Values (*Continued*)

	R_s	C_1	L_2	C_3	L_4	C_5	L_6	C_7	L_8	C_9	L_{10}
8	1.3554	0.9234	2.0454	1.6453	2.2826	1.6841	2.2300	1.5091	1.2515		
	1.4286	0.7186	2.2054	1.4350	2.5554	1.4974	2.5422	1.3882	1.6029		
	1.6667	0.5298	2.5459	1.1644	3.0567	1.2367	3.0869	1.1769	2.1477		
	2.0000	0.4042	3.0029	0.9415	3.6917	1.0118	3.7619	0.9767	2.7690		
	2.5000	0.3025	3.6859	0.7365	4.6191	0.7990	4.7388	0.7787	3.6240		
	3.3333	0.2147	4.8250	0.5421	6.1483	0.5930	6.3423	0.5820	4.9811		
	5.0000	0.1364	7.1050	0.3554	9.1917	0.3917	9.5260	0.3863	7.6164		
	10.0000	0.0652	13.9469	0.1749	18.3007	0.1942	19.0437	0.1922	15.3880		
	Inf.	1.5422	1.9106	1.9008	1.9252	1.8200	1.7231	1.3683	0.6258		
9	1.0000	1.2446	1.5017	2.2220	1.6829	2.2957	1.6829	2.2220	1.5017	1.2446	
	0.9000	1.2244	1.3765	2.3388	1.5756	2.4400	1.5870	2.3835	1.4444	1.4297	
	0.8000	1.2361	1.2276	2.5201	1.4365	2.6561	1.4572	2.6168	1.3505	1.6788	
	0.7000	1.2898	1.0670	2.7856	1.2751	2.9647	1.3019	2.9422	1.2248	2.0029	
	0.6000	1.3950	0.9035	3.1653	1.1008	3.3992	1.1304	3.3937	1.0761	2.4264	
	0.5000	1.5701	0.7419	3.7166	0.9198	4.0244	0.9494	4.0377	0.9121	3.0020	
	0.4000	1.8566	0.5840	4.5594	0.7359	4.9750	0.7630	5.0118	0.7382	3.8412	
	0.3000	2.3560	0.4307	5.9781	0.5509	6.5700	0.5736	6.6413	0.5579	5.2068	
	0.2000	3.3781	0.2822	8.8291	0.3661	9.7699	0.3827	9.9047	0.3737	7.8891	
	0.1000	6.4777	0.1386	17.3994	0.1823	19.3816	0.1912	19.6976	0.1873	15.8393	
	Inf.	1.5804	1.8727	1.9584	1.9094	1.9229	1.8136	1.7150	1.3611	0.6223	
10	1.3554	0.9146	2.0279	1.6346	2.2777	1.6963	2.2991	1.6805	2.2155	1.4962	1.2397
	1.4286	0.7110	2.1837	1.4231	2.5425	1.5002	2.5915	1.5000	2.5322	1.3789	1.5903
	1.6667	0.5240	2.5194	1.1536	3.0362	1.2349	3.1229	1.2444	3.0839	1.1717	2.1351
	2.0000	0.3998	2.9713	0.9326	3.6647	1.0089	3.7923	1.0214	3.7669	0.9741	2.7572
	2.5000	0.2993	3.6476	0.7295	4.5843	0.7962	4.7673	0.8090	4.7547	0.7779	3.6136
	3.3333	0.2124	4.7758	0.5370	6.1022	0.5907	6.3734	0.6020	6.3758	0.5822	4.9735
	5.0000	0.1350	7.0347	0.3522	9.1248	0.3902	9.5681	0.3987	9.5942	0.3871	7.6148
	10.0000	0.0646	13.8141	0.1734	18.1739	0.1935	19.1282	0.1981	19.2158	0.1929	15.4052
	Inf.	1.5460	1.9201	1.9216	1.9700	1.9102	1.9194	1.8083	1.7090	1.3559	0.6198
	$1/R_s$	L_1	C_2	L_3	C_4	L_5	C_6	L_7	C_8	L_9	C_{10}

TABLE 10-28 0.1-dB Chebyshev Element Values (*Continued*)

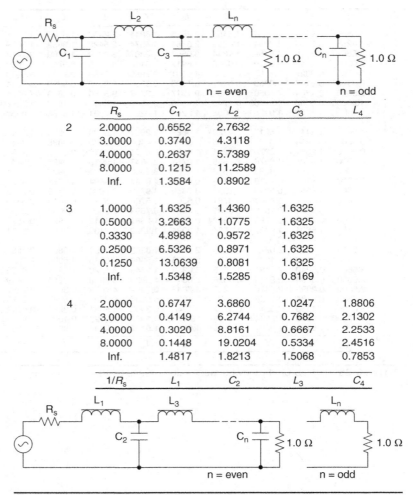

	R_s	C_1	L_2	C_3	L_4
2	2.0000	0.6552	2.7632		
	3.0000	0.3740	4.3118		
	4.0000	0.2637	5.7389		
	8.0000	0.1215	11.2589		
	Inf.	1.3584	0.8902		
3	1.0000	1.6325	1.4360	1.6325	
	0.5000	3.2663	1.0775	1.6325	
	0.3330	4.8988	0.9572	1.6325	
	0.2500	6.5326	0.8971	1.6325	
	0.1250	13.0639	0.8081	1.6325	
	Inf.	1.5348	1.5285	0.8169	
4	2.0000	0.6747	3.6860	1.0247	1.8806
	3.0000	0.4149	6.2744	0.7682	2.1302
	4.0000	0.3020	8.8161	0.6667	2.2533
	8.0000	0.1448	19.0204	0.5334	2.4516
	Inf.	1.4817	1.8213	1.5068	0.7853
	$1/R_s$	L_1	C_2	L_3	C_4

TABLE 10-29   0.25-dB Chebyshev Element Values

	R_s	C_1	L_2	C_3	L_4	C_5	L_6	C_7
5	1.0000	1.5046	1.4436	2.4050	1.4436	1.5046		
	0.5000	3.0103	0.7218	3.6080	1.4436	1.5046		
	0.3330	4.5149	0.4812	4.8100	1.4436	1.5046		
	0.2500	6.0196	0.3615	6.0130	1.4436	1.5046		
	0.1250	12.0402	0.1807	10.8230	1.4436	1.5046		
	Inf.	1.5765	1.7822	1.8225	1.4741	0.7523		
6	2.0000	0.6867	3.2074	0.9308	3.8102	1.2163	1.7088	
	3.0000	0.4330	5.0976	0.5392	6.0963	1.0804	1.8393	
	4.0000	0.3173	6.9486	0.3821	8.2530	1.0221	1.8987	
	8.0000	0.1539	14.3100	0.1762	16.7193	0.9393	1.9868	
	Inf.	1.5060	1.9221	1.8191	1.8329	1.4721	0.7610	
7	1.0000	1.5120	1.4169	2.4535	1.5350	2.4535	1.4169	1.5120
	0.5000	3.0240	0.7085	4.9069	1.1515	2.4535	1.4169	1.5120
	0.3330	4.5361	0.4723	7.3596	1.0230	2.4535	1.4169	1.5120
	0.2500	6.0471	0.3542	9.8120	0.9593	2.4535	1.4169	1.5120
	0.1250	12.0952	0.1776	19.6251	0.8631	2.4535	1.4169	1.5120
	Inf.	1.6009	1.8287	1.9666	1.8234	1.8266	1.4629	0.7555
	$1/R_s$	L_1	C_2	L_3	C_4	L_5	C_6	L_7

TABLE 10-29   0.25-dB Chebyshev Element Values (*Continued*)

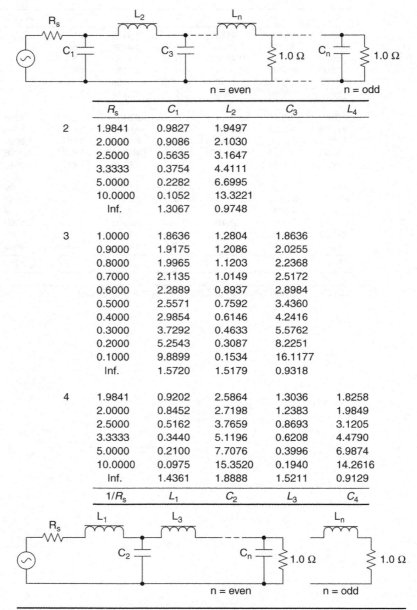

	R_s	C_1	L_2	C_3	L_4
2	1.9841	0.9827	1.9497		
	2.0000	0.9086	2.1030		
	2.5000	0.5635	3.1647		
	3.3333	0.3754	4.4111		
	5.0000	0.2282	6.6995		
	10.0000	0.1052	13.3221		
	Inf.	1.3067	0.9748		
3	1.0000	1.8636	1.2804	1.8636	
	0.9000	1.9175	1.2086	2.0255	
	0.8000	1.9965	1.1203	2.2368	
	0.7000	2.1135	1.0149	2.5172	
	0.6000	2.2889	0.8937	2.8984	
	0.5000	2.5571	0.7592	3.4360	
	0.4000	2.9854	0.6146	4.2416	
	0.3000	3.7292	0.4633	5.5762	
	0.2000	5.2543	0.3087	8.2251	
	0.1000	9.8899	0.1534	16.1177	
	Inf.	1.5720	1.5179	0.9318	
4	1.9841	0.9202	2.5864	1.3036	1.8258
	2.0000	0.8452	2.7198	1.2383	1.9849
	2.5000	0.5162	3.7659	0.8693	3.1205
	3.3333	0.3440	5.1196	0.6208	4.4790
	5.0000	0.2100	7.7076	0.3996	6.9874
	10.0000	0.0975	15.3520	0.1940	14.2616
	Inf.	1.4361	1.8888	1.5211	0.9129
	$1/R_s$	L_1	C_2	L_3	C_4

TABLE 10-30 0.50-dB Chebyshev Element Values

	R_s	C_1	L_2	C_3	L_4	C_5	L_6
5	1.0000	1.8068	1.3025	2.6914	1.3025	1.8068	
	0.9000	1.8540	1.2220	2.8478	1.2379	1.9701	
	0.8000	1.9257	1.1261	3.0599	1.1569	2.1845	
	0.7000	2.0347	1.0150	3.3525	1.0582	2.4704	
	0.6000	2.2006	0.8901	3.7651	0.9420	2.8609	
	0.5000	2.4571	0.7537	4.3672	0.8098	3.4137	
	0.4000	2.8692	0.6091	5.2960	0.6640	4.2447	
	0.3000	3.5877	0.4590	6.8714	0.5075	5.6245	
	0.2000	5.0639	0.3060	10.0537	0.3430	8.3674	
	0.1000	9.5560	0.1525	19.6465	0.1731	16.5474	
	Inf.	1.6299	1.7400	1.9217	1.5138	0.9034	
6	1.9841	0.9053	2.5774	1.3675	2.7133	1.2991	1.7961
	2.0000	0.8303	2.7042	1.2912	2.8721	1.2372	1.9557
	2.5000	0.5056	3.7219	0.8900	4.1092	0.8808	3.1025
	3.3333	0.3370	5.0554	0.6323	5.6994	0.6348	4.4810
	5.0000	0.2059	7.6145	0.4063	8.7319	0.4121	7.0310
	10.0000	0.0958	15.1862	0.1974	17.6806	0.2017	14.4328
	Inf.	1.4618	1.9799	1.7803	1.9253	1.5077	0.8981
	$1/R_s$	L_1	C_2	L_3	C_4	L_5	C_6

TABLE 10-30   0.50-dB Chebyshev Element Values (*Continued*)

	R_s	C_1	L_2	C_3	L_4	C_5	L_6	C_7	L_8	C_9	L_{10}
7	1.0000	1.7896	1.2961	2.7177	1.3848	2.7177	1.2961	1.7896			
	0.9000	1.8348	1.2146	2.8691	1.3080	2.8829	1.2335	1.9531			
	0.8000	1.9045	1.1182	3.0761	1.2149	3.1071	1.1546	2.1681			
	0.7000	2.0112	1.0070	3.3638	1.1050	3.4163	1.0582	2.4554			
	0.6000	2.1744	0.8824	3.7717	0.9786	3.8524	0.9441	2.8481			
	0.5000	2.4275	0.7470	4.3695	0.8377	4.4886	0.8137	3.4050			
	0.4000	2.8348	0.6035	5.2947	0.6846	5.4698	0.6690	4.2428			
	0.3000	3.5456	0.4548	6.8674	0.5221	7.1341	0.5129	5.6350			
	0.2000	5.0070	0.3034	10.0491	0.3524	10.4959	0.3478	8.4041			
	0.1000	9.4555	0.1513	19.6486	0.1778	20.6314	0.1761	16.6654			
	Inf.	1.6464	1.7772	2.0306	1.7892	1.9239	1.5034	0.8948			
8	1.9841	0.8998	2.5670	1.3697	2.7585	1.3903	2.7175	1.2938	1.7852		
	2.0000	0.8249	2.6916	1.2919	2.9134	1.3160	2.8800	1.2331	1.9449		
	2.5000	0.5017	3.6988	0.8878	4.1404	0.9184	4.1470	0.8815	3.0953		
	3.3333	0.3344	5.0234	0.6304	5.7323	0.6577	5.7761	0.6370	4.4807		
	5.0000	0.2044	7.5682	0.4052	8.7771	0.4257	8.8833	0.4146	7.0453		
	10.0000	0.0951	15.1014	0.1969	17.7747	0.2081	18.0544	0.2035	14.4924		
	Inf.	1.4710	2.0022	1.8248	2.0440	1.7911	1.9218	1.5003	0.8926		
9	1.0000	1.7822	1.2921	2.7162	1.3922	2.7734	1.3922	2.7162	1.2921	1.7822	
	0.9000	1.8267	1.2103	2.8658	1.3135	2.9353	1.3165	2.8834	1.2302	1.9458	
	0.8000	1.8955	1.1139	3.0709	1.2189	3.1565	1.2246	3.1102	1.1523	2.1611	
	0.7000	2.0013	1.0028	3.3565	1.1075	3.4635	1.1157	3.4232	1.0568	2.4489	
	0.6000	2.1634	0.8786	3.7621	0.9801	3.8985	0.9900	3.8647	0.9436	2.8426	
	0.5000	2.4150	0.7436	4.3573	0.8385	4.5355	0.8493	4.5087	0.8140	3.4010	
	0.4000	2.8203	0.6008	5.2792	0.6850	5.5207	0.6957	5.5023	0.6700	4.2416	
	0.3000	3.5279	0.4528	6.8474	0.5223	7.1951	0.5318	7.1876	0.5142	5.6390	
	0.2000	4.9830	0.3021	10.0212	0.3526	10.5818	0.3600	10.5925	0.3491	8.4189	
	0.1000	9.4131	0.1507	19.5995	0.1779	20.8006	0.1822	20.8588	0.1770	16.7140	
	Inf.	1.6533	1.7890	2.0570	1.8383	2.0481	1.7910	1.9199	1.4981	0.8911	
10	1.9841	0.8972	2.5610	1.3683	2.7631	1.4009	2.7795	1.3927	2.7148	1.2908	1.7801
	2.0000	0.8223	2.6845	1.2901	2.9166	1.3246	2.9390	1.3191	2.8783	1.2306	1.9398
	2.5000	0.4999	3.6868	0.8858	4.1383	0.9216	4.2020	0.9238	4.1540	0.8812	3.0919
	3.3333	0.3332	5.0071	0.6289	5.7274	0.6594	5.8399	0.6631	5.7948	0.6376	4.4804
	5.0000	0.2037	7.5446	0.4042	8.7695	0.4266	8.9727	0.4300	8.9249	0.4154	7.0518
	10.0000	0.0948	15.0578	0.1965	17.7624	0.2086	18.2313	0.2107	18.1644	0.2041	14.5199
	Inf.	1.4753	2.0107	1.8386	2.0733	1.8432	2.0494	1.7904	1.9183	1.4965	0.8900
	$1/R_s$	L_1	C_2	L_3	C_4	L_5	C_6	L_7	C_8	L_9	C_{10}

TABLE 10-30   0.50-dB Chebyshev Element Values (*Continued*)

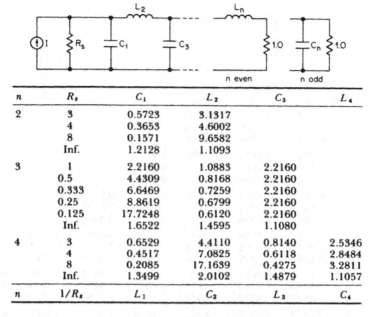

n	R_s	C_1	L_2	C_3	L_4
2	3	0.5723	3.1317		
	4	0.3653	4.6002		
	8	0.1571	9.6582		
	Inf.	1.2128	1.1093		
3	1	2.2160	1.0883	2.2160	
	0.5	4.4309	0.8168	2.2160	
	0.333	6.6469	0.7259	2.2160	
	0.25	8.8619	0.6799	2.2160	
	0.125	17.7248	0.6120	2.2160	
	Inf.	1.6522	1.4595	1.1080	
4	3	0.6529	4.4110	0.8140	2.5346
	4	0.4517	7.0825	0.6118	2.8484
	8	0.2085	17.1639	0.4275	3.2811
	Inf.	1.3499	2.0102	1.4879	1.1057
n	$1/R_s$	L_1	C_2	L_3	C_4

n	R_s	C_1	L_2	C_3	L_4	C_5	L_6	C_7
5	1	2.2072	1.1279	3.1025	1.1279	2.2072		
	0.5	4.4144	0.5645	4.6532	1.1279	2.2072		
	0.333	6.6216	0.3763	6.2050	1.1279	2.2072		
	0.25	8.8288	0.2822	7.7557	1.1279	2.2072		
	0.125	17.65565	0.1406	13.9606	1.1279	2.2072		
	Inf.	1.7213	1.6448	2.0614	1.4928	1.1031		
6	3	0.6785	3.8725	0.7706	4.7107	0.9692	2.4060	
	4	0.4810	5.6441	0.4759	7.3511	0.8494	2.5820	
	8	0.2272	12.3095	0.1975	16.740	0.7256	2.7990	
	Inf.	1.3775	2.0969	1.6896	2.0744	1.4942	1.1022	
7	1	2.2043	1.1311	3.1472	1.1942	3.1472	1.1311	2.2043
	0.5	4.4075	0.5656	6.2934	0.8951	3.1472	1.1311	2.2043
	0.333	6.6118	0.3774	9.4406	0.7955	3.1472	1.1311	2.2043
	0.25	8.8151	0.2828	12.5879	0.7466	3.1472	1.1311	2.2043
	0.125	17.6311	0.1414	25.175	0.6714	3.1472	1.1311	2.2043
	Inf.	1.7414	1.6774	2.1554	1.7028	2.0792	1.4943	1.1016
n	$1/R_s$	L_1	C_2	L_3	C_4	L_5	C_6	L_7

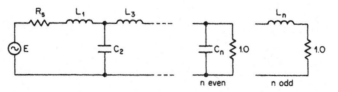

TABLE 10-31   1-dB Chebyshev *LC* Element Values

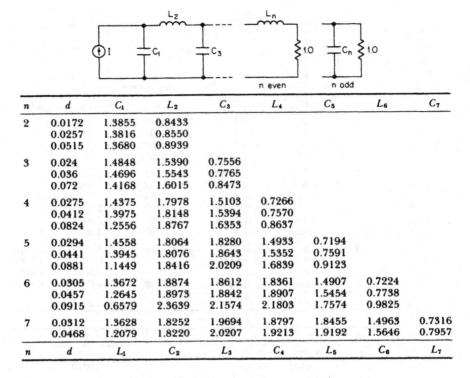

n	d	C_1	L_2	C_3	L_4	C_5	L_6	C_7
2	0.0172	1.3855	0.8433					
	0.0257	1.3816	0.8550					
	0.0515	1.3680	0.8939					
3	0.024	1.4848	1.5390	0.7556				
	0.036	1.4696	1.5543	0.7765				
	0.072	1.4168	1.6015	0.8473				
4	0.0275	1.4375	1.7978	1.5103	0.7266			
	0.0412	1.3975	1.8148	1.5394	0.7570			
	0.0824	1.2556	1.8767	1.6353	0.8637			
5	0.0294	1.4558	1.8064	1.8280	1.4933	0.7194		
	0.0441	1.3945	1.8076	1.8643	1.5352	0.7591		
	0.0881	1.1449	1.8416	2.0209	1.6839	0.9123		
6	0.0305	1.3672	1.8874	1.8612	1.8361	1.4907	0.7224	
	0.0457	1.2645	1.8973	1.8842	1.8907	1.5454	0.7738	
	0.0915	0.6579	2.3639	2.1574	2.1803	1.7574	0.9825	
7	0.0312	1.3628	1.8252	1.9694	1.8797	1.8455	1.4963	0.7316
	0.0468	1.2079	1.8220	2.0207	1.9213	1.9192	1.5646	0.7957

n	d	L_1	C_2	L_3	C_4	L_5	C_6	L_7

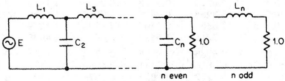

TABLE 10-32 0.1-dB Chebyshev Uniform Dissipation Network

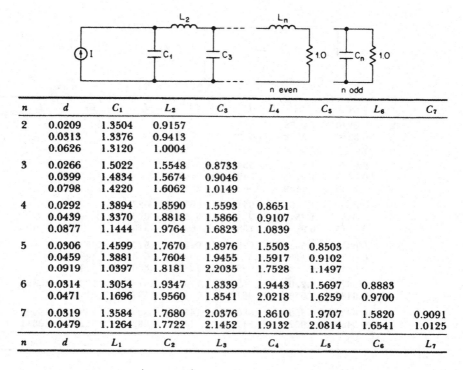

n	d	C_1	L_2	C_3	L_4	C_5	L_6	C_7
2	0.0209	1.3504	0.9157					
	0.0313	1.3376	0.9413					
	0.0626	1.3120	1.0004					
3	0.0266	1.5022	1.5548	0.8733				
	0.0399	1.4834	1.5674	0.9046				
	0.0798	1.4220	1.6062	1.0149				
4	0.0292	1.3894	1.8590	1.5593	0.8651			
	0.0439	1.3370	1.8818	1.5866	0.9107			
	0.0877	1.1444	1.9764	1.6823	1.0839			
5	0.0306	1.4599	1.7670	1.8976	1.5503	0.8503		
	0.0459	1.3881	1.7604	1.9455	1.5917	0.9102		
	0.0919	1.0397	1.8181	2.2035	1.7528	1.1497		
6	0.0314	1.3054	1.9347	1.8339	1.9443	1.5697	0.8883	
	0.0471	1.1696	1.9560	1.8541	2.0218	1.6259	0.9700	
7	0.0319	1.3584	1.7680	2.0376	1.8610	1.9707	1.5820	0.9091
	0.0479	1.1264	1.7722	2.1452	1.9132	2.0814	1.6541	1.0125
n	d	L_1	C_2	L_3	C_4	L_5	C_6	L_7

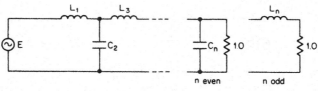

TABLE 10-33 0.25-dB Chebyshev Uniform Dissipation Network

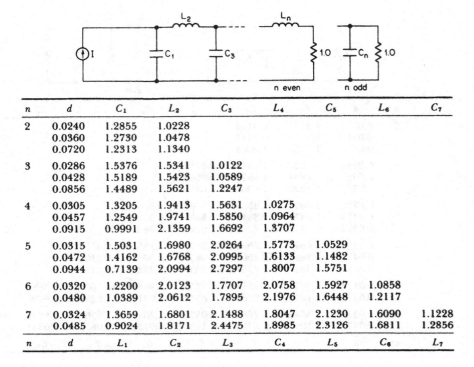

n	d	C_1	L_2	C_3	L_4	C_5	L_6	C_7
2	0.0240	1.2855	1.0228					
	0.0360	1.2730	1.0478					
	0.0720	1.2313	1.1340					
3	0.0286	1.5376	1.5341	1.0122				
	0.0428	1.5189	1.5423	1.0589				
	0.0856	1.4489	1.5621	1.2247				
4	0.0305	1.3205	1.9413	1.5631	1.0275			
	0.0457	1.2549	1.9741	1.5850	1.0964			
	0.0915	0.9991	2.1359	1.6692	1.3707			
5	0.0315	1.5031	1.6980	2.0264	1.5773	1.0529		
	0.0472	1.4162	1.6768	2.0995	1.6133	1.1482		
	0.0944	0.7139	2.0994	2.7297	1.8007	1.5751		
6	0.0320	1.2200	2.0123	1.7707	2.0758	1.5927	1.0858	
	0.0480	1.0389	2.0612	1.7895	2.1976	1.6448	1.2117	
7	0.0324	1.3659	1.6801	2.1488	1.8047	2.1230	1.6090	1.1228
	0.0485	0.9024	1.8171	2.4475	1.8985	2.3126	1.6811	1.2856
n	d	L_1	C_2	L_3	C_4	L_5	C_6	L_7

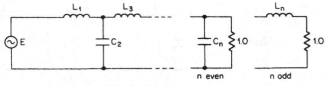

TABLE 10-34 0.50-dB Chebyshev Uniform Dissipation Network

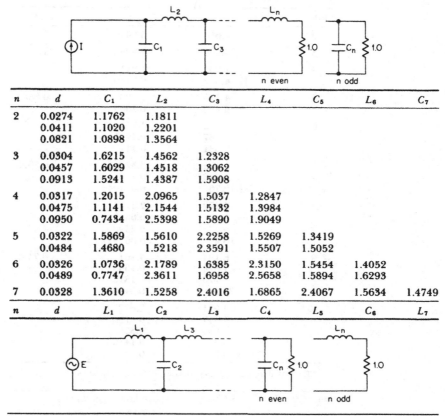

n	d	C_1	L_2	C_3	L_4	C_5	L_6	C_7
2	0.0274	1.1762	1.1811					
	0.0411	1.1020	1.2201					
	0.0821	1.0898	1.3564					
3	0.0304	1.6215	1.4562	1.2328				
	0.0457	1.6029	1.4518	1.3062				
	0.0913	1.5241	1.4387	1.5908				
4	0.0317	1.2015	2.0965	1.5037	1.2847			
	0.0475	1.1141	2.1544	1.5132	1.3984			
	0.0950	0.7434	2.5398	1.5890	1.9049			
5	0.0322	1.5869	1.5610	2.2258	1.5269	1.3419		
	0.0484	1.4680	1.5218	2.3591	1.5507	1.5052		
6	0.0326	1.0736	2.1789	1.6385	2.3150	1.5454	1.4052	
	0.0489	0.7747	2.3611	1.6958	2.5658	1.5894	1.6293	
7	0.0328	1.3610	1.5258	2.4016	1.6865	2.4067	1.5634	1.4749
n	d	L_1	C_2	L_3	C_4	L_5	C_6	L_7

TABLE 10-35   1-dB Chebyshev Uniform Dissipation Network

Order n	C_1	C_2
2	1.4826	0.7042
4	1.4874	1.1228
	3.5920	0.2985
6	1.8900	1.5249
	2.5820	0.5953
	7.0522	0.1486
8	2.3652	1.9493
	2.7894	0.8196
	4.1754	0.3197
	11.8920	0.08672

TABLE 10-36   0.01-dB Chebyshev Active
Low-Pass Values

Order n	C_1	C_2	C_3
2	1.638	0.6955	
3	6.653	1.825	0.1345
4	1.900	1.241	
	4.592	0.2410	
5	4.446	2.520	0.3804
	6.810	0.1580	
6	2.553	1.776	
	3.487	0.4917	
	9.531	0.1110	
7	5.175	3.322	0.5693
	4.546	0.3331	
	12.73	0.08194	
8	3.270	2.323	
	3.857	0.6890	
	5.773	0.2398	
	16.44	0.06292	
9	6.194	4.161	0.7483
	4.678	0.4655	
	7.170	0.1812	
	20.64	0.04980	
10	4.011	2.877	
	4.447	0.8756	
	5.603	0.3353	
	8.727	0.1419	
	25.32	0.04037	

* Reprinted from *Electronics,* McGraw-Hill, Inc., August 18, 1969.

TABLE 10-37   0.1-dB Chebyshev Active Low-Pass Values

Order n	C_1	C_2	C_3
2	1.778	0.6789	
3	8.551	2.018	0.1109
4	2.221	1.285	
	5.363	0.2084	
5	5.543	2.898	0.3425
	8.061	0.1341	
6	3.044	1.875	
	4.159	0.4296	
	11.36	0.09323	
7	6.471	3.876	0.5223
	5.448	0.2839	
	15.26	0.06844	
8	3.932	2.474	
	4.638	0.6062	
	6.942	0.2019	
	19.76	0.05234	
9	7.766	4.891	0.6919
	5.637	0.3983	
	8.639	0.1514	
	24.87	0.04131	
10	4.843	3.075	
	5.368	0.7725	
	6.766	0.2830	
	10.53	0.1181	
	30.57	0.03344	

* Reprinted from *Electronics,* McGraw-Hill, Inc., August 18, 1969.

TABLE **10-38**   0.25-dB Chebyshev Active Low-Pass Values

Order n	C_1	C_2	C_3
2	1.950	0.6533	
3	11.23	2.250	0.0895
4	2.582	1.300	
	6.233	0.1802	
5	6.842	3.317	0.3033
	9.462	0.1144	
6	3.592	1.921	
	4.907	0.3743	
	13.40	0.07902	
7	7.973	4.483	0.4700
	6.446	0.2429	
	18.07	0.05778	
8	4.665	2.547	
	5.502	0.5303	
	8.237	0.1714	
	23.45	0.04409	
9	9.563	5.680	0.6260
	6.697	0.3419	
	10.26	0.1279	
	29.54	0.03475	
10	5.760	3.175	
	6.383	0.6773	
	8.048	0.2406	
	12.53	0.09952	
	36.36	0.02810	

* Reprinted from *Electronics,* McGraw-Hill, Inc., August 18, 1969.

TABLE 10-39 0.50-dB Chebyshev Active Low-Pass Values

Order n	C_1	C_2	C_3
2	2.218	0.6061	
3	16.18	2.567	0.06428
4	3.125	1.269	
	7.546	0.1489	
5	8.884	3.935	0.2540
	11.55	0.09355	
6	4.410	1.904	
	6.024	0.3117	
	16.46	0.06425	
7	10.29	5.382	0.4012
	7.941	0.1993	
	22.25	0.04684	
8	5.756	2.538	
	6.792	0.4435	
	10.15	0.1395	
	28.94	0.03568	
9	12.33	6.853	0.5382
	8.281	0.2813	
	12.68	0.1038	
	36.51	0.02808	
10	7.125	3.170	
	7.897	0.5630	
	9.952	0.1962	
	15.50	0.08054	
	44.98	0.02269	

* Reprinted from *Electronics,* McGraw-Hill, Inc., August 18, 1969.

TABLE **10-40** 1-dB Chebyshev Active Low-Pass Values

Order n	Real Part $-\alpha$	Imaginary Part $\pm j\beta$
2	1.1030	0.6368
3	1.0509	1.0025
	1.3270	
4	1.3596	0.4071
	0.9877	1.2476
5	1.3851	0.7201
	0.9606	1.4756
	1.5069	
6	1.5735	0.3213
	1.3836	0.9727
	0.9318	1.6640
7	1.6130	0.5896
	1.3797	1.1923
	0.9104	1.8375
	1.6853	
8	1.7627	0.2737
	0.8955	2.0044
	1.3780	1.3926
	1.6419	0.8253
9	1.8081	0.5126
	1.6532	1.0319
	1.3683	1.5685
	0.8788	2.1509
	1.8575	

TABLE 10-41 Bessel Pole Locations

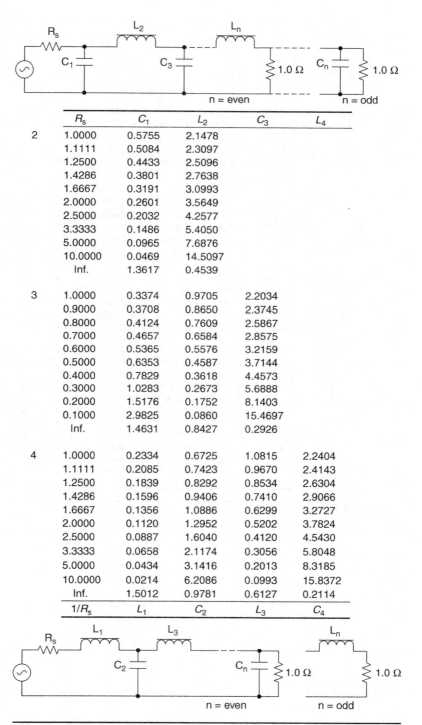

	R_s	C_1	L_2	C_3	L_4
2	1.0000	0.5755	2.1478		
	1.1111	0.5084	2.3097		
	1.2500	0.4433	2.5096		
	1.4286	0.3801	2.7638		
	1.6667	0.3191	3.0993		
	2.0000	0.2601	3.5649		
	2.5000	0.2032	4.2577		
	3.3333	0.1486	5.4050		
	5.0000	0.0965	7.6876		
	10.0000	0.0469	14.5097		
	Inf.	1.3617	0.4539		
3	1.0000	0.3374	0.9705	2.2034	
	0.9000	0.3708	0.8650	2.3745	
	0.8000	0.4124	0.7609	2.5867	
	0.7000	0.4657	0.6584	2.8575	
	0.6000	0.5365	0.5576	3.2159	
	0.5000	0.6353	0.4587	3.7144	
	0.4000	0.7829	0.3618	4.4573	
	0.3000	1.0283	0.2673	5.6888	
	0.2000	1.5176	0.1752	8.1403	
	0.1000	2.9825	0.0860	15.4697	
	Inf.	1.4631	0.8427	0.2926	
4	1.0000	0.2334	0.6725	1.0815	2.2404
	1.1111	0.2085	0.7423	0.9670	2.4143
	1.2500	0.1839	0.8292	0.8534	2.6304
	1.4286	0.1596	0.9406	0.7410	2.9066
	1.6667	0.1356	1.0886	0.6299	3.2727
	2.0000	0.1120	1.2952	0.5202	3.7824
	2.5000	0.0887	1.6040	0.4120	4.5430
	3.3333	0.0658	2.1174	0.3056	5.8048
	5.0000	0.0434	3.1416	0.2013	8.3185
	10.0000	0.0214	6.2086	0.0993	15.8372
	Inf.	1.5012	0.9781	0.6127	0.2114
	$1/R_s$	L_1	C_2	L_3	C_4

TABLE 10-42   Bessel *LC* Element Values

	R_s	C_1	L_2	C_3	L_4	C_5	L_6	C_7
5	1.0000	0.1743	0.5072	0.8040	1.1110	2.2582		
	0.9000	0.1926	0.4542	0.8894	0.9945	2.4328		
	0.8000	0.2154	0.4016	0.9959	0.8789	2.6497		
	0.7000	0.2447	0.3494	1.1323	0.7642	2.9272		
	0.6000	0.2836	0.2977	1.3138	0.6506	3.2952		
	0.5000	0.3380	0.2465	1.5672	0.5382	3.8077		
	0.4000	0.4194	0.1958	1.9464	0.4270	4.5731		
	0.3000	0.5548	0.1457	2.5768	0.3174	5.8433		
	0.2000	0.8251	0.0964	3.8352	0.2095	8.3747		
	0.1000	1.6349	0.0478	7.6043	0.1036	15.9487		
	Inf.	1.5125	1.0232	0.7531	0.4729	0.1618		
6	1.0000	0.1365	0.4002	0.6392	0.8538	1.1126	2.2645	
	1.1111	0.1223	0.4429	0.5732	0.9456	0.9964	2.4388	
	1.2500	0.1082	0.4961	0.5076	1.0600	0.8810	2.6554	
	1.4286	0.0943	0.5644	0.4424	1.2069	0.7665	2.9325	
	1.6667	0.0804	0.6553	0.3775	1.4022	0.6530	3.3001	
	2.0000	0.0666	0.7824	0.3131	1.6752	0.5405	3.8122	
	2.5000	0.0530	0.9725	0.2492	2.0837	0.4292	4.5770	
	3.3333	0.0395	1.2890	0.1859	2.7633	0.3193	5.8467	
	5.0000	0.0261	1.9209	0.1232	4.1204	0.2110	8.3775	
	10.0000	0.0130	3.8146	0.0612	8.1860	0.1045	15.9506	
	Inf.	1.5124	1.0329	0.8125	0.6072	0.3785	0.1287	
7	1.0000	0.1106	0.3259	0.5249	0.7020	0.8690	1.1052	2.2659
	0.8000	0.1372	0.2589	0.6521	0.5586	1.0803	0.8754	2.6556
	0.7000	0.1562	0.2257	0.7428	0.4873	1.2308	0.7618	2.9319
	0.6000	0.1815	0.1927	0.8634	0.4163	1.4312	0.6491	3.2984
	0.5000	0.2168	0.1599	1.0321	0.3457	1.7111	0.5374	3.8090
	0.4000	0.2698	0.1274	1.2847	0.2755	2.1304	0.4269	4.5718
	0.3000	0.3579	0.0951	1.7051	0.2058	2.8280	0.3177	5.8380
	0.2000	0.5338	0.0630	2.5448	0.1365	4.2214	0.2100	8.3623
	0.1000	1.0612	0.0313	5.0616	0.0679	8.3967	0.1040	15.9166
	Inf.	1.5087	1.0293	0.8345	0.6752	0.5031	0.3113	0.1054
	$1/R_s$	L_1	C_2	L_3	C_4	L_5	C_6	L_7

TABLE 10-42   Bessel *LC* Element Values (*Continued*)

	R_s	C_1	L_2	C_3	L_4	C_5	L_6	C_7	L_8	C_9	L_{10}
8	1.0000	0.0919	0.2719	0.4409	0.5936	0.7303	0.8695	1.0956	2.2656		
	1.1111	0.0825	0.3013	0.3958	0.6580	0.6559	0.9639	0.9813	2.4388		
	1.2500	0.0731	0.3380	0.3509	0.7385	0.5817	1.0816	0.8678	2.6541		
	1.4286	0.0637	0.3850	0.3061	0.8418	0.5078	1.2328	0.7552	2.9295		
	1.6667	0.0545	0.4477	0.2616	0.9794	0.4342	1.4340	0.6435	3.2949		
	2.0000	0.0452	0.5354	0.2173	1.1718	0.3608	1.7153	0.5329	3.8041		
	2.5000	0.0360	0.6667	0.1732	1.4599	0.2878	2.1367	0.4233	4.5645		
	3.3333	0.0269	0.8852	0.1294	1.9396	0.2151	2.8380	0.3151	5.8271		
	5.0000	0.0179	1.3218	0.0859	2.8981	0.1429	4.2389	0.2083	8.3441		
	10.0000	0.0089	2.6307	0.0427	5.7710	0.0711	8.4376	0.1032	15.8768		
	Inf.	1.5044	1.0214	0.8392	0.7081	0.5743	0.4253	0.2616	0.0883		
9	1.0000	0.0780	0.2313	0.3770	0.5108	0.6306	0.7407	0.8639	1.0863	2.2649	
	0.9000	0.0864	0.2077	0.4180	0.4588	0.6994	0.6655	0.9578	0.9730	2.4376	
	0.8000	0.0970	0.1841	0.4691	0.4069	0.7854	0.5905	1.075	0.8604	2.6524	
	0.7000	0.1105	0.1607	0.5348	0.3553	0.8957	0.5157	1.2255	0.7488	2.9271	
	0.6000	0.1286	0.1373	0.6222	0.3038	1.0427	0.4411	1.4258	0.6380	3.2915	
	0.5000	0.1538	0.1141	0.7445	0.2525	1.2483	0.3667	1.7059	0.5283	3.7993	
	0.4000	0.1916	0.0910	0.9278	0.2014	1.5563	0.2926	2.1256	0.4197	4.5578	
	0.3000	0.2545	0.0680	1.2329	0.1506	2.0692	0.2189	2.8241	0.3124	5.8171	
	0.2000	0.3803	0.0452	1.8426	0.1000	3.0941	0.1455	4.2196	0.2065	8.3276	
	0.1000	0.7573	0.0225	3.6704	0.0498	6.1666	0.0725	8.4023	0.1023	15.8408	
	Inf.	1.5006	1.0127	0.8361	0.7220	0.6142	0.4963	0.3654	0.2238	0.0754	
10	1.0000	0.0672	0.1998	0.3270	0.4454	0.5528	0.6493	0.7420	0.8561	1.0781	2.2641
	1.1111	0.0604	0.2216	0.2937	0.4941	0.4967	0.7205	0.6668	0.9492	0.9656	2.4365
	1.2500	0.0536	0.2488	0.2606	0.5548	0.4408	0.8093	0.5918	1.0654	0.8539	2.6508
	1.4286	0.0467	0.2836	0.2275	0.6327	0.3850	0.9233	0.5170	1.2147	0.7430	2.9249
	1.6667	0.0400	0.3301	0.1945	0.7366	0.3294	1.0753	0.4423	1.4134	0.6331	3.2885
	2.0000	0.0332	0.3951	0.1617	0.8818	0.2739	1.2879	0.3678	1.6913	0.5242	3.7953
	2.5000	0.0265	0.4924	0.1290	1.0995	0.2186	1.6064	0.2936	2.1076	0.4164	4.5521
	3.3333	0.0198	0.6546	0.0965	1.4620	0.1635	2.1369	0.2197	2.8007	0.3099	5.8087
	5.0000	0.0132	0.9786	0.0641	2.1864	0.1087	3.1971	0.1461	4.1854	0.2049	8.3137
	10.0000	0.0066	1.9499	0.0319	4.3583	0.0542	6.3759	0.0728	8.3359	0.1015	15.8108
	Inf.	1.4973	1.0045	0.8297	0.7258	0.6355	0.5401	0.4342	0.3182	0.1942	0.0650
	$1/R_s$	L_1	C_2	L_3	C_4	L_5	C_6	L_7	C_8	L_9	C_{10}

TABLE 10-42 Bessel *LC* Element Values (*Continued*)

Order n	C_1	C_2	C_3
2	0.9066	0.6800	
3	1.423	0.9880	0.2538
4	0.7351	0.6746	
	1.012	0.3900	
5	1.010	0.8712	0.3095
	1.041	0.3100	
6	0.6352	0.6100	
	0.7225	0.4835	
	1.073	0.2561	
7	0.8532	0.7792	0.3027
	0.7250	0.4151	
	1.100	0.2164	
8	0.5673	0.5540	
	0.6090	0.4861	
	0.7257	0.3590	
	1.116	0.1857	
9	0.7564	0.7070	0.2851
	0.6048	0.4352	
	0.7307	0.3157	
	1.137	0.1628	
10	0.5172	0.5092	
	0.5412	0.4682	
	0.6000	0.3896	
	0.7326	0.2792	
	1.151	0.1437	

TABLE 10-43   Bessel Active Low-Pass Values

Order n	Real Part $-\alpha$	Imaginary Part $\pm j\beta$
2	1.0087	0.6680
3	0.8541	1.0725
	1.0459	
4	0.9648	0.4748
	0.7448	1.4008
5	0.8915	0.8733
	0.6731	1.7085
	0.9430	
6	0.8904	0.4111
	0.8233	1.2179
	0.6152	1.9810
7	0.8425	0.7791
	0.7708	1.5351
	0.5727	2.2456
	0.8615	
8	0.8195	0.3711
	0.7930	1.1054
	0.7213	1.8134
	0.5341	2.4761
9	0.7853	0.7125
	0.7555	1.4127
	0.6849	2.0854
	0.5060	2.7133
	0.7938	
10	0.7592	0.3413
	0.7467	1.0195
	0.7159	1.6836
	0.6475	2.3198
	0.4777	2.9128

TABLE 10-44   Linear Phase with Equiripple Error of 0.05° Pole Locations

Order n	Real Part −α	Imaginary Part ±jβ
2	0.8590	0.6981
3	0.6969	1.1318
	0.8257	
4	0.7448	0.5133
	0.6037	1.4983
5	0.6775	0.9401
	0.5412	1.8256
	0.7056	
6	0.6519	0.4374
	0.6167	1.2963
	0.4893	2.0982
7	0.6190	0.8338
	0.5816	1.6453
	0.4598	2.3994
	0.6283	
8	0.5791	0.3857
	0.5665	1.1505
	0.5303	1.8914
	0.4184	2.5780
9	0.5688	0.7595
	0.5545	1.5089
	0.5179	2.2329
	0.4080	2.9028
	0.5728	
10	0.5249	0.3487
	0.5193	1.0429
	0.5051	1.7261
	0.4711	2.3850
	0.3708	2.9940

TABLE **10-45** Linear Phase with Equiripple Error of 0.5° Pole Locations

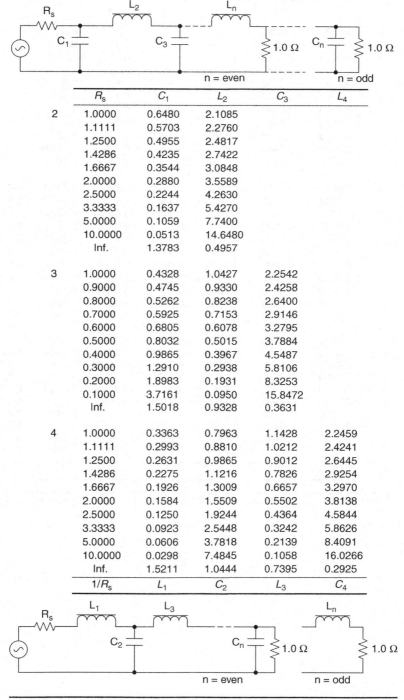

	R_s	C_1	L_2	C_3	L_4
2	1.0000	0.6480	2.1085		
	1.1111	0.5703	2.2760		
	1.2500	0.4955	2.4817		
	1.4286	0.4235	2.7422		
	1.6667	0.3544	3.0848		
	2.0000	0.2880	3.5589		
	2.5000	0.2244	4.2630		
	3.3333	0.1637	5.4270		
	5.0000	0.1059	7.7400		
	10.0000	0.0513	14.6480		
	Inf.	1.3783	0.4957		
3	1.0000	0.4328	1.0427	2.2542	
	0.9000	0.4745	0.9330	2.4258	
	0.8000	0.5262	0.8238	2.6400	
	0.7000	0.5925	0.7153	2.9146	
	0.6000	0.6805	0.6078	3.2795	
	0.5000	0.8032	0.5015	3.7884	
	0.4000	0.9865	0.3967	4.5487	
	0.3000	1.2910	0.2938	5.8106	
	0.2000	1.8983	0.1931	8.3253	
	0.1000	3.7161	0.0950	15.8472	
	Inf.	1.5018	0.9328	0.3631	
4	1.0000	0.3363	0.7963	1.1428	2.2459
	1.1111	0.2993	0.8810	1.0212	2.4241
	1.2500	0.2631	0.9865	0.9012	2.6445
	1.4286	0.2275	1.1216	0.7826	2.9254
	1.6667	0.1926	1.3009	0.6657	3.2970
	2.0000	0.1584	1.5509	0.5502	3.8138
	2.5000	0.1250	1.9244	0.4364	4.5844
	3.3333	0.0923	2.5448	0.3242	5.8626
	5.0000	0.0606	3.7818	0.2139	8.4091
	10.0000	0.0298	7.4845	0.1058	16.0266
	Inf.	1.5211	1.0444	0.7395	0.2925
	$1/R_s$	L_1	C_2	L_3	C_4

TABLE 10-46   Linear Phase with Equiriple Error of 0.05° LC Element Values

	R_s	C_1	L_2	C_3	L_4	C_5	L_6	C_7
5	1.0000	0.2751	0.6541	0.8892	1.1034	2.2873		
	0.9000	0.3031	0.5868	0.9841	0.9904	2.4589		
	0.8000	0.3380	0.5197	1.1026	0.8774	2.6733		
	0.7000	0.3827	0.4529	1.2548	0.7648	2.9484		
	0.6000	0.4420	0.3865	1.4575	0.6526	3.3144		
	0.5000	0.5248	0.3204	1.7408	0.5410	3.8254		
	0.4000	0.6486	0.2549	2.1651	0.4302	4.5896		
	0.3000	0.8544	0.1899	2.8713	0.3205	5.8595		
	0.2000	1.2649	0.1257	4.2817	0.2120	8.3922		
	0.1000	2.4940	0.0624	8.5082	0.1051	15.9739		
	Inf.	1.5144	1.0407	0.8447	0.6177	0.2456		
6	1.0000	0.2374	0.5662	0.7578	0.8760	1.1163	2.2448	
	1.1111	0.2120	0.6272	0.6799	0.9726	0.9977	2.4214	
	1.2500	0.1870	0.7032	0.6023	1.0931	0.8807	2.6396	
	1.4286	0.1622	0.8008	0.5253	1.2475	0.7652	2.9174	
	1.6667	0.1378	0.9306	0.4487	1.4530	0.6512	3.2849	
	2.0000	0.1138	1.1118	0.3725	1.7401	0.5387	3.7958	
	2.5000	0.0901	1.3830	0.2969	2.1698	0.4277	4.5579	
	3.3333	0.0669	1.8340	0.2217	2.8849	0.3182	5.8220	
	5.0000	0.0441	2.7343	0.1472	4.3129	0.2103	8.3408	
	10.0000	0.0218	5.4312	0.0732	8.5924	0.1041	15.8769	
	Inf.	1.5050	1.0306	0.8554	0.7283	0.5389	0.2147	
7	1.0000	0.2085	0.4999	0.6653	0.7521	0.8749	1.0671	2.2845
	0.8000	0.2573	0.3978	0.8274	0.6013	1.0861	0.8489	2.6655
	0.7000	0.2919	0.3470	0.9431	0.5258	1.2369	0.7400	2.9375
	0.6000	0.3380	0.2964	1.0972	0.4503	1.4381	0.6314	3.2996
	0.5000	0.4023	0.2461	1.3127	0.3749	1.7196	0.5235	3.8051
	0.4000	0.4986	0.1960	1.6356	0.2995	2.1416	0.4163	4.5613
	0.3000	0.6585	0.1463	2.1734	0.2242	2.8445	0.3101	5.8180
	0.2000	0.9778	0.0970	3.2480	0.1492	4.2496	0.2052	8.3246
	0.1000	1.9340	0.0482	6.4698	0.0744	8.4623	0.1017	15.8281
	Inf.	1.4988	1.0071	0.8422	0.7421	0.6441	0.4791	0.1910
	$1/R_s$	L_1	C_2	L_3	C_4	L_5	C_6	L_7

TABLE 10-46 Linear Phase with Equiriple Error of 0.05° *LC* Element Values (*Continued*)

	R_s	C_1	L_2	C_3	L_4	C_5	L_6	C_7	L_8	C_9	L_{10}
8	1.0000	0.1891	0.4543	0.6031	0.6750	0.7590	0.8427	1.0901	2.2415		
	1.1111	0.1691	0.5035	0.5415	0.7500	0.6813	0.9362	0.9735	2.4176		
	1.2500	0.1494	0.5650	0.4802	0.8435	0.6041	1.0527	0.8588	2.6349		
	1.4286	0.1298	0.6438	0.4191	0.9637	0.5272	1.2019	0.7459	2.9113		
	1.6667	0.1105	0.7487	0.3583	1.1237	0.4508	1.4004	0.6345	3.2767		
	2.0000	0.0914	0.8953	0.2978	1.3475	0.3748	1.6776	0.5247	3.7846		
	2.5000	0.0725	1.1148	0.2376	1.6827	0.2991	2.0927	0.4164	4.5418		
	3.3333	0.0539	1.4801	0.1776	2.2411	0.2237	2.7833	0.3096	5.7978		
	5.0000	0.0356	2.2095	0.1180	3.3568	0.1488	4.1627	0.2046	8.3004		
	10.0000	0.0176	4.3954	0.0588	6.7021	0.0742	8.2969	0.1013	15.7878		
	Inf.	1.4953	1.0018	0.8264	0.7396	0.6688	0.5858	0.4369	0.1743		
9	1.0000	0.1718	0.4146	0.5498	0.6132	0.6774	0.7252	0.8450	1.0447	2.2834	
	0.9000	0.1900	0.3724	0.6097	0.5519	0.7513	0.6529	0.9352	0.9382	2.4512	
	0.8000	0.2125	0.3302	0.6846	0.4905	0.8436	0.5805	1.0481	0.8314	2.6613	
	0.7000	0.2415	0.2882	0.7807	0.4291	0.9624	0.5079	1.1933	0.7247	2.9315	
	0.6000	0.2800	0.2463	0.9088	0.3676	1.1207	0.4352	1.3870	0.6184	3.2914	
	0.5000	0.3337	0.2046	1.0880	0.3062	1.3424	0.3624	1.6581	0.5125	3.7941	
	0.4000	0.4141	0.1631	1.3565	0.2448	1.6749	0.2897	2.0647	0.4075	4.5462	
	0.3000	0.5478	0.1219	1.8038	0.1834	2.2289	0.2170	2.7420	0.3035	5.7960	
	0.2000	0.8148	0.0809	2.6977	0.1222	3.3369	0.1445	4.0960	0.2007	8.2890	
	0.1000	1.6146	0.0403	5.3782	0.0610	6.6602	0.0721	8.1556	0.0995	15.7520	
	Inf.	1.4907	0.9845	0.8116	0.7197	0.6646	0.6089	0.5359	0.4003	0.1598	
10	1.0000	0.1601	0.3867	0.5125	0.5702	0.6243	0.6557	0.7319	0.8178	1.0767	2.2387
	1.1111	0.1433	0.4288	0.4604	0.6336	0.5609	0.7290	0.6567	0.9089	0.9608	2.4151
	1.2500	0.1267	0.4812	0.4084	0.7127	0.4977	0.8205	0.5820	1.0221	0.8471	2.6323
	1.4286	0.1102	0.5486	0.3567	0.8143	0.4348	0.9380	0.5079	1.1672	0.7354	2.9082
	1.6667	0.0939	0.6383	0.3051	0.9498	0.3721	1.0944	0.4342	1.3600	0.6254	3.2727
	2.0000	0.0778	0.7637	0.2537	1.1392	0.3096	1.3131	0.3609	1.6291	0.5170	3.7791
	2.5000	0.0618	0.9515	0.2024	1.4232	0.2473	1.6408	0.2880	2.0320	0.4102	4.5340
	3.3333	0.0460	1.2641	0.1515	1.8961	0.1852	2.1866	0.2154	2.7022	0.3049	5.7860
	5.0000	0.0304	1.8885	0.1007	2.8416	0.1232	3.2775	0.1433	4.0406	0.2014	8.2806
	10.0000	0.0151	3.7600	0.0502	5.6766	0.0615	6.5485	0.0714	8.0520	0.0997	15.7441
	Inf.	1.4905	0.9858	0.8018	0.7123	0.6540	0.6141	0.5669	0.5003	0.3741	0.1494
	$1/R_s$	L_1	C_2	L_3	C_4	L_5	C_6	L_7	C_8	L_9	C_{10}

TABLE 10-46 Linear Phase with Equiriple Error of 0.05° *LC* Element Values (*Continued*)

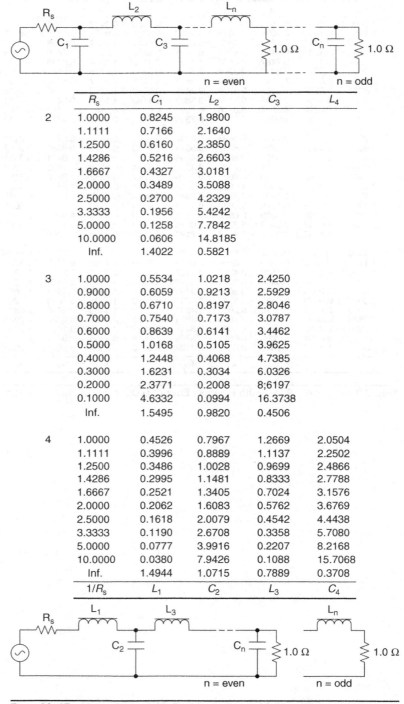

	R_s	C_1	L_2	C_3	L_4
2	1.0000	0.8245	1.9800		
	1.1111	0.7166	2.1640		
	1.2500	0.6160	2.3850		
	1.4286	0.5216	2.6603		
	1.6667	0.4327	3.0181		
	2.0000	0.3489	3.5088		
	2.5000	0.2700	4.2329		
	3.3333	0.1956	5.4242		
	5.0000	0.1258	7.7842		
	10.0000	0.0606	14.8185		
	Inf.	1.4022	0.5821		
3	1.0000	0.5534	1.0218	2.4250	
	0.9000	0.6059	0.9213	2.5929	
	0.8000	0.6710	0.8197	2.8046	
	0.7000	0.7540	0.7173	3.0787	
	0.6000	0.8639	0.6141	3.4462	
	0.5000	1.0168	0.5105	3.9625	
	0.4000	1.2448	0.4068	4.7385	
	0.3000	1.6231	0.3034	6.0326	
	0.2000	2.3771	0.2008	8;6197	
	0.1000	4.6332	0.0994	16.3738	
	Inf.	1.5495	0.9820	0.4506	
4	1.0000	0.4526	0.7967	1.2669	2.0504
	1.1111	0.3996	0.8889	1.1137	2.2502
	1.2500	0.3486	1.0028	0.9699	2.4866
	1.4286	0.2995	1.1481	0.8333	2.7788
	1.6667	0.2521	1.3405	0.7024	3.1576
	2.0000	0.2062	1.6083	0.5762	3.6769
	2.5000	0.1618	2.0079	0.4542	4.4438
	3.3333	0.1190	2.6708	0.3358	5.7080
	5.0000	0.0777	3.9916	0.2207	8.2168
	10.0000	0.0380	7.9426	0.1088	15.7068
	Inf.	1.4944	1.0715	0.7889	0.3708
	$1/R_s$	L_1	C_2	L_3	C_4

TABLE 10-47 Linear Phase with Equiripple Error of 0.5° LC Element Values

	R_s	C_1	L_2	C_3	L_4	C_5	L_6	C_7
5	1.0000	0.3658	0.6768	0.9513	1.0113	2.4446		
	0.9000	0.4027	0.6099	1.0486	0.9157	2.6062		
	0.8000	0.4485	0.5427	1.1700	0.8182	2.8114		
	0.7000	0.5069	0.4752	1.3260	0.7189	3.0787		
	0.6000	0.5843	0.4074	1.5341	0.6181	3.4387		
	0.5000	0.6921	0.3395	1.8253	0.5160	3.9462		
	0.4000	0.8530	0.2714	2.2623	0.4130	4.7108		
	0.3000	1.1201	0.2033	2.9908	0.3094	5.9881		
	0.2000	1.6524	0.1352	4.4478	0.2057	8.5444		
	0.1000	3.2454	0.0674	8.8185	0.1024	16.2117		
	Inf.	1.5327	1.0180	0.8740	0.6709	0.3182		
6	1.0000	0.3313	0.5984	0.8390	0.7964	1.2734	2.0111	
	1.1111	0.2934	0.6667	0.7446	0.8985	1.1050	2.2282	
	1.2500	0.2571	0.7515	0.6542	1.0223	0.9549	2.4742	
	1.4286	0.2219	0.8600	0.5666	1.1787	0.8164	2.7718	
	1.6667	0.1876	1.0040	0.4812	1.3848	0.6859	3.1529	
	2.0000	0.1541	1.2051	0.3976	1.6709	0.5615	3.6720	
	2.5000	0.1216	1.5058	0.3155	2.0972	0.4420	4.4362	
	3.3333	0.0898	2.0058	0.2347	2.8044	0.3266	5.6935	
	5.0000	0.0589	3.0038	0.1553	4.2137	0.2146	8.1871	
	10.0000	0.0290	5.9928	0.0771	8.4320	0.1058	15.6296	
	Inf.	1.4849	1.0430	0.8427	0.7651	0.5972	0.2844	
7	1.0000	0.2826	0.5332	0.7142	0.6988	0.9219	0.9600	2.4404
	0.9000	0.3118	0.4802	0.7896	0.6322	1.0137	0.8718	2.5953
	0.8000	0.3481	0.4271	0.8836	0.5649	1.1287	0.7809	2.7936
	0.7000	0.3945	0.3739	1.0043	0.4967	1.2768	0.6875	3.0535
	0.6000	0.4560	0.3206	1.1650	0.4277	1.4750	0.5919	3.4051
	0.5000	0.5416	0.2671	1.3899	0.3580	1.7531	0.4947	3.9025
	0.4000	0.6695	0.2136	1.7271	0.2874	2.1714	0.3961	4.6534
	0.3000	0.8819	0.1601	2.2890	0.2163	2.8700	0.2969	5.9091
	0.2000	1.3054	0.1066	3.4127	0.1445	4.2690	0.1974	8.4236
	0.1000	2.5731	0.0532	6.7835	0.0724	8.4691	0.0983	15.9666
	Inf.	1.5079	0.9763	0.8402	0.7248	0.6741	0.5305	0.2532
	$1/R_s$	L_1	C_2	L_3	C_4	L_5	C_6	L_7

TABLE 10-47 Linear Phase with Equiripple Error of 0.5° *LC* Element Values (*Continued*)

	R_s	C_1	L_2	C_3	L_4	C_5	L_6	C_7	L_8	C_9	L_{10}
8	1.0000	0.2718	0.4999	0.6800	0.6312	0.8498	0.7447	1.3174	1.9626		
	1.1111	0.2408	0.5567	0.6045	0.7116	0.7452	0.8529	1.1169	2.2146		
	1.2500	0.2114	0.6271	0.5324	0.8086	0.6506	0.9780	0.9551	2.4766		
	1.4286	0.1828	0.7173	0.4622	0.9315	0.5612	1.1331	0.8117	2.7837		
	1.6667	0.1549	0.8373	0.3934	1.0939	0.4753	1.3355	0.6795	3.1715		
	2.0000	0.1276	1.0049	0.3256	1.3201	0.3920	1.6148	0.5550	3.6960		
	2.5000	0.1009	1.2559	0.2589	1.6580	0.3107	2.0297	0.4362	4.4654		
	3.3333	0.0747	1.6734	0.1930	2.2194	0.2311	2.7164	0.3220	5.7294		
	5.0000	0.0492	2.5074	0.1279	3.3400	0.1530	4.0835	0.2114	8.2345		
	10.0000	0.0242	5.0066	0.0636	6.6971	0.0760	8.1733	0.1042	15.7101		
	Inf.	1.4915	1.0265	0.8169	0.7548	0.6709	0.6318	0.4995	0.2387		
9	1.0000	0.2347	0.4493	0.5914	0.5747	0.7027	0.6552	0.8944	0.9255	2.4332	
	0.9000	0.2594	0.4045	0.6547	0.5193	0.7754	0.5943	0.9809	0.8427	2.5822	
	0.8000	0.2900	0.3597	0.7336	0.4635	0.8662	0.5322	1.0895	0.7566	2.7745	
	0.7000	0.3291	0.3148	0.8348	0.4073	0.9829	0.4690	1.2299	0.6673	3.0283	
	0.6000	0.3810	0.2699	0.9695	0.3505	1.1388	0.4046	1.4183	0.5753	3.3734	
	0.5000	0.4533	0.2249	1.1580	0.2932	1.3572	0.3392	1.6834	0.4812	3.8629	
	0.4000	0.5613	0.1799	1.4405	0.2355	1.6854	0.2727	2.0828	0.3855	4.6032	
	0.3000	0.7407	0.1348	1.9111	0.1772	2.2331	0.2054	2.7508	0.2889	5.8424	
	0.2000	1.0986	0.0898	2.8522	0.1185	3.3299	0.1373	4.0895	0.1921	8.3246	
	0.1000	2.1702	0.0448	5.6749	0.0594	6.6230	0.0688	8.1099	0.0956	15.7718	
	Inf.	1.4888	0.9495	0.8044	0.6892	0.6589	0.5952	0.5645	0.4475	0.2141	
10	1.0000	0.2359	0.4369	0.5887	0.5428	0.7034	0.5827	0.8720	0.6869	1.4317	1.8431
	1.1111	0.2081	0.4866	0.5218	0.6141	0.6141	0.6729	0.7394	0.8187	1.1397	2.1907
	1.2500	0.1827	0.5480	0.4601	0.6972	0.5376	0.7708	0.6394	0.9483	0.9616	2.4734
	1.4286	0.1582	0.6267	0.3999	0.8024	0.4651	0.8922	0.5487	1.1042	0.8122	2.7907
	1.6667	0.1343	0.7314	0.3407	0.9416	0.3948	1.0514	0.4631	1.3052	0.6777	3.1847
	2.0000	0.1108	0.8777	0.2823	1.1356	0.3263	1.2719	0.3811	1.5809	0.5525	3.7138
	2.5000	0.0877	1.0969	0.2247	1.4258	0.2591	1.6003	0.3017	1.9888	0.4338	4.4876
	3.3333	0.0651	1.4619	0.1676	1.9085	0.1931	2.1451	0.2242	2.6628	0.3200	5.7573
	5.0000	0.0429	2.1910	0.1112	2.8724	0.1280	3.2311	0.1483	4.0033	0.2100	8.2726
	10.0000	0.0212	4.3764	0.0553	5.7612	0.0636	6.4830	0.0736	8.0118	0.1035	15.7776
	Inf.	1.4973	1.0192	0.8005	0.7312	0.6498	0.6331	0.5775	0.5501	0.4369	0.2091
	$1/R_s$	L_1	C_2	L_3	C_4	L_5	C_6	L_7	C_8	L_9	C_{10}

TABLE 10-47 Linear Phase with Equiripple Error of 0.5° *LC* Element Values (*Continued*)

Order n	C_1	C_2
2	0.9914	0.6891
4	1.0365	0.8344
	1.3426	0.2959
6	1.1231	0.9257
	1.2146	0.3810
	1.6255	0.1430
8	1.2203	1.0126
	1.2610	0.4285
	1.3864	0.1894
	1.8723	0.08324
10	1.3172	1.0957
	1.3392	0.4676
	1.3968	0.2139
	1.5444	0.1116
	2.0934	0.05483

TABLE 10-48 Linear Phase with Equiripple Error of 0.05° Active Low-Pass Values

Order n	C_1	C_2
2	1.1641	0.7011
4	1.3426	0.9103
	1.6565	0.2314
6	1.5340	1.0578
	1.6215	0.2993
	2.0437	0.1054
8	1.7268	1.1962
	1.7652	0.3445
	1.8857	0.1374
	2.3901	0.06134
10	1.9051	1.3218
	1.9257	0.3826
	1.9798	0.1562
	2.1227	0.07971
	2.6969	0.04074

TABLE 10-49 Linear Phase with Equiripple Error of 0.5° Active Low-Pass Values

Order n	Real Part −α	Imaginary Part ±jβ
3	0.9622 0.9776	1.2214
4	0.7940 0.6304	0.5029 1.5407
5	0.6190 0.3559 0.6650	0.8254 1.5688
6	0.5433 0.4672 0.2204	0.3431 0.9991 1.5067
7	0.4580 0.3649 0.1522 0.4828	0.5932 1.1286 1.4938
8	0.4222 0.3833 0.2878 0.1122	0.2640 0.7716 1.2066 1.4798
9	0.3700 0.3230 0.2309 0.08604 0.3842	0.4704 0.9068 1.2634 1.4740
10	0.3384 0.3164 0.2677 0.1849 0.06706	0.2101 0.6180 0.9852 1.2745 1.4389

TABLE 10-50 Transitional Gaussian to 6-dB Pole Locations

Order n	Real Part $-\alpha$	Imaginary Part $\pm j\beta$
3	0.9360	1.2168
	0.9630	
4	0.9278	1.6995
	0.9192	0.5560
5	0.8075	0.9973
	0.7153	2.0532
	0.8131	
6	0.7019	0.4322
	0.6667	1.2931
	0.4479	2.1363
7	0.6155	0.7703
	0.5486	1.5154
	0.2905	2.1486
	0.6291	
8	0.5441	0.3358
	0.5175	0.9962
	0.4328	1.6100
	0.1978	2.0703
9	0.4961	0.6192
	0.4568	1.2145
	0.3592	1.7429
	0.1489	2.1003
	0.5065	
10	0.4535	0.2794
	0.4352	0.8289
	0.3886	1.3448
	0.2908	1.7837
	0.1136	2.0599

TABLE 10-51   Transitional Gaussian to 12-dB Pole Locations

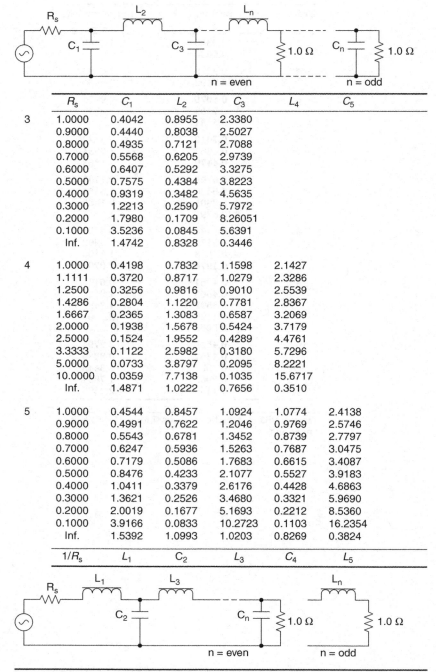

	R_s	C_1	L_2	C_3	L_4	C_5
3	1.0000	0.4042	0.8955	2.3380		
	0.9000	0.4440	0.8038	2.5027		
	0.8000	0.4935	0.7121	2.7088		
	0.7000	0.5568	0.6205	2.9739		
	0.6000	0.6407	0.5292	3.3275		
	0.5000	0.7575	0.4384	3.8223		
	0.4000	0.9319	0.3482	4.5635		
	0.3000	1.2213	0.2590	5.7972		
	0.2000	1.7980	0.1709	8.26051		
	0.1000	3.5236	0.0845	5.6391		
	Inf.	1.4742	0.8328	0.3446		
4	1.0000	0.4198	0.7832	1.1598	2.1427	
	1.1111	0.3720	0.8717	1.0279	2.3286	
	1.2500	0.3256	0.9816	0.9010	2.5539	
	1.4286	0.2804	1.1220	0.7781	2.8367	
	1.6667	0.2365	1.3083	0.6587	3.2069	
	2.0000	0.1938	1.5678	0.5424	3.7179	
	2.5000	0.1524	1.9552	0.4289	4.4761	
	3.3333	0.1122	2.5982	0.3180	5.7296	
	5.0000	0.0733	3.8797	0.2095	8.2221	
	10.0000	0.0359	7.7138	0.1035	15.6717	
	Inf.	1.4871	1.0222	0.7656	0.3510	
5	1.0000	0.4544	0.8457	1.0924	1.0774	2.4138
	0.9000	0.4991	0.7622	1.2046	0.9769	2.5746
	0.8000	0.5543	0.6781	1.3452	0.8739	2.7797
	0.7000	0.6247	0.5936	1.5263	0.7687	3.0475
	0.6000	0.7179	0.5086	1.7683	0.6615	3.4087
	0.5000	0.8476	0.4233	2.1077	0.5527	3.9183
	0.4000	1.0411	0.3379	2.6176	0.4428	4.6863
	0.3000	1.3621	0.2526	3.4680	0.3321	5.9690
	0.2000	2.0019	0.1677	5.1693	0.2212	8.5360
	0.1000	3.9166	0.0833	10.2723	0.1103	16.2354
	Inf.	1.5392	1.0993	1.0203	0.8269	0.3824
	$1/R_s$	L_1	C_2	L_3	C_4	L_5

TABLE 10-52   Transitional Gaussian to 6-dB *LC* Element Values

	R_s	C_1	L_2	C_3	L_4	C_5	L_6	C_7	L_8	C_9
6	1.0000	0.5041	0.9032	1.2159	1.0433	1.4212	2.0917			
	1.1111	0.4427	1.0079	1.0739	1.1892	1.2274	2.3324			
	1.2500	0.3853	1.1364	0.9415	1.3611	1.0620	2.5935			
	1.4286	0.3306	1.2999	0.8145	1.5753	0.9111	2.9053			
	1.6667	0.2779	1.5162	0.6914	1.8557	0.7692	**3.3032**			
	2.0000	0.2271	1.8169	0.5713	2.2433	0.6333	3.8456			
	2.5000	0.1780	2.2654	0.4534	2.8200	0.5016	4.6459			
	3.3333	0.1308	3.0091	0.3376	3.7758	0.3730	5.9662			
	5.0000	0.0853	4.4902	0.2235	5.6803	0.2468	8.5904			
	10.0000	0.0416	8.9199	0.1109	11.3810	0.1225	16.4352			
	Inf.	1.5664	1.2166	1.1389	1.1010	0.8844	0.4062			
7	1.0000	0.4918	0.9232	1.2146	1.1224	1.3154	1.1407	2.5039		
	0.9000	0.5403	0.8318	1.3393	1.0196	1.4426	1.0434	2.6575		
	0.8000	0.6001	0.7399	1.4950	0.9141	1.6040	0.9401	2.8593		
	0.7000	0.6760	0.6474	1.6952	0.8061	1.8144	0.8317	3.1285		
	0.6000	0.7763	0.5545	1.9626	0.6956	2.0986	0.7190	3.4967		
	0.5000	0.9157	0.4613	2.3373	0.5829	2.5004	0.6029	4.0203		
	0.4000	1.1236	0.3681	2.9002	0.4684	3.1072	0.4844	4.8129		
	0.3000	1.4685	0.2750	3.8389	0.3524	4.1232	0.3644	6.1397		
	0.2000	2.1560	0.1823	5.7166	0.2354	6.1604	0.2433	8.7977		
	0.1000	4.2137	0.0905	11.3483	0.1178	12.2787	0.1217	16.7743		
	Inf.	1.5950	1.2166	1.2240	1.1784	1.1260	0.8975	0.4110		
8	1.0502	0.5031	0.9699	1.2319	1.1324	1.4262	1.0449	1.6000	1.9285	
	1.1111	0.4586	1.0338	1.1286	1.2497	1.2635	1.2099	1.3372	2.2286	
	1.2500	0.3964	1.1670	0.9831	1.4404	1.0842	1.4259	1.1197	2.5453	
	1.4286	0.3392	1.3351	0.8487	1.6698	0.9299	1.6706	0.9502	2.8771	
	1.6667	0.2848	1.5571	0.7195	1.9674	0.7863	1.9808	0.7989	3.2846	
	2.0000	0.2325	1.8656	0.5939	2.3776	0.6487	2.4039	0.6569	3.8326	
	2.5000	0.1822	2.3255	0.4710	2.9870	0.5151	3.0295	0.5204	4.6374	
	3.3333	0.1337	3.0879	0.3504	3.9962	0.3840	4.0640	0.3874	5.9636	
	5.0000	0.0872	4.6062	0.2318	6.0063	0.2547	6.1243	0.2566	8.5995	
	10.0000	0.0425	9.1467	0.1150	12.0217	0.1268	12.2919	0.1276	16.4808	
	Inf.	1.5739	1.2698	1.2325	1.2633	1.2017	1.1404	0.9066	0.4148	
9	1.0000	0.4979	0.9367	1.2371	1.1589	1.3845	1.1670	1.3983	1.1422	2.5277
	0.9000	0.5475	0.8439	1.3648	1.0517	1.5194	1.0673	1.5233	1.0527	2.6698
	0.8000	0.6083	0.7505	1.5238	0.9424	1.6894	0.9625	1.6850	0.9540	2.8635
	0.7000	0.6854	0.6567	1.7278	0.8306	1.9103	0.8527	1.8996	0.8472	3.1279
	0.6000	0.7870	0.5624	1.9998	0.7165	2.2081	0.7383	2.1929	0.7342	3.4938
	0.5000	0.9280	0.4679	2.3811	0.6002	2.6288	0.6202	2.6105	0.6166	4.0174
	0.4000	1.1383	0.3732	2.9537	0.4820	3.2641	0.4993	3.2438	0.4960	4.8118
	0.3000	1.4873	0.2788	3.9087	0.3625	4.3274	0.3763	4.3062	0.3734	6.1429
	0.2000	2.1829	0.1848	5.8191	0.2421	6.4590	0.2518	6.4381	0.2496	8.8109
	0.1000	4.2652	0.0918	11.5490	0.1211	12.8601	0.1262	12.8438	0.1250	16.8186
	Inf.	1.6014	1.2508	1.2817	1.2644	1.2805	1.2103	1.1456	0.9096	0.4160
	$1/R_s$	L_1	C_2	L_3	C_4	L_5	C_6	L_7	C_8	L_9

TABLE 10-52   Transitional Gaussian to 6-dB *LC* Element Values (*Continued*)

	R_s	C_1	L_2	C_3	L_4	C_5	L_6	C_7	L_8	C_9	L_{10}
10	1.1372	0.4682	1.0839	1.1516	1.2991	1.3293	1.2748	1.4216	1.1730	1.5040	2.1225
	1.2500	0.4087	1.1987	1.0148	1.4855	1.1389	1.5155	1.1705	1.4593	1.1798	2.5537
	1.4286	0.3489	1.3718	0.8744	1.7253	0.9733	1.7813	0.9908	1.7344	0.9878	2.9155
	1.6667	0.2928	1.6000	0.7409	2.0334	0.8219	2.1124	0.8338	2.0664	0.8275	3.3380
	2.0000	0.2389	1.9169	0.6114	2.4574	0.6776	2.5622	0.6868	2.5129	0.6799	3.8995
	2.5000	0.1872	2.3893	0.4848	3.0868	0.5377	3.2264	0.5451	3.1699	0.5387	4.7218
	3.3333	0.1373	3.1723	0.3606	4.1290	0.4007	4.3241	0.4065	4.2549	0.4011	6.0762
	5.0000	0.0895	4.7317	0.2385	6.2048	0.2657	6.5094	0.2698	6.4154	0.2659	8.7681
	10.0000	0.0437	9.3953	0.1183	12.4165	0.1322	13.0503	0.1345	12.8837	0.1323	16.8178
	Inf.	1.6077	1.3178	1.2927	1.3406	1.3070	1.3160	1.2409	1.1733	0.9311	0.4257
	$1/R_s$	L_1	C_2	L_3	C_4	L_5	C_6	L_7	C_8	L_9	C_{10}

TABLE 10-52 Transitional Gaussian to 6-dB *LC* Element Values (*Continued*)

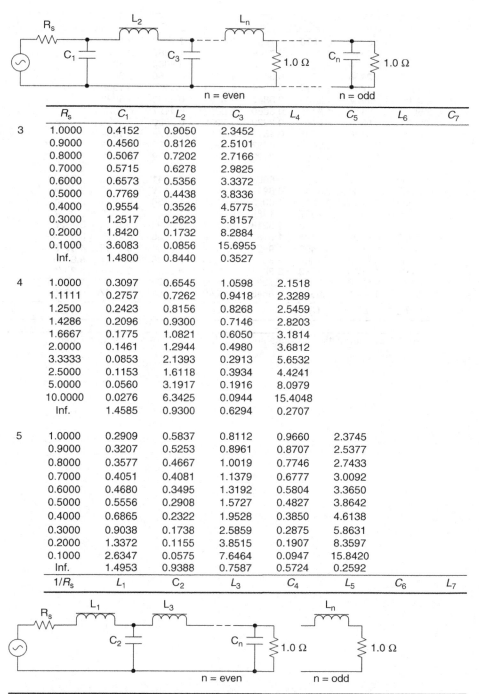

R_s		C_1	L_2	C_3	L_4	C_5	L_6	C_7
3	1.0000	0.4152	0.9050	2.3452				
	0.9000	0.4560	0.8126	2.5101				
	0.8000	0.5067	0.7202	2.7166				
	0.7000	0.5715	0.6278	2.9825				
	0.6000	0.6573	0.5356	3.3372				
	0.5000	0.7769	0.4438	3.8336				
	0.4000	0.9554	0.3526	4.5775				
	0.3000	1.2517	0.2623	5.8157				
	0.2000	1.8420	0.1732	8.2884				
	0.1000	3.6083	0.0856	15.6955				
	Inf.	1.4800	0.8440	0.3527				
4	1.0000	0.3097	0.6545	1.0598	2.1518			
	1.1111	0.2757	0.7262	0.9418	2.3289			
	1.2500	0.2423	0.8156	0.8268	2.5459			
	1.4286	0.2096	0.9300	0.7146	2.8203			
	1.6667	0.1775	1.0821	0.6050	3.1814			
	2.0000	0.1461	1.2944	0.4980	3.6812			
	3.3333	0.0853	2.1393	0.2913	5.6532			
	2.5000	0.1153	1.6118	0.3934	4.4241			
	5.0000	0.0560	3.1917	0.1916	8.0979			
	10.0000	0.0276	6.3425	0.0944	15.4048			
	Inf.	1.4585	0.9300	0.6294	0.2707			
5	1.0000	0.2909	0.5837	0.8112	0.9660	2.3745		
	0.9000	0.3207	0.5253	0.8961	0.8707	2.5377		
	0.8000	0.3577	0.4667	1.0019	0.7746	2.7433		
	0.7000	0.4051	0.4081	1.1379	0.6777	3.0092		
	0.6000	0.4680	0.3495	1.3192	0.5804	3.3650		
	0.5000	0.5556	0.2908	1.5727	0.4827	3.8642		
	0.4000	0.6865	0.2322	1.9528	0.3850	4.6138		
	0.3000	0.9038	0.1738	2.5859	0.2875	5.8631		
	0.2000	1.3372	0.1155	3.8515	0.1907	8.3597		
	0.1000	2.6347	0.0575	7.6464	0.0947	15.8420		
	Inf.	1.4953	0.9388	0.7587	0.5724	0.2592		
$1/R_s$		L_1	C_2	L_3	C_4	L_5	C_6	L_7

TABLE 10-53 Transitional Gaussian to 12-dB *LC* Element Values

	R_s	C_1	L_2	C_3	L_4	C_5	L_6	C_7	L_8	C_9
6	1.0000	0.3164	0.6070	0.7962	0.7880	1.1448	2.1154			
	1.1111	0.2813	0.6750	0.7108	0.8826	1.0087	2.3076			
	1.2500	0.2470	0.7597	0.6273	0.9994	0.8804	2.5365			
	1.4286	0.2135	0.8681	0.5452	1.1481	0.7580	2.8209			
	1.6667	0.1807	1.0123	0.4644	1.3451	0.6402	3.1908			
	2.0000	0.1487	1.2136	0.3847	1.6194	0.5263	3.6993			
	2.5000	0.1174	1.5148	0.3060	2.0292	0.4157	4.4522			
	3.3333	0.0868	2.0154	0.2282	2.7098	0.3080	5.6952			
	5.0000	0.0570	3.0146	0.1513	4.0679	0.2028	8.1654			
	10.0000	0.0280	6.0071	0.0753	8.1355	0.1002	15.5460			
	Inf.	1.4732	0.9894	0.8129	0.7484	0.5979	0.2752			
7	1.0000	0.3207	0.6267	0.8091	0.7753	0.9241	0.9649	2.3829		
	0.9000	0.3534	0.5641	0.8946	0.7016	1.0176	0.8750	2.5374		
	0.8000	0.3940	0.5015	1.0015	0.6270	1.1350	0.7824	2.7351		
	0.7000	0.4458	0.4387	1.1388	0.5513	1.2867	0.6876	2.9937		
	0.6000	0.5146	0.3758	1.3218	0.4747	1.4899	0.5910	3.3428		
	0.5000	0.6102	0.3128	1.5781	0.3972	1.7755	0.4931	3.8355		
	0.4000	0.7531	0.2498	1.9626	0.3189	2.2054	0.3943	4.5779		
	0.3000	0.9902	0.1869	2.6034	0.2399	2.9236	0.2951	5.8175		
	0.2000	1.4630	0.1243	3.8851	0.1603	4.3624	0.1961	8.2974		
	0.1000	2.8780	0.0619	7.7299	0.0803	8.6825	0.0976	15.7326		
	Inf.	1.4861	0.9693	0.8643	0.8040	0.7689	0.6157	0.2826		
	$1/R_s$	L_1	C_2	L_3	C_4	L_5	C_6	L_7	C_8	L_9

TABLE 10-53 Transitional Gaussian to 12-dB *LC* Element Values (*Continued*)

	R_s	C_1	L_2	C_3	L_4	C_5	L_6	C_7	L_8	C_9	L_{10}
8	1.0000	0.3449	0.6565	0.8686	0.8028	0.9701	0.8182	1.2503	2.0612		
	1.1111	0.3053	0.7304	0.7729	0.9044	0.8550	0.9339	1.0753	2.2930		
	1.2500	0.2674	0.8221	0.6810	1.0276	0.7489	1.0694	0.9272	2.5445		
	1.4286	0.2308	0.9394	0.5913	1.1837	0.6480	1.2378	0.7929	2.8444		
	1.6667	0.1952	1.0953	0.5034	1.3898	0.5504	1.4579	0.6673	3.2267		
	2.0000	0.1604	1.3128	0.4168	1.6764	0.4553	1.7620	0.5476	3.7473		
	2.5000	0.1265	1.6381	0.3314	2.1045	0.3621	2.2143	0.4323	4.5145		
	3.3333	0.0934	2.1788	0.2471	2.8155	0.2702	2.9640	0.3203	5.7789		
	5.0000	0.0613	3.2575	0.1638	4.2343	0.1794	4.4583	0.2111	8.2899		
	10.0000	0.0301	6.4874	0.0814	8.4843	0.0894	8.9330	0.1044	15.7917		
	Inf.	1.4974	1.0324	0.8943	0.8908	0.8494	0.8098	0.6452	0.2955		
9	1.0000	0.3318	0.6500	0.8467	0.8167	0.9426	0.8239	0.9857	0.9630	2.4140	
	0.9000	0.3657	0.5852	0.9363	0.7389	1.0390	0.7492	1.0803	0.8785	2.5608	
	0.8000	0.4078	0.5201	1.0480	0.6602	1.1599	0.6725	1.2003	0.7894	2.7524	
	0.7000	0.4614	0.4550	1.1914	0.5806	1.3160	0.5936	1.3568	0.6963	3.0070	
	0.6000	0.5324	0.3897	1.3825	0.4999	1.5251	0.5127	1.5681	0.6001	3.3538	
	0.5000	0.6312	0.3243	1.6500	0.4183	1.8192	0.4301	1.8667	0.5015	3.8460	
	0.4000	0.7787	0.2590	2.0512	0.3358	2.2620	0.3460	2.3177	0.4016	4.5897	
	0.3000	1.0234	0.1938	2.7200	0.2526	3.0021	0.2607	3.0729	0.3009	5.8332	
	0.2000	1.5113	0.1288	4.0574	0.1687	4.4850	0.1744	4.5875	0.2000	8.3219	
	0.1000	2.9716	0.0641	8.0692	0.0845	8.9384	0.0875	9.1379	0.0996	15.7849	
	Inf.	1.4917	0.9908	0.9105	0.8770	0.8910	0.8457	0.8022	0.6376	0.2917	
10	1.0139	0.3500	0.6698	0.8817	0.8148	1.0183	0.7949	1.0929	0.7508	1.4303	1.8322
	1.1111	0.3092	0.7364	0.7856	0.9200	0.8864	0.9293	0.9147	0.9187	1.1138	2.2110
	1.2500	0.2701	0.8290	0.6907	1.0477	0.7734	1.0708	0.7890	1.0712	0.9404	2.4914
	1.4286	0.2328	0.9474	0.5992	1.2075	0.6681	1.2428	0.6777	1.2503	0.7978	2.8009
	1.6667	0.1968	1.1046	0.5099	1.4179	0.5671	1.4663	0.5734	1.4791	0.6690	3.1853
	2.0000	0.1616	1.3239	0.4221	1.7103	0.4689	1.7746	0.4734	1.7920	0.5483	3.7034
	2.5000	0.1274	1.6517	0.3355	2.1467	0.3728	2.2329	0.3761	2.2552	0.4326	4.4640
	3.3333	0.0941	2.1966	0.2501	2.8714	0.2781	2.9923	0.2806	3.0215	0.3206	5.7162
	5.0000	0.0617	3.2837	0.1658	4.3171	0.1845	4.5061	0.1863	4.5478	0.2114	8.2022
	10.0000	0.0303	6.5385	0.0824	8.6473	0.0919	9.0397	0.0929	9.1181	0.1046	15.6293
	Inf.	1.4826	1.0350	0.9134	0.9263	0.9061	0.9159	0.8654	0.8190	0.6502	0.2973
	$1/R_s$	L_1	C_2	L_3	C_4	L_5	C_6	L_7	C_8	L_9	C_{10}

TABLE 10-53 Transitional Gaussian to 12-dB LC Element Values (*Continued*)

Order n	C_1	C_2
4	1.2594	0.8989
	1.5863	0.2275
6	1.8406	1.3158
	2.1404	0.3841
	4.5372	0.09505
8	2.3685	1.7028
	2.6089	0.5164
	3.4746	0.1870
	8.9127	0.05094
10	2.9551	2.1329
	3.1606	0.6564
	3.7355	0.2568
	5.4083	0.1115
	14.9120	0.03232

TABLE 10-54   Transitional Gaussian to 6-dB Active Low-Pass Values

Order n	C_1	C_2
4	1.0778	0.2475
	1.0879	0.7965
6	1.4247	1.0330
	1.5000	0.3150
	2.2326	0.09401
8	1.8379	1.3309
	1.9324	0.4106
	2.3105	0.1557
	5.0556	0.04573
10	2.2051	1.5984
	2.2978	0.4965
	2.5733	0.1983
	3.4388	0.08903
	8.8028	0.02669

TABLE 10-55   Transitional Gaussian to 12-dB Active Low-Pass Values

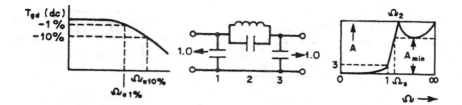

$\boxed{N = 3}$

A_{min}	Ω_s	$T_{gd}(DC)$	$\Omega_{\mu 1\%}$	$\Omega_{\mu 10\%}$	C_1	C_2	L_2	Ω_2	C_3
18	2.152	1.325	.9091	1.461	2.124	.2769	.5242	2.625	.001434
26	2.721	1.493	.8010	1.304	2.144	.1399	.6931	3.211	.1489
34	3.514	1.602	.7522	1.206	2.166	.07387	.8068	4.096	.2320
42	4.627	1.668	.7122	1.161	2.180	.03958	.8769	5.368	.2791
50	6.178	1.706	.6959	1.134	2.189	.02133	.9178	7.147	.3052
58	8.309	1.727	.6886	1.121	2.193	.01151	.9407	9.610	.3197
66	11.23	1.739	.6850	1.112	2.196	.006226	.9537	12.98	.3275
70	13.07	1.743	.6838	1.109	2.198	.004579	.9580	15.10	.3301

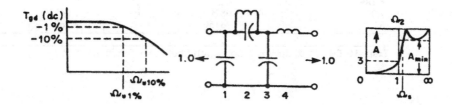

$\boxed{N = 4}$

A_{min}	Ω_s	$T_{gd}(DC)$	$\Omega_{\mu 1\%}$	$\Omega_{\mu 10\%}$	C_1	C_2	L_2	Ω_2	C_3	L_4
18	2.070	1.471	1.303	1.923	2.107	.3324	.5088	2.432	.1690	.1575
26	2.466	1.662	1.164	1.704	2.127	.1915	.6744	2.783	.3419	.1795
34	2.938	1.807	1.068	1.568	.1756	.1586	.5905	3.268	2.439	.4078
42	3.548	1.910	1.007	1.482	.2445	.09775	.6647	3.923	2.472	.4383
50	4.341	1.980	.9720	1.432	.1420	.08106	.5386	4.786	1.033	2.245
58	5.363	2.027	.9509	1.396	.1760	.04895	.5862	5.903	1.048	2.244
66	6.665	2.057	.9382	1.377	.1972	.03014	.6171	7.332	1.059	2.240
70	7.447	2.068	.9297	1.369	.2046	.02374	.6281	8.189	1.063	2.240

TABLE 10-56 Maximally Flat Delay with Chebyshev Stopband

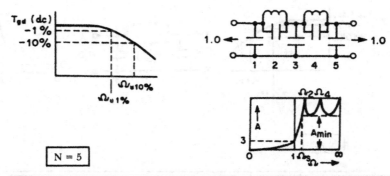

N = 5

A_{min}	Ω_s	$T_{gd}(DC)$	$\Omega_{\mu 1\%}$	$\Omega_{\mu 10\%}$	C_1	C_2	L_2
34	2.6802	1.745	1.550	2.142	.02930	.3317	.3665
42	3.0263	1.904	1.408	1.968	.1338	.2216	.4381
50	3.4467	2.035	1.316	1.840	2.1836	.08543	.8830
58	3.9701	2.139	1.304	1.750	2.203	.06060	.9418
66	4.6213	2.218	1.215	1.688	2.218	.04274	.9872
70	5.0044	2.250	1.191	1.665	2.224	.03582	1.006

Ω_2	C_3	C_4	L_4	Ω_4	C_5
2.868	2.420	.07329	.5638	4.919	.1111
3.209	2.435	.05333	.6496	5.373	.1513
3.641	.5958	.08228	.3358	6.016	.07096
4.186	.6503	.05559	.3818	6.864	.1007
4.863	.6922	.03799	.4168	7.947	.1221
5.268	.7092	.03150	.4307	8.585	.1305

TABLE 10-56 Maximally Flat Delay with Chebyshev Stopband (*Continued*)

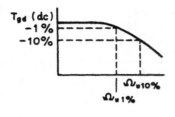

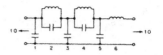

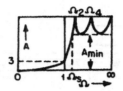

N = 6

A_{min}	Ω_s	$T_{gd}(DC)$	$\Omega_{\mu 1\%}$	$\Omega_{\mu 10\%}$	C_1	C_2	L_2
18	2.0530	1.415	2.096	—	.09332	.09400	1.907
26	2.3648	1.631	2.093	—	.1568	.07374	2.047
34	2.6429	1.820	1.910	2.604	2.128	.1820	.6982
42	2.9239	1.985	1.753	2.343	2.150	.1375	.7730
50	3.2353	2.129	1.643	2.190	2.172	.1047	.8371
58	3.6033	2.251	1.549	2.066	2.190	.07957	.8927
66	4.0446	2.352	1.485	1.982	2.206	.06014	.9396
70	4.2965	2.395	1.451	1.939	2.213	.05216	.9602

Ω_2	C_3	C_4	L_4	Ω_4	C_5	L_6
2.362	.4290	.2681	.2033	4.283	.08674	.1107
2.574	.5121	.2078	.2763	4.174	.1506	.1193
2.805	.4206	.2526	.2164	4.277	.08769	.09013
3.067	.4943	.1647	.2960	4.529	.1591	.09865
3.378	.5627	.1134	.3668	4.903	.2144	.1061
3.752	.6231	.08054	.4266	5.395	.2577	.1125
4.207	.6738	.05814	.4751	6.017	.2916	.1178
4.468	.6957	.04960	.4958	6.377	.3056	.1201

TABLE 10-56   Maximally Flat Delay with Chebyshev Stopband (*Continued*)

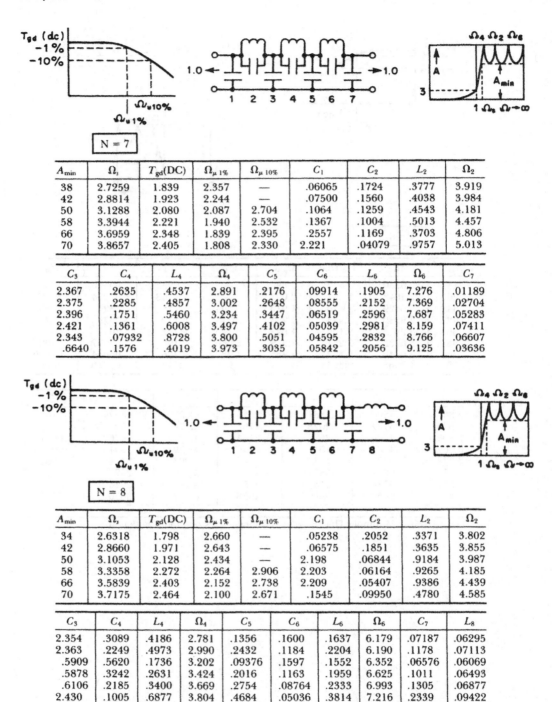

N = 7

A_{min}	Ω_s	$T_{gd}(DC)$	$\Omega_{\mu 1\%}$	$\Omega_{\mu 10\%}$	C_1	C_2	L_2	Ω_2
38	2.7259	1.839	2.357	—	.06065	.1724	.3777	3.919
42	2.8814	1.923	2.244	—	.07500	.1560	.4038	3.984
50	3.1288	2.080	2.087	2.704	.1064	.1259	.4543	4.181
58	3.3944	2.221	1.940	2.532	.1367	.1004	.5013	4.457
66	3.6959	2.348	1.839	2.395	.2557	.1169	.3703	4.806
70	3.8657	2.405	1.808	2.330	2.221	.04079	.9757	5.013

C_3	C_4	L_4	Ω_4	C_5	C_6	L_6	Ω_6	C_7
2.367	.2635	.4537	2.891	.2176	.09914	.1905	7.276	.01189
2.375	.2285	.4857	3.002	.2648	.08555	.2152	7.369	.02704
2.396	.1751	.5460	3.234	.3447	.06519	.2596	7.687	.05283
2.421	.1361	.6008	3.497	.4102	.05039	.2981	8.159	.07411
2.343	.07932	.8728	3.800	.5051	.04595	.2832	8.766	.06607
.6640	.1576	.4019	3.973	.3035	.05842	.2056	9.125	.03636

N = 8

A_{min}	Ω_s	$T_{gd}(DC)$	$\Omega_{\mu 1\%}$	$\Omega_{\mu 10\%}$	C_1	C_2	L_2	Ω_2
34	2.6318	1.798	2.660	—	.05238	.2052	.3371	3.802
42	2.8660	1.971	2.643	—	.06575	.1851	.3635	3.855
50	3.1053	2.128	2.434	—	2.198	.06844	.9184	3.987
58	3.3358	2.272	2.264	2.906	2.203	.06164	.9265	4.185
66	3.5839	2.403	2.152	2.738	2.209	.05407	.9386	4.439
70	3.7175	2.464	2.100	2.671	.1545	.09950	.4780	4.585

C_3	C_4	L_4	Ω_4	C_5	C_6	L_6	Ω_6	C_7	L_8
2.354	.3089	.4186	2.781	.1356	.1600	.1637	6.179	.07187	.06295
2.363	.2249	.4973	2.990	.2432	.1184	.2204	6.190	.1178	.07113
.5909	.5620	.1736	3.202	.09376	.1597	.1552	6.352	.06576	.06069
.5878	.3242	.2631	3.424	.2016	.1163	.1959	6.625	.1011	.06493
.6106	.2185	.3400	3.669	.2754	.08764	.2333	6.993	.1305	.06877
2.430	.1005	.6877	3.804	.4684	.05036	.3814	7.216	.2339	.09422

Table 10-56 Maximally Flat Delay with Chebyshev Stopband (*Continued*)

Order n	Real Part $-\alpha$	Imaginary Part $\pm j\omega$
1	1.0000	
2	0.7071	0.7071
3	0.3452	0.9009
	0.6203	0.0000
4	0.2317	0.9455
	0.5497	0.3586
5	0.1536	0.9681
	0.3881	0.5886
	0.4681	
6	0.1152	0.9779
	0.3090	0.6982
	0.4389	0.2400
7	0.0862	0.9844
	0.2374	0.7783
	0.3492	0.4290
	0.3821	
8	0.0689	0.9880
	0.1943	0.8248
	0.3003	0.5410
	0.3672	0.1809
9	0.0551	0.9907
	0.1573	0.8613
	0.2486	0.6338
	0.3094	0.3365
	0.3257	

TABLE **10-57**   Papoulis Filter Pole Locations

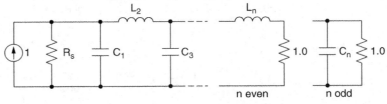

Order n	Designation	Value
1	C_1	2.0000
2	C_1	1.4142
	L_2	1.4142
3	C_1	2.1801
	L_2	1.3538
	C_3	1.1737
4	C_1	1.5644
	L_2	1.9585
	C_3	1.4768
	L_4	1.0827
5	C_1	1.9990
	L_2	1.5395
	C_3	2.0673
	L_4	1.4780
	C_5	0.9512
6	C_1	1.5763
	L_2	1.9040
	C_3	1.7442
	L_4	1.9857
	C_5	1.4852
	L_6	0.9160

7	C_1	1.8640
	L_2	1.5895
	C_3	2.1506
	L_4	1.7270
	C_5	1.9394
	L_6	1.4770
	C_7	0.8394
8	C_1	1.5564
	L_2	1.8500
	C_3	1.8411
	L_4	2.0515
	C_5	1.7672
	L_6	1.9115
	C_7	1.4688
	L_8	0.8205
9	C_1	1.7645
	L_2	1.6134
	C_3	2.1585
	L_4	1.7816
	C_5	2.0662
	L_6	1.7755
	C_7	1.8674
	L_8	1.4555
	C_9	0.7695

TABLE 10-58 Papoulis Filter LC Element Values

CHAPTER 11

Switched-Capacitor Filters

11.1 Introduction

Switched-capacitor filters have gained widespread acceptance because they require no external reactive components, are precisely controlled by a clock derived from a crystal or other precise oscillator, and can be programmed easily. They come close to truly digital filters without requiring the mathematical complexity of computation and the sophistication necessary for implementation. Miniaturization is achieved by implementing these devices on a chip, while stability is a function of the external components, which are typically low-temperature–coefficient resistors and a crystal-derived clock. They can be used for frequencies ranging from a small fraction of a hertz to a few hundred kilohertz.

11.2 The Theory of Switched-Capacitor Filters

The theory of switched-capacitor filters is based on the ability to very closely match capacitor ratios on a metal oxide semiconductor (MOS) integrated circuit. The actual magnitudes of the capacitors can vary significantly from device to device due to differences in the dielectric constant of the MOS process. The ratios between capacitors on the same device, however, are extremely precise since the areas allocated to each capacitor are very closely controlled by the manufacturing process.

11.2.1 The Switched Resistor

Figure 11-1 illustrates a capacitor C connected to ground, two switches S_1 and S_2, and an applied voltage E. The charge on a capacitor is given by

$$Q = CE \tag{11-1}$$

where Q is the charge in coulombs and E is the voltage applied to the capacitor.

The current through a capacitor is Q/T since an ampere is defined as a coulomb per second. The AC current through the capacitor of Fig. 11-1 is

$$I_{ac} = \frac{\Delta Q}{\Delta T} = \frac{C\Delta E}{\Delta T} \tag{11-2}$$

where ΔT is the time between the closing of S_1 and S_2, which are operated alternately, and where ΔE is the variation in voltage across the capacitor. The capacitor is thus

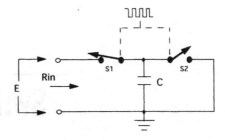

Figure 11-1   A resistor simulated by a switched capacitor.

switched between voltage E and ground during alternate parts of the switching cycle, making ΔE equal to E.

Since ΔT is determined by a clock frequency F_{clock}, we can say

$$I_{\text{ac}} = C \Delta E \, F_{\text{clock}} \qquad (11\text{-}3)$$

If we simply transpose and solve for the equivalent resistance R_{in} in terms of voltage divided by current, we obtain

$$R_{\text{in}} = \frac{1}{C F_{\text{clock}}} \qquad (11\text{-}4)$$

Equation (11-4) indicates that a capacitor can be made to behave like a resistor whose value is inversely proportional to the capacitance and to the clock-switching frequency. The switching rate should be much higher than any short-term variation in E.

11.2.2   The Basic Integrator as a Building Block

Figure 11-1 illustrates how a resistor can be implemented using clock-controlled switches and a capacitor. This is exactly the technique used in monolithic switched-capacitor filters. The switches are actually metal-oxide-semiconductor field-effect transistors (MOSFETs) and have relatively low resistance compared with the resistor values implemented. The capacitors are controlled by the geometry of the surface of the substrate.

The circuit of Fig. 11-2a shows a conventional inverting integrator. The transfer function of this circuit is

$$T(s) = -\frac{1}{sRC} \qquad (11\text{-}5)$$

If we use a switched-capacitor resistor for R, the circuit of Fig. 11-2b is obtained. This is an inverting integrator implemented using a switched-capacitor approach. We can substitute Eq. (11-4) for R in Eq. (11-5) and obtain the following transfer function:

$$T(s) = -\frac{1}{SC_b \dfrac{1}{C_a F_{\text{clock}}}}$$

Figure 11-2 (*a*) A basic integrator; and (*b*) a switched-capacitor integrator.

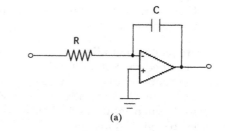

(a)

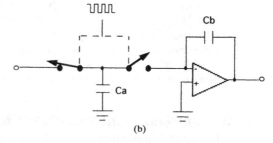

(b)

Simplifying Eq. (11-5) results in

$$T(s) = - \frac{F_{clock} C_a}{S C_b} \qquad (11\text{-}6)$$

Equation (11-6) indicates that the transfer function of a switched-capacitor integrator can be scaled proportionally by a clock frequency F_{clock} and is determined by the constant C_a / C_b, which is a *ratio* of two capacitors. Since this ratio can be precisely determined by chip geometries, we can build a filter with properties that are directly proportional to an externally supplied clock and independent of absolute magnitudes of chip capacitors. Variations of the circuit of Fig. 11-2*b* are used as basic building blocks for switched-capacitor filters.

11.2.3 The Limitations of Switched-Capacitor Filters

Noise

A complex switched-capacitor filter chip contains many capacitors within the device. In order to minimize the absolute magnitude of these capacitors to keep the devices small, internal resistor values are relatively higher than in most complementary metal oxide semiconductor (CMOS) integrated circuits (ICs). As a result, the output noise of switched-capacitor filters is higher than that produced by a conventional active filter. The noise generated by switched-capacitor filters consists of two types: clock feedthrough and thermal noise. The clock of a switched-capacitor filter is normally 50 or 100 times the operating frequency range, so it can be easily filtered by a simple *RC* at the output. Thermal noise is in-band and is typically 80 dB below operating signal levels. It can be minimized by optimizing gain distribution for maximum signal-to-noise ratio. Careful PC layout and decoupling should occur as well.

Charge injection of the CMOS switches into the integrating capacitors and high internal impedance levels result in larger DC offsets than with active filters.

However, some capacitive coupled architectures are available that can eliminate this problem.

Frequency Limitations

Switched-capacitor filters have a limiting figure of merit, which is the product of *Q times* the center frequency. In other words, the higher the center frequency, the lower the realizable *Q*. For relatively low *Q*s, traditional switched-capacitor filters are realizable for frequencies up to 200 kHz or so. With newer submicron technologies, the range has been extended to a few MHz.

Since we are dealing with a sampled system, aliasing effects can occur. If input signals occur above $F_S/2$, where F_S is the internal sampling frequency (typically 50 or 100 times the cutoff), we get foldover, which could make the signals of interest unusable if the foldover components are significant in amplitude. Sometimes an input *RC* network is useful to prevent this phenomenon, especially if the input signal consists of rectangular pulses rich in harmonic content.

11.3 Universal Switched-Capacitor Second-Order Filters

Figure 11-3 illustrates the basic architecture for a universal switched-capacitor building block using external resistors for determining characteristics in conjunction with a clock. This structure was first used in a device called an MF10 developed by National, which contains two such circuits. It was followed by the National LMF100 and has become the fundamental architecture for most complex switched-capacitor filters. Maxim also makes an MF10 and a next-generation dual-filter building block with improved performance and single supply operation as well as many other features—what they call their MAX7490/7491 series.

Linear Technology has a device called an LTC1067, which is a next-generation version of the MF10 but with significantly new features, such as single supply operation, low noise, and wide dynamic range. Both Maxim and Linear Technology have quad versions of this architecture for more complex filters. Linear Technology has a single and triple version as well.

These devices, as well as others, are listed in the selection guide of Sec. 11.5, and more details are provided in the manufacturers' data sheets, which are available at their websites. In addition, software is available from the manufacturers' websites to perform design computations.

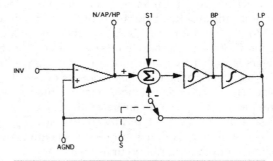

Figure 11-3 A block diagram of a universal second-order filter.

Although the following discussion and example is based on the MF10, the general design equations and methodology still apply to the other versions of universal switched-capacitor filters, as they all maintain essentially the same architecture.

This circuit realizes a second-order transfer function and can provide a band-pass, low-pass, high-pass, notch, or all-pass output by adding external resistors. The cascading of multiple sections results in switched-capacitor filters of very high complexity.

The basic configuration consists of an operational amplifier, a summing node, and two integrators. An MOS switch under the control of input S connects one of the summing node inputs either to the second integrator or to ground. The output of the operational amplifier can be a notch, all-pass, or high-pass response, depending on how the device is configured. The two integrators, on the other hand, can provide band-pass and low-pass outputs, and in most cases, these outputs can be provided simultaneously.

Although not shown, a clock at either 50 or 100 times the center (pole) frequency must be provided for the National MF10 and LMF100. A logic input to the chip determines whether the clock is 50 or 100 times F_0. Some newer devices can also internally generate clock frequencies of a few MHz proportional to a single external passive component.

11.3.1 Modes of Operation

The universal filter structure of Fig. 11-3 is quite flexible and can be reconfigured into various forms by adding external resistors and an amplifier. Each form, or *mode,* has unique properties that determine the filters' characteristics. Figures 11-4 to 11-10 show the most popular modes of operation. The design equations for each mode are given in Table 11-1.

11.3.2 Operating Mode Features

Mode 1

The circuit of mode 1 provides simultaneous band-pass and notch outputs at center frequencies that are equal to each other and directly determined by $F_{clock} \div 50$ or $F_{clock} \div 100$. As a result, we can design clock-tunable band-pass/band-stop networks with frequency

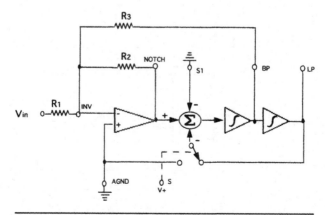

FIGURE 11-4   Mode 1: Band-pass, low-pass, $F_{notch} = F_0$.

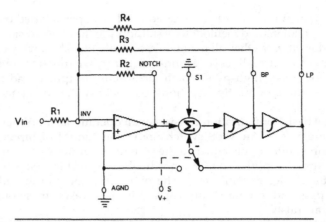

Figure 11-5   Mode 2: Band-pass, low-pass, $F_{notch} \leq F_0$.

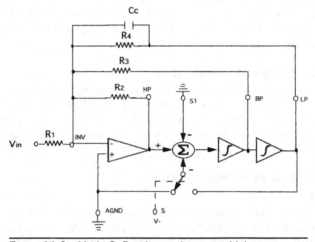

Figure 11-6   Mode 3: Band-pass, low-pass, high-pass.

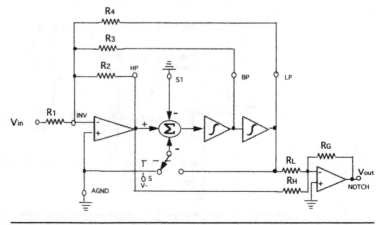

Figure 11-7   Mode 3a: Band-pass, low-pass, F_{notch} using an external operational amplifier.

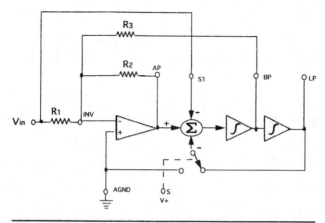

Figure 11-8   Mode 4: All-pass, band-pass, low-pass.

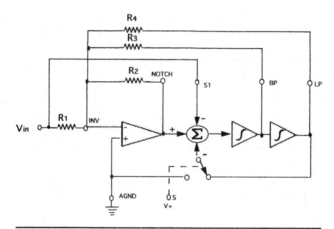

Figure 11-9   Mode 5: Numerator complex zeros, band-pass, low-pass.

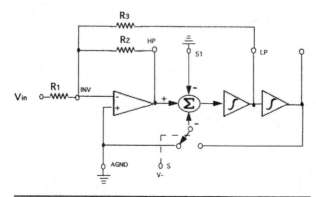

Figure 11-10   Mode 6a: Single-pole, low-pass, high-pass.

Table 11-1 Design Equations for Modes of Operation

Mode	F_0	Q	F_{notch}	LP Gain	BP Gain	HP Gain	Notch Gain	All-Pass Gain
1	$\dfrac{F_{clock}}{100(50)}$	$\dfrac{R_3}{R_2}$	F_0	$\dfrac{R_2}{R_1}$	$-\dfrac{R_3}{R_1}$		$-\dfrac{R_2}{R_1}$ (as $F \to 0$)	
2	$\dfrac{F_{clock}}{100(50)} \times \sqrt{1+\dfrac{R_2}{R_4}}$	$\dfrac{\sqrt{\dfrac{R_2}{R_4}+1}}{R_2/R_3}$	$\dfrac{F_{clock}}{100(50)}$	$-\dfrac{R_2/R_1}{R_2/R_4+1}$	$-\dfrac{R_3}{R_1}$		$-\dfrac{R_2/R_1}{R_2/R_4+1}$ (as $F \to 0$)	
3	$\dfrac{F_{clock}}{100(50)} \times \sqrt{\dfrac{R_2}{R_4}}$	$\sqrt{\dfrac{R_2}{R_4}} \times \dfrac{R_3}{R_2}$		$-\dfrac{R_4}{R_1}$	$-\dfrac{R_3}{R_1}$	$-\dfrac{R_2}{R_1}$		
3a	$\dfrac{F_{clock}}{100(50)} \times \sqrt{\dfrac{R_2}{R_4}}$	$\sqrt{\dfrac{R_2}{R_4}} \times \dfrac{R_3}{R_2}$	$\dfrac{F_{clock}}{100(50)} \times \sqrt{\dfrac{R_H}{R_L}}$	$-\dfrac{R_4}{R_1}$	$-\dfrac{R_3}{R_1}$	$-\dfrac{R_2}{R_1}$	$Q\left[\dfrac{R_G R_2}{R_H R_1} - \dfrac{R_G R_4}{R_L R_1}\right]$ (at F_0)	
4	$\dfrac{F_{clock}}{100(50)}$	$Q_{pole}\ \dfrac{R_3}{R_2}$ $Q_{zero}\ \dfrac{R_3}{R_1}$		$-\left[\dfrac{R_2}{R_1}+1\right]$	$-\dfrac{R_3}{R_2}\left(1+\dfrac{R_2}{R_1}\right)$			$-\dfrac{R_2}{R_1}$ ($R_2 = R_1$ for all-pass)
5	$\dfrac{F_{clock}}{100(50)} \times \sqrt{1+\dfrac{R_2}{R_4}}$	$Q_{pole}\ \dfrac{\sqrt{1+\dfrac{R_2}{R_4}}}{R_2/R_3}$ $Q_{zero}\ \dfrac{\sqrt{1-\dfrac{R_1}{R_4}}}{R_1/R_3}$	$\dfrac{F_{clock}}{100(50)} \times \sqrt{1-\dfrac{R_1}{R_4}}$	$-\left(\dfrac{R_2+R_1}{R_2+R_4}\right) \times \dfrac{R_4}{R_1}$	$-\left(\dfrac{R_2}{R_1}+1\right) \times \dfrac{R_3}{R_2}$			
6a	$F_c = \dfrac{F_{clock}}{100(50)}\ \dfrac{R_2}{R_3}$			$-\dfrac{R_3}{R_1}$		$-\dfrac{R_2}{R_1}$		

458

stability and accuracy as a direct function of the clock. Circuit Q is determined by the ratio R_3/R_2. A low-pass output is available as well.

Mode 2

The mode 2 configuration can provide band-pass and low-pass outputs as well as a notch output, providing that $F_{notch} \le F_0$. Unlike the mode 1 configuration, center frequency F_0 can be different from F_{notch}. This circuit is useful for cascading second-order sections to create an elliptic-function high-pass, band-reject, or band-pass response requiring transmission zeros below the passband poles.

Mode 3

The circuit of mode 3 allows us to use a single resistor ratio (R_2/R_4) to tune F_0. Because the feedback loop is closed around the input-summing amplifier, a slight Q enhancement occurs where the circuit Q is slightly above the actual design Q. A small capacitor across R_4 (10 to 100 pF) will result in a compensation phase lead, which eliminates the Q enhancement.

Mode 3a

Mode 3a is probably the most versatile of all modes. An external operational amplifier sums together the low-pass and band-pass outputs of mode 3 to form a notch output, which can be on either side of center frequency F_0. This is especially useful for an elliptic-function low-pass, high-pass, band-reject, or band-pass response requiring transmission zeros both above and below the passband poles. Gains can be distributed between the internal circuitry and the external operational amplifier to achieve maximum dynamic range.

Mode 4

An all-pass output can be provided by using mode 4 in addition to low-pass and band-pass outputs. As in the case of mode 1, the center frequency F_0 is directly determined by $F_{clock} \div 50$ or $F_{clock} \div 100$. Resistor R_1 must be made equal to R_2 for all-pass characteristics.

Mode 5

This mode can be used for implementing complex numerator zeros for a transfer function and offers low-pass and high-pass outputs as well.

Mode 6a

A single low-pass or high-pass real pole can be implemented using the mode 6a configuration. The major benefit of the added complexity of this implementation over a single RC is that the 3-dB cutoff F_c is clock-dependent. This makes this approach very useful for filters requiring variable characteristics controlled by an external clock frequency. If the cutoff is fixed, an ordinary RC implementation of the real pole is more cost-effective.

11.3.3   Using the MF10 and LMF100 Dual Universal Second-Order Filter

Figure 11-11 is a block diagram of an MF10 in a 20-pin package available as either a DIP (dual inline package) or a surface-mount version. In addition to the pins that make up the basic switched-capacitor filter, other pin functions are provided as follows.

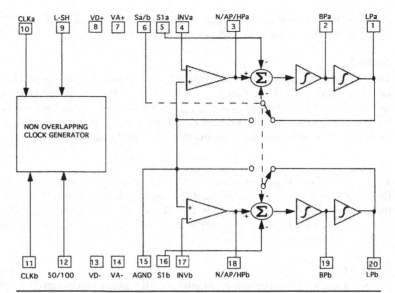

FIGURE 11-11 The block diagram of the MF10 dual universal second-order filter.

Power Supply Connections

Typically, the MF10 operates from ±5-V supplies. Separate digital and analog positive and negative power supply pins are provided, which are VD+ (pin 8), VA+ (pin 7), VD– (pin 13), and VA– (pin 14). Pins 7 and 8 can be tied together, provided that adequate bypassing exists. The same applies to pins 13 and 14. Analog ground (pin 15) is normally connected to ground, and the level shift function (pin 9) is also connected to ground.

The MF10 is also capable of operating from a single positive supply. In that case, the VD– and VA– pins, as well as level shift, are connected to system ground. The analog ground (pin 15) should be biased at half of the positive rail and adequately decoupled to system ground.

50/100 Clock Select

The 50/100 function (pin 17) determines whether the clock is either in the 50:1 mode or the 100:1 mode. When pin 17 is connected to VD+ and VA+, the mode is 50:1. If pin 17 is connected to analog ground, the mode is 100:1. Some devices also allow pin 17 to provide a sample-and-hold capability to freeze the outputs when connecting pin 17 to VD– and VA–.

Clock Input Pins

Two separate TTL–compatible clock input pins for the A section and B section filters are provided on pins 10 and 11, respectively. The clock duty cycle should be close to 50 percent.

$S_{a/b}$ Control

This function controls the two CMOS switches that connect the summer input to either analog ground or the low-pass output function. When $S_{a/b}$ is connected to the negative rail, both switches go to analog ground. If $S_{a/b}$ is connected to the positive rail, both switches are connected to the two low-pass outputs. The switch positions are determined by the requirements of the operating mode.

Example 11-1   Design of a Switched-Capacitor Elliptic-Function Low-Pass Filter

Required:

Design a switched-capacitor low-pass filter that has a 0.5-dB maximum ripple below 100 Hz and a greater than 18-dB rejection above 155.6 Hz. In addition, the filter should be programmable for a cutoff at any frequency in the range of 100 Hz to 10 kHz, while maintaining the frequency ratio of 1.6 for a greater than 18-dB attenuation.

Results:

(*a*) Open *Filter Solutions.*
    Check the *Stop Band Freq* box.
    Enter **0.177** in the *Pass Band Ripple (dB)* box.
    Enter **100** in the *Pass Band Freq* box.
    Enter **155.6** in *Stop Band Freq* box.
    Check the *Frequency Scale Hertz* box.

(*b*) Click the *Set Order* control button to open the second panel.
    Enter **18** for the *Stop Band Attenuation (dB).*
    Click the *Set Minimum Order* button and then click *Close.*
    3 *Order* is displayed on the main control panel.

(*c*) Click the *Transfer Function* button.
    Check the *Casc* box.

    The following is then displayed:

Continuous Transfer Function

Wn = 1095

$$\frac{306.6 \quad (S^2 + 1.199e{+}06)}{(S^2 + 358.2{*}S + 5.529e{+}05) \quad (S + 664.8)}$$

Wo = 743.6
Q = 2.076

3rd Order Low Pass Elliptic

Pass Band Frequency = 100.0 Hz   Stop Band Ratio = 1.556
Pass Band Ripple = 177.0 mdB   Stop Band Frequency = 155.6 Hz
                                  Stop Band Attenuation = 18.56 dB

(*d*) The design parameters are summarized as follows:

$$\text{Section } Q = 2.076$$
$$\text{Section } \omega_0 = 743.6$$
$$\text{Section } \omega_\infty = 1{,}095$$
$$\alpha_0 = 664.8 \text{ (from the denominator)}$$

(e) Convert rad/sec to Hz (divide by 2π) and restate the design parameters:

$$\text{Section } Q = 2.076$$
$$\text{Section } f_0 = 118.3 \text{ Hz}$$
$$\text{Section } f_\infty = 174.1 \text{ Hz}$$
$$\text{Real pole } f_c = 105.7 \text{ Hz}$$

(f) Design a two-section switched-capacitor low-pass filter as follows:

Section 1:

$$f_0 = 118.3 \text{ Hz}$$
$$Q = 2.076$$
$$f_\infty = 174.1 \text{ Hz}$$

Section 2:

Real pole where $f_c = 105.7$ Hz

Use a single MF10 dual second-order filter IC. The A section can be used to implement S.1 and the B section for S.2. The design proceeds as follows:

Section 1 (A Section):

Use mode 3a.

Let $F_{\text{clock}} = 5,000$ Hz

Select 50:1 clock mode.

Let $R_1 = R_4 = 10$ kΩ

Using the f_0 equation for mode 3a from Table 11-1 (where $f_0 = 118.3$ Hz):

$$R_2 = 13.99 \text{ k}\Omega$$

Using the Q equation for mode 3a from Table 11-1 (where $Q = 2.076$):

$$R_3 = 24.57 \text{ k}\Omega$$

Using the f_{notch} equation for mode 3a from Table 11-1 (where $f_{\text{notch}} = f_\infty = 174.1$ Hz):

Let $R_L = R_G = 10$ kΩ

then $R_H = 30.31$ kΩ

Section 2 (B Section):

Use mode 6a.

Let $F_{\text{clock}} = 5,000$ Hz

Using the f_c equation for mode 6a from Table 11-1 (where $f_c = 105.7$ Hz):

Let $R_1 = R_3 = 10$ kΩ

then $R_2 = 10.57$ kΩ

(g) The schematic of the resulting filter is shown in Fig. 11-12. The response curves for clock frequencies of 5,000 Hz and 500 kHz are given in Fig. 11-13, where the cutoffs are 100 Hz and 10 kHz. For any clock frequency between 5,000 Hz and 500 kHz, the filter cutoff will be exactly one-fiftieth of the clock frequency. All resistor values computed have been rounded off to their nearest 1 percent standard value.

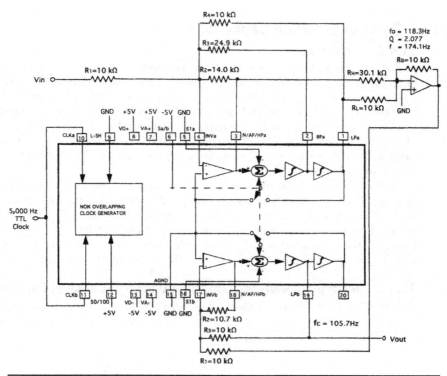

FIGURE 11-12   The low-pass filter of Example 11-1.

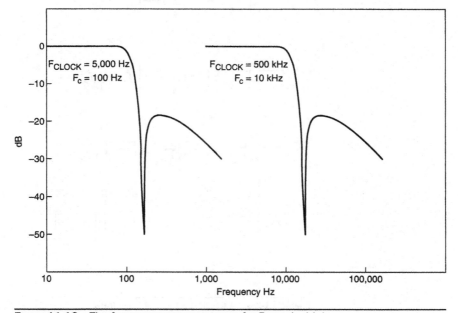

FIGURE 11-13   The frequency response curve for Example 11-1.

11.4   Types of Switched-Capacitor Filters

11.4.1   Universal

Universal switched-capacitor building blocks consist of independent universal second-order filters using the configuration of Fig. 11-3. Up to four sections can be contained within a single package. This architecture allows adjustment of filter parameters such as center frequency, Q, notch location, and gain. These parameters are also all controlled by relationships between resistors and the applied clock frequency. Figure 11-14 shows a block diagram of the Linear Technology LTC1064 and LTC1068 containing four second-order sections. Refer to the References and the manufacturers' websites for detailed application information on these devices.

Filters of varying complexity can be designed using these building blocks. The design procedure consists of transforming a complex filter transfer function into individual first- or second-order sections as required. Each section is defined in terms of parameters, such as center or cutoff frequency, Q, and notch frequency, and then implemented using cascaded ICs. This was illustrated in Example 11-1 for a third-order elliptic-function low-pass filter using the MF10 type of dual switched-capacitor filter.

11.4.2   Microprocessor-Programmable Universal Switched-Capacitor Filters

This family of switched-capacitor filters uses an external program to load in the filter mode of operation, center/corner frequency f_0 as a function of a clock division ratio,

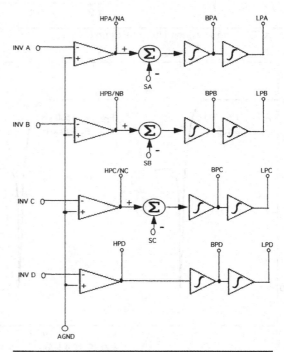

FIGURE 11-14   A block diagram of the Linear Technology LTC1064 and LTC1068.

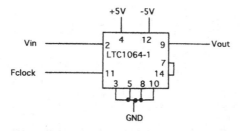

FIGURE 11-15   An eighth-order elliptic-function low-pass filter using the LTC1064-1.

and Q for each section in the form of coefficients in the ASCII format. No external components are required. MAXIM MAX260/261/262 devices fall into this family and are dual second-order universal switched-capacitor filters.

11.4.3   Pin-Programmable Universal Switched-Capacitor Filters

Unlike the microprocessor versions, these filters are defined by hard-wiring dedicated pins to either positive or negative voltages to define f_0 as a function of a clock division ratio and Q, as well as the mode of operation. This would be the MAXIM MAX263/264 product. The MAX267/268 is dedicated to the band-pass mode only, while the MAX 265/266 combines pin programming for f_0 and the use of resistors to set Q.

11.4.4   Dedicated Switched-Capacitor Filters

Dedicated switched-capacitor filters are already configured to provide a defined transfer function, and as a result do not require external resistors. Typically, they provide low-pass responses and are available with standard shapes such as Butterworth, Chebyshev, Bessel, and elliptic-function. Designing with these devices is simply a matter of connecting a clock with a frequency that is a defined multiple of the cutoff frequency and connecting power and ground to the device.

In some cases, the manufacturer simply customizes one of the universal filters by adding some components internally to the IC. A representative example of this approach is the Linear Technology LTC1064-1, which is shown in Fig. 11-15. This device is an eighth-order elliptic-function low-pass filter created from an LTC1064 and requires no external components. It has a passband ripple of ±0.15 dB and makes the transition from passband to stopband within a steepness factor of 1.5, achieving over 68 dB of stopband attenuation, where the clock-to-cutoff frequency ratio is 100:1.

11.5   The Switched-Capacitor Filter Selection Guide

Switched-capacitor filters offer circuit designers a solution to filtering needs while utilizing a minimal printed circuit board area as compared with passive and active filters. They feature high stability and precision, as well as the ability to change frequency characteristics under the control of an external clock. A selection guide to switched-capacitor filters is provided in Table 11-2.

Universal Switched-Capacitor Filters

Device	Type	Response	Order	Freq. Accur.	Max Center Freq.	Ratio Clock: F_0	Special Features
LMF100	Universal	Universal	Dual N = 2	±0.2%	100 kHz ($f_0 \times Q$ up to 1.8 MHz)	50:1, 100:1	MF10 pin-compatible
MF10	Universal	Universal	Dual N = 2	±0.6%	30 kHz ($f_0 \times Q$ up to 200 kHz)	50:1, 100:1	Industry standard
LTC1059	Universal	Universal	Single N = 2	±0.3%	40 kHz	50:1, 100:1	Low noise
LTC1060	Universal	Universal	Dual N = 2	±0.3%	30 kHz ($f_0 \times Q$ up to 1.6 MHz)	50:1, 100:1	MF-10 pin-compatible
LTC1061	Universal	Universal	Triple N = 2	±1.2%	35 kHz	50:1, 100:1	Low noise, single supply
LTC1064	Universal	Universal	Quad N = 2	±0.3%	140 kHz	50:1, 100:1	Low noise
LTC1067	Universal	Universal	Dual N = 2	±0.2%	20 kHz	100:1	Low noise, single supply
LTC1067-50	Universal	Universal	Dual N = 2	±0.2%	40 kHz	50:1	Low noise, single supply
LTC1068-200	Universal	Universal	Quad N = 2	±0.3%	25 kHz	200:1	Low noise, single supply
LTC1068	Universal	Universal	Quad N = 2	±0.3%	50 kHz	100:1	Low noise, single supply
LTC1068-50	Universal	Universal	Quad N = 2	±0.3%	50 kHz	50:1	Low noise, single supply
LTC1068-25	Universal	Universal	Quad N = 2	±0.3%	200 kHz	25:1	Low noise, single supply
LTC1164	Universal	Universal	Quad N = 2	±0.5%	20 KHz	50:1, 100:1	Low noise
MAX260	Universal	Universal	Dual N = 2	±1%	7.5 kHz	Programmable	Microprocessor programmable
MAX261	Universal	Universal	Dual N = 2	±1%	57 kHz	Programmable	Microprocessor programmable
MAX262	Universal	Universal	Dual N = 2	±1%	140 kHz	Programmable	Microprocessor programmable
MAX263	Universal	Universal	Dual N = 2	±1%	57 kHz	Programmable	Pin programmable
MAX264	Universal	Universal	Dual N = 2	±1%	140 kHz	Programmable	Pin programmable
MAX265	Universal	Universal	Dual N = 2	±0.2%	40 kHz	Programmable	Pin and resistor programmable
MAX266	Universal	Universal	Dual N = 2	±0.2%	140 kHz	Programmable	Pin and resistor programmable
MAX7490	Universal	Universal	Dual N = 2	±0.2%	40 kHz	100:1	Single supply +5V
MAX7491	Universal	Universal	Dual N = 2	±0.2%	40 kHz	100:1	Single supply +3V

TABLE 11-2 Switched-Capacitor Filter Selection Guide

	Dedicated Switched-Capacitor Filters[1]						
Device	Type	Response	Order	Freq Accur.	Max Cutoff Freq.	Ratio Clock: F_0	Special Features
LTC1062	Low-pass	Butterworth	5th	±1%	20 kHz	100:1	Zero DC offset
LTC1064-1	Low-pass	Elliptic	8th	±1%	20 kHz	100:1	Low noise
LTC1064-2	Low-pass	Butterworth	8th	±1%	140 kHz	50:1, 100:1	Low noise
LTC1064-3	Low-pass	Bessel	8th	±1%	95 kHz	75:1, 120:1, 150:1	Low noise
LTC1064-4	Low-pass	Elliptic	8th	–	100 kHz	50:1, 100:1	Low noise
LTC1069-1	Low-pass	Elliptic	8th	–	12 kHz	100:1	Single supply
LTC1069-6	Low-pass	Elliptic	8th	–	20 kHz	50:1	Single supply, low power
LTC1069-7	Low-pass	Linear phase	8th	–	200 kHz	25:1	Single supply
LTC1164-5	Low-pass	Butterworth or Bessel	8th	–	20 kHz	50:1, 100:1	Low harmonic distortion
LTC1164-6	Low-pass	Elliptic or Bessel	8th	–	30 KHz	50:1, 100:1	Low noise
LTC1164-7	Low-pass	Linear phase	8th	–	20 KHz	50:1, 100:1	
LTC1569-6	Low-pass	Linear phase	10th	±3.5%	64 kHz	1:1, 4:1,16:1	Cutoff frequency is resistor programmable
LTC1569-7	Low-pass	Linear Phase	10th	±3.5%	300 kHz	1:1, 4:1,16:1	Cutoff frequency is resistor programmable
MAX280-281	Low-pass	Butterworth/Bessel	5th	–	20 kHz	100:1	No DC error, internal or external clock
MAX291-294	Low-pass	Butterworth/Bessel/ elliptic	8th	–	25 KHz	100:1	Internal or external clocks
MAX295-297	Low-pass	Butterworth/Bessel/ elliptic	8th	–	50 KHz	50:1	Internal or external clocks
MAX7400/7403/ 7404/7407	Low pass	Elliptic	8th	–	10 KHz	100:1	Single supply, low power, internal or external clocks
MAX7401/7405	Low pass	Bessel	8th	–	5 kHz	100:1	Single supply, low power, internal or external clocks
MAX7418-7425 series	Low pass	Butterworth/Bessel/ elliptic	5th	–	45 kHz	100:1	Single supply, low power, internal or external clocks, offset adj.

[1]Note: Many of these dedicated devices are simply a customization of a mask of an existing universal product. This table is not all-inclusive since other customized products exist or can be created for a set of requirements.

TABLE 11-2 Switched-Capacitor Filter Selection Guide (*Continued*)

References

Linear Technology. (1990). *Monolithic Filter Handbook*. Milpitas, California: Linear Technology.

Linear Technology. (1990). *Linear Databook Supplement*. Milpitas, California: Linear Technology.

Maxim Integrated Circuit Products, Inc. (2005). *Application Note: The Basics of Anti-Aliasing: Using Switched Capacitor Filters*. Sunnyvale, California: Maxim Integrated Circuit Products, Inc.

National Semiconductor Corporation. (1985). *Switched-Capacitor Handbook*. Santa Clara, California: National Semiconductor Corporation.

National Semiconductor Corporation. (1988). *Linear Databook, Rev 1*. Santa Clara, California: National Semiconductor Corporation.

CHAPTER 12

Adjustable and Fixed Delay and Amplitude Equalizers

12.1 The Need for Equalization

12.1.1 Delay and Amplitude Equalization

In data communications over a channel, a phenomenon called intersymbol interference (ISI) can occur. Literally, this means that the transmission of a symbol representing a group of data bits can interfere with subsequent symbols, possibly resulting in data errors at the decision-making receiver. This is caused by delay and amplitude variations that occur over the band of interest, i.e., the channel is not ideal. Sometimes, delay equalization is required to compensate for the late arrival of some channels in a multichannel system. The process of delay and amplitude equalization of a communications channel is called line conditioning.

Modern communications technology has resulted in devices known as automatic equalizers, which compensate for this phenomenon in an automatic manner and converge into an equalized mode. However, there are some unique applications, such as private-line data systems or lines beyond the capabilities of automatic equalization, that require the use of adjustable equalizers to achieve optimum performance. Also, the frequency band of interest may be beyond the capabilities of automatic equalizers.

In the audio world, frequency response is a major concern, so at times the equalization of an audio channel is required to obtain maximum flatness. Typically, a graphic equalizer is used. The reason it is called a "graphic" equalizer is that it contains a cascade of band-pass type equalizers separated in frequency. Each equalizer has a slide switch that can add boost or amplitude reduction at its center frequency. So if you were to look at the slide switches and draw an imaginary line through the knobs, you would see a graph of the frequency response of the equalizer. In addition, a graphical equalizer may be used in audio to enhance an instrument's sound or to compensate for room acoustics.

For the remainder of this chapter we will discuss delay and amplitude equalizers. It is important to recognize that a delay equalizer can be designed that remains all-pass in its band of operation. In other words, as the delay is adjusted over a range, the amplitude versus frequency response will remain flat or all-pass. It will pass all frequencies over its band equally. However, an amplitude equalizer will affect delay because as its response is altered, the phase shift curve will vary as well.

Note that in Chap. 7 delay and amplitude equalizers were presented in the context of equalizing filters. In this chapter we will expand the design of equalizers in a more general-purpose manner for multiple applications.

12.2   The Equalization Process

12.2.1   Amplitude Equalization

Equalization is the process of altering a frequency response by adding a network, which when added to the existing frequency response, results in the desired shape. This applies to both amplitude and delay as a function of frequency. An idealized example of amplitude equalization is presented in Fig. 12-1 showing roll-off of an audio channel.

In the real world, it is not always easy to design a complementary equalizer that precisely flattens the response as shown in Fig. 12-1. As a result, adjustable equalizers offer a significant advantage over fixed equalizers since they can be tailored to optimally match the complementary characteristic. If the initial response changes, the equalizers can be readjusted to match the new requirements. In order to achieve a resulting flat response of 0 dB over the entire band, the amplitude equalizers must be capable of introducing gain as well as loss, so they must be active.

Figure 12-2 shows the response curve of a typical active amplitude equalizer. The amplitude at the center frequency of 2 KHz can be varied over a ±10-dB range.

A graphic equalizer contains a number of these sections spread out over the frequency band. By adjusting each section using a slider switch, which is actually a potentiometer (variable resistor), an equalization curve can be created that is shaped like a line running through the positions of the switch. An example is shown in Fig. 12-3.

It is often difficult to precisely create a matching equalization curve, so the net result may not be perfectly flat as shown in Fig. 12-1. It may more resemble the curve of Fig. 12-4 containing ripples.

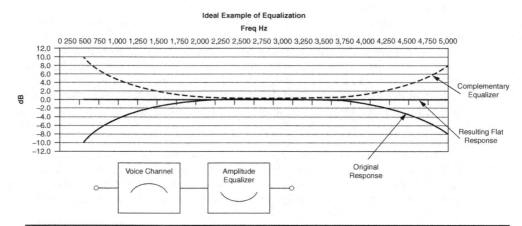

FIGURE 12-1   Ideal example of amplitude equalization of an audio channel.

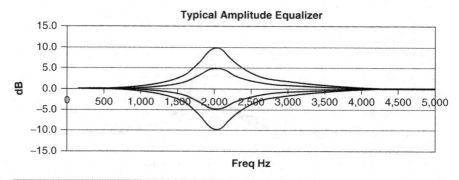

FIGURE 12-2   Frequency response of typical amplitude equalizer.

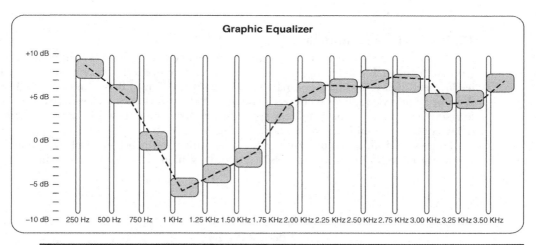

FIGURE 12-3   Graphic equalizer.

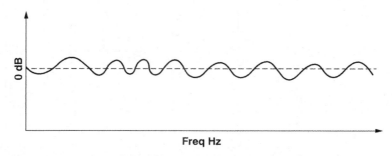

FIGURE 12-4   Equalized curve may contain ripples.

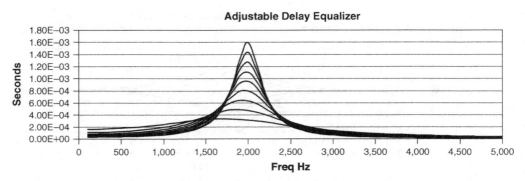

FIGURE 12-5   Adjustable delay equalizer, single section.

The magnitude of the ripples depends on the effort involved in equalization and the number of equalizers.

12.2.2   Delay Equalization

In a similar manner to amplitude equalizers, a delay equalizer section can be designed having an adjustable delay. Figure 12-5 shows a delay equalizer section that can be varied in peak delay from approximately 330 μS up to 1.6 mS. It is centered at 2 KHz. A group of delay equalizer sections can be placed at various frequencies to cover a band in a similar manner to the graphic amplitude equalizer previously discussed. A typical delay equalizer would appear as shown in Fig. 12-6.

High-speed modems use an automatic equalization process involving sophisticated digital signal processing (DSP) methods, which are beyond the scope of this book.

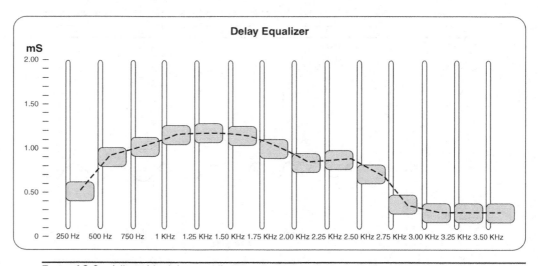

FIGURE 12-6   Adjustable-delay equalizer.

12.3 Pole-Zero Concept Applied to Amplitude and Delay Equalizers

Figure 12-7 shows some representative pole-zero patterns for an amplitude equalizer and delay equalizer. If we look at Fig. 12-7a, the poles and zeros are equidistant from the $j\omega$ axis since $\alpha_1 = \alpha_2$. Since the imaginary parts are equal as well, the effects of the poles and zeros on the amplitude of signals on the $j\omega$ axis cancel each other. Therefore, this is known as an all-pass circuit, as the frequency response is flat. Also, since the poles and zeros are symmetrical, this pole-zero constellation has "quadrantal symmetry."

In Fig. 12-7b the poles are closer to the $j\omega$ axis than the zeros, so there is no cancellation and the poles are dominant. The frequency response is peaked in the vicinity of the poles. In a similar manner, Fig. 12-7c shows the zeros closer to the $j\omega$ axis than the poles, so the response would show a dip in the vicinity of the zero location on the $j\omega$ axis.

So from an amplitude equalizer perspective, moving either poles or zeros closer to the axis results in a boost or dip in the frequency response near the pole or zero location on the $j\omega$ axis. This is what occurs as an amplitude equalizer is adjusted.

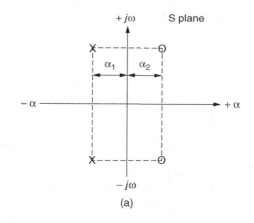

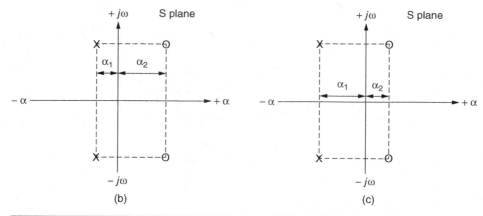

FIGURE 12-7 Pole-zero patterns: (a) all-pass $\alpha_1 = \alpha_2$; (b) amplitude boost $\alpha_1 < \alpha_2$; and (c) amplitude reduction $\alpha_1 > \alpha_2$.

If both poles and zeros are moved closer to the axis equally so that α_1 is always equal to α_2, the circuit will always remain all-pass. However, the effect of the poles and zeros on the phase shift near the center frequency increases the phase slope, so the delay will increase as shown in Fig. 12-5, so this becomes our delay equalizer.

No circuit can provide a change in amplitude without affecting its phase characteristic. Therefore, an amplitude equalizer will change its delay in the vicinity of its center frequency as the amount of amplitude equalization is adjusted. However, a delay equalizer remains all-pass as its delay is adjusted. Some circuits discussed in this section will allow the adjustment of both delay and amplitude in the same section.

12.4   Adjustable-Delay and Amplitude Equalizer Circuits

12.4.1   *LC* Delay Equalizers

The circuit of Fig. 12-8a illustrates a simplified adjustable *LC* delay equalizer section. The emitter and collector load resistors R_e and R_c are equal, so Q_1 serves as a phase splitter. Transistor Q_2 is an emitter follower output stage.

The equivalent circuit is shown in Fig. 12-8b. The transfer function can be determined by superposition as

$$T(s) = \frac{s^2 - \dfrac{1}{RC}s + \dfrac{1}{LC}}{s^2 + \dfrac{1}{RC}s + \dfrac{1}{LC}} \tag{12-1}$$

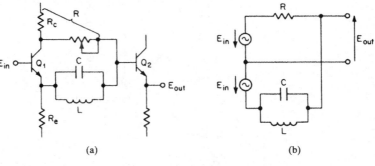

(a)   (b)

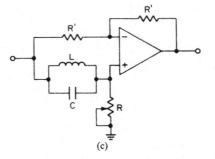

(c)

Figure 12-8   Adjustable *LC* delay equalizer: (a) the adjustable equalizer; (b) equivalent circuit; and (c) operational-amplifier realization.

This expression is of the same form as the general second-order all-pass transfer function of Eq. (7-6). By equating coefficients, we obtain

$$\omega_r = \frac{1}{\sqrt{LC}} \tag{12-2}$$

and

$$Q = \omega_r RC \tag{12-3}$$

Equation 12-3 can be substituted in Eq. (7-13) for the maximum delay of a second-order section, resulting in

$$T_{gd,\,max} = 4RC \tag{12-4}$$

By making R variable, the delay can be directly controlled while retaining the all-pass properties. The peak delay will occur at or near the LC resonant frequency.

The all-pass transfer function of Eq. 12-1 can also be implemented using an operational amplifier. This configuration is shown in Fig. 12-8c, where R' is arbitrary. Design Eqs. (12-2) to (12-4) still apply.

Example 12-1   Design of an Adjustable LC Delay Equalizer Using the Two-Transistor Circuit

Required:

Design an adjustable LC delay equalizer using the two-transistor circuit of Fig. 12-8a. The delay should be variable from 0.5 to 2.5 ms with a center frequency of 1,700 Hz.

Result:

Using a capacitor C of 0.05 μF, the range of resistance R is given by Eq. (12-4) as shown:

$$R_{min} = \frac{T_{gd,\,max}}{4C} = \frac{0.5 \times 10^{-3}}{4.05 \times 10^{-6}} = 2,500\ \Omega$$

$$R_{max} = \frac{2.5 \times 10^{-3}}{4 \times 0.05 \times 10^{-6}} = 12.5\ \text{k}\Omega$$

The inductor is computed by the general formula for resonance, $\omega^2 LC = 1$, resulting in an inductance of 175 mH. The circuit is shown in Fig. 12-9a. The emitter resistor R_e is composed of two resistors for proper biasing of phase splitter Q_1, and electrolytic capacitors are used for DC blocking. The delay extremes are shown in the curves of Fig. 12-9b.

12.4.2   *LC* Delay and Amplitude Equalizers

Frequently, the magnitude response of a transmission channel must be equalized along with the delay. An equalizer circuit featuring both adjustable amplitude and delay is shown in Fig. 12-10a. Transistor Q_1 serves as a phase splitter where the signal applied to the emitter follower Q_2 is K times the input signal. The equivalent circuit is illustrated in Fig. 12-10b. The transfer function can be determined by superposition as

$$T(s) = \frac{s^2 - \dfrac{K}{RC}s + \dfrac{1}{LC}}{s^2 + \dfrac{K}{RC}s + \dfrac{1}{LC}} \tag{12-5}$$

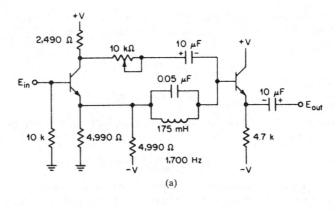

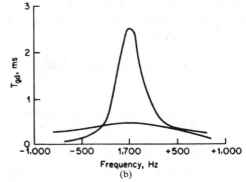

Figure 12-9 The adjustable delay equalizer of Example 12-1: (a) equalizer circuit; and (b) the delay adjustment range.

If K is set equal to unity, the expression is then equivalent to Eq. (12-1) corresponding to a second-order all-pass transfer function. As K increases or decreases from unity, a boost or null occurs at midfrequency with an asymptotic return to unity gain at DC and infinity.

The amount of amplitude equalization at midfrequency in decibels is given by

$$A_{dB} = 20 \log K \tag{12-6}$$

The maximum delay occurs at the LC resonant frequency and can be derived as

$$T_{gd,\,max} = \frac{2RC}{K} + 2RC \tag{12-7}$$

If K is unity, Eq. (12-7) reduces to $4\,RC$, which is equivalent to Eq. (12-4) for the all-pass circuit of Fig. 12-8.

An operational-amplifier implementation is also shown in Fig. 12-10c. The value of R' is arbitrary, and Eqs. 12-6 and 12-7 are still applicable.

The following conclusions may be reached based on the evaluation of Eqs. (12-5) to (12-7):

1. The maximum delay is equal to $4\,RC$ for $K = 1$, so R is a delay magnitude control.

2. The maximum delay will be minimally affected by a nonunity K, as is evident from Eq. 12-7.

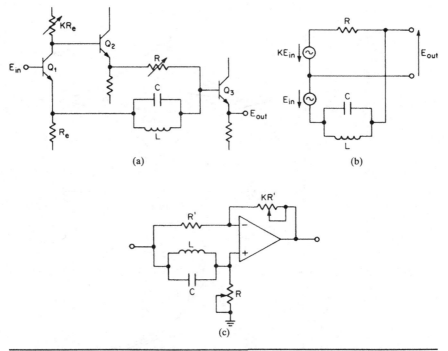

FIGURE 12-10 An adjustable *LC* delay and amplitude equalizer: (*a*) adjustable delay and amplitude equalizer; (*b*) equivalent circuit; and (*c*) an operational-amplifier realization.

3. The amount of amplitude equalization at the *LC* resonant frequency is independent of the delay setting and is strictly a function of *K*. However, the selectivity of the amplitude response is a function of the delay setting and becomes more selective with increased delay.

The curves of Fig. 12-11 show some typical delay and amplitude characteristics. The interaction between delay and amplitude is not restricted to *LC* equalizers and will occur whenever the same resonant element, either passive or active, is used to provide both the amplitude and delay equalization. However, for small amounts of amplitude correction, such as ±3 dB, the effect on the delay is minimal.

12.4.3 Active Delay and Amplitude Equalizers

The dual-amplifier band-pass (DABP) delay equalizer structure of Sec. 7.2, under "Active All-Pass Structures," and Fig. 7-13 has a fixed gain and remains all-pass regardless of the design *Q*. If resistor R_1 is made variable, the *Q*, and therefore the delay, can be directly adjusted with no effect on resonant frequency or the all-pass behavior. The adjustable delay equalizer is shown in Fig. 12-12*a*. The design equations are

$$T_{gd,\,max} = 4R_1C \tag{12-8}$$

and

$$R_2 = R_3 = \frac{1}{\omega_r C} \tag{12-9}$$

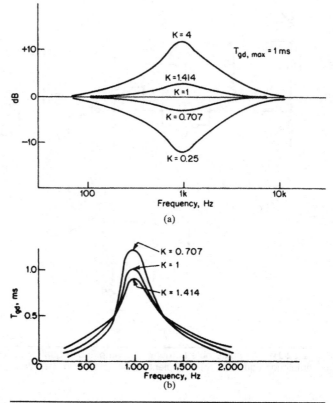

FIGURE 12-11 A typical delay and amplitude response for an *LC* delay and amplitude equalizer: (*a*) amplitude characteristics for a fixed delay; and (*b*) the delay variation for ±3 dB of amplitude equalization.

where *C*, *R*, *R'*, and *R''*, can be conveniently chosen. Resistor R_2 can be made variable for frequency trimming.

If amplitude equalization capability is also desired, a potentiometer can be introduced, resulting in the circuit of Fig. 12-12*b*. The amplitude equation at ω_r is given by

$$A_{dB} = 20 \log (4K - 1) \qquad (12\text{-}10)$$

where a *K* variation of 0.25 to 1 covers an amplitude equalization range of $-\infty$ to +9.5 dB.

To extend the equalization range above +9.5 dB, an additional amplifier can be introduced, as illustrated in Fig. 12-12*c*. The amplitude equalization at ω_r is then obtained from

$$A_{dB} = 20 \log (2K - 1) \qquad (12\text{-}11)$$

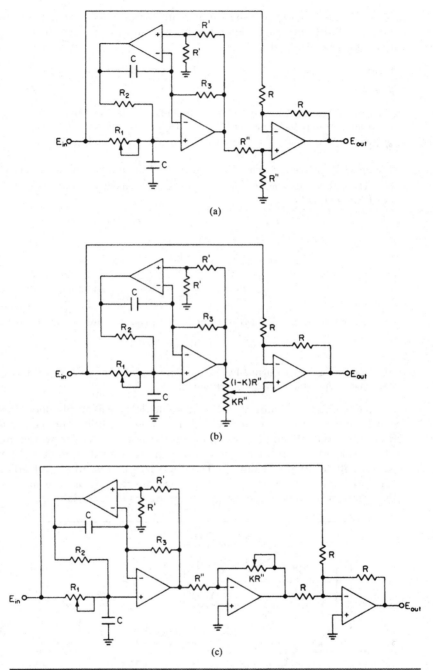

FIGURE 12-12 A DABP delay and amplitude equalizer: (*a*) adjustable delay equalizer; (*b*) adjustable delay and amplitude equalizer; and (*c*) adjustable delay and amplitude equalizer with extended amplitude range.

where a K variation of 0.5 to ∞ results in an infinite range of equalization capability. In reality, a ±15-dB maximum range has been found to be more than adequate for most equalization requirements.

Example 12-2   Design of an Active Adjustable Delay and Amplitude Equalizer

Required:

Design an adjustable active delay and amplitude equalizer that has a delay adjustment range of 0.5 to 3 ms, an amplitude range of ±12 dB, and a center frequency of 1,000 Hz.

Result:

The circuit of Fig. 12-12c will provide the required delay and amplitude adjustment capability.

If we choose $C = 0.01$ μF and $R = R' = R'' = 10$ kΩ, the element values are computed as follows using Eqs. (12-8) and (12-9):

$$R_{1,\,min} = \frac{T_{gd,\,max}}{4C} = \frac{0.5 \times 10^{-3}}{4 \times 10^{-8}} = 12.5 \text{ k}\Omega$$

$$R_{1,\,max} = \frac{3 \times 10^{-3}}{4 \times 10^{-8}} = 75 \text{ k}\Omega$$

$$R_2 = R_3 = \frac{1}{\omega_r C} = \frac{1}{2\pi \times 1,000 \times 10^{-8}} = 15.9 \text{ k}\Omega$$

The extreme values of K for ±12 dB of amplitude equalization are found from Eq. (12-11) as shown:

$$K = \frac{1}{2}\left[\log^{-1}\left(\frac{A_{dB}}{20} \right) + 1 \right]$$

The range of K is then 0.626 to 2.49. The equalizer section is shown in Fig. 12-13. Resistor R_2 has also been made adjustable for frequency trimming.

An active delay equalizer with adjustable delay was implemented by combining a second-order band-pass section with a summing amplifier. The band-pass section was required to have a fixed gain and a resonant frequency, which were both independent of the Q setting. If amplitude equalization alone is needed, the band-pass section can operate with a fixed design Q. The low-complexity multiple-feedback band-pass (MFBP) delay equalizer section of Fig. 7-12a can then be used as an adjustable amplitude equalizer by making one of the summing resistors variable.

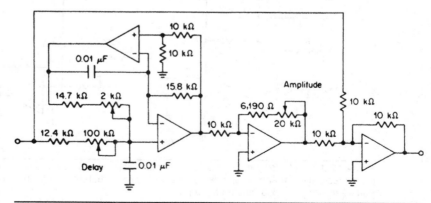

Figure 12-13   The adjustable delay and amplitude equalizer of Example 12-2.

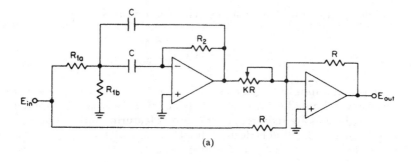

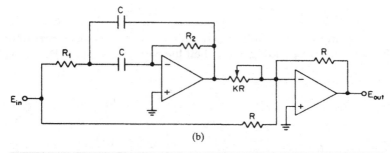

Figure 12-14   MFBP amplitude equalizer: (a) amplitude equalizer 0.707 < Q < 20; and (b) amplitude equalizer 0 < Q < 20.

This circuit is shown in Fig. 12-14a. The design equations are given by

$$R_2 = \frac{2Q}{\omega_r C} \tag{12-12}$$

$$R_{1a} = \frac{R_2}{2} \tag{12-13}$$

$$R_{1b} = \frac{R_{1a}}{2Q^2 - 1} \tag{12-14}$$

The amount of amplitude equalization at ω_r is computed from

$$A_{dB} = 20 \log\left(\frac{1}{K} - 1\right) \tag{12-15}$$

where K will range from 0 to 1 for an infinite range of amplitude equalization.

If the Q is below 0.707, the value of R_{1b} becomes negative, so the circuit of Fig. 12-14b is used. R_2 is given by Eq. (12-12), and R_1 is found from

$$R_1 = \frac{R_2}{4Q^2} \tag{12-16}$$

The attenuation or boost at resonance is computed from

$$A_{dB} = 20 \log\left(\frac{2Q^2}{K} - 1\right) \tag{12-17}$$

The magnitude of Q determines the selectivity of the response in the region of resonance and is limited to values typically below 20 because of amplifier limitations.

If higher Qs are required, or if a circuit featuring independently adjustable Q and amplitude equalization is desired, the DABP circuits of Fig. 12-12 may be used, where R_1 becomes the Q adjustment and is given by

$$R_1 = QR_2 \tag{12-18}$$

To compute the required Q of an amplitude equalizer, first define f_b, which is the frequency corresponding to one-half the pad loss (in decibels). The Q is then given by

$$Q = \frac{f_b b^2 \sqrt{K_r}}{f_r(b^2 - 1)} \tag{12-19}$$

where
$$K_r = \log^{-1}\left(\frac{A_{dB}}{20}\right) = 10^{A_{dB}/20} \tag{12-20}$$

and
$$b = \frac{f_b}{f_r} \tag{12-21}$$

or
$$b = \frac{f_r}{f_b} \tag{12-22}$$

whichever b is greater than unity.

Example 12-3   Design of an Active Fixed Amplitude Equalizer

Required:

Design a fixed active amplitude equalizer that provides a +12-dB boost at 3,200 Hz and has a boost of +6 dB at 2,500 Hz.

Result:

(a) First compute the following using Eqs. (12-20) and (12-22):

$$K_r = 10^{A_{dB}/20} = 10^{12/20} = 3.98$$

and
$$b = \frac{f_r}{f_b} = \frac{3,200 \text{ Hz}}{2,500 \text{ Hz}} = 1.28$$

The Q is then found from Eq. (12-19) as shown:

$$Q = \frac{f_b b^2 \sqrt{K_r}}{f_r(b^2 - 1)} = \frac{2,500 \times 1.28^2 \sqrt{3.98}}{3,200(1.28^2 - 1)} = 4.00$$

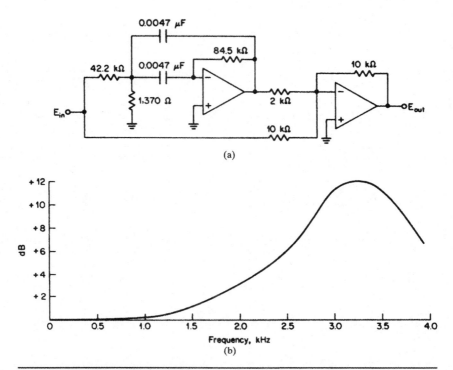

(a)

(b)

FIGURE 12-15   The amplitude equalizer of Example 12-3: (a) amplitude equalizer circuit; and (b) frequency response.

(b) The MFBP amplitude equalizer circuit of Fig. 12-14a will be used. Using a C of 0.0047 μF and an R of 10 kΩ, the element values are given by Eqs. (12-12) to (12-15) as shown:

$$R_2 = \frac{2Q}{\omega_r C} = \frac{2 \times 4}{2\pi 3,200 \times 4.7 \times 10^{-9}} = 84.6 \text{ k}\Omega$$

$$R_{1a} = \frac{R_2}{2} = 42.3 \text{ k}\Omega$$

$$R_{1b} = \frac{R_{1a}}{2Q^2 - 1} = \frac{42.3 \times 10^3}{2 \times 4^2 - 1} = 1,365 \ \Omega$$

$$K = \frac{1}{1 + 10^{A_{dB}/20}} = \frac{1}{1 + 10^{12/20}} = 0.200$$

The equalizer circuit and corresponding frequency response are shown in Fig. 12-15.

References

Lindquist, C. S. (1977). *Active Network Design*. Long Beach, California: Steward and Sons.

Williams, A. B. "An Active Equalizer with Adjustable Amplitude and Delay," *IEEE Transactions on Circuit Theory* CT-16, November 1969.

CHAPTER 13

Voltage Feedback Operational Amplifiers

13.1 Review of Basic Op-Amp Theory

Analog design is heavily based on using the versatile operational amplifier as the major building block. The versatility and low cost of integrated-circuit (IC) operational amplifiers (op amps) have made them one of the most popular building blocks in the industry. The op amp is capable of performing many mathematical processes upon signals. For active filters, op amps are specifically used to provide gain and isolation.

IC op amps have evolved from the invention of the Fairchild μA 709 in the mid-1960s to the many different types available today with a variety of special features. This section reviews some of the essential characteristics, discusses some important considerations, and provides a survey of some popular IC amplifier types. The reader is encouraged to visit the many websites of op-amp manufacturers for detailed technical data, selection guides, and applications assistance if needed.

There are two basic categories of operational amplifiers: *voltage feedback* and *current feedback*. This chapter covers voltage feedback, and Chap. 18 discusses current feedback. The differences will be obvious as the material is covered. This does not mean they are not interchangeable in many cases, but the user has to know the properties and limitations of each type to make that decision.

Throughout the book most schematics involving ICs are generic, where their function is to convey a concept. The individual data sheet for the device chosen will define in detail a recommended range of component values and specifics on device operation.

13.1.1 The Ideal Amplifier

A simplified equivalent circuit of an operational amplifier is shown in Fig. 13-1. The open circuit output voltage e_0 is the difference of the input voltages at the two input terminals amplified by amplifier gain A_o. A positive changing signal applied to the positive (+) input terminal results in a positive change at the output, whereas a positive changing signal applied to the negative input terminal (−) results in a negative change at the output. Hence, the positive input terminal is called the *noninverting* input, and the negative input terminal is referred to as the *inverting* input. If we consider the amplifier ideal, the input impedance R_i is infinite, the output impedance R_0 is zero, the voltage gain A_0 is infinite, and the bandwidth is infinite. This, of

Figure **13-1** Equivalent circuit of
operational amplifier.

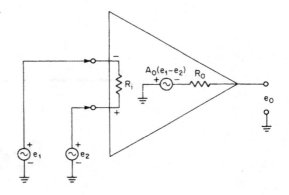

Figure **13-1** Equivalent circuit of
operational amplifier.

course, will not be the case in the real world, but let us assume so for the purposes
of this explanation.

Let us assume the positive input to the op amp is connected to ground. Negative
feedback is applied from the output to the inverting input. Since the gain A_0 is assumed
infinite, an infinitesimally small (almost zero) differential input between the – and +
terminals of the op amp results in an output determined by the feedback.

So the negative terminal is at the same potential as the positive grounded terminal
as an infinitesimally small signal is present at the terminal. Hence, the negative terminal is at ground. This effect is called the *virtual ground effect*. Although the negative terminal is effectively at ground, no current can flow into the op amp from this terminal,
as the input impedance is assumed infinite ($R_i = \infty$).

This property is extremely useful, as it permits a wide variety of different amplifier
configurations and greatly simplifies circuit analysis.

13.1.2 Inverting Amplifier

Let us first consider the basic inverting amplifier circuit of Fig. 13-2. If we consider the
amplifier ideal, the differential input voltage becomes zero because of the negative
feedback path through R_2, resulting in the virtual ground effect. Therefore, the inverting input terminal is at ground potential. The currents through resistors R_1 and R_2 are

$$I_1 = \frac{E_{in}}{R_1} \tag{13-1}$$

Figure **13-2** Inverting amplifier.

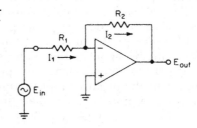

FIGURE 13-3 Intuitive representation of an ideal inverting amplifier.

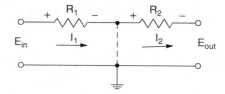

and

$$I_2 = -\frac{E_{out}}{R_2} \qquad (13\text{-}2)$$

If the amplifier input impedance is infinite, no current can flow into the inverting terminal, so $I_1 = I_2$. If we equate Eqs. (13-1) and (13-2) and solve for the overall transfer function, we obtain

$$\frac{E_{out}}{E_{in}} = -\frac{R_2}{R_1} \qquad (13\text{-}3)$$

The circuit amplification is determined directly from the ratio of two resistors and is *independent* of the amplifier itself. Also, the input impedance is R_1 and the output impedance is zero.

The diagram of Fig. 13-3 is more intuitive. The virtual ground is represented by the dashed line. The sign inversion of Eq. (13-3) occurs because of the current being applied to the inverting input of the amplifier, so E_{out} is inverted relative to E_{in}.

The input impedance, or impedance seen by E_{in}, is R_1 because of the virtual ground effect.

Multiple inputs can be summed at the inverting input terminal as a direct result of the virtual ground effect. The triple-input summing amplifier of Fig. 13-4 has the following output based on superposition:

$$E_{out} = -\frac{R_2}{R_{1a}} E_a - \frac{R_2}{R_{1b}} E_b - \frac{R_2}{R_{1c}} E_c \qquad (13\text{-}4)$$

It is important to note that the virtual ground effect occurs as a result of the feedback resistor R_2. Because the output is fed back to the *inverting* input, this is called *negative feedback*.

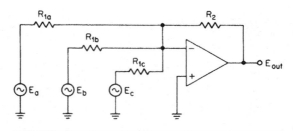

FIGURE 13-4 Summing amplifier.

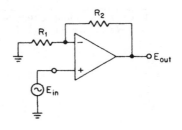

FIGURE **13-5** Noninverting amplifier.

13.1.3 Noninverting Amplifier

A noninverting amplifier can be configured using the circuit of Fig. 13-5. Since the differential voltage between the amplifier input terminals is zero, the voltage across R_1 is E_{in}. Since R_1 and R_2 form a voltage divider, we can state

$$\frac{E_{out}}{E_{in}} = \frac{R_1 + R_2}{R_1} \qquad (13\text{-}5)$$

or the more popular form

$$\frac{E_{out}}{E_{in}} = 1 + \frac{R_2}{R_1} \qquad (13\text{-}6)$$

where the input impedance is infinite and the output impedance is zero.

Figure 13-6 is an intuitive representation of Fig. 13-5. E_{in}, although applied to the + input of the amplifier, appears at the junction of R_1 and R_2 because of the virtual ground effect between the − and + terminals. R_1 and R_2 form a simple voltage divider, so

$$E_{out} = E_{in} \frac{R_1 + R_2}{R_1} \qquad (13\text{-}7)$$

which is another form of Eq. (13-5). Note that the input impedance is infinite as E_{in} is applied to a node that draws no current.

If we set R_1 to infinity and R_2 to zero, the gain becomes unity, which corresponds to the voltage follower configuration of Fig. 13-7.

This circuit would be analogous to an emitter follower in terms of transistors, and for those who remember vacuum tubes, a cathode follower. For an ideal op amp, this circuit

FIGURE **13-6** Intuitive representation of ideal noninverting amplifier.

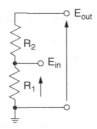

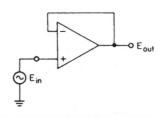

FIGURE **13-7** Voltage follower.

FIGURE 13-8 Voltage follower with feedback resistor.

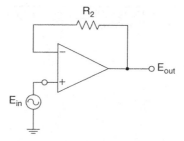

would have infinite input impedance, zero output impedance, infinite bandwidth, and be capable of providing infinite current. Of course, no such ideal amplifier exists.

For the circuit of Fig. 13-5, we can just set R_1 to infinity and leave R_2 at a convenient value. That would give us the voltage follower circuit of Fig. 13-8. The gain will still be +1. Including R_2 rather than zero ohms has a benefit with a practical amplifier, as it would limit the current in the feedback path under conditions where the amplifier output may go to extremes.

13.1.4 Differential Input Amplifier

The circuit of Fig. 13-9a is that of a differential input amplifier. The output is proportional to the *difference* of the two input voltages. Using superposition, let us compute the output. Setting E_2 to zero, the output is

$$E_{out} = -E_1 \frac{R_2}{R_1} \qquad (13\text{-}8)$$

Setting E_1 to zero results in

$$E_{out} = E_2 \frac{R_4}{R_3 + R_4} (1 + R_2/R_1) \qquad (13\text{-}9)$$

Combining Eqs. (13-8) and (13-9) results in

$$E_{out} = -E_1 \frac{R_2}{R_1} + E_2 \frac{R_4}{R_3 + R_4} (1 + R_2/R_1) \qquad (13\text{-}10)$$

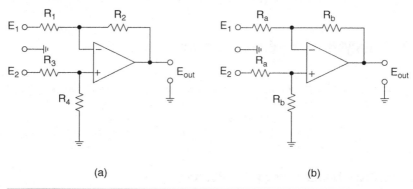

(a)   (b)

FIGURE 13-9 Differential input amplifier: (a) basic configuration and (b) gain becomes ratio of R_b/R_a.

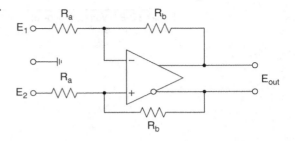

FIGURE **13-10** Differential input and output amplifier.

This is a good example of how superposition can be applied to a multiple-input circuit by taking one source at a time while the others are disabled and combining all the outputs. (Note that voltage sources are disabled by being set to zero and current sources are disabled by being set to an open circuit.)

If we let $R_1 = R_3 = R_a$ and $R_2 = R_4 = R_b$, with a bit of algebraic manipulation, we obtain

$$E_{\text{out}} = (E_2 - E_1)\, R_b / R_a \qquad (13\text{-}11)$$

This would be the circuit of Fig. 13-9*b*.

If we let all four resistors have the same value R, the gain would be unity.

13.1.5   Differential Input and Output Amplifier

An amplifier that has both differential inputs and outputs is shown in Fig. 13-10. Note that the lower output has a small circle, which implies an inversion relative to the other output. The gain of this circuit would be

$$E_{\text{out}} = (E_2 - E_1)\, R_b / R_a \qquad (13\text{-}12)$$

In this case, E_{out} is a differential signal between the two output terminals rather than relative to ground as in the previous cases.

The feedback elements in the previous figures were all resistors. They are by no means limited to resistors and can be reactive elements as well. The applications of operational amplifiers are not restricted to summing and amplification. If R_2 in Fig. 13-2, for example, were replaced by a capacitor, the circuit would serve as an integrator. Alternatively, a capacitor for R_1 would result in a differentiator. Nonlinear functions can be performed by introducing nonlinear elements into the feedback paths.

13.2   Analysis of Nonideal Amplifiers

NOTE   *This section discusses the implications of nonideal amplifiers. If an appropriate amplifier for the application is selected, the more complicated equations for the gain of noninverting and inverting amplifiers developed in this section need not be used. The simplified versions of Eqs. (13-3) and (13-6) will apply. However, it is useful to understand the implications of an amplifier that is not ideal, so that is the purpose of this section.*

13.2.1   Noninverting Amplifier Analysis

Let us revisit the noninverting amplifier of Fig. 13-5. The circuit is again illustrated in Fig. 13-11.

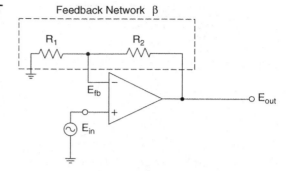

FIGURE 13-11 Noninverting amplifier showing feedback network.

Resistors R_1 and R_2 form a voltage divider where a portion β of the output is fed back to the inverting input. The output can be expressed as

$$E_{\text{out}} = A_o\,(E_{\text{in}} - E_{fb}) = A_o\,(E_{\text{in}} - \beta E_{\text{out}}) \tag{13-13}$$

where A_o is the open-loop gain of the amplifier.

We can then solve for the circuit gain, which is

$$A_c = \frac{E_{\text{out}}}{E_{\text{in}}} = \frac{1}{\beta}\,\frac{1}{1 + 1/A_o\beta} \tag{13-14}$$

A more convenient form of Eq. (13-14) is given by

$$A_c = \frac{A_0}{1 + A_0\beta} \tag{13-15}$$

This is the fundamental equation for the closed-loop gain of a noninverting amplifier where A_0 is the amplifier's open-loop gain and β is the feedback factor. The product $A_o\,\beta$ is traditionally known as *loop-gain*. This expression should be familiar to those who have studied feedback systems or servo theory. For an infinite gain and a feedback factor of 1, the circuit gain A_c becomes precisely 1. As A_o deteriorates to less than infinity, the circuit gain is reduced to less than 1.

13.2.2 Inverting Amplifier Analysis

Let us derive the expression for the gain of the inverting amplifier shown in Fig. 13-12. We can use superposition, so we first set E_{out} to zero. Then

$$E_{fb} = E_{\text{in}}\,\frac{R_2}{R_1 + R_2} \tag{13-16}$$

Next we set E_{in} to zero. Then

$$E_{fb} = E_{\text{out}}\,\frac{R_1}{R_1 + R_2} \tag{13-17}$$

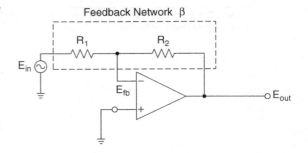

FIGURE 13-12   Inverting amplifier showing feedback network.

Combining Eq. (13-17) with Eq. (13-16) results in

$$E_{fb} = E_{out} \frac{R_1}{R_1 + R_2} + E_{in} \frac{R_2}{R_1 + R_2} \tag{13-18}$$

Equation 13-18 can be algebraically rearranged into the form

$$E_{fb} = (1 - \beta) E_{in} + \beta E_{out} \tag{13-19}$$

where

$$\beta = \frac{R_1}{R_1 + R_2}$$

Since the output voltage E_{out} is the open-loop gain $-A_o$ times the input, which in this case is E_{fb} (note the negative sign since the inverting input is used with the noninverting input at ground), the output voltage E_{out} in terms of the feedback voltage is given by

$$E_{out} = A_o (-E_{fb}) \tag{13-20}$$

Combining Eqs. (13-19) and (13-20) results in

$$A_c = \frac{E_{out}}{E_{in}} = \left(1 - \frac{1}{\beta} \right) \frac{1}{1 + 1/A_o \beta} \tag{13-21}$$

Note that β is always less than 1, so the gain is negative.
The more intuitive form of Eq. 13-21 is

$$A_c = \frac{A_0 (\beta - 1)}{1 + A_0 \beta} \tag{13-22}$$

13.2.3   Stability

In both cases, the noninverting amplifier and the inverting amplifier, the feedback factor, which corresponds to the portion of the output that is fed back to the input, is determined by

$$\beta = \frac{R_1}{R_1 + R_2} \tag{13-23}$$

Let us examine the term $1 + A_0 \beta$ corresponding to the denominator of the closed-loop gain expressions. The open-loop gain of practical amplifiers is neither infinite nor real (having zero phase shift). The magnitude and phase of A_0 will be a function of frequency.

If at some frequency $A_0\beta$ were equal to –1 (180° phase shift causing the sign inversion), the denominator of Eqs. (13-15) and (13-22) would become zero. The closed-loop gain then becomes *infinite,* which implies an oscillatory condition. In other words, an output would occur with no input.

To prevent oscillations, the amplifier open-loop gain must be band-limited so that the product $A_0\beta$ is less than 1 *below* the frequency where the amplifier phase shift reaches 180°. This is achieved by introducing a gain roll-off beginning at low-frequencies and continuing at approximately a 6 dB per octave rate. This technique of ensuring stability is called *frequency compensation.* It is evident from the closed-loop gain equations that for high closed-loop gains, β is diminished (the higher the closed-loop gain, the lower the β) so that less frequency compensation will be required to ensure $A_0\beta$ is less than 1 at the 180° phase shift frequency. Conversely, the voltage follower will need the most compensation, as it has the most feedback. An amplifier that will be stable with unity gain is referred to as *unity-gain stable.*

Note that in Fig. 13-13 the open-loop gain rolls off at approximately 20 dB/decade (which is the same as 6 dB/octave) above 5 KHz. At the frequency where the phase shift reaches 180°, the open-loop gain is about –20 dB (0.1), which, of course, is much less than 1. Therefore, stability is ensured even in the case of the voltage follower, where all of the output is fed back to the input since $\beta = 1$. The loop gain ($A_0\beta$) at the 180° phase shift point would be 0.1 for this example. The voltage follower would be the worst case as far as maintaining stability because β is at its maximum value. If oscillation were to occur, it would appear at a relatively high frequency and appear as "fuzz" if observed on an oscilloscope. However, with loop gains less than 1, the circuit would be unconditionally stable. Above 200 MHz, the curve of Fig. 13-13 has a second break point with increased slope. This region is unusable.

In both the inverting and noninverting circuits, the feedback is always connected to the negative terminal of the op amp. Hence, it is called *negative feedback.*

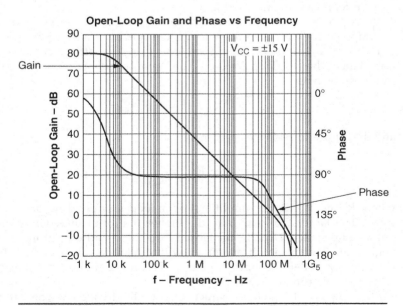

FIGURE 13-13 Open-loop gain (Texas Instruments THS4001).

13.2.4 Effects of Open-Loop Gain

To determine the effects of open-loop gain on closed-loop gain, let us substitute Eq. (13-23) into Eqs. (13-15) and (13-22) for the noninverting and inverting amplifiers. The resulting gain expressions are

for a noninverting amplifier
$$A_c = \frac{1 + \dfrac{R_2}{R_1}}{\dfrac{1}{A_0}\left(1 + \dfrac{R_2}{R_1}\right) + 1} \qquad (13\text{-}24)$$

and for an inverting amplifier
$$A_c = -\frac{\dfrac{R_2}{R_1}}{\dfrac{1}{A_0}\left(1 + \dfrac{R_2}{R_1}\right) + 1} \qquad (13\text{-}25)$$

If A_0 were infinite, both denominators would reduce to unity. Equations (13-24) and (13-25) would then simplify to Eqs. (13-3) and (13-6), the fundamental and simplified expressions for the gain of a noninverting and inverting amplifier.

To reduce the degrading effect of open-loop gain upon closed-loop gain, A_0 should be much higher than the desired A_c. Since the open-loop phase shift is usually 90° over most of the band of interest, the error term in the denominator of Eqs. (13-24) and (13-25) is in *quadrature* with unity. This quadrature (90°) relationship minimizes the effect of open-loop gain upon closed-loop gain when the feedback network is purely resistive. *An open-loop to closed-loop gain ratio of 10:1 (20 dB) will result in an error of only 0.5 percent, which is more than adequate for most requirements.* Figure 13-14 shows the effect of open-loop gain upon closed-loop gain.

An op amp has a finite open-loop input impedance and a non-zero open-loop output impedance. Typical numbers for input impedance may range around 10 meg ohms plus 1.5 pF in parallel. Finite open-loop gain also affects circuit input and output impedance when the loop is closed. The input impedance of the noninverting amplifier can be derived as

$$R_{\text{in}} = (1 + A_0\beta)\, R_i \qquad (13\text{-}26)$$

and the output impedance is given by

$$R_{\text{out}} = \frac{R_0}{A_0\beta} \qquad (13\text{-}27)$$

where R_i and R_0 are the amplifiers' open-loop input and output impedance, respectively. The larger the loop gain, the higher the input impedance and the lower the output impedance. Usually, the closed-loop input and output impedance will have a negligible effect on circuit operation, with moderate values of $A_0\,\beta$. In the case of the inverting amplifier, the input impedance is equal to the value of the resistor connected to the virtual ground negative input.

Op amps are frequently characterized by the frequency where the open-loop gain is 0 dB.

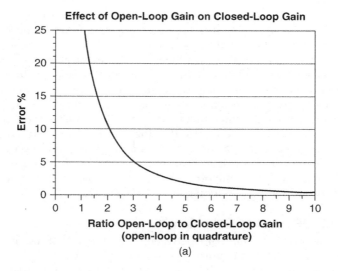

Effect of Open-Loop Gain on Closed-Loop Gain

(a)

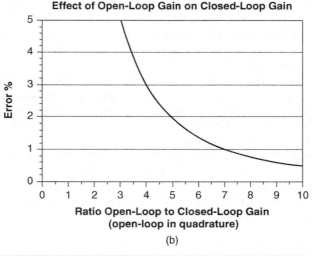

Effect of Open-Loop Gain on Closed-Loop Gain

(b)

FIGURE 13-14 (a) Effect of open-loop gain on closed-loop gain; (b) expanded scale.

13.3 Understanding Op-Amp Specifications

13.3.1 Bandwidth and Gain

The bandwidth of an op amp is frequently specified as the frequency where the open-loop gain is 1—the *unity-gain bandwidth*. Bandwidth is also specified for the closed-loop case with a design gain of +1 or –1 as the 3-dB bandwidth. Finally, the bandwidth can be specified for 0.1-dB flatness with a design gain of +1 (voltage follower).

It is important to have sufficient bandwidth of the closed-loop circuit so that the loss of high-frequency components of the input signal does not occur. Otherwise, the waveform will be spectrally distorted with associated information loss. Figure 13-14 illustrates the accuracy impact of open-loop gain upon closed-loop gain.

As a rule of thumb, the op-amp bandwidth should be at least 10 times the highest frequency of interest. The open-loop gain should be at least 20 dB more than the closed-loop gain desired to minimize the limiting effects of the amplifier itself. However, excessive gain or bandwidth can result in more sensitivity of the amplifier to PC board layout and instability with reactive loads. In many applications, the more general-purpose moderate gain and limited bandwidth amplifiers are best suited and more robust.

It is important to recognize that the gain (and many other parameters) shown on manufacturers' data sheets are for small signal conditions. Larger signal swings degrade performance. Data sheets must be read carefully. Open-loop gain is sometimes given in volts per millivolt (V/mV). This usually applies to DC where a spec of 10V/mV means an open-loop gain of 10,000 at DC.

13.3.2   Phase and Gain Margin

Phase margin is the *difference* between the phase shift of the signal passing through the op amp and 180° at the unity gain frequency. For example, a phase shift of 160° at the unity gain frequency implies a phase margin of 20°. Of course, the more the margin, the better for avoiding instability. Gain margin is the separation between unity gain and the gain (should be a loss) at the 180° point. For example, if the gain at the 180° point were −6 dB, the gain margin would be 6 dB.

13.3.3   DC Offsets

An inverting amplifier is shown in Fig. 13-15 with the addition of two bias currents I_a and I_b and an input offset voltage V_{dc}. Bipolar transistors in the input stage always draw some bias current in order to operate. A CMOS-type input has a much higher input impedance and draws significantly less bias current but has a higher offset voltage than the bipolar structure.

The input offset voltage results in an offset voltage at the output equal to V_{dc} times the closed-loop gain. The polarity of V_{dc} is random and is typically less than 10 mV.

The two bias currents I_a and I_b are nearly equal except for a small difference in currents or offset current I_0. In the circuit of Fig. 13-15, I_a produces an additional error voltage at the input given by $I_a R_{eq}$ where

$$R_{eq} = \frac{R_1 R_2}{R_1 + R_2} \tag{13-28}$$

The noninverting input bias current I_b has no effect since the positive input is grounded. To minimize the effect of I_a, high values of R_{eq} should be avoided.

FIGURE 13-15   Inverting amplifier with offsets.

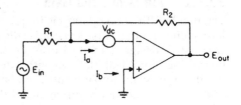

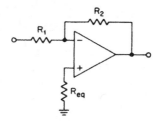

FIGURE **13-16** Minimization of DC offsets due to bias currents.

A more commonly used approach involves introducing a resistor with the value R_{eq} between the noninverting input and ground, as shown in Fig. 13-16. This has no effect on the overall gain. However, a DC offset voltage of $I_b R_{eq}$ is introduced at the noninverting input. Since the amplifier is a differential device, the net input error voltage due to the offset currents is $(I_a - I_b) R_{eq}$, or $I_0 R_{eq}$. Since I_a and I_b may each be $2\mu A$, for example, and the offset current I_0 $(I_a - I_b)$ is in the range of 200 nA, a 10:1 reduction in the effect of the offset current can occur as a result of having R_{eq} between the noninverting input and ground.

In the case of the inverting amplifier, the ratio R_2/R_1 is determined by gain as shown in Eq. (13-6). However, the actual values of R_1 and R_2 are nearly arbitrary and can be chosen so that their parallel combination (R_{eq}) is approximately equal to the DC loading on the noninverting input, i.e., the parallel combination of all resistance directly connected between the noninverting input and ground.

In general, for moderate closed-loop gains or AC-coupled circuits, the effects of DC offsets are of little or no consequence. For critical applications, such as precision active low-pass filters for the recovery of low-level DC components, the methods discussed can be implemented. Some amplifiers will provide an input terminal for the nulling of output DC offsets. Chapter 14 will show some methods for eliminating DC offset from an AC-only amplifier.

13.3.4 Slew-Rate Limiting

When an operational amplifier is used to provide a high-level output signal at a high frequency, the output will appear distorted in one way or another. For a step, the output could have a significant slope to the edge due to a long rise time. For a sine wave, the output may appear triangular. Significant harmonic distortion is the result. This effect is called *slew-rate limiting,* and the slope of the triangular waveform is referred to as the *slew rate.* It is an internally limited rate of change of the output voltage for a large amplitude input signal. Typical values can range from 10 to over 1,000 V/μs, depending upon the amplifier type. It is usually due to parasitic capacity or an internal or external compensation capacitor. If the peak-to-peak output voltage is small, the effects of slew rate will be minimized. Bandwidth can also be extended by using the minimum frequency compensation required for the given closed-loop gain when the compensation is external. However, stability would be a concern.

Figure 13-17 illustrates the effect of slew-rate limiting on a voltage step applied to a voltage follower. The units are in volts per microsecond (V/μS).

13.3.5 Settling Time

When a signal is applied to the input of an op amp, the output does not follow the input precisely. Figure 13-18 shows the output waveform of an op amp for an input step.

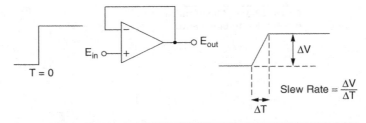

FIGURE 13-17   Effect of slew-rate limiting.

Three effects can be observed. The first effect is overshoot, where the output "over-shoots" the final value by some percentage. The second effect is slew rate, which was just discussed. The last effect is *settling time*. This is how long it takes before the output settles to within a given percent of its final value.

13.3.6   Common-Mode Rejection Ratio (CMRR)

This parameter is the rejection to a signal simultaneously present on both the positive and negative inputs; in other words, common mode. It must be the exact same signal in phase. The CMRR is given in dB. It degrades with increasing frequency. The test signal must be inside the rails, which means it must be within the power supply voltages. Associated with CMRR is the common-mode input voltage range. This is the maximum range of common-mode voltage present on both inverting and noninverting inputs to avoid damaging the op amp.

13.3.7   Output Voltage Swing

The output voltage swing of an amplifier is limited by the available supply voltages, the load on the output, and the drop from the rails (supply voltages) internal to the amplifier. Because of slew-rate limiting, this parameter is a function of frequency. An amplifier that can swing between supply voltages with almost no internal drop is said to have a *rail-to-rail* output.

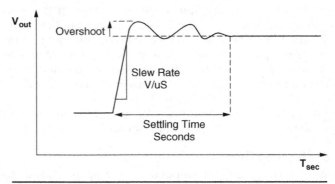

FIGURE 13-18   Settling time.

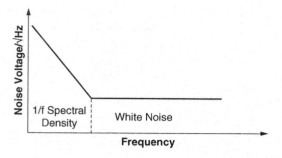

FIGURE **13-19** Noise spectral distribution.

13.3.8 Noise

There are two noise sources of an op amp: noise voltage and noise current. These sources are always reflected to the input; in other words, what value would they have if they were only present at the input. Noise voltage and current are typically in $nV/\sqrt{Hz}$ and $pA/\sqrt{Hz}$, respectively. The noise current results in a noise voltage by multiplication with the Thevenin (equivalent) impedance looking back from the op-amp input. Both of these noise sources have to be reflected to the output of the op amp by the transfer function of the entire circuit, including the effect of the feedback elements as a function of frequency, to determine if the output noise floor is acceptable. There are classes of op amps that have low noise features. Also, keeping the resistor values as low as possible will minimize their Johnson (thermal) noise contribution (see Fig. 9-19).

Op-amp noise follows the spectral shape of Fig. 13-19. Below a low-frequency break point, the noise will increase in a $1/f$ manner. Above this break point, the noise remains flat (white noise).

Noise voltage and current are specified typically in $nV/\sqrt{Hz}$ or $pA/\sqrt{Hz}$. They are also rms values. If converted to power, the $nV/\sqrt{Hz}$ or $pA/\sqrt{Hz}$ terms are squared, so the result is in watts/Hz. Of course, the load impedance would come into the picture, so the power per Hz would be nV^2/R_L or $pA^2 R_L$.

Example 13-1 Computing the output noise of an amplifier

Required:

For Fig. 13-20, compute the output noise delivered to a load of 100 Ω.

Result:

If we set E_{in} to zero, the Thevenin equivalent looking back from the inverting input is 500 Ω. The noise current I_n of $5pA/\sqrt{Hz}$ results in a noise voltage of $2.5nV/\sqrt{Hz}$. Since the gain is –1, the contribution of I_n measured at the output is $2.5nV/\sqrt{Hz}$.

The input noise voltage V_n of $10nV/\sqrt{Hz}$ results in an output noise voltage also of $10nV/\sqrt{Hz}$ due to the circuit gain of –1. To compute the total output noise, combine the two noise voltages, assuming they are not correlated (originating from the same source), so the total noise output voltage V_{nt} is found by

$$V_{nt} = \sqrt{10nV^2 + 2.5nV^2} = 10.3nV/\sqrt{Hz}$$

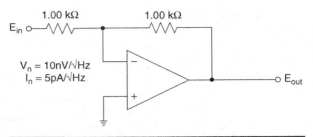

$V_n = 10nV/\sqrt{Hz}$
$I_n = 5pA/\sqrt{Hz}$

FIGURE 13-20   Inverting op amp showing noise sources.

The output power with a load of 100 Ω is 1.061×10^{-18} watts/Hz.
This can be expressed in dBm, where 0mW is the zero reference, as

$$10 \, Log(1.061 \times 10^{-18}/0.001) = -149.7 \, dBm/Hz$$

13.3.9   Total Harmonic Distortion (THD)

Harmonic distortion will occur mainly as result of slew-rate limiting and limited output swing of the op amp for high signal levels. Harmonics can occur not only at multiples of the input signal but also at interactions (sums and differences, etc.) due to nonlinearities. THD is usually expressed as a percentage.

On a ratio of power basis, THD can be defined as

$$TDH(\%) = \frac{P_2 + P_3 + P_4 + P_n}{P_1} \times 100 \tag{13-29}$$

where P_1 is a pure sine wave and P_2 to P_n are harmonics, all measured at the output of the amplifier.

On a voltage ratio basis rather than power

$$THD(\%) = \frac{\sqrt{E_2^2 + E_3^2 + E_4^2 \dots + E_n^2}}{V_1} \times 100 \tag{13-30}$$

where V_1 is a pure sine wave.

13.4   Power Supply Considerations

Most IC op amps specify dual supply voltages, typically ranging from ±5V to ±18V. The actual voltage magnitude is not critical, provided that the output swing from the amplifier is a few volts less than the supply voltages to avoid clipping. Maximum supply-voltage ratings should not be exceeded, of course. CMOS-type amplifiers can have rail-to-rail output swings and provide single-supply operation. In reality, op-amp power supply voltages do not have to be bipolar or symmetrical. A ±15V supply voltage to an op amp is the same as a single 30V supply with the op amp's negative supply terminal grounded and +30V applied to the positive terminal. The inputs have to be biased within the common-mode range of the amplifier. However, since most signals fluctuate around a ground reference, dual symmetrical-around-ground supplies are most convenient since no biasing is required. Op amps that use voltages that can be as low as ±1.5V are also called single-supply devices since they can operate off battery power.

To ensure stability, both the positive and negative power-supply voltages should be adequately bypassed to ground for high frequencies. Bypass capacitors can be 0.1 μF ceramic or 10 μF tantalum or combinations of both. Usually the op-amp data sheet will provide decoupling recommendations. Regulated supplies are desirable but not required. Most amplifiers have a typical supply-voltage sensitivity of 30 μV/V (90 dB), so a 1V power supply ripple would result in only a 30-μV change reflected at the amplifier input. This parameter is commonly known as the *power supply rejection ratio (PSRR)*.

In many cases, only a single positive supply voltage is available. Dual-voltage-type op amps can still be used by generating a reference voltage V_r, which replaces the circuit's ground connections. The amplifier's negative power terminal is returned to ground, and the positive power terminal is connected to the positive supply voltage.

The reference voltage should be midway between the positive supply voltage and ground, and should be provided from a low-impedance source. A convenient means of generating V_r directly from the positive supply is shown in Fig. 13-21 using the voltage follower configuration. The circuit of Fig. 13-21b is improved over the circuit of Fig. 13-21a by having additional bypass capacitors at the output. Typically, the 10uF is tantalum and the 0.1μF is ceramic. The 100-Ω resistor is inside the feedback loop, so it appears as a very low impedance to the circuit. Its purpose is to prevent oscillations, as some op amps do not like capacitive loads.

Input signals referenced to ground must be AC-coupled and then superimposed upon V_r. The signals are decoupled at the output to restore the ground reference. A low-pass filter design using this method will not pass low-frequency components near DC.

A very simple rule can be used to convert a circuit from dual supplies to a single supply. All points shown in the schematic as being connected to ground should be biased at a voltage midway between ground and the single supply. This bias voltage can be V_r as shown in Fig. 13-21. However, if it is clearly determined that this node does not draw any DC current—such as the positive input to an op amp, for example—you can use the voltage divider, including the decoupling capacitor of Fig. 13-21a as a reference. The voltage follower can be omitted.

This technique is illustrated in Fig. 13-22 as applied to some previous design examples. The circuit of Fig. 13-21a is used to generate V_r, and all amplifiers are powered using single-ended power supplies.

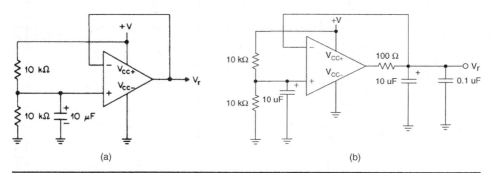

FIGURE 13-21 Generation of a reference voltage: (a) basic circuit and (b) improved circuit.

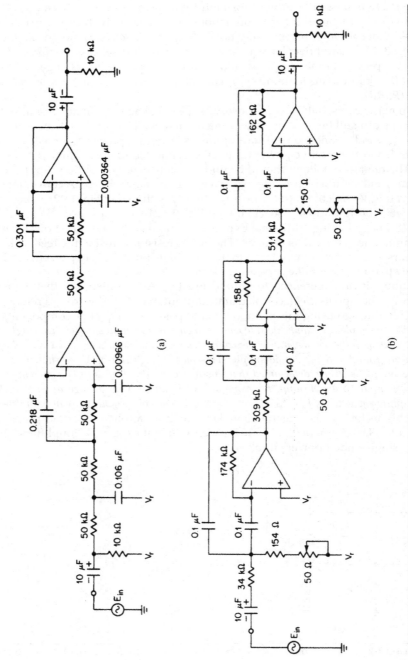

FIGURE 13-22 Designing filters for single supply operation: (a) low-pass filter of Example 3-8; (b) band-pass filter of Example 5-12.

For the sake of clarity, power supply connections and decoupling capacitors may not be shown in many of the diagrams in this book. Typical decoupling capacitor combinations are a 10-μF tantalum capacitor and a 0.1-μF capacitor in parallel from the power connection to ground. Other recommendations may be present in the data sheets.

13.5   Operational Amplifier Selection

13.5.1   Op-Amp Types

Bipolar Input

There are a variety of families of operational amplifiers. Some are differentiated by the type of input stage used within the monolithic integrated circuit. Bipolar inputs are the most commonly used type where the entire device, including the input stage, consists of bipolar transistors. Input bias and offset currents are a few hundred nanoamperes, and offset voltages are typically under 10 mV. The open-loop input impedance is a few hundred kilo-ohms.

CMOS Op Amps

CMOS op amps need very little supply current, have high input impedances, and very low bias currents. Offset voltages are somewhat higher than bipolar amplifiers. CMOS amplifiers can operate rail-to-rail and are suitable for single-supply operation and low-voltage battery-powered applications since they draw little power. CMOS amplifiers are typically more noisy then the bipolar type.

BIFET Op Amp

BIFET is an acronym for bipolar-field-effect transistor. It combines two technologies where the front end or input stages use FETs and the remainder of the circuit is bipolar. The result is wider bandwidths than bipolar devices, lower input offset currents, higher input impedances, and more output drive capability. The input offset voltages are typically higher than bipolar op amps.

13.5.2   Op-Amp Packaging

Op amps are single, dual, or quad. Dual unit packages will save space and cost. Another useful feature is that both amplifiers in the same package are virtually identical electrically—in other words, they are matched because they are on the same substrate. They will typically track very closely with temperature. Quad op amps offer four op amps in the same package. This is very useful for high-density applications, but the designer should be aware of limited flexibility when designing the PC board since all signal traces have to be routed to a single area to connect to all four op amps, which may result in crosstalk.

Virtually all IC operational amplifiers are available in surface-mount packages. Some are still available in the more traditional dual inline package. In most cases, the same die or chip is also used in a through-hole part. Figure 13-23 shows some typical packages. These packages consist of a circuit attached to a frame and encapsulated with a plastic compound. This compound can withstand the high temperatures resulting from soldering without deforming and provides a hermetic seal.

FIGURE 13-23   Typical op-amp packages (Texas Instruments).

13.5.3   Survey of Popular Amplifiers

IC operational amplifiers range from economical general-purpose devices to the more costly high-performance units. A selection guide is provided in Table 13-1 listing some of the numerous op amps that are available from multiple manufacturers and a few parameters. They are sorted by descending bandwidth and whether single, dual, or quad. Although manufacturers are listed for each device, many devices are available from multiple manufacturers.

The parameters of this table are intended as a general guideline and correspond to commercial-grade op amps at a 25°C ambient temperature. This table is not meant to be all-inclusive, and some op amps are better suited than others for specific applications. The manufacturers' data sheets will provide more detailed and specific information. *It is imperative that the designer read the applicable device data sheets.* Helpful hints are given that may be specific for that particular operational amplifier.

Each manufacturer has an applications engineering department that can provide guidance in selecting the best op amp for a given set of requirements. The designer should keep in mind that excessive gain or bandwidth can result in more sensitivity of the amplifier to PC board layout and instability, so in many applications, the more general-purpose moderate gain and limited bandwidth amplifiers are better suited and more robust. Op-amp data sheets will provide recommendations for both power decoupling and PCB layout considerations. It is good practice to have a 0.1-μF ceramic capacitor and a 10-μF tantalum capacitor as close as possible to the IC power connections.

Single Op Amps

Single		Bias Current nA max	Offset Current nA max	Offset Voltage mV max	Unity Gain BW MHz Type	Slew Rate V/µS Type	Supply Voltage	Manuf. Code	Features Code
	OPA843	37 µA	1 µA	1.5	800	600	±4 to ±6	TI	WB, LN, LH
	MAX4450	20 µA	4 µA	26	210	485	+4.5 to +11	MAX	WB, SS, LH, RR
	AD8065	6 pA	10 pA	1.5	145	180	±2.5 to ±12	AD	JF, WB, LN, SS, RR
	AD8033	11 pA	—	2	80	80	±2.5 to ±13	AD.	JF, WB, LN, SS, RR
	AD8610	10 pA	10 pA	100 µV	25	50	±5 to ±13	AD	JF, WB, LN, OA
	OPA725	200	50	3	20	30	±2 to ±6	TI	JF, WB, LN, LH
	OPA727	100 pA	100 pA	150 µV	20	30	±2 to ±6	TI	WB, LN, SS, RR
	LM318	250	50	10	15	70	±5 to ±20	TI, NA	GP, WB
	NE5534	1500	500	4	10	13	±3 to ±20	TI	GP, WB, LN, LH
	OPA134	100 pA	50 pA	2	8	20	±2.5 to ±18	TI	JF, LN, LH
	MAX4237	500 pA	1 pA	50 µV	7.5	1.3	+2.4 to +5.5	MAX	SS, RR
	LF351	200 pA	100 pA	10	4	16	±5 to ±18	ST	GP, JF, OA
	TL071	200 pA	100 pA	6	3	13	±3.5 to ±18	TI	GP, JF, LH
	TL081	400 pA	100 pA	15	3	13	±3.5 to ±18	TI	GP, JF, LH
	LF356	200 pA	50 pA	10	2.5	15	±5 to ±18	NA	GP, JF
	MAX4236	500 pA	1 pA	50 µV	1.7	0.3	+2.4 to +5.5	MAX	SS, RR
	TL061	200 pA	200 pA	15	1	3.5	±3.5 to ±18	TI	GP, JF

TABLE 13-1 Operational Amplifier Selection Guide

		Bias Current nA max	Offset Current nA max	Offset Voltage mV max	Unity Gain BW MHz Type	Slew Rate V/µS Type	Supply Voltage	Manuf. Code	Features Code
Dual	MAX4451	20 µA	4 µA	26	210	485	+4.5 to +11	MAX	WB, SS, LH, RR
	AD8066	6 pA	10 pA	1.5	145	180	±2.5 to ±12	AD	JF, WB, LN, SS, RR
	LT1813	4 µA	400 nA	1.5	100	750	±2 to ±6	LT	WB, LN, SS
	AD8034	11 pA	—	2	80	80	±2.5 to ±13	AD	JF, WB, LN, SS, LH, RR
	OPA2725	200	50	3	20	30	±2 to ±6	TI	JF, WB, LN, SS, RR
	OPA2727	100 pA	100 pA	150 µV	20	30	±2 to ±6	TI	JF, WB, LN, SS, LH, RR
	NE5532	800	500	4	10	9	±3 to ±20	TI	GP, WB, LN
	OPA2134	100 pA	50 pA	2	8	20	±2.5 to ±18	TI	JF, LN, LH
	LF353	200 pA	100 pA	10	4	13	±5 to ±18	TI	GP, JF
	TL072	200 pA	100 pA	6	3	13	±3.5 to ±18	TI	GP, JF, LH
	TL082	400 pA	100 pA	15	3	13	±3.5 to ±18	TI	GP, JF, LH
	MC4558	500	200	6	2.8	1.6	±3 to ±18	ST	GP
	MC1458	700	300	10	1.1	0.8	±3 to ±18	TI	GP
	TL062	200 pA	200 pA	15	1	3.5	±3.5 to ±18	TI	GP, JF
	LM358	250	50	6	1	0.6	±1.5 to ±18	TI	GP, SS

TABLE 13-1 Operational Amplifier Selection Guide (Continued)

506

Single Op Amps

		Bias Current nA max	Offset Current nA max	Offset Voltage mV max	Unity Gain BW MHz Type	Slew Rate V/ µS Type	Supply Voltage	Manuf. Code	Features Code
Quad	OPA4727	100 pA	100 pA	150 µV	20	30	±2 to ±6	TI	JF, WB, LN, SS, LH, RR
	LF347	200 pA	100 pA	10	4	13	±5 to ±18	TI	GP, JF, LH
	TL074	200 pA	100 pA	6	3	13	±3.5 to ±18	TI	GP, JF, LN, LH
	TL084	400 pA	100 pA	15	3	13	±3.5 to ±18	TI	GP, JF, LH
	TL054	200 pA	1	4	2.7	16	±3.5 to ±18	TI	GP, JF
	TL064	200 pA	200 pA	15	1	3.5	±3.5 to ±18	TI	GP, JF
	LM324	250	50	6	1	0.6	±1.5 to ±16	TI	GP, SS
	LM348	200	50	6	1	0.5	±3 to ±18	TI	GP

Table 13-1 Operational Amplifier Selection Guide (*Continued*)

Manufacturers' Code

Code	Manufacturer
TI	Texas Instruments
NA	National Semiconductor
ST	ST Microelectronics
LT	Linear Technology
AD	Analog Devices
MAX	Maxim

Feature Code

Code	Feature
GP	General Purpose
JF	Junction FET Input
WB	Wideband
LN	Low Noise
SS	Single Supply
LH	Low Harmonic Distortion
RR	Rail-to-Rail Output
OA	Output Offset Adjust
UN	Uncompensated

13.6   General Manufacturing Considerations

Specific mechanical considerations determine the robustness of the connections to the circuit board for surface-mount parts. In most cases, the coefficient of expansion of the components differs from that of the board itself. For example, the expansion coefficient of the ceramic dielectric of a chip capacitor is different from that of an epoxy glass printed circuit board. Since the part is held in place on the top of the pads with solder only, the absence of a further mechanical connection could cause the part to break free with time.

Pad preparation is dependent upon which of two soldering processes is used: *reflow* or *wave*. In the reflow process, the pad is coated with a dried solder paste. The chip is placed on these pads and the board is gradually heated until this solder melts and re-flows, joining the part to the pad. With the wave solder method, the parts are first held in place with glue and then the board is covered with a wave of solder. Each method imposes unique requirements on the pads' design to ensure reliability and longevity of the connections.

Although use of surface-mount technology results in high-circuit densities, there is a penalty to be paid. Through-hole parts allow the use of the component leads to convey signals from the surface of the board to the inner layers, as well as to the other outside surface of the board. Since the leads of surface-mount components lie flat on the board surface attached to pads, extra feed-throughs (referred to as *vias*) are required on the board to interconnect layers, and sometimes this results in extra layers as well. The end result is a costlier PC board.

Restriction of Hazardous Substances (RoHS) is a European Union (EU) directive (2002/95/EG) that went into effect July 1, 2006. This directive is also known as the lead-free directive in the electronics industry, even though it also restricts usage of other hazardous substances, such as mercury, etc. The major impact in terms of component selection is that you must ensure selected parts meet this directive, which should be stated on the data sheet. It also impacts manufacturing processes by dictating lead-free solder requirements.

References

Electronic Design Magazine, Gene Heftman, editor, and Analog Devices. (2003). *Basics of Op Amps Design.* New York, Penton Media.

Texas Instruments, *Handbook of Operational Amplifier Applications*, Application Report SBOAs092A, October 2001.

Franco, S., (1988). *Design with Operational Amplifiers and Analog Integrated Circuits.* McGraw-Hill, New York.

Texas Instruments. *Op Amps for Everyone*, Advanced Analog products SLOD006B, August 2002.

CHAPTER 14

Linear Amplifier Applications

14.1 Resistive Feedback Networks

Chapter 13 introduced basic inverting and noninverting op-amp configurations. This chapter will expand on these basic configurations and introduce new circuits. They all will operate in the *linear* mode. This implies that if the input amplitude were multiplied by a constant, the output will be multiplied by the same constant. No clipping or any type of nonlinear distortion will occur.

14.1.1 Adding and Subtracting Signals

Let us assume we want to add multiple signals together as follows:

$$E_{\text{out}} = -(K_1 E_1 + K_2 E_2 + K_3 E_3 + \cdots + K_n E_n) \tag{14-1}$$

Each K can be any positive multiplication factor. This approach is implemented in Fig. 14-1. All inputs are isolated from each other and combined at the output.

The minus sign at the output indicates that in addition to the gain or loss, a 180° phase shift will occur.

The circuit can be expanded to apply both positive and negative multiplication factors to the inputs. Signals can be algebraically added and subtracted. This configuration is shown in Fig. 14-2.

If a circuit that sums without the 180° phase shift is desired, use the circuit of Fig. 14-2 with only the noninverting inputs (E'_n).

A circuit that was covered in Chap. 13 is repeated in Fig. 14-3. This circuit amplifies the difference between two input voltages E_1 and E_2. It is sometimes called a subtractor or difference amplifier. The output is given by

$$E_{\text{out}} = (E_2 - E_1) R_b / R_a \tag{14-2}$$

It is clear that the input impedance to the circuits of Figs. 14-1 and 14-2 is the value of the input resistors as they are connected to virtual grounds. In Fig. 14-3, the inverting input sees an impedance of R_a and the noninverting input impedance is $R_a + R_b$. Because the impedances are different, an unbalance in the loading of the source impedances will occur, which degrades the common-mode rejection ratio (CMRR). Nevertheless, this circuit is often used as an analog building block.

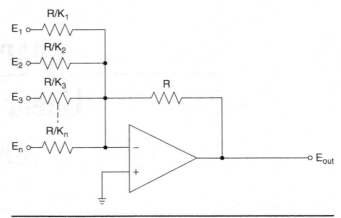

Figure 14-1 Summing amplifier.

A brute-force solution to the unequal impedance loading problem would be to add two voltage followers as buffers at the front end, as shown in Fig. 14-4. The input impedance becomes very high since the voltage follower has the highest input impedance of any configuration.

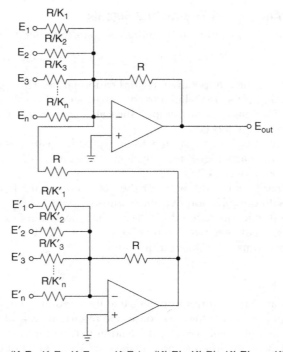

$$E_{out} = -(K_1E_1 + K_2E_2 + K_3E_3 + \cdots + K_nE_n) + (K'_1E'_1 + K'_2E'_2 + K'_3E'_3 + \cdots + K'_nE'_n)$$

Figure 14-2 Summing amplifier with both positive and negative inputs.

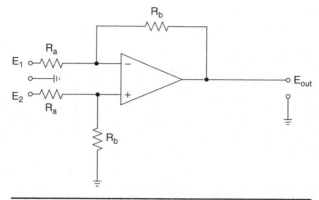

FIGURE 14-3 Difference amplifier.

A solution that provides high-input impedance and uses only two amplifiers is shown in Fig. 14-5.

To compute the output, let us first consider the amplification of E_1. The output of the first stage is

$$E_1(R_a/R_b + 1)$$

This signal gets amplified by the gain of the second stage, which results in

$$E_1(R_a/R_b + 1)(-R_b/R_a)$$

The other input voltage E_2 appears at the output as

$$E_2(R_b/R_a + 1)$$

If we then add both output signals, we obtain

$$E_{out} = E_1(R_a/R_b + 1)(-R_b/R_a) + E_2(R_b/R_a + 1) \qquad (14\text{-}3)$$

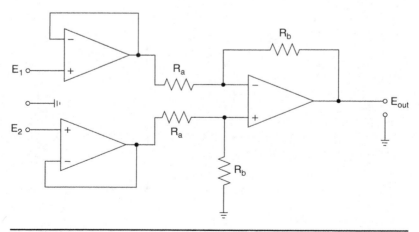

FIGURE 14-4 Buffered difference amplifier.

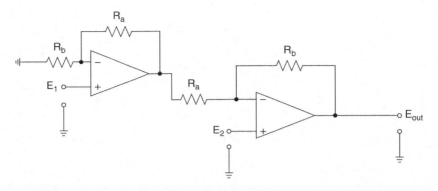

Figure 14-5   Two-op-amp difference amplifier.

Equation (14-3) can be algebraically manipulated into the following form:

$$E_{out} = (E_2 - E_1)(R_b / R_a + 1) \tag{14-4}$$

A negative of the two-amplifier design compared to the previous three-amplifier version is that the common-mode AC signal path is not symmetrical. A common-mode signal on E_1 will not follow the same path as one on E_2, so the CMRR will suffer.

In general, to achieve a high common-mode rejection ratio, resistor matching is very critical. It is best to use 0.1 percent metal film resistors.

14.1.2   The Instrumentation Amplifier

An instrumentation amplifier is a difference or differential amplifier that provides high-input impedance, a high CMRR, and the ability to change gain by varying the value of a single resistor.

The classic three-op-amp instrumentation amplifier is shown in Fig. 14-6. If we remove R_{gain} we have two voltage follower (buffer) amplifiers on the left feeding a standard differential amplifier. The gain would be R_3/R_2, and the input impedance would

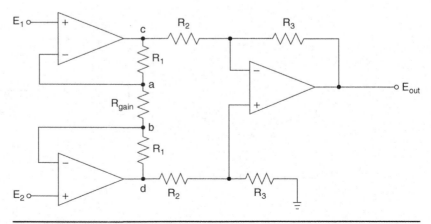

Figure 14-6   Classic instrumentation amplifier.

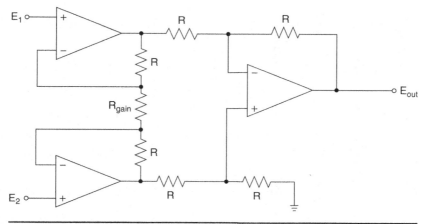

FIGURE 14-7 Equal value of resistors.

be high as a result of the two input buffers. If we add R_{gain}, the full differential input voltage appears across R_{gain} because of the virtual ground effect at the input to the two buffers. The voltage drop between a and b is then $E_1 - E_2$. The current through R_{gain} also goes through R_1 and R_2 since no current can go through the amplifier. So the voltage between c and d is

$$E_{cd} = (E_2 - E_1)(1 + 2R_1/R_{gain}) \tag{14-5}$$

If we then amplify this signal by the difference amplifier, the result becomes

$$\frac{E_{out}}{E_2 - E_1} = R_3/R_2(1 + 2R_1/R_{gain}) \tag{14-6}$$

In an ideal instrumentation amplifier, CMRR is infinite. In reality, common-mode rejection is highly dependent on the matching of the two R_1 resistors, the two R_2 resistors, and the two R_3 resistors. They should also track with temperature as well.

In the circuit of Fig. 14-7, all resistors are of equal value except for R_{gain}. This is the most commonly used approach. Resistor networks are available where all resistors are equal in value and very closely matched. They will also track each other with temperature since they are on the same substrate. The gain equation then becomes

$$\frac{E_{out}}{E_2 - E_1} = 1 + 2R/R_{gain} \tag{14-7}$$

If all resistors, including R_{gain}, are equal to R, the overall gain becomes 3.

Figure 14-8 shows an instrumentation amplifier operating from a single supply. The addition of capacitors limits the low-frequency response. If we have an RC network that effectively performs a high-pass response, the minimum value of capacitance is calculated by

$$C_{min} \geq \frac{1}{\omega_{3\,dB}R_l} \tag{14-8}$$

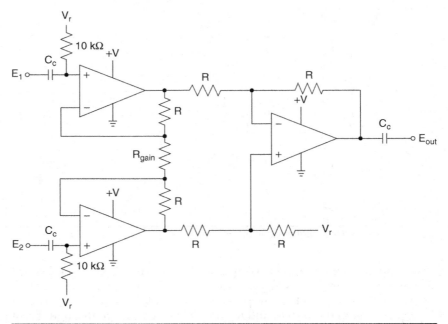

Figure 14-8 Instrumentation amplifier using a single supply.

where $\omega_{3\,dB}$ is $2\pi f_{3\,db}$, which is the lower 3-dB point of the response, and R_l is the load seen by the capacitor.

For example, for a lower 3-dB point of 100 Hz, C_C should be a minimum of 0.159 μF in the circuit of Fig. 14-8. The minimum value of C_o is computed based on the load it is terminated with. If source and load resistors apply, they should be added together, as they are effectively in series.

It should be pointed out that many IC manufacturers offer a complete instrumentation amplifier in a single package. These can be found on their websites. A common application of an instrumentation amplifier is to drive an analog-to-digital converter.

14.1.3 AC Coupling of Amplifiers

AC coupling of amplifier circuits has a number of features, providing that operation down to DC is not required. In the circuit of Fig. 14-9, AC coupling at the input of a voltage follower allows operation from a single supply. A reference voltage midway between ground and +V is generated at the plus input to the voltage follower by the two resistors labeled R. So no separate V_r reference source is needed. The capacitor C_C is calculated using Eq. (14-8), where R_1 is $R/2$ with any source resistance added to it. The output can be AC-coupled in a similar manner, or it will be riding on the midway DC voltage.

If we add the components inside the dashed box, the result is a noninverting amplifier where the gain is determined by Eq. (13-6) as shown:

$$\frac{E_{out}}{E_{in}} = 1 + \frac{R_2}{R_1}$$

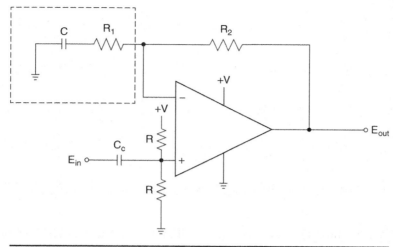

FIGURE 14-9 AC-coupled voltage follower.

The value of C in the dashed box is calculated by Eq. (14-8).

If we use the conventional dual-supply configuration, the circuit of Fig. 14-9 greatly simplifies to that of Fig. 14-10.

Note that power supply voltages must be decoupled to ground at every op amp as good practice. Recommended values and guidelines are given on the op-amp data sheet.

Chapter 13 illustrated that various DC offset voltages can occur at the op-amp output resulting from input bias and offset currents and input offset voltages. This could have a significant effect with high circuit gains. For example, if the circuit of Fig. 14-10 has a gain of 100, a 10mV offset at the input can result in a 1V DC offset at the output. If we add a capacitor as shown in Fig. 14-11, the circuit becomes a voltage follower at DC with a gain of zero since the capacitor becomes an open circuit. The value of C should be determined by Eq. (14-8).

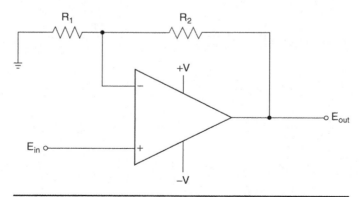

FIGURE 14-10 Noninverting amplifier with dual supplies.

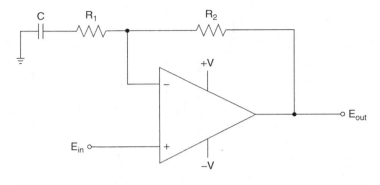

Figure 14-11 Noninverting amplifier with gain of 1 at DC.

The same approach can be taken for an inverting amplifier, as the same DC offset issue applies. The circuit is shown in Fig. 14-12, where the gain is calculated by Eq. (13-13) as shown:

$$\frac{E_{\text{out}}}{E_{\text{in}}} = -\frac{R_2}{R_1}$$

Unlike the noninverting configuration of Fig. 14-11, the gain at DC is zero so there is no output.

14.1.4 Bootstrapping a Voltage Follower for High-Input Impedance

The highest input impedance can be obtained using a voltage follower, as the most negative feedback occurs ($\beta = 1$). However, a resistor to ground must be present or the input will float and be subject to issues such as electrostatic discharge (ESD) charges building up with no input signal present. The amplifier could fail. In addition, if a large value resistor such as 10 MΩ were present, this could be a source of unacceptable Johnson (thermal) noise.

Figure 14-13 illustrates a technique known as "bootstrapping." Since the input signal appears almost precisely at the output and this signal is fed back to the bottom of the

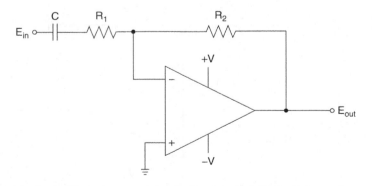

Figure 14-12 Inverting amplifier with DC blocking.

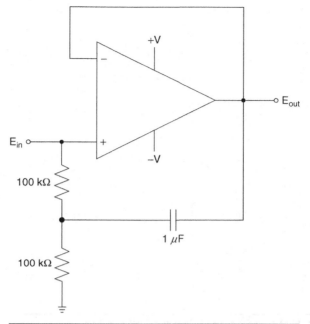

Figure 14-13   Bootstrapped voltage follower.

upper 100 kΩ, it prevents any current from flowing into the resistor, so the impedance is effectively infinite. Of course, nothing is perfect, so gain roll-off with frequency of the op amp will degrade the effect of bootstrapping. Also, the coupling capacitor will affect the bootstrapping at the low-frequency end, so a large enough value for C must be used. In the example shown, $R = 50$ kΩ and C is 1 μF. The required C is determined by Eq. (14-8).

14.1.5   T-Network in Inverting Amplifier Feedback Loop to Reduce Resistor Values

Figure 14-14a shows a conventional inverting amplifier where the gain is given by Eq. (13-3) as shown:

$$\frac{E_{\text{out}}}{E_{\text{in}}} = -\frac{R_2}{R_1}$$

Using the values of the circuit in Fig. 14-14b, the gain is −100 and the input impedance is 100 kΩ. The feedback resistor R_2 is much too high and can result in high noise, unavailability, sensitivity to moisture, etc. Also, parasitic capacitance across high-value resistors limits frequency response.

The feedback network of Fig. 14-14c can allow us to lower the feedback resistor values. The formula for the calculation of values is

$$\frac{E_{\text{out}}}{E_{\text{in}}} = -\frac{R_2 + R_3 + \dfrac{R_2 R_3}{R_4}}{R_1} \tag{14-9}$$

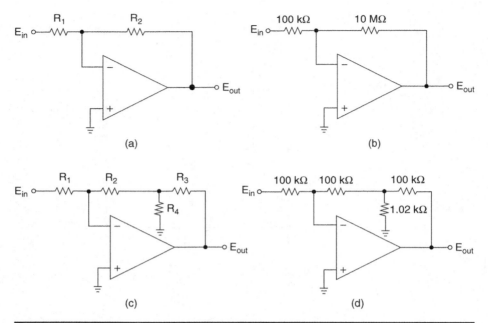

Figure 14-14   T-network to reduce resistor values.

In the example of Fig. 14-14*d*, the values shown represent a feedback resistor of 10 MΩ but do not exceed 100 kΩ.

14.1.6   Bootstrapped Inverting Amplifier for High-Input Impedance

In the circuit of Fig. 14-15, an inverting amplifier is shown with a second amplifier added to provide a controlled positive feedback.

The circuit gain is given by Eq. (13-3) as shown: $\dfrac{E_{out}}{E_{in}} = -\dfrac{R_2}{R_1}$

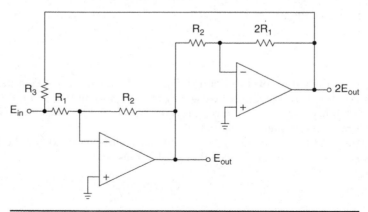

Figure 14-15   Bootstrapped inverting amplifier.

The input impedance is calculated from

$$Z_{\text{in}} = -\frac{R_3 R_1}{R_3 - R_1} \qquad (14\text{-}10)$$

Note that if $R_3 = R_1$, Z_{in} becomes infinite. However, if the denominator becomes negative, the circuit will oscillate. So R_3 should always remain reasonably greater than R_1.

14.2 Current-to-Voltage and Voltage-to-Current Converters

14.2.1 Current-to-Voltage Converter

The circuit of Fig. 14-16a converts an input current to a voltage by virtue of the virtual ground effect. The input current is forced to go through resistor R due to the high-input impedance of the op amp. The current does not see any opposition, since the negative input is at virtual ground, so all of it flows through R. Therefore,

$$E_{\text{out}} = -I_{\text{in}}R \qquad (14\text{-}11)$$

The input current may be AC or DC. An error due to the bias current can occur, so proper selection of the op amp is critical. This circuit is sometimes called a "transimpedance amplifier." It is frequently used as a photodiode amplifier, where light impinging on the diode is converted to a small current. Typically, the current could be 1 μA or less. This would mean that for high sensitivity, R must be a high value in the megohm range. This is not desirable for reasons mentioned earlier, such as high noise, unavailability, sensitivity to moisture, etc. We can then use the T-network as shown in Fig. 14-16b. The equivalent feedback resistor R is calculated from

$$R = R_2 + R_3 + \frac{R_2 R_3}{R_4} \qquad (14\text{-}12)$$

For example, assuming we need a 10-MΩ equivalent feedback resistor, let $R_2 = R_3 = 100$ kΩ. Then R_4 becomes 1.02 kΩ. All these values are much more practical than 10 MΩ.

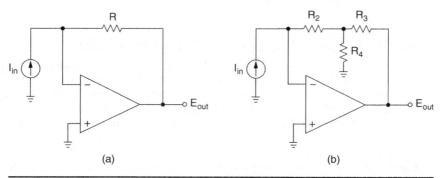

(a)   (b)

Figure 14-16 Current-to-voltage converter: (a) basic circuit and (b) high-sensitivity circuit.

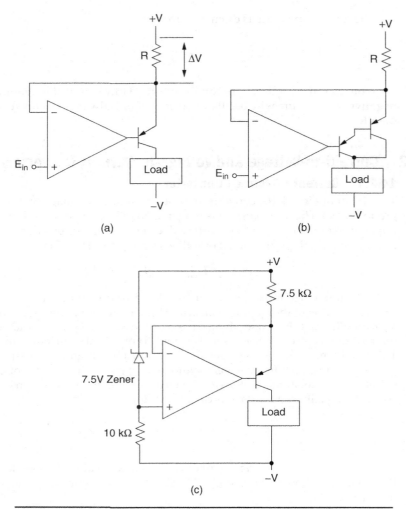

FIGURE 14-17   Unipolar current sources: *(a)* single transistor; *(b)* Darlington; and *(c)* Zener added for precision.

14.2.2 Voltage-to-Current Converter (Current Source)

An ideal voltage-to-current converter (or current source) maintains a constant load current that is directly determined by an input voltage and independent of load impedance.

The circuit of Fig. 14-17*a* maintains a constant voltage ΔV equal to the difference between $+V$ and E_{in} across resistor R. (The lower side of R is at E_{in} because of the virtual ground effect.) This forces a current into the emitter of the positive negative positive (PNP) transistor equal to $\Delta V/R$. This current is then supplied from the transistor collector to the load. However, not all the current arrives at the collector since $(1 - \alpha)$ times the emitter current flows into the base. The error is usually negligible since transistor α's are typically 0.99. Replacing the transistor with the Darlington structure of Fig. 14-17*b* further reduces this error. Circuit output impedance is quite high and

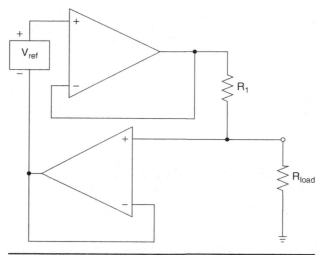

Figure 14-18   Current source using two op amps.

typically a few megohms. Higher impedances can be obtained by using a field effect transistor (FET) instead of a transistor.

Care should be taken that after subtracting ΔV and the voltage across the load from the *difference* between $+V$ and $-V$ that enough voltage is left to operate the transistor. The transistor requires a minimum V_{ce} (voltage between collector and emitter) to operate properly. Both $+V$ and $-V$ must remain stable for the current to remain stable. They should be regulated.

Figure 14-17c uses a Zener diode to provide a fixed voltage of 7.5V across a 7.5-kΩ resistor. The value of the current is then 1 mA. Because of the Zener diode, the requirement for stability of $+V$ and $-V$ no longer applies since the diode provides the regulated voltage.

The circuits of Fig. 14-17 are unipolar, meaning that only a single polarity of current can be provided. They are best suited for DC applications.

The circuit of Fig. 14-18 is a current source using only op amps. Because of the virtual ground effect at the inverting inputs of both op amps, the voltage across R_1 is always V_{ref}. The current through the load is then maintained at V_{ref}/R_1. As with all current sources, the limitations are determined by the power supply voltages, since an infinite load would require an infinite voltage.

A voltage-to-current converter can simply be an op amp with the load in the feedback path, as shown in Fig. 14-19. Both inverting and noninverting configurations are shown. These circuits are bipolar, so with both positive and negative supplies, the circuits can be used for AC applications. The loads are floating instead of being connected to ground.

14.2.3   The Howland Current Pump

The circuit of Fig. 14-20 is called a Howland current pump and was invented by Professor Bradford Howland of MIT around 1962. It is capable of extremely high impedances, which means that it approaches the characteristics of a theoretical current source. As shown in the figure, the input signal can be bipolar, so it is generally used for AC signals and with dual supplies.

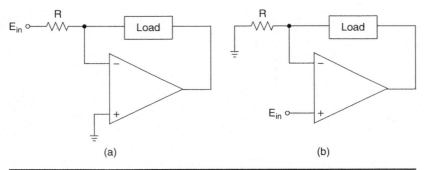

Figure 14-19   Voltage-to-current converter: *(a)* inverting and *(b)* noninverting.

The circuit output impedance is given by

$$Z_{out} = \frac{R_4}{\dfrac{R_4}{R_2} - \dfrac{R_3}{R_1}}$$ (14-13)

If the ratios of R_4/R_2 and R_3/R_1 are equal to each other, the denominator becomes zero, so the impedance becomes infinite. The higher the impedance of the current source, the less of an effect it will have when connected across a load, with other AC signals present at that load. In other words, it would not load down (reduce) any of the external AC circuitry signals.

The output current is given by

$$I_{load} = -E_{in} \frac{R_3}{R_1 R_4}$$ (14-14)

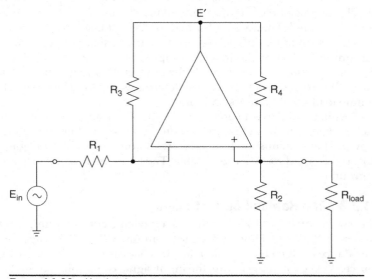

Figure 14-20   Howland current pump.

where E_{in} can be a sinusoid, DC, or any complex signal.

If we let R_1, R_2, R_3, and R_4 all equal R, the current output expression simplifies to

$$I_{load} = -\frac{E_{in}}{R} \qquad (14\text{-}15)$$

To maintain a high output impedance, low values of R should be avoided and resistor matching is critical. For a tolerance of ±1 percent, the output impedance would exceed 50 R. Resistors are available with a ±0.1 percent accuracy.

The voltage at node E' is given by

$$E' = -E_{in}\left(2\,\frac{R_{load}}{R} + 1\right) \qquad (14\text{-}16)$$

If R_{load} is too high, the amplifier will clip, with large output swings, so this must always be considered when selecting values and supply voltages. Like all analog linear circuits, the op amps must be operated well within their linear range, away from the nonlinearities that occur as the amplifier output approaches the external supply voltages.

A small capacitor can be added across R_3 to prevent oscillations if they occur.

Improved Howland Current Pump

The Howland current pump of Fig. 14-20 could be inefficient in terms of power. Let's say all the resistors are 100 kΩ. The load current being supplied from the amplifier has to flow through R_4 on its way to the load. This could result in a significant voltage drop across R_4 and a waste of power. The circuit of Fig. 14-21 has split up R_4 into R_{4a} and R_{4b}. In this new circuit, the ratio $(R_{4a} + R_{4b})/R_2$ must be equal to R_3/R_1.

In a typical application, $R_1 = R_2 = R_3 = 100$ kΩ. R_{4a} would be 99 kΩ and R_{4b} would then be 1.00 kΩ. Observe that the current from the op-amp output goes through a 1-kΩ resistor first before encountering the load, whereas if we used the circuit of Fig. 14-20, it would have gone through a 100-kΩ resistor, resulting in more I^2R power.

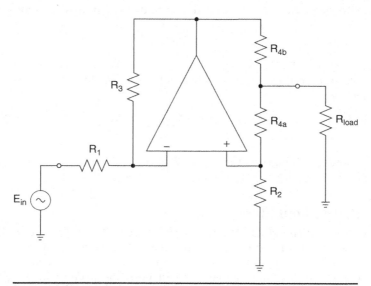

Figure 14-21 Improved Howland current pump.

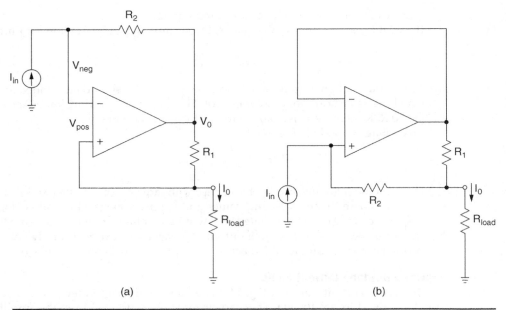

Figure 14-22 Current-mode amplifiers: (a) inverting and (b) noninverting.

14.2.4 Current-Mode Amplifiers

Inverting and noninverting current-mode amplifiers are shown in Fig. 14-22. These circuits amplify current rather than voltage. Figure 14-22a shows an inverting current-mode amplifier. Because no current flows into the op amp, $V_{neg} - V_o = R_2 I_{in}$ and $V_o - V_{pos} = R_1 I_o$. Since the virtual ground effect forces $V_{neg} = V_{pos}$, this results in $R_2 I_{in} = -R_1 I_o$. The final equation for the current gain of this circuit then becomes

$$\frac{I_o}{I_{in}} = -\frac{R_2}{R_1} \tag{14-17}$$

A noninverting current-mode amplifier is shown in Fig. 14-22b. The current gain is given by

$$\frac{I_o}{I_{in}} = 1 + \frac{R_2}{R_1} \tag{14-18}$$

Note the similarity of Eqs. (14-17) and (14-18) with the basic voltage gain equations for an inverting and noninverting op amp.

14.3 Bridge Amplifiers

The circuit of Fig. 14-23a illustrates a traditional Wheatstone bridge. This type of architecture is useful for measuring small resistance changes that occur in devices such as temperature sensors (thermistors), strain gauges, pressure sensors, and similar devices whose resistance changes as a function of a variable. In effect, these devices are transducers that convert the variable under measurement to a resistance.

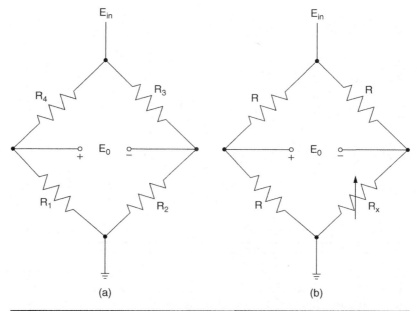

Figure 14-23   Bridge amplifier: (a) basic circuit and (b) variable element.

The output voltage can be computed by

$$E_o = \frac{R_1}{R_1 + R_4} E_{\text{in}} - \frac{R_2}{R_2 + R_3} E_{\text{in}} \qquad (14\text{-}19)$$

Equation (14-19) can be rewritten as

$$E_o = \frac{\dfrac{R_1}{R_4} - \dfrac{R_2}{R_3}}{\left(1 + \dfrac{R_1}{R_4}\right)\left(1 + \dfrac{R_2}{R_3}\right)} E_{\text{in}} \qquad (14\text{-}20)$$

Inspection of Eq. 14-20 indicates that if the ratios R_1/R_4 are equal to R_2/R_3, the bridge is at a *null* or *is balanced.* The numerator is equal to zero, so there would be no output voltage. E_o would equal zero.

In many applications, this architecture is used in a feedback system to force a null when a particular condition is reached, such as a temperature. If the bridge is not in a null, the polarity of E_o would be such as to steer the variable, such as temperature, in a direction toward the null.

In some cases, the bridge is used to make a measurement, such as the value of a resistor. In Fig. 14-23b, the value R would be the nominal value of R_x. When R_x is exactly equal to R, the bridge would be at a null, having no output.

The bridge architecture is not limited to resistors. If the sensing element is a capacitor, for example, a bridge can be constructed using only capacitors.

The output signal of a bridge must be amplified. This amplifier must also act as a buffer so it does not disturb the balance of the bridge. The best approach

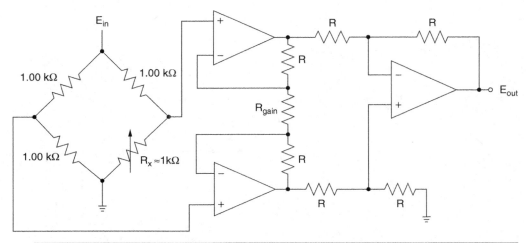

Figure 14-24   Bridge circuit feeding an instrumentation amplifier.

would be to use an instrumentation amplifier, as this would provide both buffering and gain. This is illustrated in fig 14-24.

If the bridge is used to simply detect a null, the precision and stability of E_{in} is not critical. If the bridge is used to provide a voltage indicative of the value of R_x, then E_{in} is critical and must remain stable as well as accurate.

References

Jung, Walt, editor. (2006). *Op Amp Applications Handbook*, Analog Devices. Norwood, Ma.

Graeme, Jerald G. (1973). *Applications of Operational Amplifiers: Third Generation Techniques*, McGraw-Hill. New York.

Texas Instruments, *Handbook of Operational Amplifier Applications*, Application Report SBOA092A. October 2001, Dallas, Texas.

Franco, S. (1988). *Design with Operational Amplifiers and Analog Integrated Circuits*, McGraw-Hill. New York.

Texas Instruments. *Op Amps for Everyone*, Advanced Analog Products SLOD006B, August 2002. Dallas, Texas.

Williams, Arthur B. (1984). *Designers' Handbook of Integrated Circuits*, McGraw-Hill. New York.

CHAPTER 15

Nonlinear Circuits

15.1 Ideal Rectifiers and Their Applications

An ideal diode has a transfer function where it would appear as an open circuit in one direction of current flow and a short circuit with zero resistance and no voltage drop in the other direction. However, this is not the case. The voltage-current characteristic of a practical diode appears in Fig. 15-1. In the forward direction, the diode "on voltage" must be reached before the diode exhibits conduction. In the reverse direction, no current will flow until a breakdown voltage is reached, so the diode will not behave like an ideal switch.

When a diode is used as a rectifier, the output signal is offset by the diode "on" voltage, which has to be overcome before current will flow. Figure 15-2 illustrates this effect.

In Fig. 15-2a, a half-wave rectifier is shown. It consists simply of a diode and a resistor. Its function is to remove half of the input sine wave, the portion going below zero. However, as illustrated in Fig. 15-2b, the output is offset by the 0.7V drop across the diode, which must be overcome before current can flow. So some of the positive-going waveform is lost. For low-voltage signals, the rectification would not occur at all.

An ideal rectifier circuit uses the characteristics of an op amp to reproduce only the positive (or negative) portion of the input signal without the error of the diode drop. It provides a break point that is sharp and well defined, which is not the case when just using an ordinary diode rectifier circuit. Note that for all circuits shown, the polarity of the output can be changed by simply reversing all the diodes. These circuits are alternatively called precision absolute value circuits or *super diodes*.

15.1.1 Half-Wave Precision Rectifier

Let us look at the simple half-wave precision rectifier shown in Fig. 15-3a. For a positive input, the op amp will try to maintain its positive and negative terminals equal due to the virtual ground effect. As a result, the op amp will forward-bias the diode D_1 and overcome its $\approx 0.7V$ voltage drop so the feedback voltage to its negative terminal will be equal to that at its positive terminal. Therefore, the op-amp output will follow the input for positive-going signals. For a negative-going signal, the op-amp output will swing negative, but output current will be blocked by D_1. As a result, the output signal would be zero. The op-amp output would be *saturated* at its most negative value since there is no virtual ground effect.

When the op-amp input switches from a negative input signal to a positive one, the output has to switch from its most negative value to this positive signal. Because of slewing-rate limitations, this cannot happen in zero time, so some waveform distortion

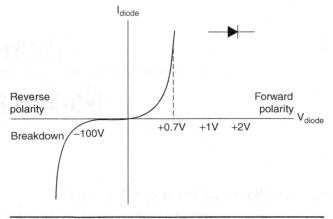

FIGURE 15-1   Diode characteristics.

will occur. As a result, this circuit has speed limitations, so amplifiers with good slewing-rate characteristics must be chosen. Also, for all precision rectifiers, the speed of the diodes will affect circuit operation, so signal diodes would be preferred to "power rectifiers," which are intended for higher-power applications and are slower.

Figure 15-3*b* shows another version of an ideal half-wave rectifier, which is improved over the circuit of Fig. 15-3*a*. When the input goes positive, the input current will flow from R_1 through D_2, causing the cathode of D_2 to be $\approx -0.7V$ since the anode is connected to a virtual ground. This negative voltage will prevent D_1 from conducting current, so the output will be zero volts. When the input goes negative, the output of the

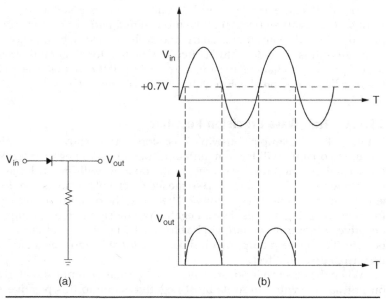

FIGURE 15-2   Diode half-wave rectifier: (*a*) input and (*b*) output.

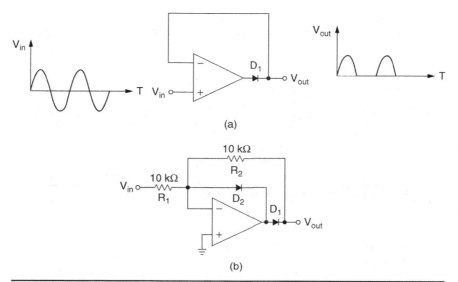

Figure 15-3   Half-wave ideal rectifier: (a) basic circuit and (b) improved circuit.

op amp swings positive. Since D_1 is inside the feedback loop, the output of the op amp itself will be about +0.7V higher than V_{out}. In order for the virtual ground to be maintained, the current through R_1 must equal the current through R_2 since no current can flow into the inverting terminal of the op amp. So

$$\frac{V_{in}}{R_1} = \frac{-V_{out}}{R_2} \qquad (15\text{-}1)$$

Therefore, the circuit gain is

$$\frac{V_{out}}{V_{in}} = \frac{-R_2}{R_1} \quad \text{for } V_{in} > 0V \qquad (15\text{-}2)$$

The major advantage of this circuit is that the op amp never goes into saturation. Diode D_2 clamps the output to $\approx -0.7V$ during the output off state, so the deleterious effects of slewing do not occur, resulting in improved speed. In addition, this circuit can provide gain.

15.1.2   Full-Wave Precision Rectifier

The circuit of Fig. 15-4 shows a precision *full-wave* rectifier.

When the input goes positive, a negative-going half-wave signal is generated at the junction of D_1 and R_3 as shown. This signal is then amplified by 2 (the ratio of $R_5/R_3 = 2$) and again inverted so it is positive going at V_{out}. Simultaneously, the input signal is inverted once through the $R_4\,R_5$ path. So we have two signals combined together at V_{out}, one that is the input *inverted* and the other that is a half-wave rectified version of the input amplified by 2 but *not inverted*. Combined, we obtain a full-wave rectified signal at the output.

Figure 15-5 illustrates the operation of Fig. 15-4 by showing the various waveforms all on the same time axis.

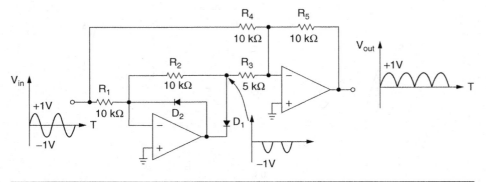

FIGURE 15-4   Full-wave precision rectifier.

FIGURE 15-5   Analysis of ideal full-wave rectifier of Fig. 15-4.

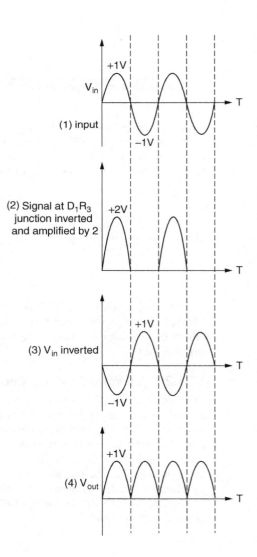

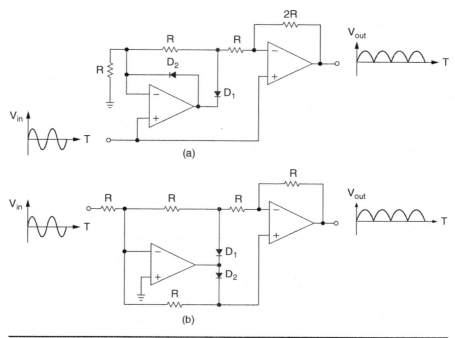

FIGURE 15-6 Two alternative forms of ideal full-wave rectifiers.

Note how the output waveform 4 is a result of the algebraic addition of waveforms 2 and 3 by the second summing amplifier.

With $R_5 = R_4$, the gain is 1. However, other gains can be achieved by letting $R_5 = AR_4$, where A is the desired gain. Resistors R_1, R_2, and R_4 should be matched as closely as possible. R_3 should be precisely half of R_4. The op amp should have sufficient slewing and bandwidth capability to accommodate the signals applied. The diodes should all be of the high-speed type.

Figure 15-6 shows two alternative forms of ideal full-wave rectifiers. The circuit of Fig. 15-6a has a high input impedance, which is not the case for the circuit of Fig. 15-4. Otherwise, it is very similar to Fig. 15-4. The circuit of Fig. 15-6b allows the summing of multiple signals at the input by the addition of summing resistors. In addition, gain or loss can be provided by reducing or increasing the input resistor.

15.1.3 Peak Detector

A peak detector is a circuit that tracks an input signal, retaining its peak value until a larger signal occurs, at which point it will capture the larger value. This will go on until a capacitor is discharged (reset), at which point this process will start all over again. Peak detectors are used in applications such as amplitude measurement, automatic gain control, and a variety of instrumentation equipment. Figure 15-7a shows a precision peak detector. A representative waveform is shown in Fig. 15-7b. A storage element C is combined with an ideal half-wave rectifier followed by a voltage follower buffer. Capacitor C must be a low leakage–type capacitor, or the signal will start to exhibit a droop before the next larger peak occurs, and charge would leak off. Some typical capacitors that exhibit low leakage are polystyrene, polypropylene, and mica. Electrolytic capacitors of all types should be avoided, if possible, as they typically have high

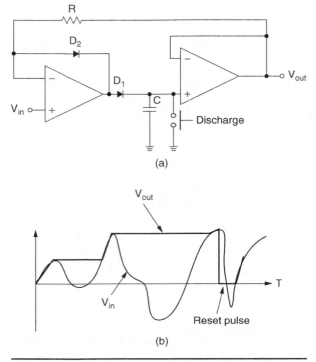

FIGURE 15-7 Peak detector: (a) circuit and (b) waveforms.

leakage, although tantalums are superior to aluminum electrolytics. Note that the switch shown in Fig. 15-7a is not really a mechanical switch but a solid-state switch. Typically, a field effect transistor (FET) or transistor would be used. Depending on the application, periodically the capacitor will be discharged and the peak detection sequence reinitiated.

15.1.4 Sample and Hold Circuit

A sample and hold circuit is normally used to capture a sample of an analog signal and hold it stable for a sufficient time so that an analog-to-digital converter (ADC) can convert the sample into a digital word. Once the sample is processed, a new sample is captured and the process is repeated. A very basic sample and hold circuit is shown in Fig. 15-8a. Periodically, the solid-state switch is closed to acquire a sample of the signal and then opened. Since the sampled signal must remain stable at the input to the ADC, the time the switch is closed must be very short. As in the case of the peak detector, the capacitor must be a low-leakage type.

The circuit of Fig. 15-8b is somewhat similar to the peak detector circuit of Fig. 15-7a. While the switch is closed, the input amplifier A_1 followed by A_2 acts as a voltage follower forcing the inverting and noninverting inputs of A_1 to be equal. Resistor R closes the feedback loop. The voltage captured on capacitor C is the input voltage during the duration of the switch being closed. When the switch is opened, the output of voltage follower A_2 is the captured or sampled voltage. The function of the diodes is to prevent A_1 from saturating when the switch is opened (hold mode) so A_1 can make a rapid

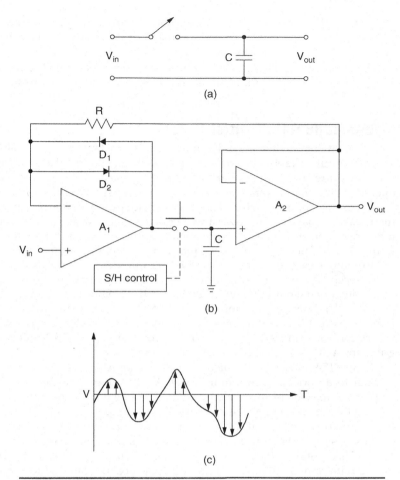

Figure 15-8 Sample and hold circuit: (a) simplified circuit; (b) implementation; and (c) typical pulse train.

recovery when a new sample is acquired. Figure 15-8c shows a typical pulse train consisting of samples of an input analog waveform.

There are a number of important definitions associated with a sample and hold circuit, as follows:

Sampling Rate To accurately reproduce the analog signal sampled, the sample rate (number of samples per second) must equal or exceed *twice* the highest frequency present in the sampled signal. This is called the *Nyquist criterion*. If this is not met, additional spectrums will be generated when the D/A process occurs, which overlap the wanted spectrum and cause a form of distortion called *aliasing*. Typically, a low-pass filter precedes the ADC to ensure the Nyquist criterion is met.

Aperture time This is the time delay between the application of the command to the solid-state switch to hold the sample and the actual point at which the circuit output to the ADC will stop tracking the input. This is typically caused by various propagation delays, so to compensate, the hold or "open switch" command could be advanced by this delay if greater precision is desired.

Acquisition time   The acquisition time is the delay between the command to sample the input signal and the point at which it appears at the output.

15.2   Automatic Gain Control

An automatic gain control (AGC) maintains a constant output amplitude for a given range of input variation. It is an adaptive feedback system where the output level is fed back to control the gain so that the output is forced to remain constant in amplitude over a given dynamic range. Essentially, it is a negative feedback system.

Let us take the application of an AM radio, for example. Considering that the received modulated signal can vary over a wide range due to propagation effects, distance from the transmitter, etc., one would not want the output audio to vary simultaneously in the same manner. This is just one of the many applications of an AGC amplifier.

In many cases, an AGC is already built into devices such as radio frequency (RF) radio receivers, analog-to-digital converters, modem analog front ends (AFEs), and other integrated circuit (IC) building blocks that would require an AGC function. However, there are some applications where an AGC must be constructed as a unique circuit. In both cases, it is useful to understand the basic principles of operation so component values can be optimized for maximum performance. Table 15-1 will provide a selection guide for AGC amplifiers and related building blocks.

Figure 15-9 illustrates the block diagram of an AGC system. The input signal is first applied to a variable gain amplifier (VGA). The output of the VGA is applied to an amplitude detector in the feedback loop. The function of the amplitude detector is to convert the signal coming from the variable gain amplifier to a DC voltage, which reflects the amplitude of the signal. This conversion can be performed in a number of ways. Frequently, a simple diode rectifier is used for this function, which half-wave rectifies the signal. This is quite primitive, and diode drops can cause errors and variations in performance with temperature. A more accurate approach would be to use an ideal rectifier

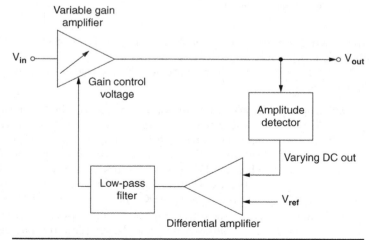

FIGURE 15-9   Basic AGC system.

circuit, which was discussed in Sec. 15.1. A peak detector can be used, although allowing the circuit to discharge so it can track a change in signal level is an added complication. A more elegant and superior solution would be to use an RMS-to-DC converter. This approach would provide a more accurate representation of the signal amplitude and take the complexity of the waveform into consideration by computing the true root mean square (RMS) value. The peak detector approach would only consider the waveform peaks, which is by no means a representation of the waveform from a power point of view.

The RMS representation of the signal amplitude is compared to a reference voltage V_{ref} using a differential amplifier to establish whether the signal is too strong or too weak.

Since the input waveform will vary with time, its short-term RMS representation can vary. Therefore, the output of the differential amplifier must be sufficiently filtered to create a smooth control voltage for the VGA. This is extremely critical. We don't want the AGC to follow the short-term variations of the signal. If that were to happen, the output signal would become distorted since it will be multiplied by the *envelope* of the input signal. However, we don't want too long a time constant either or the system will not quickly track a longer-term drift in incoming signal amplitude. This is one of the more critical design parameters.

Figure 15-10 shows a modulated waveform that maintains a constant power when averaged, but short term, the signal strength will vary as a result of the modulation. The dashed line represents the derived gain control voltage with *too short* a time constant. In this case, the waveform would be distorted by, in effect, being multiplied by the control voltage, producing unwanted sidebands. The AGC should not be responsive to short-term amplitude modulation. A steady control voltage would be desired, as shown by the solid line.

It should be pointed out that the term "time constant" is really the RC product of a simple one-pole RC network used for filtering. Lowering the 3-dB point of a low-pass filter is analogous to increasing the RC time constant. Since this is a closed-loop system, to ensure stability, the bandwidth of the low-pass filter should be limited, and the complexity of the low-pass filter limited as well. However, a compromise is desired so the AGC does not appear "sluggish" in its response to a longer-term change in input signal amplitude. Typically, the filter is a first-order filter, resulting in a 3-dB bandwidth of the AGC loop limited to a value lower than the lowest modulating frequency.

The VGA is probably the most critical building block. The simplest configuration would consist of a voltage variable attenuator, sometimes called a *variolosser*, followed by an amplifier. The voltage variable resistor can be an FET, where the resistance between drain and source is controlled by the gate-source voltage. This resistance parameter is called r_{DS}. A simple implementation is shown in Fig. 15-11a. The FET resistance r_{DS} is controlled by V_{GS}, the voltage between gate and source.

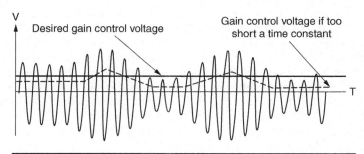

FIGURE 15-10 Gain control voltage.

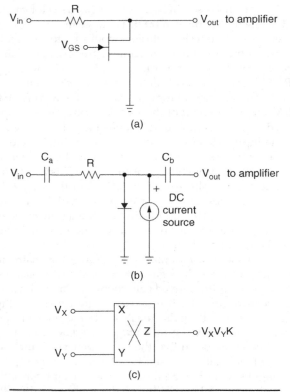

Figure 15-11 Variable attenuator: (a) FET approach; (b) using a diode; and (c) analog multiplier.

In Fig. 15-11b, a diode is used as the variable resistor where the forward-biasing current controls the diode resistance r_d. The capacitors C_a and C_b block the DC developed across the diode from being fed back to the input and output.

Although the two circuits of Figs. 15-11a and 15-11b are quite simple, they have a number of shortcomings. With too large an input signal, the instantaneous value of the waveform will interact with the FET or diode and the r_{DS} or r_d will not be a function of the control voltage only. The signal itself will *modulate* the resistance. Variations will occur among FETs and diodes from unit to unit. Stability with temperature can become an issue too. An analog multiplier can perform the function of a variable attenuator, as shown in Fig. 15-11c. An analog multiplier is a building block that accepts two analog signal inputs and produces an output that is the product of the two inputs times a constant K. In this application, the DC control voltage is one input and the analog signal applied to the AGC is the other. The control voltage will then linearly amplify or attenuate the analog signal. In effect, the control voltage will scale the analog signal as a function of its DC value.

Table 15-1 provides a list of devices that perform the AGC function and a selection of individual building blocks. This list is not exhaustive, so other devices can be available. The manufacturers' data sheets will provide a wealth of information on their usage and properties.

Variable Gain Amplifiers				
Manufacturer	**P/N**	**Bandwidth**	**Gain Range**	**Features**
Analog Devices	AD600	35 MHz	0 dB to +40 dB	dual, low distortion, low noise, low power
Analog Devices	AD602	35 MHz	−10 dB to +30 dB	dual, low distortion, low noise, low power
Analog Devices	AD603	90 MHz	−11 dB to +31 dB	low noise, programmable gain ranges
Analog Devices	AD604	40 MHz	0 dB to +48 dB	dual, ultra-low noise, programmable gain ranges
Analog Devices	AD605	40 MHz	−14 dB to +34 dB	low noise, programmable gain ranges, single supply
Analog Devices	AD8336	115 MHz	−14 dB to +46 dB	low noise, general purpose
Linear Technology	LTC6412	800 MHz	−14 dB to +17 dB	wide band, single supply, balanced in and out
Texas Instruments	VCA610	15 MHz	−38.5 dB to +38.5 dB	77 dB range, low noise, small package
Texas Instruments	VCA810	25 MHz	±40 dB	low noise, differential-in, single-ended out, high output current
Texas Instruments	VCA820	150 MHz	>40 dB	low noise, differential-in, single-ended out, high output current
Texas Instruments	VCA821	710 MHz	>40 dB	low noise, differential-in, single-ended out, high output current
Texas Instruments	VCA822	150 MHz	>40 dB	low noise, differential-in, single-ended out, high output current
Texas Instruments	VCA824	710 MHz	>40 dB	low noise, differential-in, single-ended out, high output current
Texas Instruments	LMH6502	130 MHz	70 dB	70 dB range, high slewing, high output current
Texas Instruments	LMH6503	135 MHz	70 dB	70 dB range, high slewing, high output current, low power
Texas Instruments	LMH6505	150 MHz	80 dB	80 dB range, high slewing, high output current, low power
RMS to DC Converter				
Manufacturer	**P/N**	**Bandwidth**	**Features**	
Analog Devices	AD636	1 MHz	high crest factor, 50 dB dynamtic range	
Analog Devices	AD736	460 KHz	high crest factor, negative going output, low power, low cost	
Analog Devices	AD737	8 MHz	high accuracy, positive going output, 60 dB range	
Analog Devices	AD8361	2.5 HGz	wide band, 30 dB dynamic range, low power, single supply	
Analog Devices	AD8436	1 MHz	high crest factor, very high accuracy, low power	
Linear Technology	LTC1966	1 KHz	simple to use, high accuracy, single supply	
Linear Technology	LTC1967	40 KHz	simple to use, high accuracy, single supply	
Linear Technology	LTC1968	500 KHz	simple to use, high accuracy, single supply	

TABLE 15-1 AGC Building Blocks (*Continued*)

	AGC			
Manufacturer	**P/N**	**Bandwidth**	**Range**	**Features**
Analog Devices	AD8367	500 MHz	−2.5 dB to +42.5 dB	high performance, includes AGC detector
Maxim	MAX9814	20 KHz	0 dB to +20 dB	microphone amplifier with AGC
MITEL	SL6140	400 MHz	70 dB	wide range of AGC, wide bandwidth
Texas Instruments	TL026C	50 MHz	50 dB	wide bandwidth, low distortion

TABLE 15-1   AGC Building Blocks (*Continued*)

15.3   Log and Antilog Circuits

Log and antilog amplifiers are nonlinear building blocks that have an output that is either the log or antilog of the input. By combining the log or antilog of signals, mathematical functions such as multiplication, division, and squaring can be performed on input signals. The end result is the implementation of analog multipliers, AGC circuits, modulators, and many other nonlinear functions.

As was the case with AGC amplifiers and their building blocks, log and antilog amplifiers are available as standard integrated circuits, so a selection table will be provided at the end of this section.

Figure 15-12*a* shows the logarithmic relationship between a silicon diode current and voltage. It is this property that is exploited in log and antilog circuits. A basic op-amp diode logarithmic converter is shown in Fig. 15-12*b*. This logarithmic property also applies to a transistor. A more popular implementation is shown in Fig. 15-12*c* and is called a transdiode configuration. The base-emitter voltage V_{BE} has the predictable logarithmic relationship with the collector current I_C.

For the circuit of Fig. 15-12*b*, V_{out} is given by

$$V_{out} = -V_T \ln\left(\frac{V_{in}}{I_S R}\right) \tag{15-3}$$

where I_S is the reverse-bias saturation current, which is typically a few picoamps, and V_T is the thermal voltage of the diode, which at room temperature is typically 26mV.

For the transdiode configuration of Fig. 15-12*c*

$$V_{out} = -V_T \ln\left(\frac{V_{in}}{I_{SO} R}\right) \tag{15-4}$$

where I_{SO} is the saturation current of the emitter-base diode. In both circuits, the output voltage is expressed as a natural log of the input voltage.

The input voltage polarity must be positive. This can be reversed by reversing the diode or going to a PNP transistor instead of an NPN.

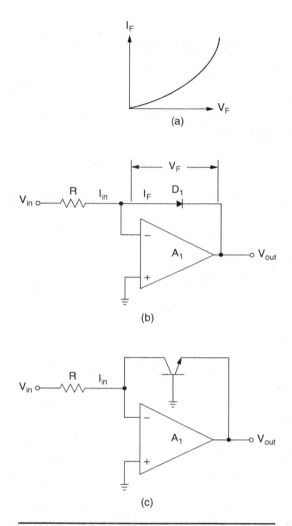

Figure 15-12 Logarithmic amplifier: (a) diode characteristics; (b) diode log amplifier; and (c) transdiode log amplifier.

Note that neither implementation considers temperature stability since V_T, I_S, and I_{SO} are a function of temperature as well as other nonideal effects and limitations that occur. These issues are addressed in many of the ICs listed in Table 15-2.

Figure 15-13 shows the antilog version of the two basic configurations. The semiconductor device is simply moved to the input, so the input voltage produces a current in an antilog proportionality.

Figure 15-14a illustrates how a multiplier can be constructed using log and antilog amplifiers. The configuration is based on the principle that the log of the product of two terms is equivalent to the summation of the individual logs of the terms. In other

Manufacturer	P/N	Bandwidth	Dynamic Range	Features
Analog Devices	AD538	400 KHz		log and antilog, multiplication, division, squaring, square rooting, resistor programmable
Analog Devices	AD640	120 MHz	50 dB	log, 5 stages
Analog Devices	AD641	250 MHz	44 dB	log, ±2 dB accuracy, built-in attenuator
Analog Devices	AD8304	10 MHz	160 dB	log, optimized for fiber optic diode interfacing, extremely wide dynamic range
Analog Devices	AD8307	500 MHz	92 dB	log, single supply operation, nine stages, ±1 dB linearity
Analog Devices	AD8310	440 MHz	95 dB	log, single supply operation, nine stages, ±0.4 dB linearity
Maxim	MAX4206	5 MHz	100 dB	log, single or dual supply operation, uncommited amplifier
Maxim	MAX4207	1 MHz	120 dB	log, dual supply operation, uncommited amplifier
Texas Instruments	LOG101	45K Hz	150 dB	log, wide dynamic range, 0.01% accuracy, dual supplies
Texas Instruments	LOG102	45 KHz	120 dB	log, wide dynamic range, 0.15% accuracy, dual supplies
Texas Instruments	LOG104	45 KHz	150 dB	log, wide dynamic range, 0.01% accuracy, dual supplies
Texas Instruments	LOG112	45 KHz	150 dB	log, wide dynamic range, 0.2% accuracy, dual supplies
Texas Instruments	LOG114	120 MHz	160 dB	log, wide dynamic range, dual supplies or single +5V supply
Texas Instruments	LH0094	10 KHz		log, antilog, multiple functions, 0.05% accuracy

TABLE 15-2 Log and Antilog Integrated Circuits

words, $\ln(V_{inA}) + \ln(V_{inB}) = \ln(V_{inA} \times V_{inB})$, so we add the natural log of the two inputs together, take the antilog, and we have the product. A constant K is shown representing gain or loss.

The configuration of Fig. 15-14*b* can implement division of two numbers. Instead of summing the log of the two inputs, they are subtracted from each other.

By combining multipliers with log and antilog building blocks, a number of additional mathematical functions can be performed on input signals, such as raising to a power or determining a root.

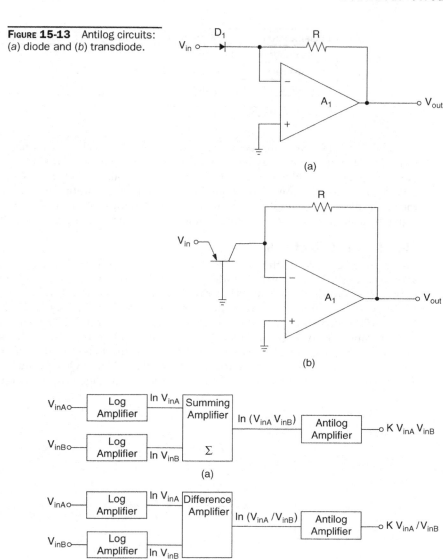

FIGURE **15-13** Antilog circuits: (a) diode and (b) transdiode.

(a)

(b)

FIGURE **15-14** Applications of log and antilog circuits: (a) multiplier and (b) divider.

15.4 Multipliers

An analog multiplier is a building block with two input ports and one output port. Its function is to accept two inputs and provide an output that is the product of the two inputs times a constant scaling factor K. Figure 15-14a showed one implementation using log and antilog circuits.

Analog multipliers are classified in terms of quadrants of operation. Table 15-3 defines these classifications.

Classification	X Input	Y Input	X × Y × K Output
Single quadrant	Unipolar	Unipolar	Unipolar
Two quadrant	Bipolar	Unipolar	Bipolar
Four quadrant	Bipolar	Bipolar	Bipolar

TABLE 15-3   Multiplier Classifications

If both input signals must be of one defined polarity, the classification is "single quadrant." If one of the signals may be of either positive or negative polarity and the other must be of a given polarity, the device is "two quadrant." Finally, if both inputs may be of either polarity, the classification is "four quadrant." The type of multiplier chosen depends on the application. For example, if a multiplier is needed for an AGC that multiplies a bipolar signal by a single polarity of control voltage, a two-quadrant multiplier can be used.

15.4.1   The Gilbert Cell

There are two basic methods of implementing multipliers. The first method is that shown in Fig. 15-14a, which is adding together the log of each input signal and then taking the antilog of the sum. The second method is using a building block called a "Gilbert cell" invented by Larry Gilbert in the 1960s. The basic Gilbert cell is shown in Fig. 15-15. It consists of two cross-connected differential amplifiers, each consisting of a pair of transistors: Q1, Q2 and Q4, Q5. The emitters of both pairs are driven by a third pair consisting of Q3 and Q6. Therefore, the currents (or bias) in both differential amplifiers is a function of the emitter current generated by Q3 and Q6 and the input voltages. The result is that the combining of the two differential stage output currents results in

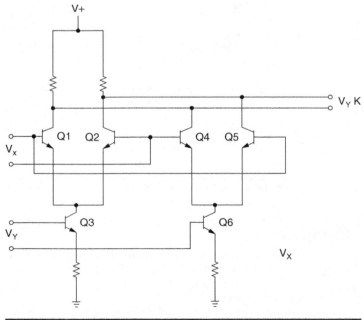

FIGURE 15-15   The Gilbert cell.

four-quadrant multiplication. Note that both inputs and the output are differential (balanced) signals.

15.4.2 Multiplier Parameters

When selecting a multiplier, there are several key parameters of importance. The first is the *number of quadrants*: single, two, or four. Next comes *accuracy*. The accuracy is defined as the deviation of the actual output from the ideal for any range of X and Y inputs within the specified limits. It is typically given as a percentage of the full-scale output. *Linearity* is the variation from a best fit of a straight line, and is expressed as a percentage of a full-scale output. *Bandwidth* is defined as the frequency range over which the device is specified, typically the 3-dB point.

15.4.3 Multiplier Math Functions

Multipliers can perform a variety of mathematical functions on input signals. Figure 15-16 illustrates a number of these functions. The circuit of Fig. 15-16*a* is a

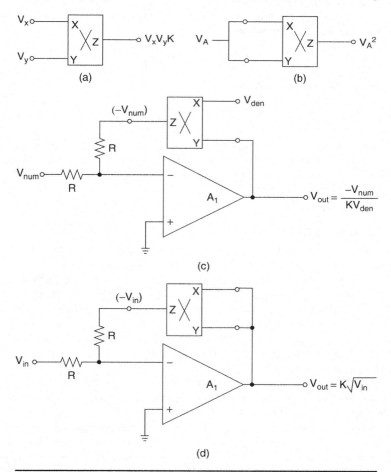

FIGURE 15-16 Mathematical applications of multipliers: (*a*) basic multiplier; (*b*) squaring function; (*c*) division; and (*d*) square root.

basic multiplier that provides the product of two input signals times a constant multiplier K. One application shown previously was to control signal level in an AGC. It also has numerous applications as a modulator and for frequency translation (also called a mixer).

The circuit of Fig. 15-16b provides the square of the input signal. It can also function as a frequency doubler. Let us consider the identity

$$\cos 2\Phi = 1 - \sin^2\Phi \tag{15-5}$$

For a sine wave, we can say $\Phi = \omega t$ so

$$\cos 2\omega t = 1 - \sin^2 \omega t \tag{15-6}$$

We can eliminate the "1" term with a capacitor, as it represents a DC component, so by squaring a sine wave, we can double its frequency. A constant multiplier K will affect the amplitude of the output signal.

Division can be performed using a multiplier in a op-amp feedback loop. In the circuit of Fig. 15-16c, the output of the multiplier must be $-V_{num}$ so that the negative input to A_1 remains a virtual ground. Therefore

$$-V_{num} = KV_{den}V_{out} \tag{15-7}$$

So the final result becomes

$$V_{out} = \frac{-V_{num}}{KV_{den}} \tag{15-8}$$

The *square root* of a signal can be determined using the circuit of Fig. 15-16d. Since the output of the multiplier is equal to $-V_{in}$ to maintain the virtual ground and $-V_{in}$ is equal to V_{out} squared, V_{out} then is proportional to the square root of the input times a constant.

15.5 Modulators

A modulator modifies a carrier signal using a waveform that contains information such as audio or data. At the opposite end, this information is recovered by a demodulator. In Fig. 15-17a, a carrier is modulated (multiplied) by a modulating waveform. The envelope of the modulated waveform takes on the shape of the modulating signal. This is an example of *amplitude modulation*. Of course, there are many other forms of modulation, such as pulse, frequency, phase, etc., but that is best left for a course or book on communications. Sometimes modulators are called *mixers*, but that usage should be differentiated from audio mixers, which simply add signals together.

In many cases, the carrier can be much higher in frequency than the modulating signal (AM radio, for example), so the carrier can be a square wave as there is no need to retain any linear information of the carrier.

Therefore, when using a multiplier for this application, one of the inputs is the carrier, which is constant in amplitude, so that particular input does not have to operate in the linear region. Using a Gilbert cell architecture, the upper pairs of transistors (Q1, Q2, Q4, and Q5 in Fig. 15-15) operate in a switching mode rather than in

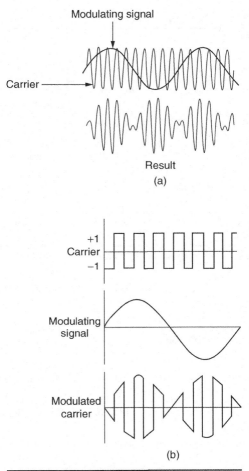

FIGURE 15-17 Modulator waveforms: (a) linear-
linear mode and (b) switching-linear mode.

their linear mode. The lower pair operates in a linear mode. The carrier drives the
upper pairs, which then switch the currents from the lower pairs, which is the
modulating signal. The carrier alternately inverts segments of the modulating
signal based on the instantaneous polarity of the carrier. This is shown in
Fig. 15-17b.

The switching-linear modulator, by its nature, is not fully linear, so the output will
contain harmonics of the switching carrier frequency, with each harmonic replicating a
modulated waveform although lower in amplitude. Therefore, filtering is usually
required to eliminate these harmonics. The larger the ratio of the carrier to the modulating frequencies, the less complex will be the filtering.

Table 15-4 provides a selection guide for multipliers.

Manufacturer	P/N	Bandwidth	Quadrants	Features
Analog Devices	AD532	1 MHz	four-quadrant	Internal amplifier for mulitple functions
Analog Devices	AD534	1 MHz	four-quadrant	adjustable scale factor
Analog Devices	AD538	400 KHz	one or two quadrant	wide dynamic range
Analog Devices	AD539	60 MHz	two-quadrant	2 independent multipliers in one package, low distortion
Analog Devices	AD632	1 MHz	four-quadrant	internal amplifier for multiple functions
Analog Devices	AD633	1 MHz	four-quadrant	internal amplifier for multiple functions
Analog Devices	AD734	10 MHz	four-quadrant	high performance replacement for AD534
Analog Devices	AD834	500 MHz	four-quadrant	wide bandwidth, very low distortion
Analog Devices	AD835	250 MHz	four-quadrant	wide bandwidth, fast low setting
Intersil	HA-2556	57 MHz	four-quadrant	high slewing
Texas Instruments	MPY634	10 MHz	four-quadrant	internal amplifier for multiple functions
Texas Instruments	4413	550 KHz	four-quadrant	internal amplifier for multiple functions
Texas Instruments	MPY100	550 KHz	four-quadrant	internal amplifier for multiple functions

TABLE 15-4   Multiplier Selection Guide

References

Jung, W., ed. (2006). *Op Amp Applications Handbook*, Analog Devices, Norwood, Ma.

Graeme, J. G. (1973). *Applications of Operational Amplifiers: Third-Generation Techniques*, McGraw-Hill, New York.

Texas Instruments, *Handbook of Operational Amplifier Applications*, Application Report SBOA092A, October 2001, Dallas, Texas.

Franco, S. (1988). *Design with Operational Amplifiers and Analog Integrated Circuits*, McGraw-Hill, New York.

Texas Instruments. *Op Amps for Everyone*, Advanced Analog products SLOD006B, August 2002, Dallas, Texas.

Williams, A. B. (1984). *Designer's Handbook of Integrated Circuits*, McGraw-Hill, New York.

CHAPTER 16

Waveform Shaping

16.1 Integrators and Differentiators

16.1.1 The Ideal Integrator

Figure 16-1a shows an ideal integrator. An integrator is defined as a circuit that provides an output proportional to the integral with time of the input. It can be described as a circuit whose output reflects the area under the input waveform. Mathematically, this can be expressed as

$$V_{out} = K \int V_{in} dt \tag{16-1}$$

where K is a constant.

In the circuit of Fig. 16-1a the output voltage is the voltage across capacitor C since one end is connected to the virtual ground of the op amp at the negative terminal. The input current is given by V_{in}/R due to the virtual ground. This current flows into the capacitor, building up a charge. Since the voltage across a capacitor is defined by

$$V = \frac{1}{C} \int I dt \tag{16-2}$$

The output of the circuit of Fig. 16-1a can be expressed as

$$V_{out} = -\frac{1}{C} \int \frac{V_{in}}{R} dt = -\frac{1}{RC} \int V_{in} dt \tag{16-3}$$

Equation (16-3) is a bit awkward to work with since it could involve mathematically integrating complex waveforms. Let us start with the fundamental equation for the charge on a capacitor: the familiar $Q = CV$. Since the charge Q is equal to the product of current and time, we can state IT = CV. Replacing I with V_{in}/R and V_{out} for V, the output of the circuit of Fig. 16-1a can be expressed as

$$V_{out} = -\frac{V_{in}T}{RC} \tag{16-4}$$

The negative sign is a result of the inversion that occurs since the current is injected into the negative input of the op amp.

Equation (16-4) allows us to mathematically integrate a waveform. Figure 16-1b shows the integration of a square wave resulting in a triangular waveform.

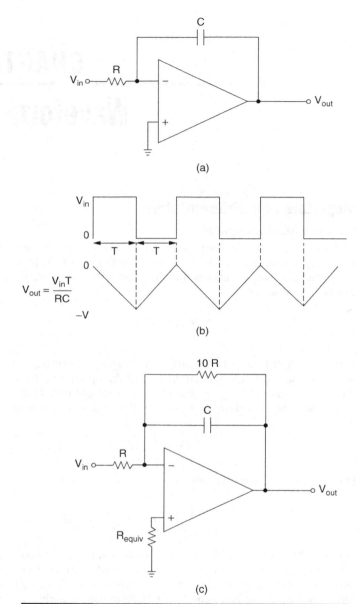

FIGURE 16-1 Integration with op amp: (a) basic circuit of inverting integrator; (b) integration of a square wave; and (c) practical integrator.

In Fig. 16-1b for the first half-cycle of the square wave, the positive input will cause the capacitor to charge negative, as determined by Eq. (16-4). Since the square wave is symmetrical, for the second half-cycle, the output will charge positive exactly the same amount. The result is a symmetrical triangular wave. This implies that the integral of a square wave is a triangular wave. Equation (16-4) can also be used for unsymmetrical waveforms where the pulse duty cycle is not 50 percent of a full period.

The previous analysis of an integrator was in the time domain. It is also useful to analyze an integrator in the frequency domain. The transfer function of the op-amp integrator of Fig. 16-1a is

$$T(s) = -\frac{Z_{feedback}}{Z_{input}} = -\frac{\frac{1}{sC}}{R} = -\frac{1}{sCR} \tag{16-5}$$

where $s = j\omega$ and $\omega = 2\pi f$.

The response will roll off monotonically at 6 dB per octave (20 dB per decade). It is the response of a one-pole network, where the pole is at DC. The phase shift is constant at +90°.

The minus sign of Eq. (16-5) indicates an inversion at the output. A one-pole low-pass normally has a negative phase shift, but the inversion converts it to positive (−90° gets converted to +90°.)

16.1.2   A Practical Integrator

In the circuit of Fig. 16-1a, with no input to the circuit, any offset present at the inverting input of the op amp will cause a continuous charging of the capacitor until the op-amp output saturates at either the positive or negative rails (maximum output), depending on the polarity of the offset. The circuit of Fig. 16-1c limits the amount of saturation to no more than a factor of 10 times the input offset. The resistor in parallel with the capacitor can be even more than 10 times the input resistor. The higher it is, the more linear will be the integration, but a factor of 10 is a reasonable number.

It is generally good practice to connect a resistor from the noninverting input to ground equal to the parallel combination of the other two resistors to minimize the effects of offset. That would be R_{equv} in Fig. 16-1c.

16.1.3   Differentiators

An ideal differentiator is shown in Fig. 16-2a. This circuit performs the opposite function of the integrator in the previous section. It performs the mathematical function of differentiation rather than integration of the input signal. In this ideal circuit, the current through the capacitor is defined by the classic equation $I = C\,dV/dt$. Since the inverting input terminal is at virtual ground, the output voltage can be given by the product of current and resistor R (with an inversion), as follows:

$$V_{out} = -RC\frac{dV_{in}}{dt} \tag{16-6}$$

If the input were to change linearly over the region of interest, we could define V_{out} by

$$V_{out} = -RC\frac{\Delta V_{in}}{\Delta t} \tag{16-7}$$

Figure 16-2b shows how a triangular wave is converted into a square wave. This circuit performs the opposite of integration; it reverses the process.

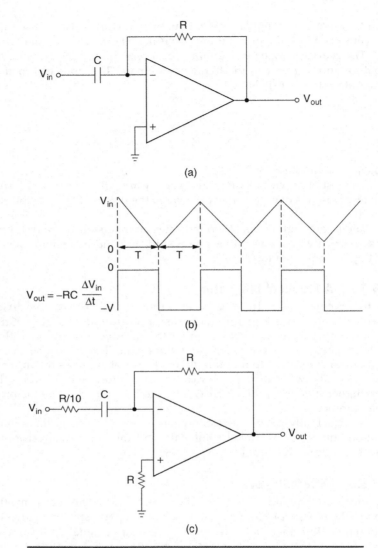

FIGURE **16-2** Op-amp differentiator: (a) ideal differentiator; (b) differentiating a triangular waveform; and (c) practical differentiator.

In the frequency domain, the response of the differentiator would be

$$T(s) = -\frac{Z_{feedback}}{Z_{input}} = -\frac{R}{\dfrac{1}{sC}} = -sCR \qquad (16\text{-}8)$$

where $s = j\omega$ and $\omega = 2\pi f$.

This implies that the output will increase at 6 dB per octave (20 dB per decade) and the phase shift would be a constant –90° (+90° combined with the 180° phase shift resulting from the inversion).

Because of the 6 dB per octave increase with frequency, any extraneous high-frequency noise at the input could be amplified to the point where it becomes an issue. So a maximum high-frequency amplification limit must be established. Figure 16-2c illustrates how this is accomplished with the addition of a resistor. The maximum gain would be 10. The resistor from the noninverting input to ground is for offset cancelation.

16.2   Comparators

16.2.1   Basic Comparator

A comparator compares two analog signals, where one is typically a fixed reference voltage and the other is an analog signal. The output of the comparator has a binary (two state) output, which indicates whether the analog signal is more positive or more negative than the reference. There are specific integrated circuits (ICs) designed for comparator applications. A selection chart is provided in Table 16-1. There are numerous comparators available, so it is always desirable to consult with the various manufacturers' applications engineers to make the best choice for your application. In addition, the individual data sheets are a valuable resource.

An op amp with no negative feedback can be used as a comparator since the high open-loop gain can compare small differences in input voltages and saturate the output either positively or negatively. However, when used as a comparator, the op amp has a few disadvantages. The first is that the op-amp response time may be too slow, whereas a comparator is designed specifically for this purpose and would be significantly faster. The second is that an op amp will not have a precisely defined output voltage due to internal voltage drops. Most comparators have an open-collector output, which means that a resistor to a precisely defined supply voltage allows level conversion between different logic families. A third advantage over op amps is that some comparators have built-in hysteresis to avoid additional transitions at their output. Nevertheless, there are applications where an op amp is acceptable for use as a comparator.

Figure 16-3a shows a general-purpose comparator—the Texas Instruments LM397. Let us assume its function is to convert an analog waveform into logic level signals. Typically, a reference voltage is established at a desired switching threshold. When the analog signal exceeds this threshold (V_{ref}), the output goes low since the analog waveform is at the inverting input. Conversely, when the analog signal is below V_{ref}, the output would go high. Note that the output stage is the collector of a transistor. A pull-up resistor is connected to the desired maximum output voltage for a logic-high, which in this case is +3.3V.

Since the resistor is external, it doesn't have to be connected to V_{CC}, the comparator power. In this example, V_{CC} is +5V and the pull-up goes to +3.3V, so the output would be compatible with +3.3V logic-level devices. Operation of this circuit is shown in Fig. 16-3b where a triangular-like waveform is converted into a +3.3V logic-level signal. The reference voltage V_{ref} must be inside the range of the analog input signal. If the input were a logic-level signal—let's say precisely between 0V and +5V—the reference should be +2.5V.

Manufacturer	P/N	Supply Voltage Range	Propogation Delay	Features
Analog Devices	ADCMP370	+2.5V to +5.5V	5uS	open-drain output, 22V tolerance on inputs
Analog Devices	ADCMP371	+2.5V to +5.5V	5uS	push-pull output, 22V tolerance on inputs
Analog Devices	ADCMP608	+2.5V to +5.5V	40nS	ultra fast, rail-to-rail output, shut-down pin, tiny package
Analog Devices	AD8561	+3V to +10V	7nS	ultra fast, differential output, low power
Analog Devices	AD8468	+2.5V to +5.5V	40nS	ultra fast, rail-to-rail output, shut-down pin, tiny package
Analog Devices	ADCMP609	+2.5V to +5.5V	40nS	ultra fast, rail-to-rail output, shut-down pin
Linear Technology	LT1016	+/-5V	10nS	ultra fast, TTL compatibility, output latrching capability
Linear Technology	LT1716	+2.7V to +44V	3uS	wide voltage range, internal output pullup
Linear Technology	LT1719	+2.7V to +10.5V	4.5nS	ultra fast, rail-to-rail output
Linear Technology	LT6703	+1.4V to +18V	18uS	built-in reference, built-in hysteresis, open-collector output
Maxim	MAX917	+1.8V to +5.5V	22uS	built-in +1.245V reference, push-pull output
Maxim	MAX918	+1.8V to +5.5V	22uS	built-in +1.245V reference
Maxim	MAX919	+1.8V to +5.5V	22uS	low supply current, push-pull output
Maxim	MAX920	+1.8V to +5.5V	22uS	low supply current
Maxim	MAX9021	+2.5V to +5.5V	3uS	single, low power, built-in hysteresis, ultra small package
Maxim	MAX9022	+2.5V to +5.5V	3uS	dual, low power, built-in hysteresis, ultra small package
Maxim	MAX9024	+2.5V to +5.5V	3uS	quad, low power, built-in hysteresis, ultra small package
Maxim	MAX9030	+2.5V to +5.5V	188nS	shutdown, built-in hysteresis
Maxim	MAX9031	+2.5V to +5.5V	188nS	single, built-in hysteresis
Maxim	MAX9032	+2.5V to +5.5V	188nS	dual, built-in hysteresis
Maxim	MAX9034	+2.5V to +5.5V	188nS	quad, built-in hysteresis
Maxim	MAX9060	−0.3V to +6V	15uS	open drain output, ultra low power

TABLE 16-1 General Purpose Comparators

Manufacturer	P/N	Supply Voltage Range	Propogation Delay	Features
Maxim	MAX9061	−0.3V to +6V	15uS	open drain output, ultra low power
Maxim	MA9062	−0.3V to +6V	15uS	built in reference, ulta low power
Maxim	MAX9063	−0.3V to +6V	15uS	built in reference, ulta low power
Maxim	MAX9064	−0.3V to +6V	15uS	built in reference, ulta low power
Texas Instruments	LM111	+/−18V	115nS	open collector output
Texas Instruments	LM193	+2V to +36V	1.3uS	dual, ultra low power, open-collector output
Texas Instruments	LM397	+5V to +30V	440nS	open collector output
Texas Instruments	LM2903	+2V to +36V	1.3uS	dual, ultra low power, open-collector output
Texas Instruments	LMV331	+2.7V to +5.5V	1uS	single, open collector output
Texas Instruments	LMV339	+2.7V to +5.5V	1uS	quad, open-collector output
Texas Instruments	LMV393	+2.7V to +5.5V	1uS	dual, open-collector output
Texas Instruments	LMC7211	16V differential	4uS	specified for +2.7V, +5V and +15V operation, push-pull output
Texas Instruments	LMV7219	+2.7V to +5V	1.3nS	ultra fast, built-in hysteresis, push-pull output
Texas Instruments	LMV7235	+2.7V to +5.5V	75nS	low power, open-drain output
Texas Instruments	LMV7239	+2.7V to +5.5V	75nS	low power, push-pull output

TABLE 16-1 General Purpose Comparators (Continued)

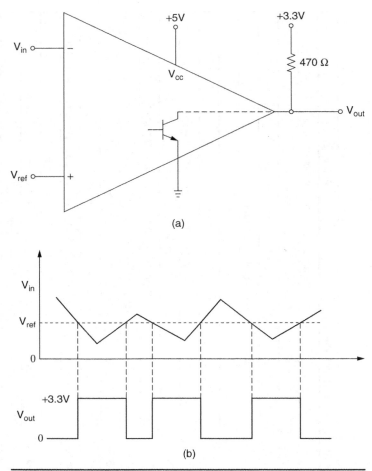

Figure 16-3   General-purpose comparator: (*a*) schematic showing internal open collector output and (*b*) example waveforms.

16.2.2   Window Comparator

Comparators can be used to see if a voltage is between two boundaries. A typical application might be to monitor a critical supply voltage and indicate when it is outside of preset limits. Another might be for production line testing.

Figure 16-4*a* shows a circuit for a comparator with two level thresholds: $V_{\text{ref HIGH}}$ and $V_{\text{ref LOW}}$. If the input voltage V_{in} exceeds $V_{\text{ref HIGH}}$, the output of A_1 will go high (the collector of the internal transistor will open). If V_{in} goes below the lower limit $V_{\text{ref LOW}}$, the output of A_2 will go high. A high coming from either A_1 or A_2 will cause the output of the NOR gate V_{out} to go low. Figure 16-4*b* shows a set of waveforms that illustrates this behavior.

If we reverse the input polarity of A_1 and A_2 we can use the "Wired-OR" property of the two open-collector outputs to perform like the NOR gate. A signal outside of the desired range will cause a low to occur at the associated op-amp output. Either amplifier output going low would result in the desired output, thus eliminating the NOR gate. The circuit is shown in Fig 16-4*c*.

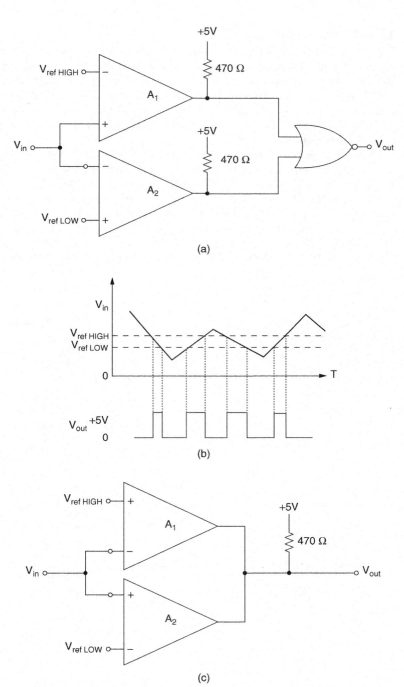

Figure 16-4   Window comparator: (a) window comparator circuit:
(b) waveforms; and (c) simplified circuit using wired-OR function.

16.2.3   Hysteresis

Figure 16-5a illustrates a basic comparator configuration, and Fig. 16-5b shows a very noisy pulse-like waveform. Because of the noise, the output waveform contains a number of extraneous narrow pulses on both the leading and trailing edges. This effect could wreak havoc, especially if the waveform is used in a digital manner. It could represent extra counts, for example. When a signal that varies slowly compared to the speed of the comparator is applied, multiple output transitions can occur for each edge. The solution is *hysteresis*.

Hysteresis is essentially positive feedback. Once the input signal crosses the reference threshold, the output switches and changes the reference voltage in a direction to ensure that a small variation of the input signal in the *opposite* direction does not cause the output to change state. The *amount* of hysteresis is equal to the voltage difference between the two references. In effect, hysteresis reinforces the state transition that just occurred.

Let us analyze the inverting comparator circuit of Fig. 16-5c. A three-resistor network has been added to the circuit of Fig. 16-5a. Let us first take the case where the V_{in} is lower than the reference, so the output will be high. As the signal rises, it will

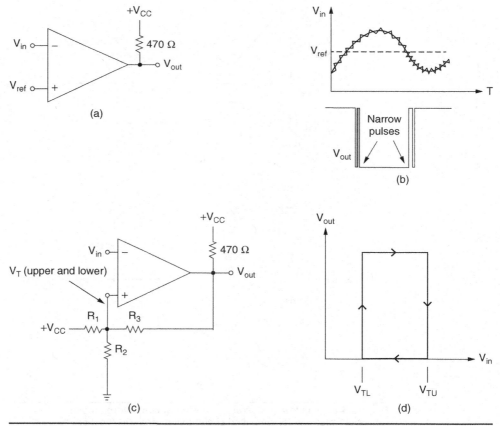

FIGURE 16-5   Adding hysteresis: (a) original circuit; (b) noisy input waveform resulting in extra edges (pulses); (c) circuit with added hysteresis; and (d) hysteresis path.

approach the upper threshold V_{TU}, which is computed by first calculating the parallel combination of R_1 and R_3

$$R_{P1-3} = \frac{R_1 R_3}{R_1 + R_3} \qquad (16\text{-}9)$$

then

$$V_{TU} = V_{CC} \frac{R_2}{R_2 + R_{P1-3}} \qquad (16\text{-}10)$$

When V_{in} starts off greater than the reference, then V_{out} is low, so R_2 is in parallel with R_3. So

$$R_{P2-3} = \frac{R_2 R_3}{R_2 + R_3} \qquad (16\text{-}11)$$

and

$$V_{TL} = V_{CC} \frac{R_{P2-3}}{R_1 + R_{P2-3}} \qquad (16\text{-}12)$$

The total amount of hysteresis is given by $V_{TU} - V_{TL}$.

Figure 16-5d illustrates the hysteresis path where the threshold will alternate between the two states based on V_{in}.

All of the previous resistance calculations are based on the assumption that the resistor values are at least 10 times larger than the output pull-up resistor, so it has little effect.

It is important to remember that slowly changing signals applied to a high-gain device such as a comparator can take very low-level noise signals and greatly amplify them as the signal passes through the threshold level, which becomes a linear region. This can result in extra transitions and even oscillation, so the addition of hysteresis becomes mandatory. Also, decoupling of the power supply voltages and clean ground planes are a necessary requirement.

Digital circuits such as CMOS and TTL require fast edges for input transitions, so they are subject to similar effects where rounded edges due to trace lengths or stray capacitance occur. There are digital circuits with built-in hysteresis at their inputs, are called *Schmitt triggers*, that can resolve these issues.

16.2.4 Limiters

A limiter is a circuit that converts an analog signal into a digital one, where a one or zero is determined by whether the analog signal is positive or negative with respect to ground. The only information contained in this digital signal is where in time a zero crossing occurred and in which direction. A typical application would be in an FM receiver where all the frequency information is contained in the zero crossings. If the limiter circuit has enough open-loop gain, the output signal would not be input-level independent over a wide dynamic range. So a high-gain amplifier or comparator would be needed, similar to the circuit of Fig. 16-5a, except V_{ref} would be ground.

It is important to minimize any DC offset at the input, which would distort the zero crossings, so this should be considered when selecting a device with low offset. The input should be capacitive-coupled to eliminate any DC offset that may be present in the input signal resulting from previous stages of amplification. This type of circuit is called

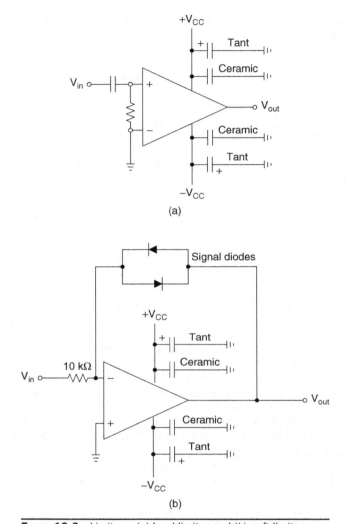

FIGURE **16-6**   Limiters: (a) hard limiter and (b) soft limiter.

a *hard limiter*, as the output has only two states. An example is shown in Fig. 16-6a. Power supply decoupling is important. Typically, a 0.1-µF ceramic and a 10-uF tantalum are used. Frequently, the device data sheet will have recommendations.

You will sometimes come across the term *soft limiter*. This type of circuit would be used for dynamic range compression. A typical limiter in this family is one that performs logarithmic compression. A representative circuit is shown in Fig. 16-6b.

16.2.5   Time-Delay Circuits Using Comparators

A time-delay circuit using op amps or comparators is one that can produce a predefined delay, typically to a digital signal, but the timer operates in an analog fashion. The delay results from using the time constant of an *RC* network to delay an edge (or both edges) of the signal. Figure 16-7a shows the results from closing SW1 to charge a capacitor to 1V and then closing switch SW2 to discharge the capacitor. A value of 1V was chosen for convenience. The time axis is given in RC units, where RC is called the time constant.

In approximately 5 time constants, the capacitor will be fully charged for the SW1 case or fully discharged for the SW2 case. In approximately 0.7 time constants, the 50 percent point occurs for both charging and discharging.

The general formula for the charging of a capacitor is

$$V_C = V_{DC}\left(1 - e^{-\frac{\tau}{RC}}\right) \tag{16-13}$$

For the discharging of a capacitor, it would be

$$V_C = V_{DC}e^{-\frac{\tau}{RC}} \tag{16-14}$$

where V_{DC} is the DC voltage applied to the RC circuit and τ is the time. The time constant is RC.

Let us assume we want to delay a digital signal by two seconds. The height of the signal is +5V. If we set a V_r of +2.5V, it would take approximately 0.7 time-constant units to get to the 50 percent voltage point. So

$$2 \text{ seconds} = 0.7\, RC$$

$$RC = \frac{2}{0.7} = 2.86 \tag{16-15}$$

So if we used a 10-uF capacitor, the resistor value would be 287 kΩ (closest standard value to 286 kΩ). The circuit is shown in Fig. 16-7b. The delay to each edge is shown in Fig. 16-7c. For this application, we could use either an amplifier or comparator.

The capacitor can be nonpolarized or a polarized tantalum. If a polarized capacitor is used, observe polarity. In the case shown in Fig. 16-7b, the upper capacitor terminal would be positive.

Longer time delays are possible, but large capacitor values, which are typically tantalum, do have leakage in the form of an equivalent parallel resistor, which can form a voltage divider with the external charging resistor and provide errors in the results. Avoid electrolytics due to much higher leakage than tantalum. The input V_{in} should come from a source impedance that is less than one-tenth the value of the charging resistor.

The circuit of Fig. 16-7 has symmetrical delay. Both the positive- and negative-going edges are delayed equally. It is possible to have unequal delays to both edges. Figure 16-8 demonstrates how to accomplish this.

In the circuit of Fig. 16-8a, the charge time would be 0.7 RC, where R is the parallel combination of R_1 and R_2. The discharge time would also be 0.7 RC, where R is just R_2. So the charge time would be faster than the discharge time.

For the circuit of Fig. 16-8b, the opposite would be true. The charge time would be determined only by R_2 and the discharge time by the parallel combination of R_1 and R_2, so the discharge time would be faster than the charge time.

For the cases shown in Fig. 16-8, it is assumed that the input voltage swings from ground to +5V, and also the diode drop is ignored. If the input were, let's say, +10V and the reference were +5V, the diode drop would be even less significant.

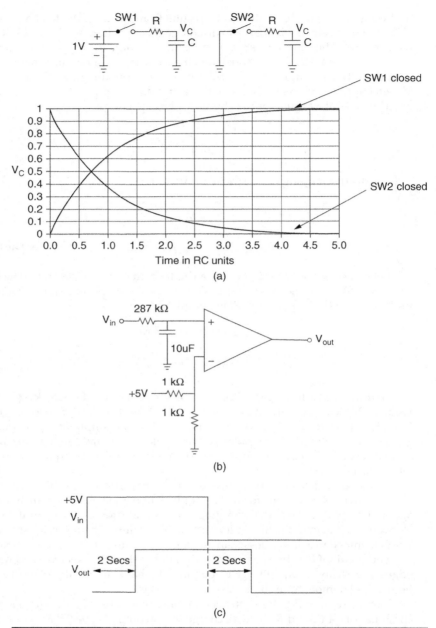

FIGURE 16-7   Time-delay circuit: (a) charging and discharging of a capacitor; (b) two-second delay circuit; and (c) waveforms.

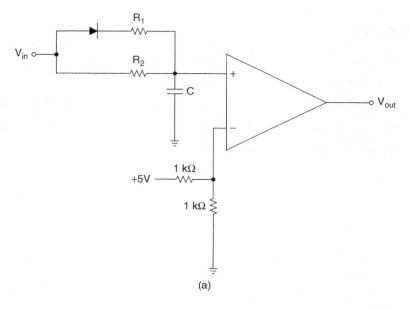

(a)

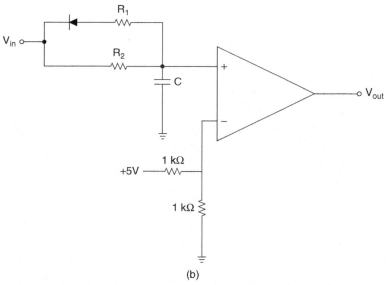

(b)

FIGURE 16-8   Unsymmetrical delay: (a) faster charge time and (b) faster discharge time.

References

Jung, W., ed. (2006). *Op Amp Applications Handbook*, Analog Devices, Norwood, Ma.

Graeme, J. G. (1973). *Applications of Operational Amplifiers: Third-Generation Techniques*, McGraw-Hill, New York.

Texas Instruments. Application Report SBOA092A, "Handbook of Operational Amplifier Applications," October 2001, Dallas, Texas.

Franco, S. (1988). *Design with Operational Amplifiers and Analog Integrated Circuits*, McGraw-Hill, New York.

Texas Instruments. Advanced Analog products SLOD006B, "Op Amps for Everyone," August 2002, Dallas, Texas.

Williams, A. B. (1984). *Designer's Handbook of Integrated Circuits*, McGraw-Hill, New York.

CHAPTER 17

Waveform Generation

17.1 Sine Wave Generators

In previous chapters the condition of oscillation in op amps was treated as a mode to avoid, and great measures were taken in this regard, such as adequate decoupling, control of circuit Qs, gains to prevent poles on the $j\omega$ axis, etc. However, let us take the case where we want to produce a sine wave where the frequency and amplitude are controlled. Let us consider the basic circuit of Fig. 17-1 showing the canonical form of a basic feedback circuit. In Chap. 13 we developed the fundamental equation for the closed-loop gain of a noninverting amplifier, which was shown in Eq. (13-15) to be

$$A_c = \frac{A_0}{1 + A_0\beta}$$

So applying this formula to Fig. 17-1 we obtain

$$\frac{V_{\text{out}}}{V_{\text{in}}} = \frac{A}{1 + A\beta} \tag{17-1}$$

The condition for *instability* occurs when the denominator is zero. This will happen when $A\beta = -1$. By precisely controlling gain A and feedback factor β so $A\beta$ is -1 at a single frequency, the circuit will oscillate at that frequency, producing a sine wave output *with no input*. This effect is known as the *Barkhausen criterion* for oscillation.

This would be a linear oscillator, whereas oscillators that produce signals by the op amp saturating alternately at the supply rails are nonlinear, and are called relaxation oscillators. Even when the gain is too high, as long as the phase shift is $180°$ the nonlinearities of the op amps will seek out the condition for oscillation.

17.1.1 Phase Shift Oscillators

In order for $A\beta$ to become -1, a $180°$ phase shift must occur. If β consists of a phase shift network that produces $180°$ at a specific frequency, and the product of the absolute magnitude of $A\beta$ is 1 (at that frequency), the circuit will oscillate. The main objective in designing an oscillator is to force $A\beta$ to equal -1 at the frequency of interest. This is the Barkhausen criterion, where the denominator of Eq. (17-1) is equal to zero.

The steeper the change in phase around the $180°$ point ($\Delta\Phi/\Delta\omega$), the more stable the frequency will be as long as the components that produce the phase shift remain stable. A large $\Delta\Phi/\Delta\omega$ occurs with high Q circuits, which is not the case when the phase shift is produced by RC networks.

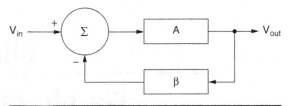

FIGURE 17-1 Canonic feedback diagram.

A phase shift oscillator is based on the principle that a cascade of RC networks can produce a phase shift of 180°at a particular frequency. A single high-pass RC network has a phase shift of +45° at $\omega = 1/RC$ and asymptotically approaches +90° at DC. A cascade of three RC networks will asymptotically approach the +270° point, but in doing so it must pass through a +180° point. Controlling where this 180° occurs will determine the frequency of a sine wave oscillator in the circuits of Fig. 17-2. In Fig. 17-2a, all resistors (except for the feedback resistor) are equal and all capacitors are equal. This greatly simplifies the mathematics of calculating values.

The frequency of the oscillation is then given by

$$f_{osc} = \frac{1}{2\pi RC\sqrt{6}} \tag{17-2}$$

and
$$A = 29 \tag{17-3}$$

In Fig. 17-2a, the resistor connected to the inverting input of the amplifier is part of the RC phase shift circuit. That is not the case in the alternative configuration shown in Fig. 17-2b. R_1 must be at least 10 times R so as not to significantly affect the calculated oscillator frequency.

It must be pointed out that the circuits of Figs. 17-2a and 17-2b are not just a simple cascade of three RC networks since each one is loaded down by the next one.

A phase shift of 180° can also be obtained by a cascade of low-pass networks rather than high-pass networks. This is shown in Fig. 17-2c. The design equations would then be

$$f_{osc} = \frac{1.732}{2\pi RC} \tag{17-4}$$

and
$$A = 8 \tag{17-5}$$

where $R_1 \gg R$.

However, because of the interaction between the three RC networks, each loading down the previous one, the required gain can be somewhat more than 8 and as high as 30 to start oscillation. To prevent interaction between RC networks, individual RC networks can be buffered from each other with voltage followers so the results will be closer to the calculated frequency and gain. Also, they can be tapered in impedance so each RC circuit would be loaded by a much higher impedance, thus minimizing interaction.

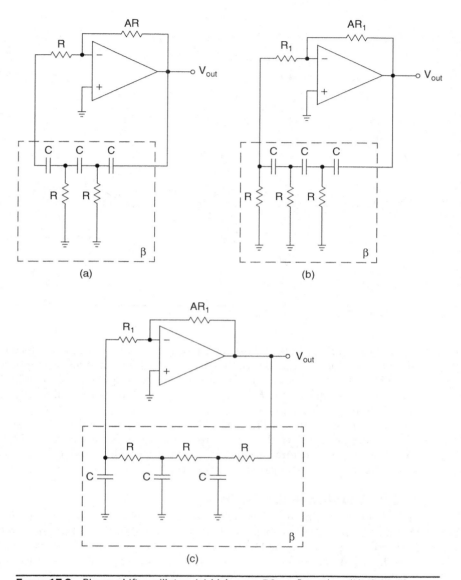

FIGURE 17-2 Phase shift oscillator: (a) high-pass *RC* configuration; (b) alternative high-pass *RC* configuration; and (c) low-pass version.

The circuit of Fig. 17-3 is called a Bubba oscillator. It consists of four buffered *RC* networks, each one contributing $-45°$ of phase shift at $f = \dfrac{1}{2\pi RC}$. Since the attenuation is 3 dB, or 0.707, at the $-45°$ point, the total loss in going through four stages is 0.707^4, or 0.250, so the required minimum gain A is 4 with $R_1 \gg R$. This circuit can produce two outputs in quadrature (90°) with each other. Since op amps are available in quads, this circuit would be easy to implement.

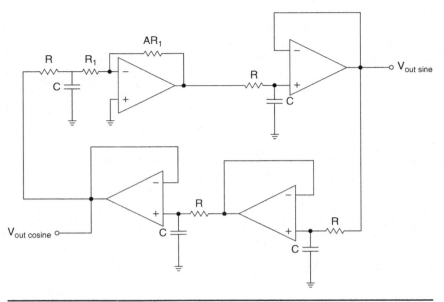

FIGURE 17-3 Bubba oscillator.

A phase shift oscillator is best suited for low-frequency applications (below 1 MHz) where large values of R and C can be used. At higher frequencies, the phase shift of the amplifier itself will have an effect on the oscillation frequency and stability. Also, we are not restricted to the simple RC networks shown. Phase shift oscillators in general produce a relatively clean sine wave.

The all-pass circuits of Fig. 7-11 can produce a phase shift of $-90°$ at $\omega = 1/RC$, so two cascaded circuits would be required. The circuit is shown in Fig. 17-4. By making one or two of the resistors variable, the frequency at which $180°$ occurs can be changed, so an adjustable frequency oscillator can result.

17.1.2 The Wien Bridge Oscillator

The circuit shown in Fig. 17-5a is referred to as a Wien bridge and was first developed by Max Wien in 1891. It has survived since then and was the one used in Hewlett Packard's first product, the HP200A Precision Wein Bridge Sine Wave Oscillator. Let us consider what happens at the frequency where the capacitive reactance is equal to the resistance, or

$$R = \frac{1}{2\pi fC} \tag{17-6}$$

The impedance of the series RC is then equal to

$$Z_{series} = 1.414R\angle - 45° \tag{17-7}$$

The impedance of the parallel RC is

$$Z_{parallel} = 0.707R\angle - 45° \tag{17-8}$$

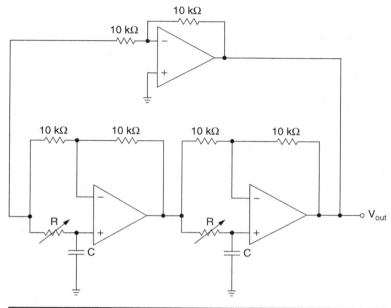

FIGURE 17-4 All-pass oscillator circuit.

The voltage divider formed by Z_{series} and $Z_{parallel}$ is

$$divider = \frac{Z_{parallel}}{Z_{series} + Z_{parallel}} = \frac{0.707R\angle - 45°}{1.414R\angle - 45° + 0.707R\angle - 45°} = \frac{1}{3} \qquad (17\text{-}9)$$

With an amplifier gain of 3, this voltage loss due to the divider is compensated for, so the feedback to the positive terminal is 1. To obtain a noninverting gain of 3, the ratio of R_2/R_1 should be 2.

Although the Wein bridge maintains a stable frequency, it is very sensitive to the gain resulting from the values of R_1 and R_2. If the gain is low, the circuit will not oscillate. Too high a gain results in a distorted waveform. A type of automatic gain control (AGC) has been used, which although it appears primitive, works well. If we replace R_1 with a standard small incandescent bulb we can better control the gain. The bulb's tungsten filament has a very high positive temperature coefficient of resistance versus temperature, so as the signal tends to increase, the current through the bulb will increase, thus increasing its resistance and lowering the gain. This results in a form of AGC regulation and a low distortion sine wave. R_2 must be chosen as to provide sufficient current for the bulb. The circuit is shown in Fig. 17-5b.

If a real electronic AGC is desired, the circuit of Fig. 17-5c can be used. The diode samples the negative side of the output and produces a DC voltage, which is filtered by R_A and C_A. This negative DC voltage controls the resistance of the field effect transistor (FET) where a stronger oscillation results in a more negative DC voltage, which increases the resistance of the FET and lowers the gain.

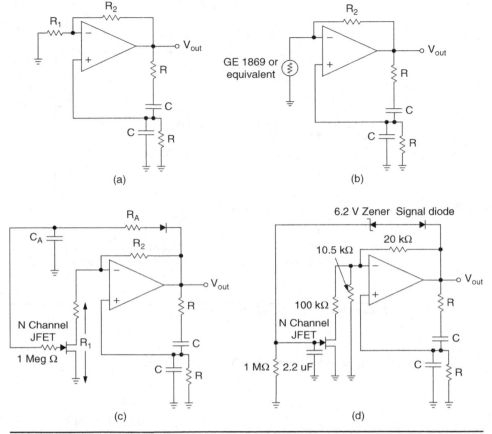

Figure 17-5 Wien bridge oscillator: (a) basic circuit; (b) addition of incandescent bulb for gain control; (c) AGC; and (d) alternative AGC.

There are multiple permutations of this approach. Another example is the circuit shown in Fig. 17-5d. As the oscillation builds up and exceeds the barrier created by the diodes, a negative voltage to the gate of the junction field effect transistor (JFET) will increase the resistance of the metal oxide semiconductor field effect transistor (MOSFET), thus reducing the gain.

17.1.3 Multiple-Feedback Band-Pass Oscillator

To implement β in the canonic feedback network of Fig. 17-1, band-pass filters can be used that will provide 180° of phase shift at their center frequency. This approach will also reduce harmonics of the sine wave by taking the output from the band-pass filter.

Section 5.2 discusses the multiple-feedback band-pass (MFBP) band-pass filter section. Figure 17-6 combines this circuit with the soft diode limiter previously discussed in Sec. 16.2. The soft limiter will adjust the circuit gain to ensure an oscillatory condition. The MFBP band-pass filter should have a Q of 10 to ensure oscillation. Also, the 10 kΩ resistor may have to be lowered to 1 kΩ at higher frequencies.

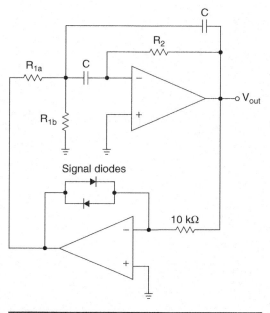

FIGURE 17-6   MFBP oscillator.

The values for the $Q = 10$ case are calculated as follows:

$$R_2 = \frac{10}{\pi f_{osc} C}$$ (17-10)

$$R_{1a} = R_2/2$$ (17-11)

$$R_{1b} = R_2/398$$ (17-12)

where f_{osc} is the frequency of oscillation. R_{1b} can be made variable over a small range to allow precise frequency adjustment.

17.2   Generating Nonsinusoidal Waveforms

17.2.1   Square Wave Relaxation Oscillator

Figure 17-7a illustrates a relaxation oscillator that produces a square wave. Let us first assume the capacitor is discharged and V_{out} is at $+V$. The capacitor will start to charge positive until it reaches $V_{feedback}$, which is positive. At that point the output will go negative and the capacitor will start to charge in the negative direction. When it reaches $V_{feedback}$, which is now negative, the output will go positive and the process will start all over again. This is shown in Fig. 17-7b.

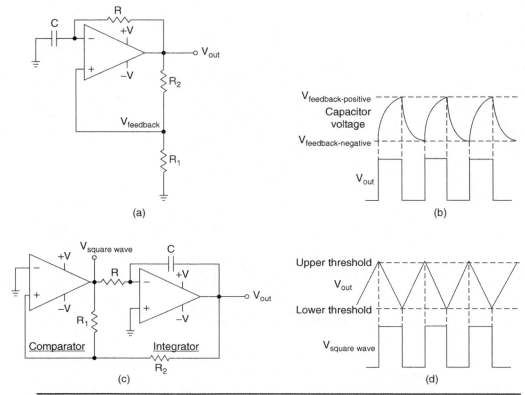

Figure 17-7 Relaxation oscillators: (*a*) square wave generator; (*b*) waveforms of square wave generator; (*c*) triangular waveform relaxation oscillator; and (*d*) waveforms of triangular waveform relaxation oscillator.

The frequency of oscillation is given by

$$f_{osc} = \frac{1}{2RC\ln\left(\dfrac{2R_1}{R_2}+1\right)} \tag{17-13}$$

If $R_1 = R_2$, then the formula simplifies to

$$f_{osc} = \frac{1}{2.2RC} \tag{17-14}$$

17.2.2 Triangular Wave Relaxation Oscillator

The comparator-integrator relaxation oscillator of Fig. 17-7*c* can generate triangular waves as well as square waves. The corresponding waveforms are in Fig. 17-7*d*. The circuit operates as follows: Let us assume that the output of the comparator is at its maximum negative level. This voltage will establish a negative bias voltage at the non-inverting input of the comparator through R_1 and will simultaneously start to charge the integrator capacitor C through resistor R in a linear manner. The output of the integrator will ramp positive through R_2 until it overcomes the negative voltage produced by R_1.

This will reverse the output of the comparator to its maximum positive level. The integrator will then start to charge in the negative direction. Once the ramp reaches the point where it overcomes the positive bias from R_1 at the noninverting input of the comparator, the comparator output will switch negative, and this cycle will continue.

It is important to recognize that although both circuits produce square waves, only the circuit of Fig. 17-7c produces a true triangular wave due to the integration of the square wave.

Since the comparator changes state once the triangular waveform reaches the upper and lower threshold, it is these voltages that determine the peak-to-peak amplitude of the triangular wave. The upper threshold is computed by

$$\text{upper threshold} = +V(R_2/R_1) \tag{17-15}$$

$$\text{lower threshold} = -V(R_2/R_1) \tag{17-16}$$

The voltages $+V$ and $-V$ are the supply voltages, so Eqs. (17-15) and (17-16) assume the comparator has a rail-to-rail output. If not, these voltages have to reflect the actual positive and negative comparator output.

The frequency of oscillation is given by

$$f_{osc} = \frac{1}{4RC}(R_1/R_2) \tag{17-17}$$

Resistor R can be made adjustable to vary the frequency.

17.2.3 The 555 Timer

The 555 timer is an amazingly versatile integrated circuit (IC) invented in 1971 by Hans Camenzind under contract to Signetics. It can operate in a variety of modes, including monostable and astable. It can perform functions such as accurate and very stable time delays ranging from microseconds to hours and produce precise square waves, pulses, and other waveforms.

A block diagram is shown in Fig. 17-8. The pin numbers indicated correspond to the eight-pin package, although this device is available in a number of packages and configurations, including dual and quad, and in multiple technologies.

A voltage divider is present consisting of three 5 kΩ resistors. (It is possible this timer got its name from the three 5 kΩ resistors, hence 555). The three resistors develop two threshold voltages at one-third and two-thirds of V_{CC}, the positive supply voltage.

The lower (trigger) comparator goes high when the trigger input goes below one-third of V_{cc}, setting the flip-flop, so Q goes high. Since $\bar{Q}$ is now low, the transistor is off (open). The upper (threshold) comparator goes high when the threshold input goes above two-thirds of V_{cc}, resetting the flip-flop so Q goes low and $\bar{Q}$ goes high, turning on the transistor and pulling pin 7 low.

To summarize, the flip-flop is set when the trigger input (pin 2) is less than one-third V_{cc} and the flip-flop is reset when the threshold voltage (pin 6) exceeds two-thirds V_{cc}. A low on pin 4 will immediately reset the circuit, overriding all other signals.

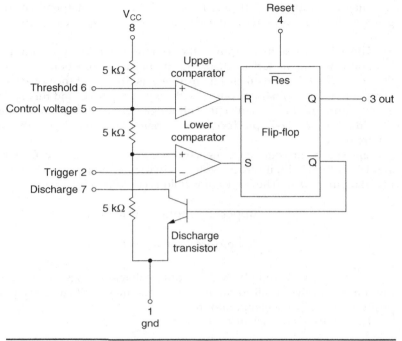

FIGURE 17-8 555 timer block diagram.

Sometimes the flip-flop is referred to as a latch, but the terms are the same in this application.

Monostable

Let us now look at the monostable configuration (sometimes called a one-shot). This circuit is shown in Fig. 17-9a and the waveforms in Fig. 17-9b. With no signal applied, the output is low. A negative-going pulse that goes lower than one-third V_{cc} allows the lower comparator to set the flip-flop, so Q goes high and $\overline{Q}$ goes low. The transistor is turned off.

The capacitor C will now charge through R_A until the voltage is high enough to exceed two-thirds V_{cc} and trip the upper comparator, which resets the flip-flop, so the Q output goes low. The $\overline{Q}$ output goes high, turning on the transistor, which discharges the capacitor. Note that the input must be restored to a high before the capacitor charge reaches two-thirds R. Typically, an RC differentiator is used to convert an edge transition to a pulse, as shown in the figure to ensure triggering.

The pulse duration is given by

$$T_{PW} = 1.1 R_A C \tag{17-18}$$

A capacitor is normally located between the control input and ground to protect against false triggering from transients. Also, reset should be held positive if not used.

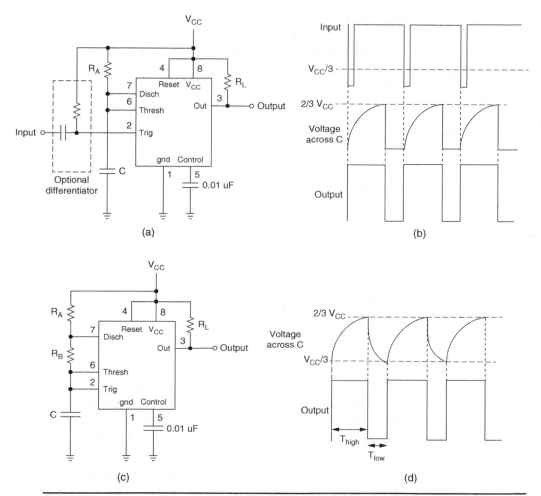

Figure 17-9 Monostable and astable configurations of the 555 timer: (a) monostable configuration; (b) waveforms of monostable; (c) astable configuration; and (d) waveforms of astable.

Since the charging rate and threshold levels are proportionally determined by V_{cc}, the timing interval is immune to supply voltage changes over the specified operating voltage range.

If, during the charging cycle, a negative-going pulse is simultaneously applied to both the reset terminal and trigger terminal, the capacitor will discharge and the charging cycle will start over again on the positive edge of the reset pulse. For the duration of the reset pulse, the output will be forced low.

Astable

Figure 17-9c shows an astable configuration of the 555 timer. By adding a second resistor and joining the trigger and threshold inputs, the circuit will free-run as a multivibrator (oscillator). The capacitor will charge through the series combination of R_A and R_B

and discharge through R_B alone. So through choosing the appropriate values for R_A and R_B, the frequency and duty cycle can be controlled. The capacitor voltage will swing between the threshold voltage and the trigger voltage levels (between two-thirds V_{cc} and one-third V_{cc}).

The charge time of C where the output is high is given by

$$T_{high} = 0.693(R_A + R_B)C \tag{17-19}$$

The discharge time of C where the output is low is defined by

$$T_{low} = 0.693R_BC \tag{17-20}$$

The period is given by

$$Period = T_{high} + T_{low} = 0.693(R_A + 2R_B)C$$

The frequency of oscillation is the reciprocal of period

$$f_{osc} = \frac{1.44}{(R_A + 2R_B)C} \tag{17-21}$$

Finally, the duty cycle is given by

$$\% \ Duty \ Cycle = \frac{R_B}{R_A + 2R_B} \times 100 \tag{17-22}$$

The waveforms are shown in Fig. 17-9d.

The value of R_A cannot be zero, as excessive current will flow through the transistor during discharge of the capacitor. To limit the discharge current to 200 mA, the minimum value for R_A should be $V_{cc}/0.2$. So a precise 50 percent duty cycle cannot be obtained, but one can come close with the proper selection of values.

Sequential Timer

The 555 timer has a multitude of applications, but only a few will be covered here to demonstrate its versatility. Data sheets and many application notes can be downloaded to illustrate other configurations.

The circuit of Fig. 17-10 shows a sequential timer that produces a sequence of pulses. The negative edge of the input triggers timer 1 for a duration of T1 seconds. The negative edge of the output from T1 triggers timer 2, which has an output of T2. Finally, the negative edge of T2 triggers timer 3, which has an output of duration T3.

Pulse-Width Modulator

Figure 17-11a shows a pulse-width modulator. A string of pulses is applied as the input. By applying a voltage to the control terminal, the charging time of capacitor C can be changed by varying the control voltage at pin 5, thus overriding the upper threshold. The result is that the width of the pulses in the pulse train can be modulated.

FSK Modulator

Figure 17-11b illustrates a frequency shift keyed (FSK) modulator that can switch from one frequency to another in a binary fashion. The modulating data is applied to a transistor, which switches a second resistor in parallel with R_{A1} to change the frequency.

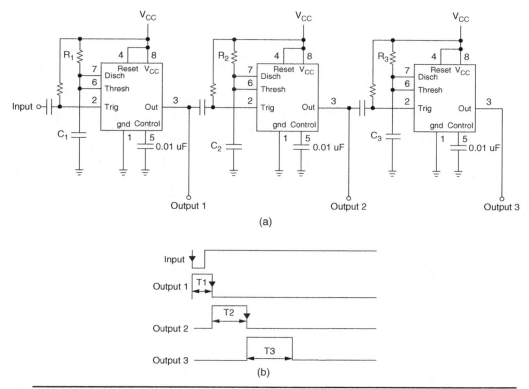

Figure 17-10 Sequential timer: (*a*) circuit and (*b*) waveforms.

Linear Ramp Waveform Generator

The monostable circuit of Fig. 17-11*c* replaces the charging resistor *R* with a transistor current source. As a result, the capacitor will charge in a linear fashion rather than exponential. Care should be taken that no capacitive load is in parallel with the capacitor *C* to avoid affecting the frequency. Buffering with a voltage follower may be a solution. The input and output waveforms are shown in Fig. 17-11*d*.

17.2.4 Hex Inverter *RC* Oscillators

A versatile building block in the digital world is the hex inverter. It consists of six inverters in a single package and is available in virtually all digital technologies, such as transistor-transistor logic (TTL), low-power Schottky, CMOS, etc. Versions are available with a Schmitt trigger interface, open collector, tri-state output, and so forth.

Although intended for inverting digital signals, they can be used for analog functions where the signals operate in the linear region. This is the area where the output transitions between 0 and 1. In effect, they can function as a high-gain amplifier where the output can saturate since there is no feedback. If we look at the circuit of Fig. 17-12*a*, a string of inverters are connected in a closed loop. As long as the number of inverters is odd, the circuit will oscillate. An intuitive way of analyzing this is by considering a

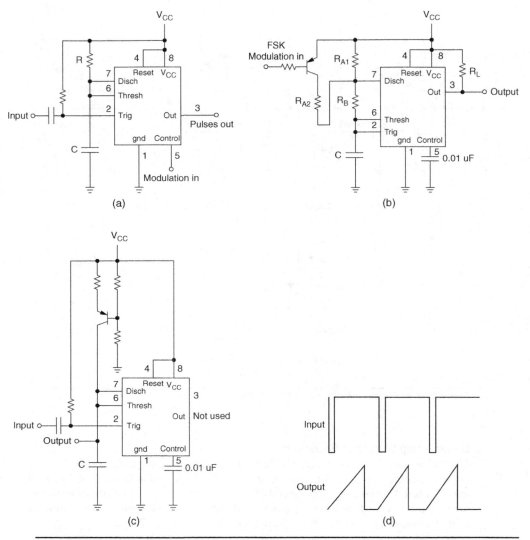

FIGURE 17-11 Some 555 timer applications: (*a*) pulse-width modulator; (*b*) FSK modulator; (*c*) linear ramp waveform generator; and (*d*) waveforms of linear ramp waveform generator.

logic "1" at some point in the ring chasing itself around the ring as a 1 input causes a 0 output, which causes a 1 output, and so forth. The frequency of oscillation can be computed as follows:

$$f_{osc} \approx \frac{1}{2NT_{PD}} \tag{17-23}$$

where N is the number of inverters and T_{PD} is the propagation delay of each inverter.

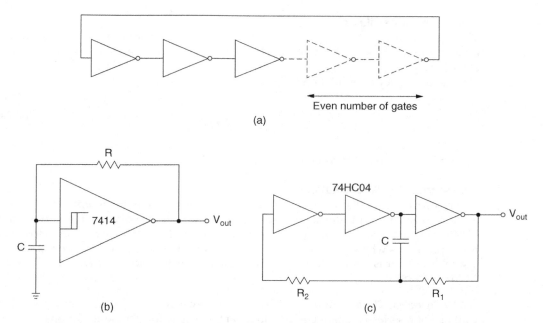

FIGURE **17-12**   Hex inverter oscillators: (*a*) ring oscillator; (*b*) relaxation oscillator; and (*c*) three-inverter oscillator.

This circuit is not a very precise or stable oscillator, as it depends on the propagation delay of the inverter as well as IC switching thresholds, but it certainly is usable. Typically, the frequency of oscillation is in the 10-MHz region, so it is not suitable for low-frequency oscillators.

Schmitt Trigger Oscillator

Square wave generators can be built by simply using hex inverter ICs, which are easily available in multiple technologies and low in cost. Figure 17-12*b* shows an extremely simple square wave oscillator using a single hex inverter 7414 of TTL technology with a Schmitt trigger input. The device has a lower threshold of typically +0.9V and an upper threshold of typically +1.7V. An initial positive output will cause the capacitor to charge until it reaches the upper threshold, at which point the output will go low. The capacitor will start to discharge until the input is at the lower threshold, at which point the output will go high and the process will start all over again. The circuit is a classic relaxation oscillator.

For values of $R = 10$ kΩ and $C = 2,700$ pF, the circuit will oscillate at nominally 35 KHz. For other technologies such as CMOS, the frequencies will be different. They can be scaled by varying R or C or both. One must keep in mind that the frequency stability and accuracy are not very precise and are sensitive to supply voltage, but the circuit's simplicity is a major feature.

As with all hex inverter oscillators, the output should be buffered from the load by one or two stages of hex inverters or other gates to ensure a waveform with rectangular edges.

Three-Inverter Oscillator

The circuit of Fig. 17-12c uses three hex inverter stages. It is not restricted to hex inverters, but will work with other CMOS gates that can produce an inversion such as NAND and NOR gates.

The frequency of oscillation is approximately given by

$$f_{osc} \approx \frac{1}{2C(0.405R_{EQ} + 0.693R_1)} \tag{17-24}$$

where R_{EQ} is the parallel combination of R_1 and R_2.

Hex Inverter Crystal Oscillators

The hex inverter oscillators described in the previous section are acceptable for many applications. They typically have guaranteed start-up, a wide voltage range of operation (CMOS versions), and the ability to easily interface with other logic families. However, for applications where stability and accuracy better than a few percent are required, a crystal oscillator is a must. It represents a quantum leap in accuracy and stability over *RC* oscillators.

Crystal Equivalent Circuit Figure 17-13a is the equivalent circuit of a quartz crystal. It contains of a series-resonant circuit and a parallel capacitor. The series-resonant circuit consisting of L_1, C_1, and R_1 represents the mechanical characteristics of the crystal. A parallel capacitor C_O is formed by the two electrodes, which effectively act like the plates of a capacitor, and the crystal as the dielectric. A parallel-resonant circuit is formed by the series capacitance of C_1 and C_0 in conjunction with R_1 and L_1.

Analytically, the resonant frequencies are

$$f_{series} = \frac{1}{2\pi\sqrt{L_1 C_1}} \tag{17-25}$$

$$f_{parallel} = \frac{1}{2\pi\sqrt{L_1 \dfrac{C_1 C_0}{C_1 + C_0}}} \tag{17-26}$$

Typically, the parallel-resonant frequency falls about 0.1 percent above the series-resonant frequency.

Figure 17-13b illustrates the impedance of the equivalent circuit. At series resonance, the impedance is resistive and approximately equal to R_1. Slightly above the series resonance frequency, parallel resonance occurs where the impedance changes from highly inductive to highly capacitive as you transition through resonance.

A crystal exhibits extremely high Qs, which are impossible to realize with a conventional inductor-capacitor circuit. Inductances are extremely high and series capacitances are extremely low, resulting in the enormous Qs. Table 17-1 has some typical values.

It is this extremely high Q and stability with time and temperature that make a crystal oscillator mandatory when high stability and precision are required. Crystals are cut from raw quartz at specific angles to the axis of the quartz. This determines not only the frequency, but also the behavior of the crystal, as some cuts result in resonances at frequencies other than the fundamental, such as harmonics. Sometimes this is

Parameter	200-KHz Fundamental Mode	2-MHz Fundamental Mode	30-MHz Fundamental Mode
R_1	2 kΩ	100 Ω	20 Ω
L_1	27 H	529 mH	11 mH
C_1	0.024 pF	0.012 pF	0.0026 pF
C_0	9 pF	4 pF	6 pF
Q	18,000	54,000	100,000

TABLE 17-1 Typical Crystal Parameters

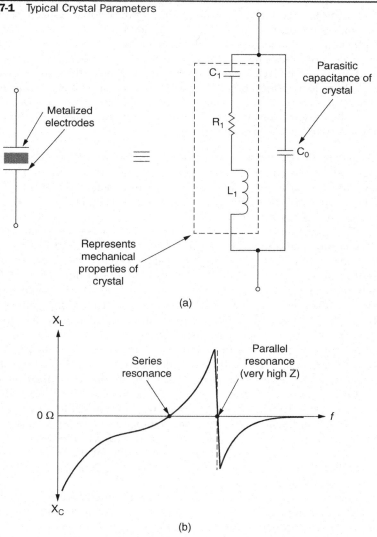

FIGURE 17-13 Quartz crystal: (*a*) equivalent circuit and (*b*) impedance of equivalent circuit.

desirable, so oscillator circuits can be optimized to operate at overtones (multiples) of the fundamental frequency. Crystals above 20 MHz are somewhat difficult to manufacture, so they are used in the overtone mode. To ensure oscillation at the higher overtone frequency, the feedback network should have more gain at the overtone frequency than the fundamental. Some crystals are optimized for the parallel-resonant mode of operation. The manufacturer of the crystal will specify a capacitive load for the crystal to obtain the exact parallel-resonant frequency.

Depending on the frequency, different axis cuts will apply. The most popular is the AT cut, but other cuts with special features are used as well. Typical accuracy and stability can be as tight as ±5 ppm (±0.0005 percent) over a temperature range of −10°C to +60°C, but for most applications ±50 ppm or ±100 ppm is acceptable. Techniques such as placing the crystal in what is called a "proportional controlled oven" to maintain a constant temperature somewhat independent of the environment are used for very high stability. The temperature of the oven is set for the point at which the frequency versus temperature curve is the flattest for least sensitivity to temperature.

Sometimes it is desirable to electronically change the frequency of a crystal, such as in a phase-locked-loop (PLL) circuit. This is accomplished by placing a device called a *varycap* in series with a crystal to "pull" its frequency so you can have a voltage-controlled oscillator. A varycap is basically a reverse-biased diode whose capacitance is a function of the DC bias on the diode.

Pierce-Gate Crystal Oscillator   The Pierce-Gate crystal oscillator circuit is shown in Fig. 17-14*a*. This circuit is the most common configuration for a crystal oscillator since the CMOS inverter itself is quite often embedded as part of a complex device, such as a microprocessor, and its pins are accessible, as shown in Fig. 17-14*b*. The crystal operates in the parallel-resonant mode.

To analyze how the circuit of Fig. 17-14*a* works, we need to again consider the Barkhausen requirement for an oscillatory condition. Let us consider the gain of the loop formed by the crystal and the inverter. The inverter provides a phase shift of 180° by the nature of its inversion. The feedback path through the crystal forms a PI-shaped network if we include C_A and C_B, which provides an additional 180° of phase shift, so the end result is 360°. The inverter provides sufficient loop gain to sustain oscillation. The phase shift will automatically be maintained at exactly 360° to sustain oscillation, so the oscillation frequency may shift slightly.

Resistor R_F provides some negative feedback and biases the inverter in the high-gain linear region. Typically, the value may be in the vicinity of 500 kΩ to 5 MΩ. Sometimes, R_F is internal to the microprocessor or application specific integrated circuit (ASIC). Capacitors C_A and C_B in series effectively form the load capacitance for the crystal. Usually, the capacitors are equal in value. The crystal manufacturer's data sheet will specify the load capacitance the crystal is designed for. Resistor R_S limits the current from the inverter so as not to overdrive the crystal. A typical resistor value would be equal to the reactance of C_B at the frequency of oscillation. For a 20-MHz AT cut crystal, some typical values would be an R_F of 1 MΩ , R_S of 390 ohms, and $C_A = C_B = 27$ pF.

A crystal oscillator design must first be qualified before going into production. The circuit must be tested for start-up to ensure it will always start in an oscillating mode, at the correct frequency, over the temperature range of operation, and with a selection of crystals, as properties may vary from crystal to crystal within a manufactured batch. Also, the circuit should be tested for operation with supply voltage variations.

It is good practice to buffer the output of the oscillator from its load with one or two stages of inverters to "square up" the waveforms if the inverter is not part of the microprocessor or ASIC. A Schmitt trigger device does this nicely.

There are two types of CMOS gates, which are either a single inverting stage or three inverting stages in cascade. The single inverting stage has been found to provide better stability. For the multiple inverter stages, the values of R_S or C_B can be increased by a factor of 10 or more since the loop gain is higher.

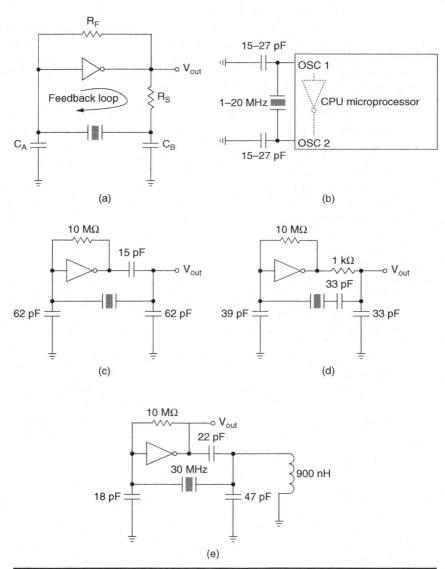

FIGURE 17-14 Hex inverter oscillators: (a) basic Pierce-Gate crystal oscillator; (b) use with a microprocessor; (c) configuration for >4 MHz; (d) circuit for improved stability; and (e) circuit for use with overtone oscillators.

Part of the phase shift that occurs in the feedback loop results from the gate propagation delay, which can be estimated by

$$\Phi(\text{degrees}) = \text{frequency (Hz)} \times \text{Propagation Delay} \times 360° \qquad (17\text{-}27)$$

At frequencies above 4 MHz, a propagation delay through the inverter of 10 nsec would represent 14.4° of phase shift. Resistor R_S should be replaced with a small capacitor to minimize additional phase shift. This is shown in Fig. 17-14c. For improved stability, the circuit of Fig. 17-14d should be used

When a crystal is used in an overtone mode, a properly located tuned circuit would ensure oscillation at the overtone frequency rather than a fundamental or other unwanted frequency. In Fig. 17-14e, an inductor forms a parallel-resonant circuit with the parallel capacitor resulting in a high impedance near the desired frequency and a low impedance above and below this frequency, thus preventing unwanted oscillation at unwanted frequencies.

It must be said that crystal oscillator design is somewhat of an inexact science that may require a bit of trial and error. A few tries may be required to optimize a circuit. The designer should work closely with the crystal manufacturer to determine a recommended circuit. If the device is to be used with a microprocessor, the device data sheet would call out the required crystal parameters.

All oscillators that have a nonsinusoidal output can be followed by a low-pass or band-pass filter to eliminate all harmonics and any other unwanted frequencies to provide a spectrally pure sine wave output.

References

Fairchild Semiconductor. "HCMOS Crystal Oscillators Application Note 340," May 1983, San Jose, CA.

Fairchild Semiconductor. "CMOS Oscillators Application Note 118," October 1974, San Jose, CA.

Raymond Cerda, Crystek Corp. "Pierce Gate Crystal Oscillator, An Introduction," *Microwave Product Digest,* March 2008, Ft Myers, Fl.

Mancini, R. "Design of Op Amp Sine Wave Oscillators," *Analog Applications Journal,* Texas Instruments, August 2000, Dallas, Texas.

Mancini, R. *Op Amps for Everyone,* Texas Instruments, August 2002, SL0D006B, Dallas, Texas.

Graeme, J. G. (1973). *Applications of Operational Amplifiers: Third-Generation Techniques,* McGraw-Hill, New York.

Jung, W., ed. (2006). *Op Amp Applications Handbook,* Analog Devices, Norwood, Ma.

Texas Instruments, "Handbook of Operational Amplifier Applications," Application Report SBOA092A, October 2001, Dallas, Texas.

Williams, A. B. (1984). *Designer's Handbook of Integrated Circuits,* McGraw-Hill, New York.

CHAPTER 18

Current Feedback Amplifiers

18.1 Introduction to Current Feedback Amplifiers

Many engineers, even those who specialize in analog design, are uncomfortable using current feedback amplifiers since they don't have the traditional structure of voltage feedback amplifiers. However, once the designer gains some familiarity with current feedback amplifiers (CFB op amps), they can become a powerful tool for many applications where voltage feedback amplifiers (VFB op amps) fall short.

Let us start with the significant differences between the VFB and CFB families of devices. CFB op amps do not have the precision of VFB op amps. However, they offer much wider bandwidths, larger output slewing, and higher output current capability than their VFB op amp counterparts by orders of magnitude.

The CFB op amps have a low inverting input impedance, but the noninverting input impedance is comparatively high, so they can't be used in a balanced differential input type application. They are vulnerable to a peaking phenomenon when stray capacitance at the inverting input or across the feedback resistor is present, and can burst into oscillation when these parasitics exist. As a result, printed circuit board (PCB) layout is more critical.

VFB amplifiers have a fixed gain-bandwidth product, so the higher the gain, the poorer the frequency response will be. CFB op amps, on the other hand, have a bandwidth that is mainly controlled by the feedback resistor *independent of gain*. So the closed-loop gain can be changed by varying the input resistor, and the closed-loop bandwidth will stay relatively constant as long as the feedback resistor remains fixed in value. The manufacturers' data sheets will provide a list of recommended feedback resistor values for various gain ranges of operation. If the feedback resistor value is increased, the bandwidth will be reduced. For lower resistor values, the bandwidth will increase but the phase margin will be lowered and instability can occur. This would rule out a voltage follower with zero feedback resistance. The data sheet will provide a recommended value for this case.

To summarize, CFB amplifiers provide much faster slewing rate, lower distortion, and much wider bandwidths than the VFB class of amplifiers, but at the expense of restriction of the feedback resistor values. Also CFB amplifiers will have higher noise and poorer DC performance, such as DC offset.

18.2   Analysis and Applications of Current Feedback Amplifiers

18.2.1   Models of Current Feedback Amplifier

Noninverting CFB Op Amp

Figure 18-1 illustrates the topology of a current feedback amplifier configured in the noninverting mode. The first observation is that the noninverting input is separated from the inverting input by a unity gain buffer. So the positive input impedance is high (that of the buffer) and the inverting input impedance is low (R_O) and near zero. The input unity gain buffer forces the negative input to follow the positive input to the amplifier. The error signal is a small current that flows from the inverting input. This error current "i" is mirrored into a high-impedance Z, which is a function of frequency. Z is called a *transimpedance*. A voltage is developed across this transimpedance of iZ.

The transimpedance in a CFB amplifier is analogous to the gain in a VFB amplifier. The higher the transimpedance, the more the amplifier's performance is dependent only on its surrounding passive components. Like the gain in a VFB op amp, the transimpedance is also a function of frequency.

Let us now analyze the circuit of Fig. 18-1.

Because of the unity gain output buffer, the output voltage is clearly

$$V_{OUT} = iZ \tag{18-1}$$

$$i = \frac{V_G}{R_G} - \frac{V_{OUT} - V_G}{R_F} \tag{18-2}$$

$$V_G = V_{IN} - iR_O \tag{18-3}$$

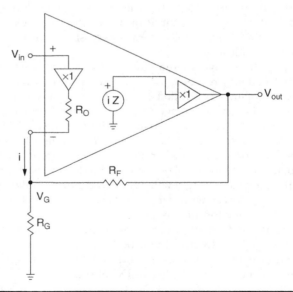

FIGURE 18-1   CFB op amp model for noninverting configuration.

If we manipulate these equations, we can obtain

$$\frac{V_{\text{OUT}}}{V_{\text{IN}}} = \frac{\dfrac{Z\left(1+\dfrac{R_F}{R_G}\right)}{R_F\left(1+\dfrac{R_O}{R_{eq}}\right)}}{1+\dfrac{Z}{R_F\left(1+\dfrac{R_O}{R_{EQ}}\right)}} \tag{18-4}$$

where R_{EQ} is the parallel combination of R_F and R_G.

With the input buffer's output impedance R_O at zero, Eq. (18-4) simplifies to

$$\frac{V_{\text{OUT}}}{V_{\text{IN}}} = \frac{Z\left(1+\dfrac{R_F}{R_G}\right)}{1+\dfrac{Z}{R_F}} \cdot \frac{1}{R_F} = \frac{1+R_F/R_G}{1+R_F/Z} \tag{18-5}$$

Note that with Z approaching ∞, Eq. (18-5) reduces to the form of the voltage feedback op amp equation for the gain of a noninverting amplifier, which would be

$$\frac{V_{\text{OUT}}}{V_{\text{IN}}} = 1+\frac{R_F}{R_G} \tag{18-6}$$

Inverting CFB Op Amp

Let us now consider the inverting topology, although the noninverting configuration is preferred because of the higher input impedance. The inverting version is shown in Fig. 18-2.

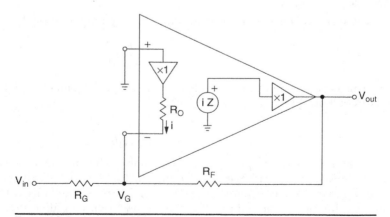

FIGURE 18-2 Inverting CFB op amp.

To derive the transfer function, let us start with

$$i + \frac{V_{IN} - V_G}{R_G} = \frac{V_G - V_{OUT}}{R_F} \tag{18-7}$$

$$iR_O = -V_G \tag{18-8}$$

and
$$iZ = V_{OUT} \tag{18-9}$$

By manipulating these equations, we can obtain

$$\frac{V_{OUT}}{V_{IN}} = -\frac{\dfrac{Z}{R_g\left(1 + \dfrac{R_O}{R_{EQ}}\right)}}{1 + \dfrac{Z}{R_F\left(1 + \dfrac{R_O}{R_{EQ}}\right)}} \tag{18-10}$$

where R_{EQ} is the parallel combination of R_F and R_G.

As we did for the noninverting configuration, let us assume R_O is very small. The resulting equation becomes

$$\frac{V_{OUT}}{V_{IN}} = -\frac{1/R_G}{\dfrac{1}{Z} + \dfrac{1}{R_F}} \tag{18-11}$$

With Z very large, Eq. (18-11) reduces to

$$\frac{V_{OUT}}{V_{IN}} = -\frac{R_F}{R_G} \tag{18-12}$$

Note that Eq. (18-12) is also the conventional expression for the gain of an inverting VFB op amp.

18.2.2   Stability

Equation (18-5), the gain of a noninverting CFB op amp, was established as

$$\frac{V_{OUT}}{V_{IN}} = \frac{1 + R_F/R_G}{1 + R_F/Z}$$

This is quite similar to the general equation for the gain of the classic noninverting feedback loop shown in Eq. (13-15), which was

$$A_c = \frac{A_0}{1 + A_0\beta}$$

In both cases, the amplifier becomes unstable when the denominator becomes zero, so either R_F/Z in Eq. (18-5) or $A_O\beta$ in Eq. (13-15) becomes -1. Since both R_F and β are typically fixed resistive components, it is Z, the transimpedance for the CFB op amp, and A_O, the open-loop gain in the voltage feedback amplifier, that are a function of frequency both in magnitude and phase, and can result in instability.

For Eq. (18-5), the loop gain R_F/Z determines circuit stability. In most cases, Z contains two or more poles, so we can conclude the CFB op amp stability is almost completely dependent on the feedback resistor R_F and the transimpedance Z, and not the closed-loop gain, as in the case of the VFB op amp.

The CFB op amp data sheet typically will provide the optimum value of R_F for different supply voltages and desired gains. These values are the result of extensive measurements by the manufacturers' engineers to determine the optimum values that will result in maximum bandwidth with minimal peaking of the response. A compromise value of R_F is suggested in the data sheets for stable operation and minimal peaking in the frequency response.

The designer has the freedom to choose R_F values other than the suggested values. However, larger values of R_F will result in a reduction in bandwidth, and lower values may result in unstable operation. Instability can also occur as a result of stray capacity around the inverting input node and across R_F.

Figure 18-3 illustrates the noninverting frequency response of a Texas Instruments THS3111 CFB op amp for a gain of 2. The recommended value of RFB is 1.15 kΩ. When decreased in value, the bandwidth increases but a peaking phenomenon occurs. This would imply approaching instability. Increasing the value reduces bandwidth. Clearly, 1.15 kΩ appears to be the optimum value.

A CFB op amp must not have a capacitor in the feedback loop, as this would reduce the feedback impedance due to decreasing capacitive reactance for higher frequencies. The end result would be oscillation.

Figure 18-4 shows the frequency response versus gain for both the noninverting and inverting case. For each gain, a specific value of R_F is recommended.

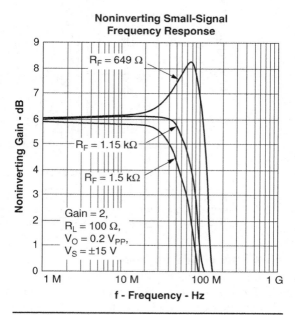

FIGURE 18-3 Noninverting small signal frequency response of Texas Instruments THS3111. Courtesy Texas Instruments.

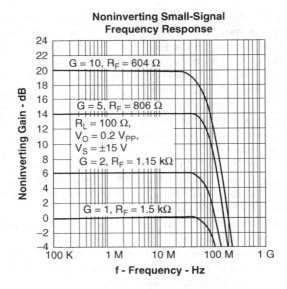

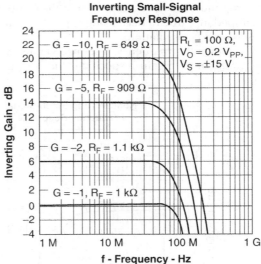

FIGURE 18-4 Noninverting and inverting small signal frequency response of Texas Instruments THS3111 versus gain. Courtesy Texas Instruments.

18.2.3 Slew Rate of CFB Op Amps

CFB op amps excel over VFB op amps in terms of slew rate capabilities in addition to bandwidth and low distortion. The output buffer, as shown in Figs. 18-1 and 18-2, has the capability of large voltage and current swings. This allows high slew rates at higher currents. However, the input buffer must be capable of following a fast-changing input signal in order to accomplish high slew rate at the output. The end result is faster rise and fall time when pulses are applied to a CFB op amp.

18.2.4 Implementing VFB Designs Using CFB Op Amps

Most VFB op amp circuits can be implemented by substituting a CFB op amp as long as care is taken to make the proper choice of the feedback resistor, which is probably the most critical decision during the design process. One can interpolate between values of R_F for different gains, but care should be taken not to reduce R_F to the point of instability or to increase R_F to where the bandwidth is overly limited. Good practice in general as far as PCB layout is required even more than for VFB amplifiers, such as avoiding long traces at the inverting input to minimize stray capacitance. The manufacturers' data sheets will provide excellent guidance, and in many cases suggested PC layouts. Because R_{FB} values are typically in the 1-kΩ range the value of R_G, the gain setting input resistor connected to the inverting input will be lower for increasing gains. So when implementing an existing VFB design using a CFB op amp, the resistor values will most likely all be reduced.

Figure 18-5 shows three basic implementations of a VFB op amp function with a CFB op amp that emphasize the difference between the two. In Fig. 18-5a, a voltage

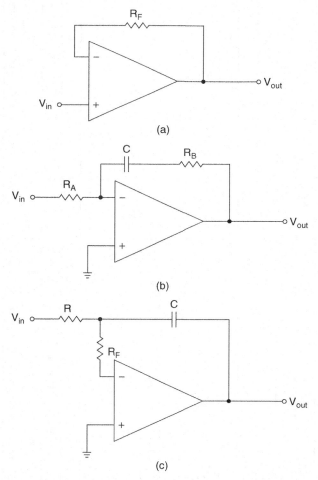

FIGURE 18-5   CFB op amp applications: (a) voltage follower; (b) integrator; and (c) improved integrator.

Manufacturer	Part Number	Small Signal Bandwidth	Slew Rate	Output Current	Supply Voltages	Features
Analog Devices	AD8000	1.5 GHz	4,100 V/μS	100 mA	+4.5 V to +12 V	ultra high speed
Analog Devices	AD8001	800 MHz	1,200 V/μS	70 mA	±5 V	low power
Analog Devices	AD8003	1.65 MHz	4,300 V/μS	100 mA	+5 V, ±5 V	low noise, low offset
Analog Devices	AD8005	270 MHz	280 V/μS	–	+5 V, ±5 V	ukltra low power
Analog Devices	AD8007/AD8008	650 MHz	1,000 V/μS	–	+5 V to +12 V	ultra low distortion
Analog Devices	AD8011	300 MHz	2,000 V/μS	–	+5 V, ±5 V	ukltra low power
Analog Devices	AD8012	350 MHz	2,250 V/μS	125 mA	+5 V, ±5 V	low power, low noise, low offset
Analog Devices	AD8014	400 MHz	4,000 V/μS	50 mA	+4.5 V to +12 V	low power, very low noise, low distortion
Analog Devices	ADA4860-1	800 MHz	790 V/μS	–	+5 V to +12 V	low power, very low noise, low offset
Texas Instruments	LM6181	100 MHz	2,000 V/μS	100 mA	±5 V, ±15 V	capacitive load tolerant
Texas Instruments	LMH6703	1.2 GHz	4,500 V/μS	90 mA	±5 V	low noise, low distortion, shut down
Texas Instruments	LMH6723, 24, 25	370 MHz	600 V/μS	110 mA	+4.5 V to +12 V	low power
Texas Instruments	OPA694	1.5 GHz	1,700 V/μS	80 mA	±5 V	low power
Texas Instruments	OPA2673	600 MHz	3,000 V/μS	700 mA	+13 V, ±6.5 V	high output current
Texas Instruments	OPA2674	220 MHz	2,000 V/μS	500 mA	+12 V, ±5 V	high output current
Texas Instruments	OPA2677	220 MHz	1,800 V/μS	500 mA	+5 V to +12 V, ±2.5 V to ±6 V	dual, high output current, low distortion
Texas Instruments	OPA2683	150 MHz	450 V/μS	110 mA	+5 V to +12 V, ±2.5 V to ±6 V	dual, ultra low power
Texas Instruments	OPA2684	170 MHz	650 V/μS	120 mA	+5 V to +12 V, ±2.5 V to ±6 V	dual, low power
Texas Instruments	OPA2695	850 MHz	2,900 V/μS	80 mA	±5 V	dual, wide bandwidth
Texas Instruments	OPA3684	170 MHz	650 V/μS	120 mA	+5 V to +12 V, ±2.5 V to ±6 V	triple, low power, disable
Texas Instruments	OPA3691	280 MHz	2,100 V/μS	190 mA	+5 V to +12 V, ±2.5 V to ±6 V	triple, low power, disable
Texas Instruments	OPA4684	170 MHz	650 V/μS	120 mA	+5 V to +12 V, ±2.5 V to ±6 V	quad, low distortion
Texas Instruments	THS3201	1.8 GHz	6,700 V/μS	100 mA	±3.3 V to ±7.5 V	wide band, high slew rate, low distortion

TABLE 18-1 Current Feedback Amplifier Selection Guide

follower is shown but a feedback resistor R_F must be present, whereas in the VFB op amp case, a short can exist between the output and the inverting input. In Fig. 18-5*b* a resistor must be present in series with the feedback capacitor in the integrator shown to maintain stability and avoid a purely capacitive feedback impedance. Figure 18-5*c* contains a feedback resistor R_F, but it is inside the feedback loop so it does not affect the integrator. In any type of integrator configuration, it is good practice to place a high-value resistor across the capacitor to avoid a DC voltage buildup at the output resulting from input DC offsets.

Table 18-1 presents a current feedback amplifier selection guide. This table is by no means all inclusive, and is limited in sources, but is meant to show a sampling of CFB op amps and the range of parameters available. A wealth of information is normally contained in the CFB op amp data sheets and should always be reviewed prior to the design process.

References

Analog Devices. "Current Feedback (CFB) Op Amps MT-034 Tutorial," October 2008, Norwood, Ma.
Texas Instruments. "Current Feedback Amplifier Analysis and Compensation Application Report," SLOA021A, March 2001, Dallas, Texas.
Williams, A. B. (1984). *Designer's Handbook of Integrated Circuits*, McGraw-Hill, New York.

CHAPTER 19

Large Signal Amplifiers

19.1 Class D Amplifiers for Audio

Amplifiers can be classified by letter, ranging from Class A to Class D. The amplifiers we have dealt with up until now have all been Class A. The op amp operates in a linear mode where, for example, if a sine wave were applied at the input, the op-amp output will track the input linearly over 360° of the sine wave.

An op amp operating in the Class B mode provides an output for the sine wave, for example, only for 180°. A second op amp operating in conjunction with the first will provide an output for the remaining 180°, and the two outputs are combined together, typically through a transformer, to reconstruct the proper output signal. However, there is a small dead zone or discontinuity where neither amplifier conducts, which results in a nonlinearity at the crossover point.

Class AB is a combination of Class A and Class B where each device conducts for slightly more than 180° so there is no discontinuity at the crossover point.

For Class C, the op amp conducts for less than 180°. Although the signal is severely clipped, this mode would be used if there was a parallel-resonant circuit as the load to reconstruct a sine wave.

Now we get to Class D, which is unique compared to all the other classes. The input signal instantaneous level is converted to the pulse width (pulse-width-modulated or PWM) of a pulse stream at a much higher frequency than the input signal. This stream is then applied to a low-pass filter for integration, which re-creates the original signal. This approach is significantly more efficient than Classes A, B, or AB. Class A is typically 20 percent efficient. For Classes B and AB, the efficiency increases to around 50 percent. Class D can result in efficiencies as high as 90 percent. The reason for this is that for Class D the brunt of the power lost is in an electronic switch, which does not have to support a simultaneous voltage and current, unlike a linear amplifier, so power losses are low. Class D amplifiers are best suited for audio applications.

19.1.1 Half-Bridge Topology

Figure 19-1*a* shows a basic half-bridge configuration of a Class D amplifier. A pulse-width modulator produces a train of pulses where the pulse-width modulation is created by comparing the input audio signal to a much higher-frequency triangular wave. Typically, for audio, the triangular waveform is between 250 KHz and 1.5 MHz. With no input, the default pulse duty cycle is 50 percent which results in a DC level after the low-pass filter of +V/2. This DC bias must be blocked by a polarized capacitor, as shown

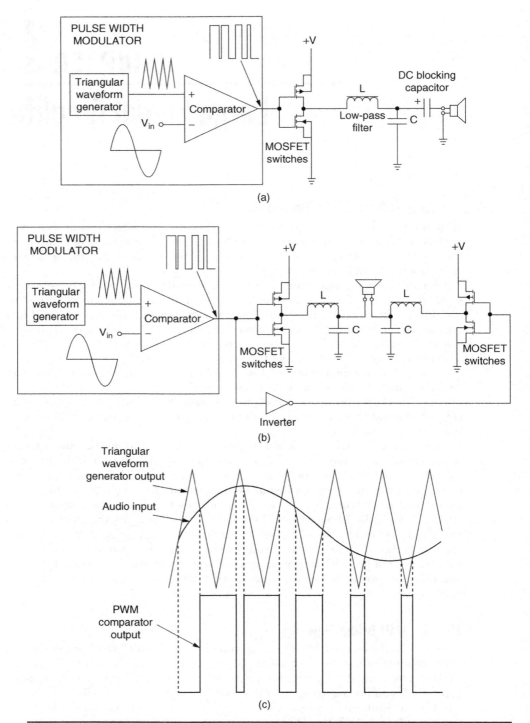

Figure 19-1 Class D amplifier topology: (a) half-bridge; (b) full-bridge; and (c) operation of PWM.

in the figure, to avoid saturating and damaging the loudspeaker. As the pulse width is modulated above and below 50 percent, the output of the low-pass filter will vary above and below +V/2, thus reconstructing the audio signal.

The P and N channel MOSFETs alternately switch between +V and ground, therefore supplying the load current. They are either open or closed—ideally when open, no current flows and when closed, there is no voltage drop—so no power should be dissipated in the MOSFETs. In reality, there will be some voltage drop but the overall circuit efficiency will be extremely high. The losses of the inductor, especially the DC resistance, should be much lower than the impedance of the loudspeaker.

19.1.2 Full-Bridge Topology

The simplified circuit of Fig. 19-1b illustrates a full-bridge topology. The loudspeaker is driven differentially by two sets of MOSFETs. For a 50 percent duty cycle, the signals on each side of the loudspeaker after filtering are both equal to +V/2, so no blocking capacitor is needed. One advantage to this is extended low-frequency response as the capacitor would have caused a low-end roll-off. The second advantage, and probably the major one, is that twice the voltage swing is developed, which results in quadrupling the output power. This circuit is sometimes referred to as an "H" configuration or a bridged tied load (BTL).

The waveforms of a PWM configuration are illustrated in Fig. 19-1c. The comparator will produce a pulse that varies in duration by comparing the input audio to the triangular waveform. Traditional PWM is sometimes called AD modulation.

19.1.3 Class D Operation Without an Output Filter

The previous section illustrated how PWM, also known as AD modulation, can be used to convert a low-frequency analog signal to a much higher-frequency pulse stream in a very efficient manner. This pulse stream can then be integrated with a low-pass filter to restore the analog signal and then applied to a loudspeaker. The low-pass filter would consist of inductors and capacitors. The inductors would have losses due to copper loss (DC resistance) and AC losses such as core losses and some skin effect. Also, they can saturate, which results in nonlinearities such as intermodulation and harmonic distortion. Finally, inductors are costly and bulky.

It would be highly desirable to minimize the need for filtering by using an alternative modulation scheme to PWM. A solution is the BD modulation method. In Fig. 19-2a the input audio is inverted and both the noninverted and inverted versions are applied to comparators, which then control two sets of MOSFET switches in a BTL configuration.

The timing diagram of Fig. 19-2b results in an output of A-B, which is differentially applied to the loudspeaker represented by R and L. This form of modulation is sometimes also called suppressed carrier PWM modulation. From the timing diagram, the A-B signal can be developed. It is a three-level signal that will significantly reduce or eliminate the need for the low-pass filter. The inductance of the loudspeaker itself is used for filtering the modulated signal to recover the audio. The end result is greater efficiency, which can run as high as 90 percent. The modulated signal must be high enough in frequency so the speaker itself can be the filter. Typically, it is up around 700 KHz.

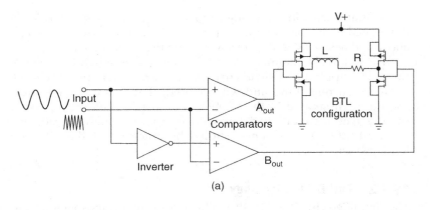

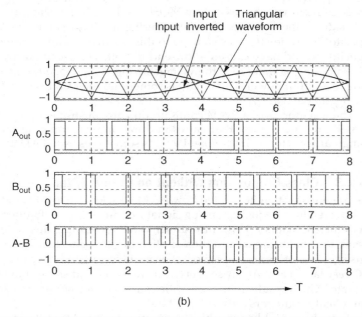

FIGURE 19-2   BD modulation: (a) circuit block diagram and (b) waveforms.

A number of other modulation schemes are used in integrated circuits (ICs) where the intent is to eliminate the external filters and thus obtain high efficiency of the overall circuit. They are all based on creating a stream of pulses, either binary (two level) or tertiary (three level), where the objective is to encode the analog signal information into the pulse width, pulse density, or pulse amplitude, which by basic integration using the loudspeaker parameters, the audio can be restored.

Not all loudspeakers are suitable for filterless operation. For example, high-frequency loudspeakers such as tweeters and piezoelectric speakers do not have sufficient inductance to allow filtering of the high-frequency components of the waveform.

Minimizing EMI

Although the various forms of pulse modulation result in very high efficiency, which implies less heat and lower power consumption, they produce a significant amount of electromagnetic interference (EMI) due to the sharp pulse edges. There are two forms of EMI: conducted and radiated. The conducted EMI travels via the loudspeaker and power supply wires and can even go back into the AC mains. The radiated EMI originates from printed circuit board (PCB) traces and amplifier components. One solution is to use spread-spectrum modulation where the pulse edges have a jitter component introduced, which spreads out the EMI energy over a wide spectrum but at a lower amplitude. Wiring between the amplifier and loudspeaker should be shielded.

Although filterless operation implies the absence of a filter, there are filtering techniques that can be used to minimize EMI without a significant impact on efficiency. RF chokes or ferrite beads can be placed in series with external wires. Careful PCB design to minimize trace length, along with low impedance ground planes, can reduce ground loops and EMI. Decoupling the amplifier output lines with capacitors at both ends to a good ground will also help to reduce EMI. Finally, shielding can be used to contain radiated electric fields.

19.1.4   Class D *LC* Filter Design

For PWM modulation, a frequently used low-pass filter design is an $N = 2$ Butterworth in each leg. The 3-dB bandwidth should be in the 40-KHz range so the audio band up to 20 KHz is well in the flat region. Let us assume the source impedance is that of the MOSFET drivers and is zero ohms. Also let us assume the load is that of the loudspeaker, which is typically 8 Ω. So we need to design a 2nd-order Butterworth low-pass filter with a 3-dB cutoff of 40 KHz that operates between a 0 ohm source and an 8-ohm load.

Table 10-2 provides the element values of the normalized filter, as shown in Fig. 19-3*a*. Let us now scale this filter to a cutoff of 40 KHz and an impedance level of 8 Ω. We can now denormalize the filter as follows using Eqs. (2-1), (2-9), and (2-10):

$$\mathrm{FSF} = \frac{2\pi 40,000 \ \mathrm{rad/s}}{1 \ \mathrm{rad/s}} = 2.513 x 10^5$$

$$L' = \frac{L \times Z}{\mathrm{FSF}} = \frac{1.414 x 8}{2.513 x 10^5} = 45 \ \mathrm{uH}$$

$$C' = \frac{C}{\mathrm{FSF} \times Z} = \frac{0.707}{2.513 x 10^5 \, x 8} = 0.352 \ \mathrm{uF}$$

The denormalized circuit is shown in Fig. 19-3*b*.

The circuit of Fig. 19-3*c* is a filter configuration for a BTL architecture. It is balanced and has decoupling to ground as well to minimize EMI. The 0.1-uF capacitors to ground will also filter common-mode unwanted signals. The values are close to the theoretical using standard value components. For a loudspeaker impedance other than 8 Ω, the Z *in* the formulas can be different. Also, a lower cut-off can be used for more high-frequency rejection at the expense of response flatness.

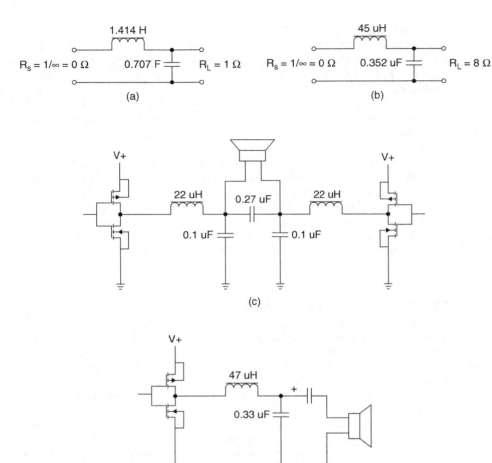

Figure 19-3 Class D amplifier filtering: (a) normalized LPF; (b) denormalized LPF; (c) BTL filtering; and (d) half-bridge filtering.

For a half-bridge configuration, the circuit of Fig. 19-3d can be used. Often, two half-bridges are used for a stereo application. A half-bridge requires a large value DC blocking capacitor to avoid damaging the loudspeaker.

Selection of inductors is very critical. The geometry will determine the degree of EMI from the inductor. The best choices would be a toroid or other self-shielding magnetic structure. The ability to support the high currents required without saturation is critical to prevent nonlinearities. Efficiency is a function of DC resistance and AC core and winding losses. Shielded inductors help reduce EMI.

A selection table for Class D power amplifiers is provided in Table 19-1. This table is simply a cross-section of what is available from a variety of manufacturers and is not meant to be all-inclusive. All devices contain internal MOSFET drivers for output interfacing unless otherwise specified.

Manufacturer	Part Number	Power	Channels	Modulation	Frequency	Features
International Rectifier	AUIRS2092S	Up to 500 watts	1	PWM	800 KHz	External MOSFETs and filter, overcurrent protection
Analog Devices	SSM2380	2 × 2 watts	2	Σ–Δ modulation	325 KHz	Filterless, Stereo
Intersil	ISL99201	1.5 watts	1	PWM	375 KHz	Filterless High Efficiency, thermal shutdown, overcurrent protection
Maxim Integrated Products	MAX9700	1.2 watts	1	spread spectrum	1.45 MHz	Filterless, reduced EMI
Maxim Integrated Products	MAX97000	725 mW	2	spread spectrum	250 KHz	Headphone and speaker amplifiers
Maxim Integrated Products	MAX97001	700 mW	2	spread spectrum	250 KHz	Headphone and speaker amplifiers
Maxim Integrated Products	MAX9708	2 × 21 watts 1 × 42 watts	2	spread spectrum	250 KHz	Filterless, high power
Maxim Integrated Products	MAX9742	2 × 16 watts	2	PWM	300 KHz	Differential inputs, low distortion
Maxim Integrated Products	MAX9744	2 × 20 watts	2	PWM	300 KHz	Volume control
Maxim Integrated Products	MAX9776	2 × 1.5 watts	2	spread spectrum	1.1 MHz	High efficiency, filterless operation, headphone compatible
Maxim Integrated Products	MAX98304	3.2 watts	1	spread spectrum	300 KHz	Filterless, reduced EMI, high efficiency
Maxim Integrated Products	MAX98306	2 × 3.7 watts	2	spread spectrum	320 KHz	Filterless, Stereo, low EMI, 5 built-in gain settings
Maxim Integrated Products	MAX98400A	2 × 20 watts 1 × 40 watts	2	PWM	330 KHz	High power, filterless operation, built in limiter
Maxim Integrated Products	MAX98500	2.2 watts	1	PWM	300 KHz	Automatic level control

TABLE 19-1 Class D Power Amplifiers (*Continued*)

Manufacturer	Part Number	Power	Channels	Modulation	Frequency	Features
NXP Semiconductors	TDA8932B	2 × 15 watts	2	PWM	320 KHz	High efficiency, filterless operation
NXP Semiconductors	TDA8933B	2 × 10 watts	2	PWM	320 KHz	High efficiency
On Semiconductor	NCP2824	2.4 watts	1	PWM	300 KHz	high efficiency, includes AGC
RoHM	BD5460GUL	2.5 watts	1	PWM	250 KHz	filterless, low voltage
STMicroelectronics	TS4999EIJT	2 × 2.8 watts	2	PWM	280 KHz	Filterless, 3D effects
Texas Instruments	LM4675	2.65 watts	1	spread spectrum	300 KHz	Filterless
Texas Instruments	LM48310	2.6 watts	1	spread spectrum	300 KHz	Ultra-Low EMI, Filterless,
Texas Instruments	LM48520	2 × 1 Watt	2	spread spectrum	300 KHz	Filterless, Stereo, low EMI, 4 built-in gain settings
Texas Instruments	TAS5176	5 × 15 watts 1 × 25 watts	6	PWM	384 KHz	High power, multiple channels
Texas Instruments	TAS5615	2 × 160 watts 1 × 300 watts	2	PWM	400 KHz	High power, low distortion
Texas Instruments	TAS5630	2 × 300 watts	2	PWM	400 KHz	High power, high efficiency
Texas Instruments	TAS5630B	2 × 300 watts 1 × 400 watts	2	PWM	400 KHz	High power, low distortion
Texas Instruments	TPA2015D1	2 watts	1	PWM	600 KHz	Built in AGC
Texas Instruments	TPA2028D1	2.65 watts	1	PWM	300 KHz	Built in AGC
Texas Instruments	TPA2100P1	19-Vpp	1	PWM	1.2 MHz	For Piezo/Ceramic speakers
Texas Instruments	TPA3110D2	2 × 15 watts	2	PWM	310 KHz	Filterless, high power
Texas Instruments	TPA3111D1	10 watts	1	PWM	310 KHz	Filterless, high power
Texas Instruments	TPA3112D1	25 watts	1	PWM	310 KHz	Filterless, high power
Texas Instruments	TPA3123D2	2 × 25 watts	2	PWM	250 KHz	High power

TABLE 19-1 Class D Power Amplifiers (*Continued*)

600

19.2 Crossover Networks

This book is by no means a discourse on audio. However, having discussed Class D amplifiers and the earlier concentration on filters, crossover networks should be introduced.

The Class D audio power amplifiers discussed have wide bandwidths covering the entire audio spectrum. However, loudspeakers are limited in bandwidth, so typically a crossover network is used to split the band appropriately to accommodate an array of a variety of speakers, which typically are low band (woofer), midrange, and high band (tweeter). Let us first consider a single channel. There are two approaches for splitting the band. The first is to use active filters followed by amplifiers. In the case just mentioned, three amplifiers would be required, one for each band. The second approach, and the one most commonly used, would be to employ passive crossover networks at the output of a single amplifier (per channel).

Tweeters typically are small loudspeakers optimized for the 2 KHz to 20 KHz band. They usually use the same structure as their larger lower-frequency counterparts and can be driven through a crossover network, but there are some that are piezoelectric devices that require a high voltage drive and use a special amplifier.

Figure 19-4a illustrates a typical configuration involving two band-separation filters—a low-pass filter to extract the band below 1 KHz and a high-pass filter to extract the band above 1 KHz—and feed the two audios to their respective loudspeakers. So the crossover frequency is 1 KHz. Figure 19-4b shows a three-way crossover system where the crossover frequencies are 800 Hz and 4 KHz.

Let us consider the design of a simple first-order low-pass filter and high-pass filter crossover network. Let the source impedance be zero ohms (assuming an ideal amplifier) and a load impedance of 8 Ω, which is the loudspeaker. The normalized values are 1 H and 1 F as shown in Fig. 19-4c. This would result in a 3-dB cutoff of 1 radian. To denormalize these values for an 8 Ω loudspeaker and a 3-dB point (crossover) of 1 KHz using Eqs. (2-9) and (2-10):

$$L' = \frac{L \times Z}{\text{FSF}} = \frac{8}{2\pi 1,000} = 1.27 \text{ mH}$$

$$C' = \frac{C}{\text{FSF} \times Z} = \frac{1}{2\pi 1,000 x 8} = 19.9 \text{ uF}$$

The final result is shown in Fig. 19-4d. For speaker impedances other than 8 Ω, use the appropriate value for Z in the equations.

If a higher order than $N = 1$ is desired to achieve a steeper slope, the tables of Chap. 10 can be utilized. A design can be chosen normalized for a 0 Ω source (on the amplifier side) and a 1 Ω load. The design can then be scaled to the impedance of the loudspeaker. Orders greater than $N = 4$ provide little benefit. The best choice would be the Butterworth type to avoid ripples and sensitivity to component values.

For the band-pass filter shown in Fig. 19-4b, a cascade of a low-pass and high-pass filters should be used. The simplest case would be a cascade of an inductor and capacitor in series, with the loudspeaker with values determined using Eqs. (2-9) and (2-10). Since the spread between the lower cutoff and the upper cutoff of the band-pass filter is more than an octave, higher-order filters can be designed independently and then cascaded.

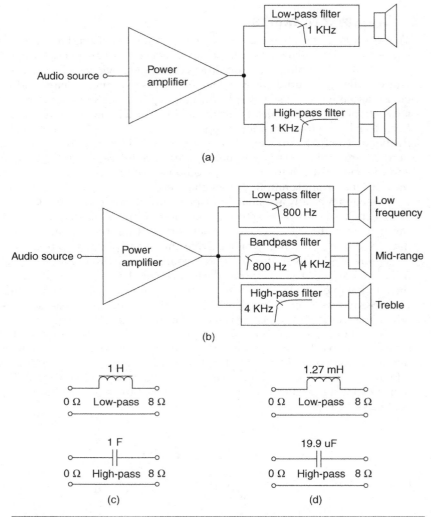

FIGURE 19-4   Crossover networks: (*a*) low band and high band; (*b*) addition of midband; (*c*) normalized N = 1 crossover networks; and (*d*) denormalized for 1-KHz crossover.

19.2.1   Component Selection

Capacitors

Although electrolytic capacitors can be used providing there is no polarity reversal across the capacitor, it is best to use plastic film type capacitors. They would provide better reliability and lower losses. Typical dielectrics would be Mylar (polyester), polycarbonate, polypropylene, and polystyrene.

Be sure to use a voltage rating of at least twice the maximum voltage that would be applied to the capacitor. If you do use an electrolytic, be sure to select one with a low dissipation factor (DF).

Inductors

Inductors are the most critical and expensive component of a crossover network. Inductors have a parameter called Q, which is $\omega L / R$, where R is both the DC and AC losses.

The lower the Q, the less efficient the inductor. Over the low end of the audio frequency band, most of the losses are in the DC resistance of the inductor, but as the frequency increases, the core losses become a consideration. Air cores have no core losses and no nonlinear behavior, but the DC resistance would tend to be high as the permeability is low (unity) requiring a large number of turns. This results in inefficiencies. If a magnetic core is used, the DC resistance is significantly reduced, but the magnetic material could cause nonlinearities such as harmonic distortion if not properly selected. For the perfectionist, an air core would be best, with a large wire size to minimize DC resistance. Although they can be purchased, some audiophiles prefer to wind their own.

19.3   Transformer-Coupled Line Driver Configurations

Chapter 18 discussed current feedback amplifiers which have high-output capability in terms of output current and speed. However, many applications require a balanced output, which dictates the addition of a transformer. For example, transmission lines for telephone and audio systems are vulnerable to interference from external stray electric fields and magnetically coupled noises.

A balanced line allows noises, which are equally induced in both wires, to be cancelled in the transformer at the opposite end. These signals would be common-mode rather than differential, and a transformer only passes differential signals applied to one winding on to the secondary winding. Usually the wires are twisted-pair to ensure each conductor is equally susceptible to the external interference. Some examples would be Ethernet cable, telephone lines, local area networks (LANs), balanced audio lines, and so forth. Additional advantages of transformer coupling are the ability to match the balanced impedance of the line and to provide DC isolation of the circuitry.

19.3.1   Traditional Transformer-Coupled Line Driver

The circuit of Fig. 19-5a is a basic CFB (current feedback amplifier) that has a transformer-coupled output. The source impedance R_S is determined by the value of R_O reflected into the secondary of the transformer by the turns ratio N squared, or

$$R_S = R_O N^2 \qquad (19\text{-}1)$$

Let us assume that the output voltage swing at $E_{\text{OP-AMP}}$ is from $-V$ to $+V$ with no internal drop within the op amp (there will always be a small internal drop). The circuit gain up to $E_{\text{OP-AMP}}$ is $R_F / R_G + 1$.

With proper impedance matching of the transformer, the maximum voltage swing after R_O (across the transformer primary) will be from $-V/2$ to $+V/2$ as a result of the 6-dB drop across R_O. If we assume a turns ratio $N = 1$, where $R_S = R_O$, E_{OUT} will also range from $-V/2$ to $+V/2$, assuming the transformer is lossless.

19.3.2   Differential Transformer-Coupled Line Driver

Figure 19-5b combines the instrumentation amplifier geometry of Chap. 14 with a transformer-coupled output. The op amp is the CFB type. Both the input and output are differential. Let us look at the maximum output available from this revised configuration.

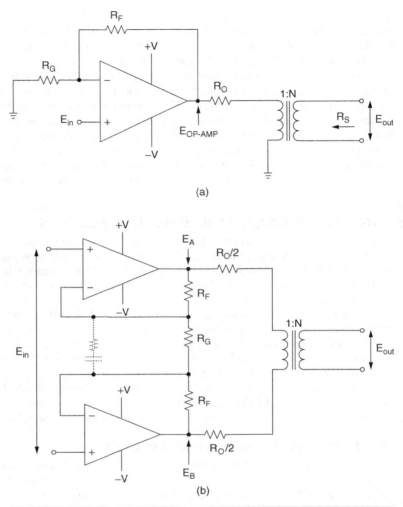

FIGURE 19-5 Transformer-coupled line driver: (a) traditional configuration and (b) differential output configuration.

The signal present at E_A will range from $-V$ to $+V$ (again assuming no internal voltage drop of the op amp). Simultaneously, the voltage swing at E_B will range from $-V$ to $+V$ but the signal will be 180° out of phase with E_A, so the *differential* between E_A and E_B will vary from $2(+V)$ to $2(-V)$. For example, if $+V = +12V$ and $-V = -12V$, then the differential will vary from $+24V$ to $-24V$. The amplifier could differentially produce a $\pm24V$ output, which is twice that of Fig. 19-5a.

A combined 6-dB drop will occur across both source resistors $(R_O/2 + R_O/2)$, so E_{OUT} will range from $-V$ to $+V$, assuming $N = 1$ and a lossless transformer. This is twice the output to the load R_L of Fig. 19-5a, or *four times the power.*

The optional RC network in parallel with R_G can be used to provide a high-frequency boost to compensate for op-amp and transformer roll-off.

In some cases, the transformer is not 1:1. This will occur if R_O is not equal to R_L, the load impedance. The turns ratio is then computed by

$$N = \sqrt{\frac{R_L}{R_O}}$$ (19-2)

19.3.3 Active Output Impedance Line Driver

One of the negatives of the circuit of Fig. 19-5b is that half of the output voltage of the amplifier is lost across the series-matching resistor R_O ($R_O/2 + R_O/2$). The voltage across the primary of the transformer becomes $(E_A - E_B)/2$. Therefore, half the output power of the amplifier itself is wasted across the series-matching resistors. The power supply voltages would have to be doubled to compensate for this.

The circuit of Fig. 19-6a is called an *active output impedance line driver*. It allows the values of the series-matching resistors R_S to be significantly lowered, which minimizes their voltage drop and hence the power wasted, and yet provides the proper impedance matching. This could be quite significant in a system that requires a large number of line drivers. To accomplish this, the two cross-coupled feedback resistors designated as R_{FB} provide positive feedback, which increases the effective values of R_S.

To see how positive feedback accomplishes this, let us examine Fig. 19-6b, which is a voltage source E_1 applied to resistor R, where the opposite side of the resistor is connected to KE_1, which is a fraction K of the applied voltage. Let us assume KE_1 is in phase with E_1. That would correspond to positive feedback. The general equation for current would be

$$I_1 = (E_1 - KE_1)/R$$ (19-3)

If $K = 0$, the current I_1 is given by E_1/R. If $K = 1$, then no current would flow.

This equation can also be expressed as

$$I_1 = \frac{E_1}{\dfrac{R}{1 - K}}$$ (19-4)

So the effective resistance becomes

$$R_{EFF} = \frac{R}{1 - K}$$ (19-5)

Therefore, as a result of the positive feedback, the effective resistance can be increased all the way up to infinity as K approaches unity. This property is used in the circuit of Fig. 19-6a, where R_S is a small value for minimal power dissipation, and positive feedback increases its effective value to provide the proper impedance toward the load.

Because of the positive feedback through R_{FB}, the effective value of R_S becomes

$$R_{EFF} = \frac{R_S}{1 - \dfrac{R_F}{R_{FB}}}$$ (19-6)

Therefore, we can use a lower value of R_S and "multiply" it by $\dfrac{1}{1 - \dfrac{R_F}{R_{FB}}}$ with proper selection of R_{FB}.

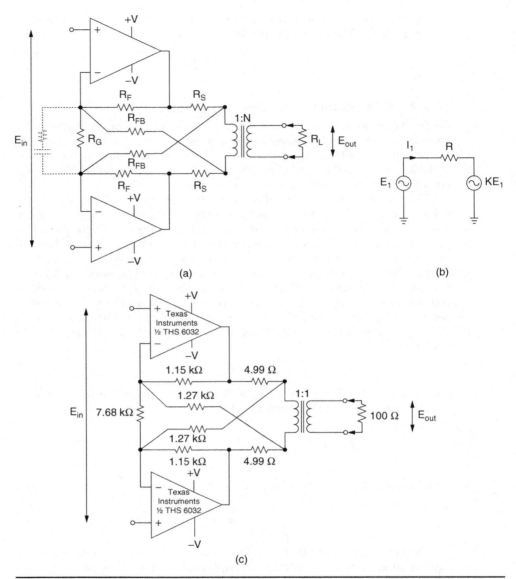

FIGURE 19-6   Active impedance: (a) active impedance differential output circuit; (b) effect of positive feedback; and (c) active impedance differential output circuit with typical values.

Let us look at an example to illustrate the reduction in R_S. Figure 19-6c shows a line driver application using a Texas Instruments THS 6032 current feedback op amp. We can compute the effective source impedance as follows:

$$R_{EFF} = \frac{R_S}{1 - \dfrac{R_F}{R_{FB}}} = \frac{4.99}{1 - \dfrac{1,150}{1,270}} = 52.8 \ \Omega \tag{19-7}$$

So a dramatic reduction in R_S has occurred where instead of using a value in the vicinity of 50 Ω, a resistor of 4.99 Ω can be used, dissipating much less power for a 100 Ω load.

The negative to this approach is that the effective impedance will change when the amplifier gain rolls off, so significantly more bandwidth would be required than with a conventional approach, such as that shown in Fig. 19-5. A good rule of thumb could be 10 or 20 times the actual required bandwidth. The series RC network across R_G shown in Fig. 19-6 will help boost the response at higher frequencies if needed. Also, it is suggested that precision resistors be used, especially for R_F and R_{FB}. Metal film resistors with tolerances of ±0.1 percent are readily available.

19.4 Thermal Management

Many of the devices presented in this chapter dissipate significant power. This results in heat buildup within the device, which can approach critical junction temperatures of the ICs. The end result is reduced life expectancy and, in many cases, component failure.

The three ways heat can be removed from a device are radiation, convection, and conduction. Radiation is something you get for "free," providing there is clearance in the vicinity of the device. Convection typically involves fans. The most often used approach is conduction.

Thermal resistance is the temperature difference across a surface for a unit of power, and is measured in degrees Celsius per watt, or °C/W. It can be defined as a measure of a material's ability to *resist* heat transfer, and is represented by the symbol θ, so, for example, θ_{JC} would be the thermal resistance from junction to case of a semiconductor device.

There are a number of ways to remove heat from a semiconductor device. Adequate ventilation is difficult at times since multiple PC boards in system will restrict air flow. The most commonly used approach is a heat sink—the larger, the better. Heat sinks have low thermal resistance to air, which allows efficient transfer of heat from the device to the surrounding environment.

Heat sinks come in a variety of forms, some of which are part of the IC. For example, Fig. 19-7 shows an exposed thermal pad, also called a "paddle," on the bottom of an IC package. The thermal pad is meant to be soldered to an appropriate copper plane on the PC board, which as a result will function as a heat sink by providing a direct thermal path from the die to the copper on the PC board. This can lower the thermal resistance by 25 percent in comparison to a plastic package. Most data sheets provide a recommended PC layout. Some devices are available both with and without the thermal pad. A negative to the thermal pad is that it becomes more difficult to remove the IC from the board if replacement is necessary. Some high-power ICs have a thermal pad on top of the IC package. It is meant to be in contact with an external heat sink mounted over the IC.

FIGURE **19-7** Thermal pad on bottom of IC package.

For high-power devices, the thermal pad is replaced or enhanced by a heat slug embedded in the top of the IC package. Its function is to make thermal contact with an external heat sink. In all cases where an external heat sink is used, thermal grease should be applied to the mating surfaces to ensure good contact. Use of a heat sink can be combined with forced air cooling.

Many integrated circuits have a built-in automatic thermal shutdown as well as current limiting. If the semiconductor junction temperature exceeds a preset limit, the device will shut down. This may prevent device destruction, but it will also cease to operate. Even if this doesn't happen, operation near the junction temperature will result in performance degradation and eventual failure.

As far as the selection of external heat sinks that make physical contact with the IC goes, the thermal resistance calculations come into the picture. The choice of a heat sink is totally based on keeping the IC's die temperature below the point at which the internal thermal shutoff circuitry is activated as much as possible. This is where the thermal resistance computations come into the picture.

To determine if a device needs a heat sink, first determine the device's thermal resistance from the junction to case, θ_{JC}. Then add to that the device's thermal resistance case to ambient θ_{CA}. Based on the dissipated wattage, compute the device's temperature rise above ambient.

The calculation is

$$\theta_{JATotal} = \theta_{JC} + \theta_{CA} = \frac{T_J - T_A}{P} \tag{19-8}$$

where T_J is the junction temperature and T_A is the ambient temperature, both in degrees C, and P is the total power dissipated in watts.

So by rearranging Eq. (19-8), the temperature rise is computed by

$$T_J - T_A = P(\theta_{JC} + \theta_{CA}) \tag{19-9}$$

Sometimes, rather than specify θ_{JC} and θ_{CA} individually, they are combined into a θ_{JA} specification.

Then
$$T_J - T_A = P\theta_{JA} \tag{19-10}$$

The effect of interposing a heat sink is to eliminate the case-to-ambient term θ_{CA} and replace it with a case-to-sink term θ_{CS} and a sink-to-ambient term θ_{SA}. So with a heat sink, the temperature rise becomes

$$T_J - T_A = P(\theta_{JC} + \theta_{CS} + \theta_{SA}) \tag{19-11}$$

These relationships are represented schematically in Fig. 19-8. Since heat flow has an analogy to current flow, the thermal resistance would be analogous to electrical resistance. The total temperature rise of the junction above ambient is computed by determining the total thermal resistance and multiplying it by the total power to be dissipated. A temperature drop in the thermal domain is analogous to a voltage drop in the electrical domain.

Keep in mind that the ambient may not be the exact temperature of the room because in a dense system the PC board will see elevated temperatures. This has to be considered when determining T_A.

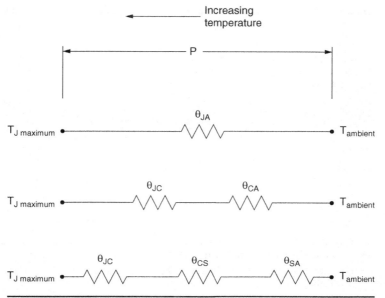

FIGURE 19-8   Thermal distribution.

To ensure good conductivity between a device and a heat sink, a thermal grease compound should be used on the mating surfaces. The thermal resistance case-to-sink θ_{CS} is usually specified by the grease compound manufacturer. In some cases, a thermally conductive insulating pad is needed between the device and the heat sink. If possible, the heat sink should be connected to a ground plane or to a chassis, as this will further enhance its ability to convey heat away from the semiconductor device.

The performance of a heat sink is enhanced by forced air if a fan is present. Heat sink data shows the effect of various velocities of forced air on their thermal resistance.

Example 19-1   Determine If a Heat Sink Is Needed Based on Junction Temperature

Required:

Desired maximum junction temperature: 85°C
Maximum ambient temperature: 55°C
$\Theta_{JA} = 8°C/W$
Power consumption of device: 1.5 watts

Result:

Using Eq. (19-10):

$$T_J - T_A = P\theta_{JA}$$

Then

$$T_J - T_A = 1.5W \times 8°C/W = 12°C$$

With an ambient of 55°C and a temperature rise of 12°C, the junction temperature would be 67°C, so no heat sink would be needed.

Let us now repeat the example with different numbers.

Example 19-2   Determine the Thermal Resistance Requirement for a Heat Sink

Required:

Desired maximum junction temperature: 85°C
Maximum ambient temperature: 55°C
$\Theta_{JA} = 8°C/W$
$\Theta_{JC} = 0.5°C/W$
Power consumption of device: 6 watts
$\Theta_{CS} = 0.5°C/W$ as specified by thermal grease manufacturer

Result:

Using Eq. (19-10):

$$T_J - T_A = P\theta_{JA}$$

Then,
$$T_J - T_A = 6W \times 8°C/W = 48°C$$

With an ambient of 55°C and a temperature rise of 48°C, the junction temperature would be 103°C, which exceeds the maximum junction temperature, so a heat sink would definitely be needed.

Compute the maximum thermal resistance of the heat sink to ambient for the junction temperature of 85°C using Eq. (19-11):

$$T_J - T_A = P(\theta_{JC} + \theta_{CS} + \theta_{SA})$$

$$\theta_{SA} = \frac{T_J - T_A}{P} - \theta_{JC} - \theta_{CS}$$

$$\theta_{SA} \le \frac{85°C - 55°C}{6W} - 0.5°C/W - 0.5°C/W = 4.0°C/W$$

A heat sink with a thermal resistance of less than 4.0°C/W is required.

References

Honda, J. and Adams, J. Application Note AN-1071, "Class D Audio Amplifier Basics," International Rectifier, El Segunda, CA.
Texas Instruments. Application Report SLOA119A, "Class D LC Filter Design," April 2006, Dallas, Texas.
Texas Instruments. Application Report, "Reducing and Eliminating the Class D Output Filter," August 1999, Dalals, Texas.
Maxim, Inc. Application Note 3977, "Class D Amplifiers: Fundamentals of Operation and Recent Developments," January 31, 2007, San Jose, CA.
Nielsen, T. A., *Loudspeaker Crossover Networks*, Technical University of Denmark, 2005.
Jung, W., ed. (2006). *Op Amp Applications Handbook*, Analog Devices, Norwood, MA.
Stephens, R. and Texas Instruments. Application Report SLOA100, "Active Output Impedance for ADSL Line Drivers," November 2002, Dallas, Texas.

APPENDIX A

Software Download and Errata

A.1 Software Download

Associated with the *Analog Filter and Circuit Design Handbook* are three software programs:

- Filter_Solutions_FB2_Install
- ELI 1.0 Program for Design of Odd-Order Elliptic Function Low-Pass Filters up to 31st Order
- FLTRFORM.XLS Spreadsheet of formulas

These three programs are all located in a single ZIP folder called AFCDH on a McGraw-Hill website. The following URL will link to the website:

www.mhprofessional.com/AFCDH

The software registration page will require you to fill out a form. Upon completion you will be sent a confirming email with a link to download the software.

After unzipping the folder containing the three programs they are installed as described in the following sections.

A.2 Installing and Using "FILTER SOLUTIONS" (Book Version) Software for Design of Elliptic Function Low-Pass Filters

Previous books on filter design have contained extensive numerical tables of normalized values which have to be scaled to the operating frequencies and impedance levels during the design process. This is no longer the case.

The "Filter Solutions Book Version" program is quite intuitive and self-explanatory. The reader is encouraged to explore its many features on his/her own. Never-the-less all design examples using this program will elaborate on its usage and provide helpful hints. The user can obtain transfer functions, pole-zero plots, reflection coefficients, and step and impulse response, as well as phase, group delay and attenuation plots versus frequency for each low-pass design up to a 10th order elliptic function low-pass filter. This program has been provided by Nuhertz Technologies LLC. (http://www.nuhertz.com/).

Extract and install this program by running Filter_Solutions_FB2_Install

The program will automatically create a shortcut on the desktop.

The reader is encouraged to obtain the full version, which in addition to passive implementations covers many filter polynomial types and includes distributed, active, switched capacitor, and digital along with many very powerful features, and integrates with the popular Microwave Office® design software from AWR® Corporation, CST Studio Suite®, and Sonnet® software.

A.3 Installing and Using "ELI 1.0" Program for Design of Odd-Order Elliptic Function Low-Pass Filters up to 31st Order

This program allows the design of odd-order elliptic function LC low-pass filters up to a complexity of 15 nulls (transmission zeros) or the 31st order.

The program inputs are Passband edge (Hz), Stopband edge (Hz), Number of nulls (up to 15), Stopband rejection in dB and Source and Load terminations (which must always be equal). The output parameters are Critical Q (theoretical minimum Q), Passband ripple (dB), Nominal 3dB cutoff and a list of component values along with resonant null frequencies.

To install the program, first copy ELI1.0 to a new folder. This folder can be located on your desktop or anywhere in your directory. If not on your desktop create a shortcut on your desktop to the folder.

To run, double-click on ELI1 and enter inputs as requested. Upon completing execution, a "dataout" file will be created in the folder which you can open using notepad, and it should contain the resulting circuit description.

If the number of nulls is excessive for the response requirements (indicated by zero passband ripple), the final capacitor may have a negative value as a result of the algorithm.[1] Reduce the number of nulls, increase the required attention, define a steeper filter, or a combination of these.

A.4 FLTRFORM.XLS Spreadsheet of Formulas

Fltrform.xls is an EXCEL spreadsheet arranged by chapter that contains all the significant formulas in the chapter to simplify some of the calculations. By entering variables, the end result is precisely calculated, in real time, so values can be optimized if desired. Simply copy the spreadsheet to your computer. Note that cells containing formulas are locked to avoid inadvertently changing or deleting the formulas.

A.5 Errata

From time to time as errors or misprints may arise they will be posted on the website.

---

[1]P. Amstutz, "Elliptic Approximation and Elliptic Filter Design on Small Computers" IEEE Transactions on Circuits ad Systems CAS-25 no. 12 1978.

Index

CPSIA information can be obtained
at www.ICGtesting.com
Printed in the USA
LVOW03*1649061115
460645LV00013BB/200/P